Fossil Reptiles of Great Britain

M.J. Benton and P.S. Spencer

Department of Geology,
University of Bristol,
Bristol, UK.

GCR Editors: W.A. Wimbledon and D. Palmer

CHAPMAN & HALL

London · Glasgow · Weinheim · New York · Tokyo · Melbourne · Madras

Published by Chapman & Hall, 2–6 Boundary Row, London SE1 8HN, UK

Chapman & Hall, 2-6 Boundary Row, London SE1 8HN, UK

Blackie Academic & Professional, Wester Cleddens Road, Bishopbriggs, Glasgow G64 2NZ, UK

Chapman & Hall GmbH, Pappelallee 3, 69469 Weinheim, Germany

Chapman & Hall USA, One Penn Plaza, 41st Floor, New York NY10119, USA

Chapman & Hall Japan, ITP-Japan, Kyoto Building, 3F, 2-2-1 Hirakawacho, Chiyoda-ku, Tokyo 102, Japan

Chapman & Hall Australia, Thomas Nelson Australia, 102 Dodds Street, South Melbourne, Victoria 3205, Australia

Chapman & Hall India, R. Seshadri, 32 Second Main Road, CIT East, Madras 600 035, India

First edition 1995

© 1995 Joint Nature Conservation Committee

Typeset in 10/12 Garamond ITC Book
Printed in Great Britain at the University Press, Cambridge

ISBN 0 412 62040 5

A catalogue record for this book is available from the British Library

Library of Congress Catalog Card Number: 94-69920

♾ Printed on acid-free text paper, manufactured in accordance with ANSI/NISO Z39.48-1992 (Permanence of paper).

Fossil Reptiles of Great Britain

THE GEOLOGICAL CONSERVATION REVIEW SERIES

The comparatively small land area of Great Britain contains an unrivalled sequence of rocks, mineral and fossil deposits, and a variety of landforms that span much of the earth's long history. Well-documented ancient volcanic episodes, famous fossil sites and sedimentary rock sections used internationally as comparative standards, have given these islands an importance out of all proportion to their size. The long sequences of strata and their organic and inorganic contents have been studied by generations of leading geologists, thus giving Britain a unique status in the development of the science. Many of the divisions of geological time used throughout the world are named after British sites or areas, for instance, the Cambrian, Ordovician and Devonian systems, the Ludlow Series and the Kimmeridgian and Portlandian stages.

The Geological Conservation Review (GCR) was initiated by the Nature Conservancy Council in 1977 to assess, document and ultimately publish accounts of the most important parts of this rich heritage. Since 1991, the task of publication has been assumed by the Joint Nature Conservation Committee on behalf of the three country agencies, English Nature, Scottish Natural Heritage and the Countryside Council for Wales. The GCR series of volumes will review the current state of knowledge of the key earth-science sites in Great Britain and provide a firm basis on which site conservation can be founded in years to come. Each GCR volume will describe and assess networks of sites of national or international importance in the context of a portion of the geological column, or a geological, palaeontological or mineralogical topic. The full series of approximately 50 volumes will be published by the year 2000.

Within each individual volume, every GCR locality is described in detail in a self-contained account, consisting of highlights (a précis of the special interest of the site), an introduction (with a concise history of previous work), a description, an interpretation (assessing the fundamentals of the site's scientific interest and importance), and a conclusion (written in simpler terms for the non-specialist). Each site report is a justification of a particular scientific interest at a locality, of its importance in a British or international setting and ultimately of its worthiness for conservation.

The aim of the Geological Conservation Review series is to provide a public record of the features of interest in sites being considered for notification as Sites of Special Scientific Interest (SSSIs). It is written to the highest scientific standards but in such a way that the assessment and conservation value of the site is clear. It is a public statement of the value given to our geological and geomorphological heritage by the earth-science community which has participated in its production, and it will be used by the Joint Nature Conservation Committee, English Nature, the Countryside Council for Wales and Scottish Natural Heritage in carrying out their conservation functions. The three country agencies are also active in helping to establish sites of local and regional importance. Regionally Important Geological/Geomorphological Sites (RIGS) augment the SSSI coverage, with local groups identifying and conserving sites which have educational, historical, research or aesthetic value, enhancing the wider earth science conservation perspective.

All the sites in this volume have been proposed for notification as SSSIs; the final decision to notify, or renotify, lies with the governing Councils of the appropriate country conservation agency.

Information about the GCR publication programme may be obtained from:

Earth Science Branch,
Joint Nature Conservation Committee,
Monkstone House,
City Road,
Peterborough PE1 1JY.

Titles in the series

Contents

Contents

Contents

Acknowledgements

We thank Bill Wimbledon (Countryside Council for Wales, Bangor) for his constant support throughout this project, and for his considerable editorial input. The whole text was read through by Andrew and Angela Milner (London), and we thank them for their input, as well as David Brown (Newcastle upon Tyne), Liz Cook (Bristol), John Cope (Cardiff), Chris Duffin (Morden), Paul Ensom (York), Susan Evans (London), Nick Fraser (Martinsville, Virginia), Malcolm Hart (Plymouth), John Hudson (Leicester), Allan Insole (Sandown, Isle of Wight), Ed Jarzembowski (Brighton), Jim Kennedy (Oxford), David Martill (Leicester), Chris McGowan (Toronto), Dick Moody (Kingston), Alec Panchen (Newcastle upon Tyne), Jon Radley (Sandown, Isle of Wight), Jean-Claude Rage (Paris), Glenn Storrs (Bristol), Mike Taylor (Leicester), David Thompson (Keele), David Unwin (Bristol), David Ward (Orpington) and Geoff Warrington (British Geological Survey, Keyworth), who read portions of the work.

Thanks are also due to the GCR Publication Production Team: Neil Ellis, Publications Manager; Nicholas D.W. Davey, Editorial Assistant (Scientific Officer) and Valerie Wyld, Text Officer. Diagrams were drafted by Chris Pamplin (R & W Publishing, Newmarket) and the photographs were developed by Simon Powell (Bristol).

Access to the countryside

This volume is not intended for use as a field guide. The description or mention of any site should not be taken as an indication that access to a site is open or that a right of way exists. Most sites described are in private ownership, and their inclusion herein is solely for the purpose of justifying their conservation. Their description or appearance on a map in this work should in no way be construed as an invitation to visit. Prior consent for visits should always be obtained from the landowner and/or occupier.

Information on conservation matters, including site ownership, relating to Sites of Special Scientific Interest (SSSIs) or National Nature Reserves (NNRs) in particular counties or districts may be obtained from the relevant country conservation agency headquarters listed below:

English Nature,
Northminster House,
Peterborough PE1 1UA.

Scottish Natural Heritage,
12 Hope Terrace,
Edinburgh EH9 2AS.

Countryside Council for Wales,
Plas Penrhos,
Ffordd Penrhos,
Bangor,
Gwynedd LL57 2LQ.

Museum abbreviations

AUGD, Aberdeen University Geology Department.

AUZD, Aberdeen University Zoology Department.

BATGM, Bath Geology Museum.

BGS(GSE), British Geological Survey, Edinburgh.

BGS(GSM), British Geological Survey, Keyworth (old Geological Survey Museum collection, London).

BMNH, Natural History Museum, London (formerly British Museum (Natural History), London).

BRSMG, Bristol City Museum and Art Gallery.

BRSUG, Bristol University Geology Department.

BUCCM, Buckinghamshire County Museum, Aylesbury.

AMMZ, Cambridge University Museum of Zoology.

CAMSM, Sedgwick Museum, Department of Earth Sciences, Cambridge University.

DORCM, Dorset County Museum, Dorchester.

ELGNM, Elgin Museum.

EXEMS, Royal Albert Memorial Museum, Exeter.

GLRCM, Gloucester City Museum and Art Gallery.

IWCMS, Isle of Wight Museum Geology, Sandown.

LEICS, Leicestershire Museums, Leicester.

LIVCM, National Museums on Merseyside, Liverpool.

MAIDM, Maidstone Museum.

MANCH, Manchester Museum.

MCZ, Museum of Comparative Zoology, Cambridge, Massachusetts.

NEWHM, Hancock Museum, Newcastle upon Tyne.

NMS, National Museums of Scotland, Edinburgh (formerly RSM).

NMW, National Museum of Wales, Cardiff.

NORCM, Norwich Castle Museum.

OUM, University Museum, Oxford.

OXFPM, Oxford Polytechnic Geology Department.

SHRBM, Shrewsbury Borough Museum.

SHRCM, Shropshire County Museum, Ludlow.

SDM, Stroud District Museum.

WARMS, Warwickshire Museum, Warwick.

WHIMS, Whitby Museum.

YORYM, Yorkshire Museum, York.

YPM, Yale Peabody Museum, New Haven, Connecticut.

Chapter 1

Introduction

Introduction

Britain is famous for its fossil reptiles, partly for historical reasons, but also because there are so many richly fossiliferous localities that have supplied, and continue to supply, excellent material. The continuing potential of British fossil reptile sites is illustrated by recent work on such internationally important localities as the Mid Triassic localities of England (e.g. Benton, 1990c, Benton, *et al.*, in press; Milner *et al.*, 1990), the Late Triassic faunas of Elgin (e.g. Benton and Walker, 1985), the Late Triassic marine bone beds of the south-west of England (Storrs, 1994; Storrs and Gower, 1993), the Late Triassic to Early Jurassic fissures around Bristol and in south Wales (e.g. Evans, 1980, 1981; Crush, 1984; Fraser, 1982, 1985, 1986, 1988a, 1988b, 1994; Fraser and Walkden, 1983; Whiteside, 1986), the Early and Late Jurassic marine faunas of Dorset and Somerset (e.g. McGowan, 1974a, 1974b, 1976, 1986, 1989a, 1989b; Brown, 1981; Padian, 1983; Galton, 1985b; Brown *et al.*, 1986; Taylor, 1992a, 1992b), the Mid Jurassic terrestrial faunas of the Cotswolds (e.g. Galton, 1980a, 1983a, 1983b, 1985b; Evans *et al.*, 1988, 1990; Evans, 1989, 1990, 1991, 1992a; Evans and Milner, 1991, 1994; Metcalf *et al.*, 1992), the diverse small reptiles from the Purbeck of Swanage (e.g. Evans and Kemp, 1975, 1976; Gaffney, 1976; Galton, 1978, 1981b; Buffetaut, 1982; Estes, 1983; Howse, 1986; Ensom *et al.*, 1991; Sereno, 1991a; Clark, in press), the Wealden of the Weald and of the Isle of Wight (e.g. Galton, 1969, 1971a, 1971b, 1971c, 1973, 1974, 1975; Buffetaut and Hutt, 1980; Norman, 1980, 1986, 1990b; Blows, 1987; Buffetaut, 1982; Charig and Milner, 1986, 1990; Unwin, 1991; Clark, in press), the pterosaurs and other reptiles from the Cambridge Greensand (e.g. Unwin, 1991), and the various Palaeogene faunas of southern England (e.g. Moody and Walker, 1970; Moody, 1974, 1980a; Walker and Moody, 1974; Meszoely and Ford, 1976; Hooker and Ward, 1980; Rage and Ford, 1980; Milner *et al.*, 1982). The main focus in selecting sites for conservation has been to choose those which have been studied recently, and which have supplied abundant reptile specimens. An attempt was also made to balance the coverage, so that each major stratigraphic unit and facies is represented.

The historical records of fossil reptiles from Britain extend back a long way. Earliest finds included fossils that we now recognize as dinosaur bones (Figure 1.1) from the Mid Jurassic of Oxfordshire (Plot, 1677; Lhuyd, 1699; Woodward, 1728; Platt, 1758; more details in

Figure 1.1 Lower end of the thigh bone of *Megalosaurus*, from Cornwell, Oxfordshire: one of the first fossil reptile bones to be illustrated from Britain, and the oldest recorded figure of a dinosaur (from Plot, 1677).

Delair and Sarjeant, 1975) and a marine crocodile from the Early Jurassic of Whitby, Yorkshire (Chapman, 1758; Wooller, 1758). More intensive collecting began only in the 19th century, and large numbers of marine ichthyosaurs and plesiosaurs were obtained from the Early Jurassic of Lyme Regis, Dorset and Whitby, Yorkshire (e.g. Home, 1814, 1819a; Conybeare, 1822, 1824; Young and Bird, 1822; more details in Benton and Taylor, 1984). More dinosaur specimens were found in the Mid Jurassic of Oxfordshire (Buckland, 1824) and in the Early Cretaceous of south-east England (Mantell, 1822, 1825), and footprints of Permian age came to light in Scotland (Buckland, 1828; Grierson, 1828; details in Sarjeant, 1974).

Throughout the remainder of the nineteenth century, large collections were amassed, and most of the localities noted in the present work were identified. Locality information for nineteenth century collections may be problematic in many cases, because of a lack of direct contact between the collectors and the palaeontologists who made the descriptions. Prolific authors such as Owen, Huxley, Seeley, Lydekker and others seem to have worked largely in their institutions on material that was sent to them from a network of local natural history and geological societies throughout the country. Only rarely did these biologically trained palaeontologists record geographic or geological details of the context of their specimens. A notable exception is the account of the discovery and excavation of a partial skeleton of the ornithopod dinosaur *Camptosaurus prestwichii* (Hulke, 1880a) by Prestwich (1879, 1880).

Sporadic collecting has been carried out during the twentieth century, much of it by amateurs and professional collectors, but the network of suppliers and describers seems to have broken down rather. This was partly because of the lack of professional palaeontologists in Britain with suitably broad interests and the desire to encourage active collecting: indeed, the most prolific describer of British fossil reptiles between 1900 and 1930 was the German palaeontologist Baron Friedrich von Huene! A further problem was the decline of local natural history societies and the loss of skilled collectors with local knowledge. Unfortunately, this has meant that many finds were recorded only rather poorly, if at all, and much of the material has been inadequately curated, or even lost altogether. In addition, many of the small local museums set up by natural history societies in the 1830s and 1840s declined into disuse and were either closed or handed over to local authorities. In most cases, there was no longer anyone with any knowledge or appreciation of the local specimens, and a tremendous amount of fossil reptile material must have been lost or damaged during this time, or abandoned in such a way that curatorial information was lost (see Torrens and Taylor, 1990 for a typical example, the sorry story of the Cheltenham museums).

It is only in the last 10 or 20 years that local museum standards in geology have improved dramatically, and that serious excavations by amateurs and professional scientists have been renewed in any numbers. These factors have led to the discovery and exploitation of several important sites, as noted above. The collections made during these years are to be seen in a large number of museums (listed at the end of this introduction).

REPTILIAN EVOLUTION

Reptiles today are readily identifiable: they are of course the turtles, crocodilians, lizards, snakes and the tuatara. However, the diversity of reptiles in the past was much greater than these surviving lineages would suggest. Without the fossil record, we could not begin to guess at the evolutionary history of the group. In phylogenetic terms, the Class Reptilia is a paraphyletic group, meaning

that it arose from a single ancestor (among the amphibians), but that the Class does not include all of its descendants, namely the birds and the mammals. The reptiles are a part of the larger monophyletic group, the Amniota (= reptiles + birds + mammals).

Modern amniotes are defined by the possession of a cleidoic (= closed) or amniotic egg, an egg that has an outer protective coating or shell, and a complex system of membranes around the embryo within the egg. Unlike the amniotic eggs of fishes and amphibians (e.g. frog spawn), the cleidoic egg can be viewed as a 'private pond' in which the embryo can develop in relative safety on land, and with all nutritional supplies (the yolk) available. Waste materials are collected in the allantois, and the embryo can breathe through the semipermeable eggshell, which may be leathery or calcareous. The cleidoic egg allows amniotes to lay their eggs away from water, and this may have been an important advantage when the group arose, in Carboniferous times, in allowing them to occupy upland and dry areas.

The oldest reptiles have been known from the early Late Carboniferous of Nova Scotia, Canada, since the 1850s, and these include 'protorothyridids' and synapsids. A major discovery in Scotland in 1988 (Smithson, 1989; Smithson and Rolfe, 1991) has pushed the origin of amniotes back even further into the Carboniferous than had been suspected: the Nova Scotia animals date from about 300–310 Ma, while the new Scottish find, dubbed 'Lizzie' by its discoverer, Mr Stan P. Wood, is dated as about 330 Ma old. The exact affinities of 'Lizzie' are not yet certain.

Over the past 100 years, it has become clear that the major lines of amniote evolution were clearly laid out during the Late Carboniferous. The amniotes split into three main lineages, the synapsids (mammal-like reptiles and ultimately, the mammals), the diapsids (early forms, dinosaurs, extinct marine reptiles, lizards, snakes, crocodilians and ultimately birds), and the anapsids (primitive groups and turtles). Traditionally, the amniotes have been divided into four groups on the basis of their skull openings (Figure 1.2). The opening(s) behind the orbit (eye socket), termed the temporal opening(s), are present in various arrangements: no temporal opening in the anapsids, two temporal openings in the diapsids, a lower temporal opening only in the synapsids, and an upper temporal opening in the euryapsids. The first three of these groups is still regarded as having taxonomic validity, but the 'euryapsids'

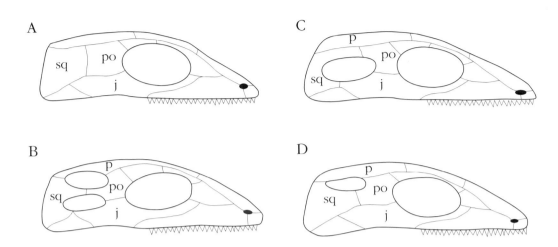

Figure 1.2 The skull patterns, in side view, of the major lineages of reptiles. The anapsid pattern (A) is plesiomorphic (primitive), being present also in fishes and amphibians, while the diapsid (B) and synapsid (C) patterns define two major clades of amniotes, the Diapsida (thecodontians, dinosaurs, pterosaurs, crocodiles, birds) and the Synapsida (mammal-like reptiles and mammals). The euryapsid pattern (D) may have arisen more than once, in different marine groups, and appears to be a derivative of the diapsid pattern. Abbreviations: j – jugal, p – parietal, po – postorbital, sq – squamosal. After Benton (1990a).

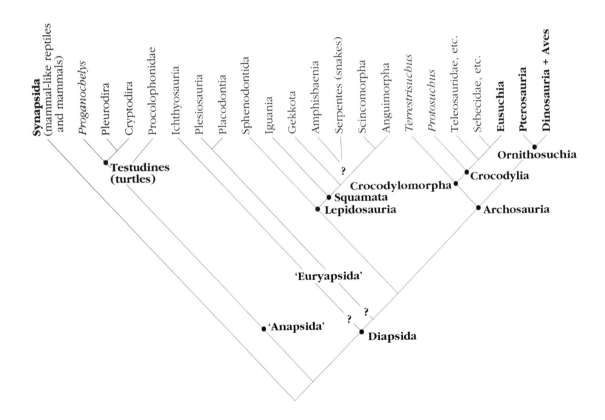

Figure 1.3 Cladogram of the major groups of reptiles, based on recent analyses (after Benton, 1990a).

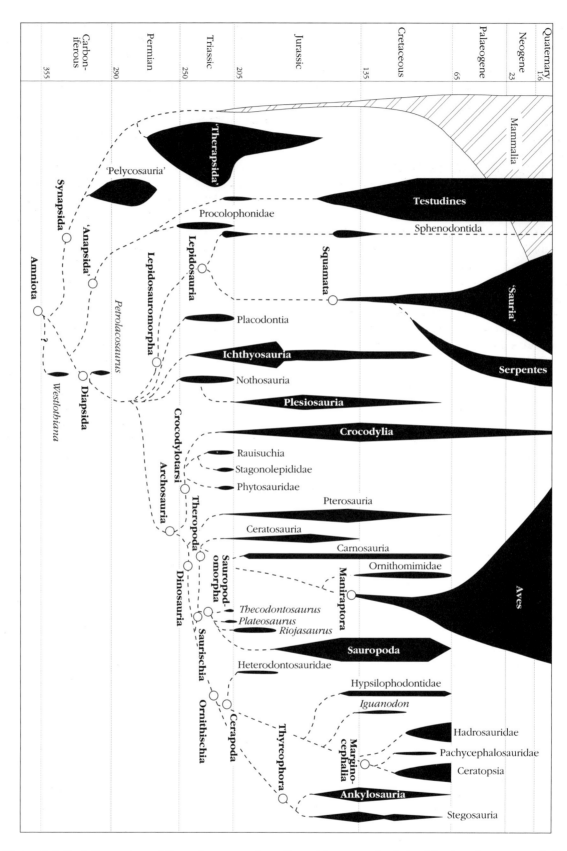

Figure 1.4 Evolutionary tree of the main groups of reptiles, with proposed relationships based on recent cladistic analyses, and the stratigraphical distributions based on global data. The width of the 'spindles' represents diversity of the groups (after Benton, 1990a).

appear to be an artificial assemblage of extinct marine reptiles that are modified diapsids.

In recent years a number of attempts have been made to disentangle the evolution of the major amniote groups by the application of cladistic analysis (e.g. Gaffney, 1980; Gardiner, 1982; Kemp, 1982; Evans, 1984, 1988a; Gauthier, 1986; Heaton and Reisz, 1986; Benton, 1985, 1990b; Benton and Clark, 1988; Gaffney and Meylan, 1988; Gauthier *et al.*, 1988a, 1988b; Massare and Callaway, 1990; Storrs, 1991; Spencer, 1994). The phylogenetic analyses are still tentative in part, and there has been disagreement over the placement of the major groups, particularly the birds and mammals. Gardiner (1982) and many molecular biologists, find strong evidence for linking birds and mammals closely as the Haematothermia (sharing a common ancestor presumably in the Triassic), while most other authors accept a more 'traditional' view, followed here also, that the phylogenetic split between birds and mammals lies in the Carboniferous (i.e. the diapsid/synapsid split). A cladogram, based on the work of the above-noted authors, and updated from those in Benton (1990a, 1990b), based on the work of Massare and Callaway (1990), Storrs (1991) and Spencer (1994), is given in Figure 1.3.

An evolutionary tree (Figure 1.4) and a classification of reptiles (Figure 1.5), both based on the cladogram, show the main features of the global evolution of the reptiles after the mid Carboniferous. Reptiles were rare animals during the Carboniferous, being restricted apparently to life in and around the trees of the great coal forests of Europe and North America. However, despite their rarity and generally modest size, the main lineages of amniote evolution, the anapsids, synapsids and diapsids, became clearly established then. During the Early Permian, as documented particularly in the mid-western United States, the pelycosaurs (mammal-like reptiles with and without 'sails') became abundant and diverse, to be followed in the Late Permian of South Africa and Russia by the radiation of various groups of therapsid mammal-like reptiles (dicynodonts, gorgonopsians, dinocephalians). During the Triassic, as indicated in many parts of the world, including Britain, there was a major turnover of faunas after the end-Permian extinction event, and new groups of synapsids (cynodonts, new dicynodonts) and diapsids (prolacertiforms, rhynchosaurs, archosaurs) came on the scene. These faunas apparently disappeared during the Late Triassic, to be replaced by a global 'modern' fauna, consisting of dinosaurs, pterosaurs, crocodilians, turtles, 'lizards', lissamphibians and mammals. These Late Triassic, as well as the Jurassic and Cretaceous dinosaur faunas are very similar worldwide, because of the conjunction of all continents in Pangaea, and because of the apparently equable climatic conditions worldwide. Faunal provinces become evident by Late Cretaceous times as a result of the opening-up of the Atlantic Ocean and the break-up of Gondwanaland, and this theme continues through the Tertiary to the present.

Fuller details of reptilian evolution may be found in textbooks such as Carroll (1988), Benton (1990a, 1990d) and Colbert and Morales (1991). Books on particular groups of fossil reptiles include Kemp (1982) and Hotton *et al.* (1986) on the mammal-like reptiles, Norman (1985), Benton (1989), Weishampel *et al.* (1990) and Carpenter and Currie (1990) on the dinosaurs, and Wellnhofer (1991) on the pterosaurs.

The British record of fossil reptiles illustrates a remarkably high proportion of the evolution of the group (Figures 1.6 and 1.7). Missing portions are the Late Carboniferous to Early Permian, known only from footprints and sporadic body fossils, virtually the whole of the evolution of mammal-like reptiles before the latest Triassic, and the Miocene and Pliocene. Otherwise, there is strong representation for the Late Permian, the Mid and Late Triassic, the marine Jurassic and terrestrial Mid Jurassic, the terrestrial Early Cretaceous and marine mid- to Late Cretaceous, and the terrestrial Palaeogene and Pleistocene. A comparison of the sequence of major reptile-bearing units in Great Britain with those from other parts of the world highlights the strengths and weaknesses. Of the major reptile lineages, British sites have produced tritylodontids among the synapsids; pareiasaurs, procolophonids and turtles, among the anapsids; and sphenodontids, lizards, snakes, rhynchosaurs, 'thecodontians', pterosaurs, dinosaurs, crocodilians, plesiosaurs and ichthyosaurs, among the diapsids.

STRATIGRAPHY

British fossil reptile sites range over the maximum time range possible, from the Early Carboniferous (Brigantian, *c.* 330 Ma) to the Pleistocene.

The stratigraphic location of most sites is relatively well-fixed by international standards. This is partly because of the mature state of local biostratigraphy in Britain. In addition, it has been

Series Amniota
  *Class Reptilia
    Subclass Synapsida
     'Family Protorothyrididae'
     †Family Mesosauridae
      *†Order Pelycosauria
      Order Therapsida
       †Suborder Biarmosuchia
       †Suborder Dinocephalia
       †Suborder Dicynodontia
       †Suborder Gorgonopsia
       Suborder Cynodontia
        †Family Procynosuchidae
        †Family Galesauridae
        †Family Cynognathidae
        †Family Diademodontidae
        †Family Chiniquodontidae
        †Family Tritylodontidae
        †Family Tritheledontidae
        Class Mammalia
    Subclass Anapsida (*sensu stricto*)
     †Family Captorhinidae
     †Family Procolophonidae
     †Family Pareiasauridae
     Order Testudines (Chelonia)
      †Family Proganochelyidae
      Suborder Pleurodira
      Suborder Cryptodira
       Superfamily Baenoidea
       †Family Meiolaniidae
       Superfamily Chelonioidea
       Superfamily Trionychoidea
       Superfamily Testudinoidea
     †Family Protorothyrididae
    Subclass Diapsida
     †Family Millerettidae
     †Family Petrolacosauridae
     †Family Weigeltisauridae
     Infraclass Lepidosauromorpha
      †Order Younginiformes
      Superorder Lepidosauria
       Order Sphenodontida
        Family Sphenodontidae
        †Family Pleurosauridae
       Order Squamata
        *Suborder Sauria (Lacertilia)
        Infraorder Gekkota
        Infraorder Iguania
        Infraorder Scincomorpha
        Infraorder Anguimorpha
        Infraorder Amphisbaenia
        Suborder Serpentes (Ophidia)
     Infraclass Archosauromorpha
      †Family Trilophosauridae
      †Family Rhynchosauridae
      †Order Prolacertiformes
      Division Archosauria
       Family Proterosuchidae
       Family Erythrosuchidae
       Family Euparkeriidae
       Subdivision Crocodylotarsi
        †Family Phytosauridae
        †Family Stagonolepididae
        †Family Rauisuchidae
        †Family Poposauridae
        Superorder Crocodylomorpha
         †Family Saltopusuchidae
         †Family Sphenosuchidae
         Order Crocodylia
          †Family Protosuchidae
          *†Suborder Mesosuchia

          Family Teleosauridae
          Family Metriorhynchidae
          Family Sebecidae, etc.
         Suborder Eusuchia
          Family Gavialidae
          Family Crocodylidae
          Family Alligatoridae
      Subdivision Ornithosuchia
       †Family Ornithosuchidae
       †Family Lagosuchidae
       †Order Pterosauria
        *Suborder Rhamphorhynchoidea
        Suborder Pterodactyloidea
       *†Superorder Dinosauria
        Family Herrerasauridae
        Order Saurischia
         Suborder Theropoda
          Infraorder Ceratosauria
          Infraorder Carnosauria
          Family Ornithomimidae
          Infraorder Maniraptora
           Family Compsognathidae
           Family Coeluridae
           Family Oviraptoridae
           Family Dromaeosauridae
           Family Troodontidae
           Class Aves
         Suborder Sauropodomorpha
         *Infraorder Prosauropoda
          Family Thecodontosauridae
          Family Plateosauridae
          Family Melanorosauridae
         Infraorder Sauropoda
          *Family Cetiosauridae
          Family Camarasauridae
          Family Brachiosauridae
          Family Diplodocidae
          Family Titanosauridae
       Order Ornithischia
        Family Pisanosauridae
        Family Fabrosauridae
        Suborder Cerapoda
         Infraorder Ornithopoda
          Family Heterodontosauridae
          Family Hypsilophodontidae
          *Family Iguanodontidae
          Family Hadrosauridae
         Infraorder Pachycephalosauria
         Infraorder Ceratopsia
          Family Psittacosauridae
          Family Protoceratopsidae
          Family Ceratopsidae
        Suborder Thyreophora
         Family Scelidosauridae
         Infraorder Stegosauria
         Infraorder Ankylosauria
          Family Nodosauridae
          Family Ankylosauridae
 Diapsida *incertae sedis*
   †Superorder Sauropterygia
    Order Placodontia
    Order Nothosauria
    Order Plesiosauria
     Family Plesiosauridae
     Family Cryptoclididae
     Family Elasmosauridae
     Family Pliosauridae
    Order Ichthyosauria

Figure 1.5 Table showing the classification of the major groups of reptiles, based on the cladograms summarized in Figure 1.3. Symbols: † extinct group; * paraphyletic group (after Benton, 1990a).

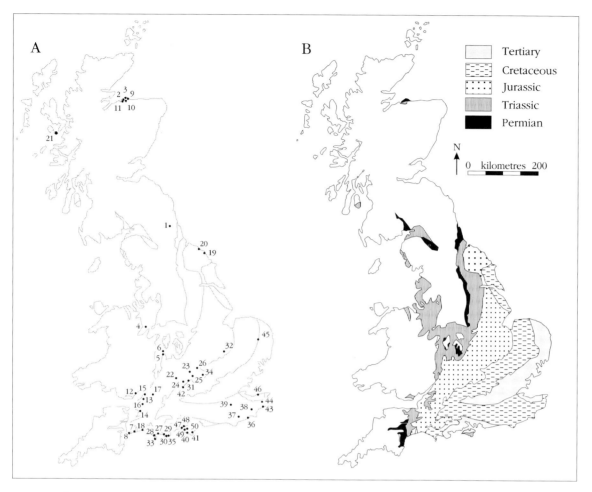

Figure 1.6 (A) Map of Great Britain showing the distribution of the 50 GCR fossil reptile sites; (B) The outcrop pattern of Permian, Triassic, Jurassic, Cretaceous and Tertiary rocks in Great Britain. After Benton (1988).

possible to correlate sites to ammonite zones, or even subzones, for most of the Jurassic and Cretaceous. Dating evidence for the terrestrial Permian, Triassic and Palaeogene sites is less secure, but is often tied to evidence from palynology, or other floral and microfossil evidence. The age of fossil reptile faunas is crucial for a proper understanding of the evolution, palaeoecology and palaeobiogeography of the group, and considerable emphasis has been placed on establishing the age of each site as precisely as possible. The evidence, and any controversial issues, are recounted in detail in the site descriptions.

HOW THE SITES WERE SELECTED

Fifty sites were selected as Geological Conservation Review (GCR) sites for their significance in representing aspects of the evolution of

reptiles (Figure 1.6). A full account of the site-selection procedure, as well as discussions of the use of sites, the detective work involved, and conservation issues are given by Benton and Wimbledon (1985) and Benton (1988). The exact procedure followed in investigating Britain's heritage of fossil reptile sites, in selecting those that should become Sites of Special Scientific Interest (SSSIs), and hence come under the protection of the Wildlife and Countryside Act (1981), and in producing the present volume were as follows (modified from Benton and Wimbledon, 1985):

1. Initial data handling (1981–2). M.J.B. examined all published papers about British fossil reptiles and noted all site information, poor as it usually was (e.g. 'Wealden, Sussex', 'Bathonian, Oxfordshire', 'Chalk, Dover'). M.J.B. then studied most of the major museum collections in Britain (listed below) and again noted site

information, especially in the very rare cases where some of the original collector's notes had been kept. M.J.B. then organized this mass of information into broad stratigraphical and geographical blocks (e.g. Triassic of Elgin, Early Jurassic of Yorkshire, Kimmeridge Clay of Dorset).

2. Site location (1981–2). M.J.B. then tried to find as many of the quarries as possible on old and new 6 inches to the mile (1:10 000) Ordnance Survey maps. This stage involved the use of geological maps, relevant stratigraphical literature, historical archives and much guess-work. Eventually, a working list of some 500 sites, with map references was drawn up. The stratigraphical distribution of those 500 sites was as follows:

Caenozoic / Pleistocene		50
Cretaceous	– Late	60
	– Early	90
Jurassic	– Late	90
	– Mid	90
	– Early	50
Triassic	– Rhaetian	20
	– Scythian–Norian	40
Permian		10
Carboniferous		1
		500

3. Preliminary site sorting (1981–2). In an ideal world, one would like to preserve all 500 sites for future scientific use, and prevent infilling or other developments. However, this would be futile, or impracticable, for many of the 500 certain sites were discarded from the list of potential SSSIs at once – those that had yielded only a few scraps, and those that had been obliterated by later developments. This was a step designed to minimize the amount of field-work required.

A

Period	Epoch	Stage	Britain	Continental Europe	North America	Southern Continents
Triassic	Late	Rhaetian	Aust Cliff; Bristol fissures	Rhät		Los Colorados
		Norian	Bendrick Rock; Bristol fissures	Stubensandstein	Upper Chinle & Dockum	Lower Elliot
		Carnian	Elgin sites	Schilfsandstein	Popo Agie; Lower Chinle & Dockum	Santa Maria; Ischigualasto; Maleri
	Mid	Ladinian		Muschelkalk		
		Anisian	Grinshill; Coten End; Devon coast	Muschelkalk; Donguz Series	Moenkopi	
	E	Scythian		Buntsandstein		Beaufort Group Karroo
Permian	Late	Tatarian	Cutties Hillock; Masonshaugh	Urals		
		Kazanian		Zechstein		
		Ufimian	Middridge	Kupferschiefer		
	Early	Kungurian			Pease River Group	
		Artinskian		Rotliegendes	Clear Fork Group	
		Sakmarian			Wichita Group	
		Asselian				
Carboniferous	Late	Gzelian				
		Kasimovian			Garnett	
		Moscovian		Nýřany	Joggins; Florence	
		Bashkirian				
	Early	Serpukhovian				
		Viséan	(East Kirkton)			
		Tournaisian				

Figure 1.7 Generalized stratigraphic column showing the major British fossil reptile sites in sequence, and comparable sites elsewhere in the world. A: Carboniferous to Triassic; B (opposite): Jurassic to Cretaceous; C (page 12) Tertiary to Quaternary.

B

			Britain	Continental Europe	North America	Southern Continents
Cretaceous	Late	Maastrichtian		Sinpetra; Rognac; Maastricht	Hell Creek; Lance; Laramie; Scollard; Frenchman	Titanosaur beds; Kronosaur chalk, Australia; Nemegt
		Campanian	Burham		Two Medicine; Milk River; Judith River; Horseshoe Canyon; Fruitland	Djadochta; Barun Goyot
		Senonian				
		Cenomanian				Quishan
	Early	Albian	Folkestone		Dakota; Paluxy; Cloverly	
		Aptian				Kukhtekskaya
		Barremian	Clock House / Brighstone	Bernissart; Nehden; Las Zabacheras	Lakota	Santana
		Hauterivian				
		Valanginian	Hastings; Telham			
		Berriasian	Durlston			
Jurassic	Late	Portlandian	Chicksgrove / Portland	Wimereux; Wimille		
		Kimmeridgian	Kimmeridge Smallmouth	Boulogne; Solenhofen; Eichstätt; Cerin; Guimarota	Morrison	Tendaguru
		Oxfordian	Peterborough	Calvados		Shangshaximiao (Sichuan)
	Mid	Callovian		Dives		
		Bathonian	Shipton; Kirtlington; Stonesfield; New Park	Caen		Xiashaximiao (Sichuan)
		Bajocian				
		Aalenian				
	Early	Toarcian	Whitby	Holzmaden	Navajo	Khota
		Pliensbachian			Kayenta	
		Sinemurian				
		Hettangian	Lyme Regis		Wingate; Moenave	Clarens; Lower Lufeng; Upper Elliot

Figure 1.7 – *contd.*

C

				Britain	Continental Europe	North America	Southern Continents
Quaternary			Holocene	Ubiquitous	Ubiquitous	Ubiquitous	Pampas; Olduvai (part)
Quaternary			Pleistocene	Ubiquitous	Ubiquitous	Ubiquitous	Omo; Olduvai (part); etc.
Tertiary	Neogene		Pliocene		Montpellier; Pikermi; Samos	Hemphill; Clarendon; Ogalalla (part)	Siwaliks (part); Montehermosa; Huayqueria
Tertiary	Neogene		Upper Miocene			Barstow; Ogalalla (part)	Chasico; Fort Ternan; Siwaliks (part)
Tertiary	Neogene		Middle Miocene		Steinheim; Oeningen	Hemingford	Chinji (Siwalik)
Tertiary	Neogene		Lower Miocene			Arikareean	Santa Cruz; Bugti; Turgai
Tertiary	Palaeogene	Oligocene	Chattian	Bouldnor Cliff		White River (part)	Colhuehuapian
Tertiary	Palaeogene	Oligocene	Rupelian		Quercy (part)	White River (part); Chadron	Deseado; Fayûm (part)
Tertiary	Palaeogene	Eocene	Priabonian	Hordle; Headon Hill	Quercy (part); Robiac		Fayûm (part)
Tertiary	Palaeogene	Eocene	Bartonian	Barton Cliff		Uinta; Washakie (part)	
Tertiary	Palaeogene	Eocene	Lutetian		Messel; Geiseltal	Bridger; Washakie (part); Green River (part)	Musters
Tertiary	Palaeogene	Eocene	Ypresian	Sheppey	Erquelinnes; Dormaal; Monte Bolca	Wind River; Willwood; Wasatch	Casa Mayor
Tertiary	Palaeogene	Palaeocene	Thanetian		Cernay	Fort Union; Tongue River; Clark Fork	Rio Chico; Itaboraí; Pernambuco
Tertiary	Palaeogene	Palaeocene	Danian			Torrejonian	

Figure 1.7 – *contd.*

4. Site visits and further site sorting (1981–2). Every site on the reduced list of 200 or so was visited, and an attempt was made to locate the fossiliferous horizon(s). At this stage, further sites were struck off the list of potential SSSIs if they were filled in, or if the relevant horizons were completely inaccessible.

5. Selection of major sites (1981–2). The selection of key sites for each unit was then made. Each of these sites had to have demonstrated potential (i.e. major finds of international importance, whether published or not), as well as the potential for more finds from known fossiliferous horizons.

6. Publication of the work (1990–2). P.S.S., in association with M.J.B., updated all the records made in 1981–2, arranged the information in a logical format, and produced the present volume. The focus of the text is on the 50 SSSIs, but all other sites that were identified as having produced any reptile fossils are also documented in the relevant places in the text. Figure 1.6 shows the distribution of these 50 sites.

Further information on the site-selection procedure, with a detailed example, based on the Oxfordian sites, is given by Benton and Wimbledon (1985). Benton (1988) lists all 50 British fossil reptile SSSIs in synoptic form, and full details and justifications are given in this volume.

Chapter 2

British Carboniferous fossil reptile sites

British Carboniferous fossil reptile sites

Within the past decade two possible Carboniferous reptiles have come to light in Britain, one from the Lower Carboniferous of West Lothian, Scotland and the other from the Upper Carboniferous of Newsham, Northumberland. The Scottish material, collected by Mr Stanley Wood and the National Museums of Scotland, came from the Lower Carboniferous (Brigantian) East Kirkton Limestone near Bathgate, West Lothian, and forms part of an important terrestrial assemblage that includes some of the earliest recorded temnospondyl amphibians, eurypterids, myriapods, scorpions and the earliest known opilionid (harvestman). Two 'reptile' specimens have been collected from different horizons; the type from bed 82, the black shale member (Smithson, 1989), and the second from bed 76 (Smithson *et al.*, 1994). There are two further specimens (A.R. Milner, pers. comm., 1994).

The type specimen (NMS G.1990.72.1), named *Westlothiana lizziae* (Smithson and Rolfe, 1991), consists of an almost complete articulated skeleton preserved in part and counterpart. The total length of the skeleton is 180–200 mm with a presacral length of about 120 mm. Assignment to the Reptilia (Division Amniota) was based on two main criteria. Firstly, a well-developed astragalus and calcaneum are present in the pes, a character shared by all extant amniotes. Secondly, the cranial remains show clear reduction of the temporal series (intertemporal, supratemporal and tabular) permitting contact between the parietal, postorbital and the squamosal. The latter character is found in all other Carboniferous tetrapods which are regarded as true amniotes.

However, recent further study of the type specimen following preparation of the palate and braincase, and detailed examination of the second specimen (Smithson *et al.*, 1994) have revealed that *Westlothiana* has a mixture of primitive tetrapod and derived amniote characters. It shares with early amniotes the pattern of bones in the temporal series, a large vertical quadrate, gastrocentrous vertebrae, gracile humerus with distinct supinator process, and hind feet with a pedal formula 23454. But, unlike other early amniotes, it lacks a tooth-bearing pterygoid flange and, contrary to the original description, it has three proximal tarsals, tibiale, intermedium and fibulare, and not an astragalus and calcaneum.

The earliest amniote fossil identified prior to the discovery of the Scottish reptile was *Hylonomus*, a protorothyridid from the lower Westphalian B (Upper Carboniferous) of Joggins, Nova Scotia (*c.* 308 Ma) (Carroll, 1964); thus the Scottish 'reptile', if that is what it is (Brigantian, 335 Ma, Lower Carboniferous) pre-dates *Hylonomus* by 27 Ma.

The only other reptilian material reported from the British Carboniferous are the supposed remains of a reptile collected during the late 19th century from the lower Westphalian B (Upper Carboniferous) of Newsham, Northumberland (NZ 306791). The specimen (NEWHM G24.84), which consists of an incomplete skull table, was referred to the 'Romeriidae', and subsequently to the Protorothyrididae by Boyd (1984, 1985). A recent examination of the specimen, however, has demonstrated that it belongs to the skull of an acanthodian fish of a variety common at the find locality (Coates and Smithson, pers. comm. to Milner, 1987, p. 500).

Footprints of amphibians and reptiles have been recovered (Sarjeant, 1974) from Butts Quarry, Aveley, Shropshire (SO 7684) in the Keele Beds, dated as Westphalian D (Smith *et al.*, 1974, p. 9) or Stephanian (Haubold and Sarjeant, 1973, p. 897).

No reptile sites are scheduled as SSSIs from the British Carboniferous because East Kirkton has so far produced very few specimens. Should more come to light there, it would be a strong candidate for scheduling as a GCR reptile site.

Chapter 3

British Permian fossil reptile sites

INTRODUCTION: PERMIAN STRATIGRAPHY AND SEDIMENTARY SETTING

British Permian time can be grouped into two broad units, an earlier phase of predominantly terrestrial deposition and a later phase characterized by greater marine influence. The Early Permian of the British Isles, and also of much of northern Europe, was mostly a time of terrestrial subaerial erosion with desiccation of newly uplifted areas that had been generated during the final phases of the Variscan Orogeny. A large number of fault-bounded sedimentary basins developed, with the creation of wide-ranging facies variations of coarse- and fine-grained sediments as well as evaporites. Later in the Early Permian, aeolian deposits and evaporites became dominant, indicating a prevailing arid climate. The sediments were deposited in a westward extension of the German–Dutch basin and are known as the Rotliegendes. By the end of the Early Permian, Britain had been reduced to a gently rolling peneplain (Smith *et al.*, 1974; Smith, 1989; Smith and Taylor, 1992), which was largely an inhospitable desert.

A major marine transgression in the Late Permian led to the development of an inland (epicontinental) sea which flooded the North Sea Basin and part of mainland Britain, leading to deposition of the thick, evaporitic Zechstein sequences. The Zechstein deposits of north-east England, the North Sea and Germany comprise five major sedimentary cycles, each commencing in shelf carbonates and grading up into evaporites. The base of each cycle is sometimes marked by widespread development of bituminous shale which passes directly into the main carbonate sequence. The earliest of these deposits contain plant remains, perhaps reflecting a temporary increase in humidity following the establishment of the Zechstein seaway. This, however, was short-lived, as the return of arid conditions led to a re-establishment of evaporite sedimentation.

Arid and semi-arid conditions continued to the end of the Permian throughout the British region, but in the isolated sedimentary basins of north-east Scotland a diverse reptilian fauna appears to have flourished in spite of the harsh conditions.

The lack of biostratigraphic indicators makes relative dating of Permian deposits in Britain very difficult (Smith *et al.*, 1974), and only parts of the succession may be dated with any degree of accuracy. The diversity of facies and their diachronous nature rule out wide-ranging lithological correlations and such correlations do not in any case necessarily amount to time-equivalence. The most important sequences bearing reptiles are the Marl Slate (Early Permian), of County Durham and the Cutties Hillock Sandstone Formation and Hopeman Sandstone Formation (Late Permian) of Scotland. Figure 3.1 shows the distribution of Permian rocks in Great Britain and the position of the GCR Permian reptile sites.

REPTILE EVOLUTION DURING THE PERMIAN

During the Early Permian, reptiles broadly replaced amphibians to become the dominant terrestrial tetrapods and, during this time, the main reptile groups that had radiated towards the close of the Carboniferous Period continued to evolve, establishing themselves in a wide variety of previously vacant ecological niches. Most significant in this respect were the synapsid reptiles (mammal-ancestors), and among these, the carnivorous and herbivorous pelycosaurs. These were the most diverse tetrapods of the Early Permian, representing upwards of 70% of the known reptiles. Pelycosaurs are best known from the mid-western United States (Texas, Oklahoma, New Mexico), but sporadic occurrences from Central Europe have shown similar faunas. Their remains are largely absent in Britain, apart from a site at Kenilworth, and a number of footprint localities (see below) which show that these reptiles did occupy the region but their skeletal remains are rarely preserved.

The Late Permian was marked by further radiation of the synapsids, with the appearance of numerous new groups belonging to the major derived group, the therapsids. Key groups include the dicynodonts (specialized herbivorous forms with reduced numbers of teeth), the dinocephalians (an assemblage of large herbivores and carnivores) and the gorgonopsians (moderate to large-sized carnivores, many with 'sabre'-teeth). The dicynodonts, and certain other groups, survived with reduced diversity beyond the end-Permian extinctions into the Triassic, whereas the dinocephalians and gorgonopsians died out. Other important reptile groups from the Permian include the anapsids (e.g. captorhinomorphs) and primitive diapsid reptiles, whose descendants were to dominate the course of reptile evolution to the present day.

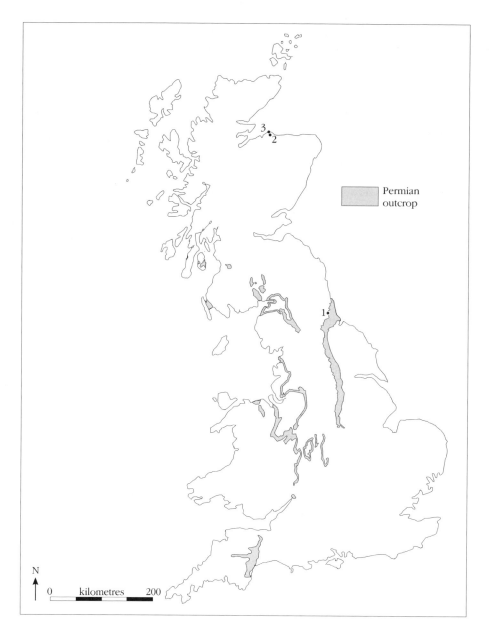

Figure 3.1 Map showing the distribution of Permian rocks in Great Britain. GCR Permian reptile sites: (1) Middridge; (2) Cutties Hillock; (3) Masonshaugh.

The best information on Late Permian reptilian evolution comes from the early parts of the Beaufort Group of the Karroo Basin in South Africa, with supplementary information from Madagascar, central Europe and eastern Russia, all of which confirm the general story. The British sites illustrate parts of this story, with typical South African/Russian-style dicynodonts from Elgin, and smaller diapsid reptiles, like those of Madagascar and central Europe, from Durham.

BRITISH PERMIAN REPTILE SITES

Early Permian fossil reptile bones have been reported from Whitemoor Brickpit, Kenilworth (SP 294717), probable source of remains of the large amphibian *Dasyceps bucklandi*, and certain source of the 'pelycosaur' mammal-like reptiles *Sphenacodon britannicus* (Huene, 1908a) (type: BGS(GSM) 22893–4) and *Haptodus grandis* Paton, 1974 (type: WARMS Gz 1071) (Huene, 1908a; Paton, 1974b; Reisz, 1986, p. 78). The

Enville Beds here contain a limited flora including the conifer *Lebachia* (= *Walchia*) dated as earliest Permian (Shotton, 1929; Smith *et al.*, 1974). A second site, 'one mile north-west of Coventry', yielded a jaw bone of *Ophiacodon* sp. in the Kenilworth Breccia (Murchison and Strickland, 1840; Paton, 1974b).

Other sites have yielded Early Permian footprints (McKeever, 1990, 1991). The key localities are in Dumfriesshire at Corncockle Muir (NY 086870), Locharbriggs (NX 907813), and Greenmills (NY 023692), all old quarries in the Locharbriggs Sandstone Formation (Brookfield, 1978a), which is the same as the Dumfries Sandstone of Smith *et al.* (1974). Footprints were reported from the 1820s onwards, these being some of the first tetrapod footprints recorded in the literature (Grierson, 1828; Harkness 1850, 1851; Hickling, 1909; Sarjeant, 1974; Delair and Sarjeant, 1985). Similar footprint faunas have been found in the Penrith Sandstone at Penrith, Cumbria (?NY 5729) (Smith, 1884; Sarjeant, 1974). None of these could be selected as a GCR site since they have all been filled to a greater or lesser extent. The main hope is that current sporadic quarrying in the Dumfries area may reveal more footprints.

Late Permian reptiles are known from the Marl Slate of the Durham area, in quarries at Eppleton, Middridge and Quarrington (Mills and Hull, 1976; Bell *et al.*, 1979; Evans and King, 1993), and from the Cutties Hillock Sandstone Formation of Cutties Hillock Quarry, near Elgin, Morayshire (Benton and Walker, 1985).

Late Permian reptile footprints have been reported from the Lower Magnesian Limestone of Rock Valley Quarry, Mansfield Nottinghamshire (SK 524613), now filled in (Hickling, 1909; Sarjeant, 1974, pp. 332–4) and from Poltimore, Devon in the Broadclyst Sandstone Member of the latest Permian Dawlish Sandstone Formation (Clayden, 1908a, 1908b; Warrington and Scrivener, 1990). Footprints are also known from Masonshaugh Quarry, and other sites, in the Hopeman Sandstone Formation of the Morayshire coast (Peacock *et al.*, 1968; Benton and Walker, 1985; McKeever, 1991). Three locations are selected as GCR sites to represent British Permian reptiles:

1. Middridge, Durham (NZ 24552535). Upper Permian (Ufimian–lowest Kazanian), Marl Slate.
2. Cutties Hillock, Grampian (NJ 185638). Upper Permian (Tatarian), Cutties Hillock Sandstone Formation.
3. Masonshaugh, Cummingstown, Grampian (NJ 125692). Upper Permian (Tatarian), Hopeman Sandstone Formation.

MIDDRIDGE, DURHAM (NZ 24552535)

Highlights

Middridge Quarry has been the source of several fossil reptile specimens from the Marl Slate. These reptiles are close to the origin of groups that became important later, such as lizards and dinosaurs. Middridge is Britain's best Upper Permian reptile locality.

Introduction

The Upper Permian Marl Slate exposed in a quarry and railway cutting 1 km south-south-west of Middridge, and close to East Thickley and Thickley Wood, has long been known (e.g. Hancock and Howse, 1870a, 1870b) for its rich fossil plant, invertebrate and vertebrate assemblages. There is another quarry, Old Towns Quarry (NZ 257246), about 1 km to the south-east, and closer to Newton Aycliffe than to Middridge. However, the reptile site is almost certainly the former, sometimes termed Thickley Quarry. Extensive collections were made in the 19th century, and these include important specimens of the reptiles *Protorosaurus*, *Adelosaurus* and the 'amphibian' *Lepidotosaurus*. The sections of the quarry that lie near the railway line and the side of the railway cutting are now rather overgrown and the Marl Slate is no longer visible. However, a new excavation in the floor of the eastern end of the old quarry exposes a good section right through the Marl Slate and gives clear access to the fossiliferous beds (Mills and Hull, 1976, pp. 137–8; Bell *et al.*, 1979). The Marl Slate here has already produced abundant fossils which include possible reptile bones (Bell *et al.*, 1979, p. 452), and there is a good chance of further discoveries.

Description

Middridge Quarry and railway cutting expose sections in the lowest portion of the Upper Permian which rests unconformably on Carboniferous sediments. Typical sections taken in the new pit at Middridge show the following sequence (Bell *et al.*, 1979, p. 445):

	Thickness (m)
Lower Magnesian Limestone	4+
Marl Slate	2.58–2.76
calcareous laminated siltstones and thin silty limestones	(1.47–1.60)
laminated limestone (upper invertebrate bed)	(0.02–0.03)
calcareous laminated siltstones and thin silty limestones	(1.09–1.13)
Basal Permian Breccia	
Calcareous breccia (lower invertebrate bed) with abundant *Lingula* in the top (0.02–0.03 m)	0.38–0.42
-------- unconformity --------	
Lower Coal Measures	
Thin-bedded micaceous sandstones and shales	1.20

The new pit exposed the Basal Breccias (?Lower Permian) which may be equivalent to the breccias observed elsewhere in Durham, Yorkshire and North Nottinghamshire lying below the Lower Permian Yellow Sands (Smith *et al.*, 1974; Smith, 1989; Smith and Taylor, 1992). The Yellow Sands are not seen at Middridge.

The Marl Slate is well represented, compared with the thicknesses of 0–3 m elsewhere in south Durham. It comprises a succession of rusty brown-weathering, thinly laminated, calcareous siltstones and thin silty limestones rich in bituminous and other organic material. There is a thin, highly fossiliferous laminated limestone (upper invertebrate bed) just over 1 m above the base of the Marl Slate. Pyrite, galena and sphalerite occur as spherulitic aggregates, small veins and as a partial replacement of some fossils (Bell *et al.*, 1979).

Numerous fossils have been found in the Marl Slate at Middridge, in addition to the reptiles and amphibians (Pattison *et al.*, 1973; Bell *et al.*, 1979). These include 12 genera of plants (Thallophyta, Pteridophyta, Pteridospermae, Coniferales), as well as a wide selection of invertebrates (foraminifers, bryozoans, brachiopods, bivalves, nautiloids and ostracods) and fish. The fishes are represented by isolated scales and fragments, as well as by a few complete flattened specimens. Typical genera are the shark *Wodnika*, the holocephalian *Janassa*, the palaeoniscoids *Acentrophorus*, *Acrolepis*, *Dorypterus*, *Palaeoniscum*, *Platysomus* and *Pygopterus*, and the coelacanth *Coelacanthus*. Some fish remains are found in coprolites deposited by other fishes or by tetrapod predators.

The reptile remains were found in the Marl Slate, and the amphibian just above (Hancock and Howse, 1870a, 1870b). Hancock and Howse (1870a, p. 556) state that 'it is, in the middle, or nearly so, of this yard of Marl-Slate that Mr. Duff has found . . . the remains of two species of reptiles . . .'. They then note (p. 557) that the amphibian *Lepidotosaurus* was found 'at about seven feet above the Marl-Slate proper'. Hancock and Howse (1870a) were referring to a section taken at Middridge by Sedgwick (1829), and it is clear that by 'Marl-Slate proper', they refer to the lower portion of the 'Marl-Slate' of Bell *et al.* (1979). A height of 7 ft (*c.* 2 m) above this 'Marl-Slate proper' would appear to lie near the base of the Lower Magnesian Limestone, an assignment noted by Pattison *et al.* (1973, p. 232). However, Hancock and Howse (1870a, p. 557) state that the *Lepidotosaurus* specimen was associated with the fossil invertebrates which suggests an assignment to the Marl Slate near the 'upper invertebrate bed' of Bell *et al.* (1979).

Fauna

The amphibian and reptile remains from Middridge are:

?Sarcopterygii/Amphibia
  Lepidotosaurus duffii (Hancock and Howse, 1870a)
    Holotype specimen: NEWHM G.55.38

Diapsida *incertae sedis*
  Adelosaurus huxleyi (Hancock and Howse, 1870b)
    Holotype specimen: NEWHM G.26.49

Diapsida: Archosauromorpha: Prolacertiformes: Protorosauridae
  Protorosaurus speneri Meyer, 1830 (described in Hancock and Howse, 1870b)
    Holotype specimen: NEWHM G.55.46

Interpretation

The Marl Slate is interpreted as a shallow-water marine deposit. It is generally reckoned to be the oldest unit in the British Late Permian, and is treated as a correlatable stratigraphic marker that stretches from north Nottinghamshire, through central and east Yorkshire, south Durham, the Durham coast and into the North Sea (Smith *et al.*, 1974; Smith, 1989; Smith and Taylor, 1992). It is correlated with the Kupferschiefer of north-west Europe (Lower Zechstein).

The specimen of *Lepidotosaurus* shows numerous ribs, large scales and a partial skull. Hancock and Howse (1870a) were convinced that it was a labyrinthodont amphibian, but the ganoid scales, ribs and skull look more like those of a bony fish (?lung fish) than an amphibian. There has been no recent work on this specimen.

Hancock and Howse (1870b) described a specimen of a small (1 m+) reptile from Middridge (Figure 3.2A) which they assigned to *Protorosaurus speneri* (Meyer, 1830), described previously from the Kupferschiefer of Germany. Seeley (1888b) further described *P. speneri*, and made speculations on its relationships. The Durham *P. speneri* is represented by a series of 35 or 36 vertebrae and casts of vertebrae, as well as a few partial ribs and a fragment of ?pelvis. *Protorosaurus* is best characterized by its long neck, perhaps developed in relation to a semi-aquatic mode of life.

The taxonomy of *Protorosaurus speneri* has been seen as problematic in the past, and it has been variously related to the euryapsids and diapsids. Recent cladistic analyses of diapsid relationships (e.g. Benton, 1985; Evans, 1988b) have shown that *Protorosaurus* is a basal prolacertiform, related to Triassic forms such as *Prolacerta*, *Macrocnemus* and *Tanystropheus*.

In the same paper as they described the Durham *Protorosaurus speneri*, Hancock and Howse (1870b) described a new species, *P. huxleyi*, distinguished from *P. speneri* on the basis of differences in rib structure and limb proportions. Watson (1914) confirmed differences between the skeletons of *P. huxleyi* and *P. speneri*, and erected the new genus *Adelosaurus*. The type specimen of *A. huxleyi* (Figure 3.2B) is a fairly complete skeleton exposed in ventral view. It is about 130 mm long, and shows the trunk, both forelimbs and one hindlimb; a large portion of the tail and the skull are missing. *Adelosaurus* is broadly similar to *Protorosaurus*, but differs in several respects, notably in the proportions of the humerus and cervical vertebrae, in the length of

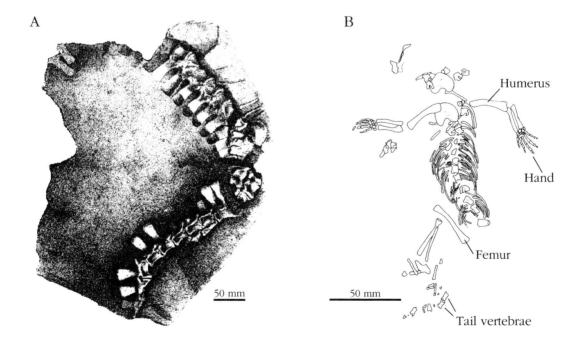

A

B

Humerus

Hand

Femur

Tail vertebrae

50 mm

50 mm

Figure 3.2 Reptile specimens from the Late Permian Marl Slate of Middridge Quarry, County Durham: (A) *Protorosaurus speneri* Meyer, 1830, part of the backbone (after Hancock and Howse, 1870b); (B) *Adelosaurus huxleyi* (Hancock and Howse, 1870), partial skeleton (after Evans, 1988b).

the neural spines, and in size (estimated length of skeleton 250 mm, as opposed to 1 m+ in *Protorosaurus*). *Adelosaurus* was probably a fully terrestrial form (Watson, 1914; Evans, 1988b), and it may represent an immature (?neotenous) individual (Evans, 1988b).

The taxonomic position of *Adelosaurus hux-leyi* has been difficult to resolve. Vaughn (1955) was unable to establish any definite relationship between *Adelosaurus* and other early amniotes and left it *incertae sedis*, while Huene (1956) and Kuhn (1969) referred it to the Broomidae, and Romer (1966) to the Younginiformes or Protorosauridae. Haubold and Schaumberg (1985) classified *Adelosaurus* as a junior synonym of *Protorosaurus speneri*. Evans (1988b) reassigned the reptile to *Adelosaurus* and concluded that, in the absence of diagnostic features such as the ankle and skull, the taxonomic position remained equivocal. Derived diapsid features noted by Evans (1988b) include the strong humerus with poorly expanded ends, the strong sigmoidal curvature of the femur, and a triangular ilium in the pelvic girdle.

Comparison with other localities

Fossil reptiles have been found in the Marl Slate of Durham at Eppleton Quarry, or High Downs Quarry (NZ 360483) near Hetton-le-Hole, which has produced a specimen of *Coelurosauravus* (= *Gracilosaurus, Weigeltisaurus*; Pettigrew, 1979; Evans, 1982; Evans and Haubold, 1987). A new specimen of *Protorosaurus* has been reported from Quarrington Quarry (NZ 329378; Evans and King, 1993).

Outside Britain, the most closely comparable tetrapod-bearing formation to the Marl Slate is the Kupferschiefer of Germany. This is a similar, fine grained, flaggy rock in which skeletons are well preserved, flattened on individual laminae. The Kupferschiefer has produced numerous specimens of *Protorosaurus* and *Coelurosauravus* as well as nearly identical fishes, invertebrates, and plants. Haubold and Schaumberg (1985), reviewing the Kupferschiefer fauna, also note the presence of a pareiasaur (*Parasaurus*) and *Nothosauravus* which they tentatively link to the aquatic diapsid *Claudiosaurus* (Late Permian, Madagascar). The glider *Coelurosauravus* is also known from the Lower Sakamena Formation of Madagascar (Carroll, 1978; Evans, 1982; Evans and Haubold, 1987).

Conclusions

Middridge Quarry is the best tetrapod locality in the Marl Slate of the British Upper Permian sequence. The remains found in the last century are well preserved, are very important in themselves, and allow correlation of the Marl Slate with the German Kupferschiefer. The diapsid reptile *Protorosaurus* lies at the base of the archosauromorph branch of reptile evolution, and is a member of a group of Upper Permian diapsids important in establishing the wider ancestry of all Mesozoic and Cenozoic groups. Relatively little is known of contemporary diapsid faunas in northern Pangaea, which adds to the value of the other diapsids, *Adelosaurus* and *Coelurosauravus*.

The palaeontological importance of the fossil reptiles from here and the potential for future discoveries with re-excavation give the site considerable conservation value.

CUTTIES HILLOCK, GRAMPIAN (NJ 185638)

Highlights

Cutties Hillock Quarry is world-famous for its fauna of dicynodonts and pareiasaurs, bulky medium-sized, plant-eating reptiles. Four or more species have been reported, and these provide unique information on the reptiles of the latest Permian, just before a major global mass extinction event at the Permian/Triassic boundary.

Introduction

The main Cutties Hillock Quarry lies concealed in Quarrywood Forest, in the eastern portion of Quarry Wood, 400 m south-east of Quarrywood School, and is reached along forest roads. The quarry, now mostly overgrown, exposes sections in aeolian units of the Cutties Hillock Sandstone Formation. Some fresh rock has been broken up at the eastern end, where access is easiest. The quarry yields an important fossil reptile fauna of latest Permian age which includes at least two genera of dicynodont, a specialized pareiasaur, and a possible procolophonid, the chief references to which are: Newton (1893), Walker (1973), Rowe (1980), Benton and Walker (1985) and Maxwell (1991). Further commercial working would doubtless yield more fossils in view of the number collected between 1885 and 1890.

Cutties Hillock quarry was opened for building stone in the early 19th century. Many of the buildings in Elgin, including the Town Hall, are built of sandstone from this quarry. The uniform nature of the stone also made it suitable for mill stones, and it is probable that this is the millstone quarry referred to by Harkness (1864) and others.

Fossil reptiles were collected around 1884, and displayed at the Aberdeen meeting of the British Association in 1885. Further nearly complete skeletons were obtained in 1884 and 1885 (Judd, 1885, 1886a, 1886b; Traquair, 1886) and these were described by Newton (1893) as species of the new dicynodont genera *Gordonia* and *Geikia*, and the new horned pareiasaur *Elginia*.

The Geological Survey drove test pits in the quarry in 1885, in an attempt to settle the contentious question of the true age of the Elgin reptile beds: most others had admitted their New Red (Permo-Triassic) age by that time. It was agreed by all that the reptiles had been found in the working portion of the quarry, and that a diagnostic Old Red Sandstone (Devonian) fish (*Holoptychius*) had been found 20–25 ft below in the trial pit. Judd (1886a, pp. 400–2) claimed he could identify a pebble band between two sandstone units, presumably marking the base of the New Red Sandstone. However, Linn, the Survey geologist, and J. Gordon Phillips, the Elgin Museum curator, did not see this pebble bed, and it might have been merely a local phenomenon (Gordon, 1892, p. 242; Peacock *et al.*, 1968, pp. 73–5).

A New Red, possibly Triassic, age was widely accepted by 1890 for the sandstones of Lossiemouth and Spynie which lie nearby, and it was assumed at first that the Cutties Hillock animals could be of the same age. A Permian age was, however, proposed early on (Taylor, 1894; Huene, 1902; Watson, 1909a), but Walker (1973) tentatively suggested a lowermost Triassic assignment. Benton and Walker (1985) opt firmly for a latest Permian (Tatarian) age on the basis of comparisons of the reptiles with independently dated faunas in southern Africa and Russia.

Description

The quarries at Cutties Hillock comprise the type locality for the Cutties Hillock Sandstone Formation (*sensu* Benton and Walker, 1985, pp.

215–16). The sandstones of Cutties Hillock were formerly supposed to represent only the lower part of the Hopeman Sandstone Formation. Because of lithological similarity and tracks found near Cutties Hillock Quarry, Watson and Hickling (1914) correlated the 'Sandstones of Cutties Hillock' (Quarry Wood) and Hopeman. This has been accepted until recently by most authors (Peacock *et al.*, 1968; Williams, 1973). Benton and Walker (1985) questioned the validity of such a correlation and erected the Cutties Hillock Sandstone Formation to include the reptile-bearing beds around Cutties Hillock, and to distinguish them from the coastal series of rocks, the Hopeman Sandstone Formation.

The Cutties Hillock Sandstone Formation is a 30–45 m thick succession of coarse- to medium-grained, predominantly aeolian sandstones which outcrop as a series of isolated fault-bounded blocks in a belt stretching south-south-west from the district of Cutties Hillock. Two main units of the formation are recognized (Peacock *et al.*, 1968; Williams, 1973), comprising a lower member of up to 4 m of pebbly sandstones and an upper member, about 30 m thick, of large-scale, yellow to light-brown, cross-bedded sandstone (Figure 3.3). The base of the formation lies discordantly on Old Red Sandstone. The lower pebbly beds have been interpreted as sheet-flood deposits, but occasional pebble beds, up to 20 m thick, may contain ventifacts, providing evidence of wind erosion during deposition of at least a part of the unit (Mackie, 1902; Watson, 1909b; Watson and Hickling, 1914). The sandstones show unidirectional foresets which indicate fossil barchan dunes and star dunes (Williams, 1973). The sandstone is reworked Old Red Sandstone and is lithologically very similar, which explains the difficulty in identifying the boundary. The petrology of most of the sandstones shows aeolian characters, such as abundant well-rounded, millet-seed quartz grains (Williams, 1973).

Judd (1886a, pp. 400–2) presented the following section, recorded from the trial pit (as summarized in Peacock *et al.*, 1968, p. 74):

Thickness (ft)

Coarse sandstone, white to pale
    yellow, often felspathic and gritty,
    becoming pebbly downwards;
    five reptiles recovered from
    one horizon, and one from the
    course below   20

Figure 3.3 Part of the worked face at the east end of the main quarry, Cutties Hillock, showing cross-bedding. The fossil reptile remains were recovered from the foot of the cross-bedded units. (Photo: M.J. Benton.)

Thickness (ft)

grades into	
Conglomerate; pebbles of white and purple quartz up to fist size	*c.* 4
sharp contact	
Finely laminated, pink and red sandstone with much false bedding. Yielded at base *Holoptychius nobilissimus*	13

The Cutties Hillock Sandstone Formation broadly correlates with the Hopeman Sandstone Formation (Peacock *et al.*, 1968; Warrington *et al.*, 1980; Glennie and Buller, 1983) on the basis of striking lithological similarities and the presence of footprints that might have been formed by reptiles like those of Cutties Hillock. An associated footprint and other trackways have been discovered on Quarry Hill near the main Cutties Hillock quarries (Linn, 1886; Huene, 1913; Watson and Hickling, 1914), and A.D. Walker (pers. comm. to M.J.B., 1990) discovered a dicynodont trackway *in situ* in the main quarry. A slab in Elgin Museum, showing footprints with a tail-drag on top of ripple marks, probably came from 'Robbies Quarry', the position of which is uncertain, but it was apparently one of the Crownhead group of quarries, on the south side of Quarry Wood Hill.

The reptiles *Elginia, Gordonia* and *Geikia* all came from Cutties Hillock Millstone Quarry (NI 185638), apparently from aeolian sandstones just above the pebbly sandstones. Judd (1886a, pp. 400-1) noted that 20 ft (6.2 m) of the 'Reptiliferous Sandstone' was to be seen above the pebbly layers, that the remains of five reptiles all came from one horizon and that a sixth came from the bed below. Phillips (1886) confirmed this. Gordon (1892, p. 242) referred to 'a portion of this conglomerate containing reptilian remains'. Newton (1893, pp. 462, 466) also noted that the specimens of *Gordonia juddiana* and *Geikia elegans* contained pebbles in the matrix like those of the 'conglomerate' bed. Similar pebbles are also preserved in the slabs containing *Gordonia duffiana*. These respectively (ELGNM, 1978.559.1, 2) show quartz pebbles up to 20 mm and up to 7 mm in diameter.

In general, the reptile skeletons are preserved in articulation, with most elements in their natural positions. Skulls are usually in position aligned with the attitude of the rest of the anterior skeleton. However, some remains lack certain elements. The type specimen of *Elginia* (BGS(GSE)4783–8) lacks its lower jaws. A record of the natural association of parts has been lost in some specimens because of poor techniques of collection, and many blocks in museum collections are no longer associated. Most of the reptiles are preserved on their sides, although the isolated pelvis (NMS, 1966.42.3) lies horizontal to the bedding. Some elements of the skeletons, however, may pass vertically through bedding (e.g. vertebrae and limb bones in *G. duffiana*; ELGNM, 1978.559.1, 2). The alignment of skulls in relation to the bedding seems to be related to skull broadness and length; narrow long skulls are usually preserved sideways on to bedding whereas broad backed skulls, such as the skull of *Elginia*, generally lie flat.

The fossil bones are preserved in the form of natural moulds from which the bone has been removed by percolating solutions. The bone/rock interface is frequently stained with black material containing iron, manganese and cobalt (Newton, 1893, p. 425). The cavities may be deformed to the extent that opposite walls may almost touch, and skulls are often vertically compressed (Newton, 1893; Walker, 1973; Rowe, 1980). Prefossilization damage is rare, but specimens lacking certain elements (e.g. the skull of *G. duffiana*) suggests disarticulation through erosional forces or through the activities of scavengers. The open cavities permit casts to be made for study, and various synthetic, flexible, rubber-like materials (e.g. RTV silicone rubber, PVC) provide excellent representatives of the original bone morphology.

Fauna

Anapsida: Pareiasauridae
  Elginia mirabilis Newton, 1893
    2 individuals: BGS(GSE) 4783–8, ELGNM, 1978.550

Anapsida: ?pareiasaurid
  'procolophonid' of Walker, 1973, p. 179
    1 individual: EM, 1978.560; ?BMNH R4807

Synapsida: Therapsida: Dicynodontia
  Gordonia traquairi Newton, 1893
    3 individuals: BGS(GSE) 4805–6, 11703, ?ELGNM, 1978.550
  Gordonia huxleyana Newton, 1893 (?= *G. traquairi*)
    2 individuals: BGS(GSE) 4799–802, 11704–5, ?ELGNM, 1978.549
  Gordonia duffiana Newton, 1893 (?= *G. traquairi*)
    1 individual: ELGNM, 1978.559
  Gordonia juddiana Newton, 1893 (?= *G. traquairi*)
    1 individual: ELGNM 1890.3
  Geikia elginensis Newton, 1893
    1 individual: BGS(GSM) 90998–1015
  'dicynodonts indet.'
    7 individuals: BMNH R4794, ELGNM, 1935.8, 1978.558, 886, NMS, 1956.8.3, 1966.42.1–3, 1984.20.7

Interpretation

In the absence of any associated fossils, the reptiles provide the only means of dating the Cutties Hillock Sandstone. Comparison with similar forms from South Africa led Walker (1973) to suggest an age in either the upper *Cistecephalus* or *Daptocephalus* Zone (uppermost Permian), or more probably in the *Lystrosaurus* Zone (lowermost Triassic). Rowe (1980) showed that the close relatives of *Geikia*, the cryptodontid dicynodonts, all come from the Late Permian of South Africa or Zambia (i.e. *Daptocephalus* Zone) and implied a similar age for the Cutties Hillock Sandstone. Benton and Walker (1985) accept a latest Permian age for the Cutties Hillock Sandstones, based on the nature of the dicynodonts and the pareiasaur, known elsewhere only from the Late Permian. These reptile-defined biozones are generally (e.g. Anderson and Cruickshank, 1978) assigned to the Tatarian Stage.

Gordonia and *Geikia* are dicynodonts, members of a group of specialized, herbivorous mammal-like reptiles with beak-like snouts, most of which had no teeth except for a pair of 'tusks' midway along the upper jaws. *Gordonia* (Figure 3.4A) is represented by the remains of skulls and skeletons of between eight and thirteen individuals; Newton (1893) established four species (*G. traquairi*, *G. huxleyana*, *G. duffiana*, *G. juddiana*), but they are probably all synonymous, the differences being the result of individual variation, age and sex differences, and the susceptibility of the Cutties Hillock fossils to early post-depositional distortion (Walker, 1973; King, 1988, p. 93). *Gordonia* has a heavy, broad skull, 100–180 mm long, modified to house a powerful musculature for mastication. *Gordonia*, the only known member of its infra-order from Europe, is a generalized dicynodont – a group of herbivorous mammal-like reptiles that have highly reduced dentition, and often only a pair of bony canine tusks. Cluver and King (1983, p. 268) stated that *Gordonia* was 'possibly related to *Kingoria* or *Dicynodon*', whereas King (1988, p. 93) synonymized *Gordonia* with *Dicynodon*, a genus known otherwise from the Late Permian of South Africa.

The single specimen of *Geikia* (Figure 3.4B) has no teeth at all, the skull is very short and the snout is square. The foreshortening of the skull, and its great breadth at the back, could both be connected with the development of a powerful biting mechanism to deal with tough vegetation. The foreshortening of the skull is like that of *Lystrosaurus* from Gondwanaland, but the similarity is only superficial. In *Geikia* the intertemporal area is broadened, the interorbital area broadened and depressed, and the premaxillae descend abruptly vertically. Rowe (1980) has redescribed the specimen and assigned it to the Cryptodontidae, a family otherwise known from the Late Permian of South Africa and Zambia. He also placed '*Dicynodon*' *locusticeps* (Huene,

1942) from the Late Permian Lower Bone-Bearing Series of Kingori, Tanzania in the genus *Geikia* and noted that the closest relative of *Geikia* is *Pelanomodon*. Cluver and King (1983) placed *Pelanomodon* in the new family Aulacephalodontidae, a view confirmed by Cruickshank and Keyser (1984) and King (1988, pp. 88–9).

Elginia was a pareiasaur with a highly spinescent skull (Figure 3.4C). The 210 mm long holotype skull is broad and covered with rough pits and spines of various lengths and sizes. Other remains of *Elginia* include vertebrae and a sacrum probably belonging to the holotype, as well as an undescribed partial skeleton and skull. The teeth are leaf-shaped, indicating that *Elginia* was probably a herbivore. The 'frill' at the back of the skull was probably to protect the neck, and the spines are also defensive structures (compare Cretaceous ceratopsian dinosaurs). The body was also covered in spinose scutes – overall an animal highly armoured against a predator that has not yet been found, but probably a large cynodont, gorgonopsian, or therocephalian, mammal-like reptile. *Elginia* shows relationship with pareiasaurs from the *Cistecephalus* and *Daptocephalus* Zones of South Africa and Zone IV of Russia, but seemingly it is cladistically more derived (Walker, 1973, p. 181; Maxwell, 1991).

A fourth reptile from Cutties Hillock is represented by a small partial skeleton described by Newton (1893, pp. 461–2, pl. 33, fig. 5) as a tail of *?Gordonia*. This specimen, consisting of seven dorsal vertebrae, the blades of two scapulae and the blade of an ilium, was later identified as a nearly complete postcranial skeleton and was assigned to the Procolophonidae by Walker (1973). A re-examination of the material by P.S.S. (1994), however, revealed characters, including a tall and narrow scapular blade, very wide, flattened neural arches of the dorsal vertebrae, and long laterally projecting ribs, that are shared by pareiasaurids, and hence *Elginia*. Thus, EM, 1978.560 may well represent an immature *Elginia*, and also one of the smallest pareiasaurid specimens known.

Comparison with other localities

Reptiles comparable to those from Cutties Hillock have been obtained from York Tower Hill (Knock of Alves) (NJ 162629) where, in 1953, Walker discovered parts of the skull and jaws of an unnamed dicynodont allied to *Geikia* (Walker, 1973). An unidentified bone in Forres Museum was found in

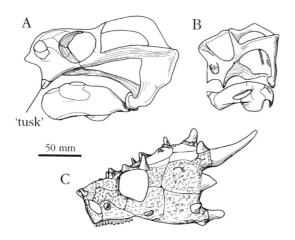

Figure 3.4 Reptile specimens from the Late Permian Cutties Hillock Sandstone Formation of Cutties Hillock Quarry, Morayshire. Skulls of (A) *Gordonia*; (B) *Geikia*; and (C) *Elginia*, all drawn to the same scale. After Benton and Walker (1985).

Crownhead Quarry (NJ 183630) on the south side of Quarry Wood Hill in sandstones of the same age. Apart from a small scrap of bone (Peacock *et al.*, 1968, p. 59), the coastal exposure of the time-equivalent Hopeman Sandstone Formation has yielded nothing except reptile tracks.

The Cutties Hillock fauna shows most similarity with uppermost Permian faunas of southern Africa, especially the *Cistecephalus* and *Daptocephalus* biozones, and Zone IV of Russia. This points to a Tatarian age. There are no comparable localities in the British Isles or in Europe.

Conclusions

Cutties Hillock Quarry is a key Permian reptile locality because of its unique fauna (Figure 3.4) which provides clear links between the Gondwana faunas of southern Africa and the mainland Eurasian faunas of western Russia. The dicynodonts *Gordonia* and *Geikia* are well preserved, and offer much useful palaeobiological information. *Elginia* is one of the most specialized pareiasaurs, being distinguished by its excessive spinescence.

The conservation value of this quarry lies in its uniqueness in Britain, its international importance and potential for future significant finds with reworking.

MASONSHAUGH QUARRY, CUMMINGSTOWN, GRAMPIAN (NZ 125692)

Highlights

Masonshaugh is famous for its fossil reptile tracks. Many complete specimens of trackways were found when the quarry was operational, and these show evidence of many reptiles, small, medium and large, trotting northwards across the sands.

Introduction

Masonshaugh Quarry, situated next to a disused railway line, comprises a 400 m long north-facing exposure on the coast. The quarry lies in aeolian units of the Hopeman Sandstone Formation, and the importance of the site lies in its abundant ichnofauna of tetrapod tracks. The trackways are extremely well preserved and provide almost the sole record of tetrapods from the formation. Masonshaugh Quarry was fully operational in the 1860s, when Martin (*c.* 1860) visited and observed tracks. It was part worked in 1912, and used as a tip in the 1930s as appears in County valuation rolls. The site is now badly weathered and much rubbish has been tipped in front of it. However, the faces are free of overgrowth, and fresh workings would doubtless yield new finds *in situ.*

An extensive quarrying industry was established in Burghead and the coastal area in the 1790s. Many of the townspeople were quarriers and stonemasons and 'five large boats, with six people in each, are also employed in transporting stones from the quarries, to different parts of the country' (Anon., *in* Sinclair, 1793, vol. 8, p. 390). The stone was used largely in 'harbour and sea-wall building' (Duff, 1842, p. 25).

In October 1850, Captain Lambart Brickenden, a fossil collector who had moved to Elgin from Sussex, obtained a slab from quarrymen at Masonshaugh (Figure 3.5A) which showed a trackway of 34 footprints of a small animal. Since the rock was considered to be of Old Red Sandstone age, the find occasioned great excitement locally (Anon., 1850a, 1850b) and in the scientific world (Lyell, 1852, p. ix; Brickenden, 1852) as evidence of the oldest tetrapod, together with the *Leplopleuron (Telerpeton)* from Spynie (Benton, 1983c).

Numerous further slabs were collected during the 1850s (Beckles, 1859; Huxley, 1859b; Hickling, 1909). Huxley (1859b, pp. 456-9, pl. 14, figs 4, 5) described and figured footprints which were larger than those found by Brickenden. At the same time, Beckles (1859) hired workmen and carpenters and extracted many tracks from the quarry, which was supplying material for the new railway from Burghead to Hopeman on the coast.

Huxley (1877, pp. 45-52, pl. 14-16) described the 'ichnites of Cummingstone' further and named them *Chelichnus megacheirus* (Figure 3.5B). Hickling (1909, pp. 12-14, pl. 2, figs 6-8) compared the Elgin tracks with those from Mansfield, Notts., and Penrith, and concluded that the Cummingstone (*sic*) beds were Permian in age. The tracks were revised by McKeever (1990, 1991).

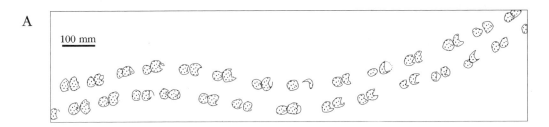

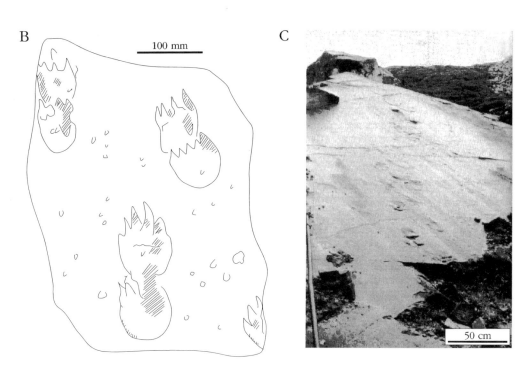

Figure 3.5 Reptile footprints from the Late Permian Hopeman Sandstone Formation of Masonshaugh Quarry, Morayshire: (A) small prints; (B) medium prints, *Chelichnus megacheirus* Huxley, 1877; and (C) large prints. After Benton and Walker (1985).

Description

Masonshaugh Quarry lies in the Hopeman Sandstone Formation, and is just east of the fault separating that unit from the Burghead Sandstone Formation. In the quarry, 8 m faces of orange-weathered, jointed sandstone are exposed, extensively silicified and hardened near the fault. The well-rounded sand grains and large-scale cross-bedding suggest that the sandstones of Hopeman are here of aeolian origin, although some water-laid pebbly beds were found at the west end of the outcrop near the old railway line.

The Hopeman Sandstone Formation is best observed on the coast between Cummingstown and Covesea Skerries and Haliman Skerries, where it is some 60 m thick. The coastal exposures show large-scale cross-bedding in sandstones generally composed of well-rounded quartz grains and feldspar, often of high sphericity, with only a little mica (Peacock *et al.*, 1968, p. 59). These are features typical of aeolian deposition. Well-developed aeolian dune features have been observed along the coast, and these include complex star dunes which indicate the major wind direction from the NNE, secondary winds from the SSE, and subordinate winds from the NW (Clemmensen, 1987). Rarer lenses of coarse sandstone and well-rounded pebbles with small-scale cross-bedding, as well as contorted beds, indicate times of fluvial deposition (Peacock, 1966). Glennie and Buller (1983) interpret the contorted beds as the result of marine flooding. Williams (1973, pp. 10–11) identified four phases of barchan and seif dune formation, each cycle being followed by water-deposited, contorted beds and sheet-flood or playa-lake deposits.

The tracks are preserved as depressions, often 'smudged', and may be associated with ripple marks and sun-cracks on sub-horizontal flagstones. McKeever (1991) studied the nature of track preservation in the Hopeman Sandstone Formation, and argued that they were not imprinted on dry loose sandy dunes. This has bearing on a debate about track formation in aeolian situations, with some workers (e.g. Brand, 1979) arguing that dune beds bearing tracks must have been laid down entirely underwater, while others (e.g. McKeever, 1991) accept the extensive evidence for aeolian deposition, but find evidence for local and short-term flooding or rainfall. McKeever (1991) found clay minerals in the footprint-bearing levels from Early Permian track sites in Dumfriesshire, clear evidence for wetting of the particular layers. No such clay minerals were found in track-bearing horizons in the Hopeman Sandstone Formation, but other evidence for the presence of water (fluvial lenses; contortion of beds) suggests that footprints may be found where the dune faces are wetted.

The age of the Hopeman Sandstone Formation has, since 1900 at least, been assumed to be Upper Permian, largely because of comparisons with the likely track-makers from Cutties Hillock. Indeed, closer comparisons of the tracks with comparative ichnofaunas in continental Europe and North America, confirm the Late Permian age. On the other hand, Glennie and Buller (1983) postulated an Early Permian age, since they considered that the Hopeman Sandstone Formation was laterally equivalent to the Early Permian Weissliegend (White sandstones) of Germany and the North Sea. In addition, they interpreted the heavily contorted beds in the Hopeman Sandstone Formation (Peacock, 1966) as the result of flooding by the Zechstein Sea (early Late Permian). These views were disputed by Benton and Walker (1985), and the Late Permian age confirmed by a comparison of the footprints with ichnofaunas elsewhere (McKeever, 1991).

Fauna

At least two kinds of footprint have been identified from the Hopeman Sandstone Formation. Footprint Type A is represented by a slab collected in 1850 and is important as the first fossil from Elgin recognized as reptilian (Brickenden, 1852;

Hickling, 1909, p. 13). The footprints consist of roughly circular impressions, 30-40 mm long, with the fore and hind feet forming tracks that nearly touch. The stride length is 110-120 mm, the width of the trackway is 80-90 mm and there is no sign of toe marks.

Footprint Type B (Huxley, 1859b, pp. 456-9, pl. 14, figs 4, 5; 1877, pp. 45-52, pl. 14-16; Hickling, 1909, p. 13, pl. 2, fig. 6; Haubold, 1971, p. 37, fig. 22 (4)) was named *Chelichnus megacheirus* by Huxley (1877) and type C15 by Hickling (1909). The fore and hind feet were clearly different. The print of the fore foot (smaller print) is semicircular, about 40 mm long and 60 mm wide, with impressions of nine or five claws at the front. The sole part of the footprint is 60 mm wide, 30-40 mm long and the claws would measure 10-15 mm. The print of the 'hind foot' (larger print) is longer: 80-90 mm long and 80mm wide, bearing five claw marks at the front 20-40 mm long. The prints overlap in pairs and show a stride length 300-400 mm, with the width of trackway (between midpoints of tracks), c. 150 mm.

There may be a third track type, like Type B, but larger. Huxley (1877, pl. 15, fig. 6) described such a track (prints 170 mm long and 140 mm broad) and with impressions of three claws. These larger tracks measure 150-250 mm long and 100-150 mm wide, and the stride length is 700-800 mm. A slab of such large prints, 100-150 mm wide, and with a stride length of 700-800 mm (Figure 3.5C), were observed *in situ* in Clashach Quarry (Benton and Walker, 1985, p. 208).

McKeever (1990, 1991) has revised the Hopeman Sandstone footprints, and notes the presence of the ichnogenera *Chelichnus, Laoporus, Herpetichnus* and *Palmichnus*. Fuller details of these determinations have yet to be published.

Footprint specimens from Masonshaugh Quarry include: BGS(GSM) 113445 (Brickenden's 1850 specimen) and BGS(GSM) several slabs (Huxley 1859b, 1877). Undescribed material includes ELGNM (six slabs), NMS (several slabs), and other material in Forres Museum, Inverness Museum and MANCH.

Interpretation

Brickenden (1850, 1852) thought tracks of Type A were produced by tortoises, and Huxley (1859b,

p. 459) thought the track type B might have been formed by *Stagonolepis*, but in 1877 (pp. 49-51) he could not ascribe them to any definite fossil amphibian or reptile then known. In a recent review of vertebrate tracks, Haubold (1971, p. 37) describes *C. megacheirus* (Type B) as possibly formed by a dicynodont, and indeed *Gordonia* from Cutties Hillock is the right size.

The footprints may be preserved on low-angle dune foresets, but this has only been observed in a few *in situ* occurrences. The slabs collected in the nineteenth century may include some from horizontal bedding planes. However, there is usually a mound of sand behind each print (Brickenden, 1852; Huene, 1913; Watson and Hickling, 1914), perhaps indicating that the producers of most trackways were moving uphill. These mounds are seen also behind the large footprints at Clashach.

Martin (*c.* 1860) gave a detailed account of the occurrence of tracks at Masonshaugh, and notes that all were heading in one direction (towards today's North Pole). He considered that the producers were moving down to the Moray Firth across the beach to find the sea! Benton and Walker (1985, p. 217), more plausibly, interpret the footprints as individual trackways probably formed by two or three species of mammal-like reptiles (?anomodonts), each displaying a range of sizes, heading across a dune-field towards the depositional basin to the north.

Comparison with other localities

Trackways and individual footprints similar to those from Masonshaugh have been observed in Greenbrae and Clashach Quarries. The 16 m working face in Greenbrae Quarry (NJ 137692) displays large-scale cross-bedding in fine- to medium-grained, yellow-brown sandstone. Evidence of water action includes fine lamination, small channels, ripple marks and small quartz pebbles. In addition to footprints, Peacock *et al.* (1968, p. 59) report an unidentified bone fragment from this locality. This quarry is still worked to some extent for ornamental stone (1990).

Clashach Quarry (NJ 163702) is also still in operation to a small extent and contains stone very like that at Greenbrae. Murchison (1859, p. 429) recorded tracks from this quarry, and some poor specimens were noted by Peacock *et al.* (1968) and Walker (pers. comm., 1990). A range of tracks from small 5 mm 'lizard-like' forms to large 100 mm dicynodont prints was seen *in situ* and on nearby spoil tips at Clashach by M.J.B. (April, 1980).

A third kind of track (Type C) was described from the coastal exposure of Hopeman Sandstone by Huxley (1877) and Hickling (1909, pl. 2, figs 7 and 8), and Watson and Hickling (1914, p. 400, fig. 1) found 'one of the typical Cummingstone footprints' (i.e. Type C prints) in a quarry '300 yards WNW from the Cutties Hillock reptile quarry', from a site that cannot now be identified. Tracks of this type were also seen at Cutties Hillock Quarry in 1878 before the reptiles there were discovered (Peacock *et al.*, 1968, p. 73) and these have occasionally been seen since (Walker, pers. comm., 1981). These footprints are similar to the Type B prints from Masonshaugh, but the toes are broader. The dimensions are: print 30 mm wide, 20 mm long; toes 8 mm long, 6 mm wide. These footprints supposedly differ from Types A and B in having broader toes, but the generally poor preservation of most specimens makes such a distinction inadvisable.

Conclusions

The best British Late Permian tetrapod trackway site. Its importance rests on the diversity of tracks observed there, and their potential in allowing stratigraphic and palaeobiological observations.

Despite the partial degradation of the site by weathering and infill, its importance in Britain and potential for re-excavation give it significant conservation value.

Chapter 4

British Triassic fossil reptile sites

INTRODUCTION: TRIASSIC STRATI-GRAPHY AND SEDIMENTARY SETTING

The British Triassic deposits have a broad U-shaped outcrop in the English Midlands, with a continuation south-westwards to south Wales and Devon (Figure 4.1). Smaller outcrops occur in northwest England, in Northern Ireland and in Scotland (Warrington *et al.*, 1980, figs 2 and 3). The sediments are almost wholly continental ter-

restrial red-beds deposited in fault-bounded basins in southern and western Britain and on the more regionally subsiding Eastern England Shelf, which formed the onshore marginal part of the Southern North Sea Basin (Audley-Charles, 1970; Holloway, 1985). In the Late Permian and Early Triassic, renewed and extensional subsidence in the Wessex Basin, Worcester Graben and the Needwood and Cheshire basins, resulted in the establishment of an axial drainage system which

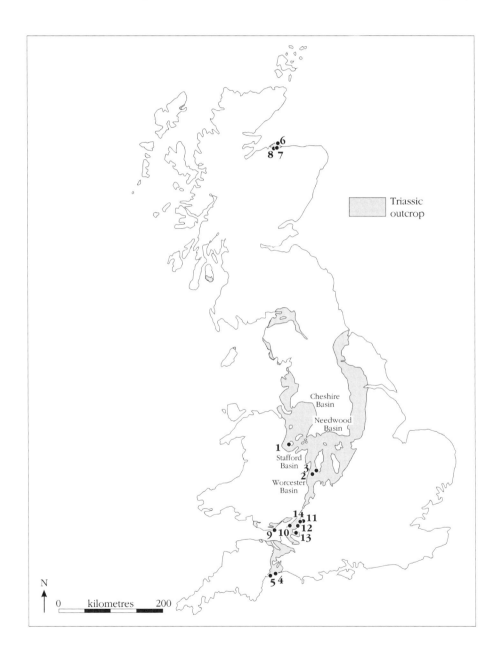

Figure 4.1 Map showing the distribution of Triassic rocks in Great Britain. GCR Triassic reptile sites: (1) Grinshill Quarries; (2) Coten End Quarry; (3) Guy's Cliffe; (4) Sidmouth coast section; (5) Otterton Point; (6) Lossiemouth East Quarry; (7) Spynie; (8) Findrassie; (9) Bendrick Rock; (10) Aust Cliff; (11) Slickstones (Cromhall) Quarry; (12) Durdham Down; (13) Emborough Quarry; (14) Tytherington Quarry.

flowed northwards from the Variscan Highlands (Holloway, 1985). The south-to-north regional palaeoslope, and the proximal to distal depositional pattern which developed, is reflected in the diachronous nature of the Sherwood Sandstone–Mercia Mudstone Group boundary (Figure 4.2), with coarse clastics being deposited in the south, while mudstones and evaporites accumulated farther north (Warrington, 1970a, 1970b; Warrington *et al.*, 1980; Warrington and Ivimey-Cook, 1992). This general sedimentary pattern was complicated locally by the introduction of coarse-grained deposits along basin margins and the deposition of marine–intertidal sediments during Mid Triassic marine incursions. The widespread occurrence of transgressive intertidal facies of Mid Triassic age indicates extremely low relief in central England and suggests that the contemporary vertebrates were disporting themselves in lowland areas close to sea level, a suggestion first offered in 1839 by Buckland, who proposed (1844) a palaeoenvironment of intertidal sandbanks.

In the British Triassic there is a general dearth of biostratigraphically useful fossils (Figure 4.2). This, combined with rather limited vertical facies variations throughout the sequence, has led to problems of correlation across Britain and between the British Isles and abroad. The standard stages of the Triassic were defined using ammonoids in the marine sequences of southern Europe, and it has been hard to correlate the continental Triassic of Britain with these type successions. In Germany the Early to Mid Triassic Buntsandstein ('mottled sandstone') and Mid to Late Triassic Keuper ('red marl') consist of continental facies which compare superficially with parts of the British sequence.

Sedgwick (1829) recognized the British New Red Sandstone as equivalent in part to the German Triassic and considered some units equivalent to the German Buntsandstein and Keuper. Hull (1869) equated the English Bunter Sandstone with the German Buntsandstein (broadly Early Triassic in age) and the Lower Keuper Sandstone with the German Lettenkohle (latest Mid Triassic to early

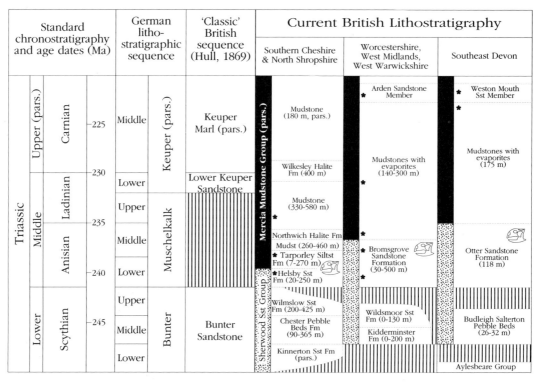

Figure 4.2 The stratigraphy of the British Triassic reptile faunas. Correlations of the standard Triassic divisions and the German Triassic sequence with the British Triassic, as proposed by Hull (1869) for the 'classical' British succession, and by Warrington *et al.* (1980) for currently recognized lithostratigraphical units. Skull symbols indicate the levels of the main tetrapod faunas, and asterisks denote palynological evidence of relative age. Age dates (Ma ± 5) after Forster and Warrington (1985). From Benton *et al.* (1994).

Late Triassic in age). He argued that a major unconformity in the British sequence corresponded to most of the Mid Triassic and represented the Muschelkalk. Warrington *et al.* (1980) advocated the abandonment of the terms 'Bunter' and 'Keuper' as applied in Britain, and established a lithostratigraphic nomenclature with correlations based on palynomorphs and other fossils where possible (Figure 4.2).

Palynological work (Warrington, 1967, 1970b; Geiger and Hopping, 1968) showed that deposits of Mid Triassic age are present in Britain (Figure 4.2), where correlatives of the Muschelkalk, including brackish-water to littoral marine facies, occur in the upper part of the Sherwood Sandstone Group and lower parts of the Mercia Mudstone Group in central and northern parts of England (Geiger and Hopping, 1968; Warrington, 1974a; Ireland *et al.*, 1978; Warrington *et al.*, 1980; Warrington and Ivimey-Cook, 1992).

The Sherwood Sandstone Group includes the former 'Bunter Sandstone' and the arenaceous (lower) parts of the former British 'Keuper'. Its boundaries are diachronous, the lower varying from Late Permian to Early Triassic and the upper from Early to Mid Triassic in age (Warrington *et al.*, 1980). The group comprises up to 1500 m of arenaceous deposits that form the lower part of British Triassic successions. The sandstones are red, yellow or brown in colour, and pebbly units occur, especially in the Midlands. Most of the deposits are of fluvial origin, but there are many aeolian units (Thompson, 1969, 1970a, 1970b).

The Mercia Mudstone Group corresponds broadly with the former 'Keuper Marl', and encompasses the dominantly argillaceous and evaporitic units that overlie the Sherwood Sandstone Group throughout much of Britain. Its lower boundary may be sharp, but there is commonly a passage upwards from predominantly sandy to predominantly silty and muddy facies at a diachronous interface that varies regionally from Early to Mid Triassic in age. The upper boundary, associated with a marine transgression which apparently occurred approximately contemporaneously throughout much of Europe, lies within the Rhaetian Stage. The group comprises dominantly red mudstones with subordinate siltstones. Extensive developments of halite and of sulphate evaporite minerals suggest deposition in hypersaline epeiric seas, connected to marine environments in associated sabkhas, and in playas (Warrington, 1974b).

The Penarth Group, which overlies the Mercia Mudstone Group, consists of argillaceous, calcareous and locally arenaceous formations, predominantly of marine and lagoonal origin. The topmost beds of the Penarth Group (Lilstock Formation, Langport Member), pass up into grey bituminous shales and limestones, which are lithologically indistinguishable from and continuous with the beds of the overlying Jurassic. The top of the British Triassic is placed above the Penarth Group within the Blue Lias, at the point of appearance of the first ammonite, *Psiloceras* (Cope *et al.*, 1980a; Warrington *et al.*, 1980).

In Scotland red-bed sequences assigned to the Permian and Triassic form fairly numerous, although scattered and usually small, outcrops (Judd, 1873) which represent the thin marginal expressions of extensive thicker successions, possibly of different facies, present in important offshore basins. Of these, the small occurrences of Late Triassic deposits in the Moray Firth area and in particular those in the Elgin district (Lossiemouth Sandstone Formation), are important for their datable vertebrate faunas.

REPTILE EVOLUTION DURING THE TRIASSIC

The Triassic Period was a time of major flux among tetrapods. The earliest Triassic faunas were of low diversity by comparison with the Late Permian faunas, a consequence of the end-Permian mass extinction event. The ecological niches left vacant by removal of the old faunas were soon filled in the Early Triassic by several new tetrapod groups and by re-radiation of surviving groups from the Permian. In several parts of Gondwanaland, Early Triassic faunas were uniquely dominated by the dicynodont *Lystrosaurus*, which often comprised 95% or more of the individuals in a fauna. The remainder of these faunas consisted of small plant- and insect-eating therapsids, and some archosaurs. The archosaurs (Triassic forms are often termed 'thecodontians'), new groups of therapsids, particularly cynodonts, and the rhynchosaurs (archosauromorphs) rose to prominence during the Early and Mid Triassic.

Mid Triassic tetrapod faunas worldwide were strikingly similar (Benton, 1983a, 1983b) being dominated by rhynchosaurs or by dicynodonts as herbivores, and with associated fish-eating temnospondyl amphibians and a variety of prolacertiforms, small- to medium-sized insectivores, and procolophonids, modest-sized plant-

and insect-eaters. These animals were preyed on by a range of cynodonts and thecodontians, some of the latter, the rauisuchians, achieving large size (up to 5 m long). These faunas continued into the early part of the Late Triassic, but were apparently decimated at the end of the Carnian (Benton, 1983a, 1983b, 1986a, 1986b, 1991, 1994a, 1994b): the rhynchosaurs and larger therapsids all died out, as well as some smaller groups.

Following this extinction event, the dinosaurs, pterosaurs, crocodilomorphs, sphenodontians (lizard relatives), and other groups radiated during the last 15–20 Ma of the Triassic, the Norian. The thecodontians, particularly the rauisuchians, ornithosuchids, fish-eating phytosaurs, and herbivorous aetosaurs dwindled during this time, and finally died out near the Triassic/Jurassic boundary, an extinction event that was also marked by major effects on marine life.

BRITISH TRIASSIC REPTILE SITES

The British Triassic is for the most part unfossiliferous, and tetrapod faunas occur only sporadically. However, some of these faunas are locally rich. The principal reptiliferous horizons lie within three rock units: the uppermost portion of the Sherwood Sandstone Group of south-west England and the Midlands (Anisian), the Lossiemouth Sandstone Formation (Carnian) of north-east Scotland, and the Penarth Group (Rhaetian), of central and south-west England. These provide excellent information on Mid Triassic terrestrial reptiles, rare elsewhere in the world, on Late Carnian pre-extinction faunas, and on terminal Triassic forms. Virtually all the Mid Triassic bone-bearing sites are selected as GCR sites, as are all the Carnian localities around Elgin, and most of the fissure localities. Aust Cliff is selected as the sole Rhaetian GCR site out of dozens of other candidates, since it has yielded most specimens in the past.

Some of the most unusual British Triassic reptile faunas are the insular assemblages from fissure-fill deposits within fossilized cave systems developed in the Carboniferous Limestone of south-west England and South Wales. The deposits range in age at least from the Late Norian, possibly the Late Carnian, to the Early Jurassic and some have been correlated lithostratigraphically and biostratigraphically with the local marginal Trias (formerly the 'Dolomitic Conglomerate'). The best examples of these fissure sites, as well as sites in the Mid Triassic of the English Midlands and Devon, the Carnian of north-east Scotland and the Penarth Group of south-west England, have been selected as GCR sites.

Reptile bones have not been reported from the British Early Triassic, but Wills and Sarjeant (1970) noted a variety of small reptilian footprints from the Bunter (=Kidderminster and Wildmoor Sandstone Formations) of a borehole at Bellington, Worcestershire. *Cheirotherium* footprints have been observed in the Wilmslow Sandstone Formation in the Wilmslow Waterworks Borehole (Thompson, pers. comm., 1993). The British record of fossil reptiles is also sporadic during Mercia Mudstone Group times: reptiles are represented by undiagnostic dissociated bones from the conglomeratic marginal Triassic of the Bristol district, and rarely from the Arden Sandstone Member of the Midlands and the Weston Mouth Sandstone Member of south-east Devon, both of which are Carnian in age. Some of the fissure deposits around Bristol and in South Wales may date from Late Carnian and Norian times, as might the footprints from Barry, South Wales (see below).

MID TRIASSIC OF THE ENGLISH MIDLANDS

Numerous localities in the English Midlands have yielded fossil reptiles of Mid Triassic age, from the upper part of the Sherwood Sandstone Group (Bromsgrove Sandstone Formation, Helsby Sandstone Formation; Figure 4.2) and the lower part of the Mercia Mudstone Group (Tarporley Siltstone Formation). These, and other neighbouring units, have also yielded significant ichnofaunas (see below).

In the Warwick area, old quarries at Guy's Cliffe, Leek Wooton, Cubbington Heath, Coten End and Leamington have yielded many fragmentary fossil reptiles, but only Guy's Cliffe and Coten End are extant. A number of localities in Leamington and Warwick (e.g. Coten End, SP 29006550; Leamington Old Quarry, SP 325666) have produced remains of the reptiles *Macrocnemus* (type specimen of Owen's *Rhombopholis scutulata* (Owen, 1842a, pp. 538–41, pl. 46, figs 1–5), *Rhynchosaurus brodiei* (Benton, 1990c), *Cladeiodon lloydi* (Owen, 1841b), *Bromsgroveia walkeri* (Galton, 1985a), a possible prosauropod tooth (Murchison and Strickland, 1840, pl. 28, fig. 7a; Huene, 1908b, figs 210–11, 265), the temnospondyls *Mastodonsaurus*, *Cyclotosaurus pachygnathus*, *C. leptognathus* and the fish *Gyrolepis* (Walker, 1969, p. 472). Cubbington Heath Quarry (SP 335694) has yielded *M. jaegeri*, *C. pachygnathus* and *C. leptognathus* (Huxley, 1859c; Woodward, 1908a; Wills, 1916, pp. 9-11, pl. 3). Guy's Cliffe (SP 293667) has produced remains of the jaws of *Mastodonsaurus* sp. (=*M. jaegeri*) (Owen, 1842a, pp. 537-8, pl. 44, figs 4-6; pl. 37, figs 1-3; Miall, 1874, p. 433), probably the first find of a tetrapod to be made in the area, having been collected in 1823 (Buckland, 1837). A ?prosauropod femur and tooth have been recorded from Leek Wooton (SP 289689), but the site of the quarry from which this specimen was collected is uncertain. Elsewhere in the Midlands good remains of *Rhynchosaurus articeps* have been obtained from a series of quarries at Grinshill, Shropshire (SJ 520237), and a fine skull of *C. leptognathus* was collected from Stanton, near Uttoxeter, Staffs (SK 126462) (Woodward, 1904).

Many localities in the Early and Mid Triassic of the English Midlands, especially in Cheshire, have yielded rhynchosauroid and *Cheirotherium* footprints (Tresise, 1993). The richest of these footprint localities is Storeton Quarry, Higher Bebington (SJ 303838), source of hundreds of slabs, but now filled in (Tresise, 1989, 1991). Other localities in Cheshire and Merseyside include Rathbone Street, Liverpool; Delamere Forest; 'Mr Leach's quarry', Runcorn; Beetle Rock Quarry, Runcorn; Overhill, Iveston and Weston Point, all near Runcorn; Runcorn Hill; Flaybrick Hill, Birkenhead; Daresbury; Oxton Heath; Moorhey, near Great Crosby; Eddisbury; Warrington; Five Crosses Quarry, Frodsham; Potbrook, Mottram St Andrews; Wazards Well, Alderley Edge; Haymans Farm Borehole, Nether Alderley, all in the Helsby Sandstone Formation and Tarporley and Lymm in the Tarporley Siltstone Formation. The only Cheshire site that has recently produced footprints is Red Brow Quarry, Daresbury (SJ 567834) and scattered occurrences in boreholes and in field walls (Thompson, 1970a; Ireland *et al.*, 1978; Sarjeant, 1974, pp. 312-13). Localities in Shropshire include Grinshill (site of reptile finds as well - see below) and Oaken Park Farm, Albrighton (Sarjeant, 1974, pp. 316-17). Localities in Derbyshire are Weston Cliff on the River Trent and Dale Abbey, Stanton-by-Dale (Sarjeant, 1974, p. 321) and in Staffordshire, Stanton; Coven, near Brewood; Burton Bridge, Burton-on-Trent; Ashby Road, Burton-on-Trent; Hollington; Townhead Quarry, Alton; Chillington;

Great Chatwell (Sarjeant, 1974, pp. 319-21; Delair and Sarjeant, 1985, pp. 131-2). Localities in Warwickshire are Birkbeck; Shrewley Common; Witley Green, near Preston Baggot; Coten End Quarry, Warwick; Rowington (Sarjeant, 1974, pp. 314-16; 1985), and in Worcestershire: Barrow Churz, Malvern (Sarjeant, 1974, p. 324). Localities in Leicestershire are Shoulder-of-Mutton Hill, Leicester; Castle Donington; and Derby Road, Kegworth, although the first two are rather uncertain (Sarjeant, 1974, pp. 317-19), and in Nottinghamshire: Colwick, Nottingham; Ollerton; Sherwood district or Mapperley Park, Nottingham (Sarjeant, 1974, pp. 321-3). Unfortunately, most of these localities, listed largely from 19th century reports, are either lost or untraceable. A strong case could not be made for listing any Mid Triassic footprint locality or localities as a GCR site. A complete overview of these sites is urgently required (King, M.J. in prep.).

At Bromsgrove, near Birmingham, three quarries near Hilltop Hospital, on Breakback or Rock Hill (SO 948698) are also known for their Middle Triassic tetrapods. These quarries (Wills, 1907, 1910, pp. 254-6) formerly showed good sections in the Finstall Member of the Bromsgrove Sandstone Formation (the former Building Stones and Waterstones). Wills (1907; 1908, pp. 29-32; 1910, pp. 254-6) described the succession as 15-20 m of alternating sandstone and shales, and a band of 'marl conglomerate'. Some lenticular beds are 'true marls', others are sandy shales, green, brown or red in colour. Individual units are lens-shaped, and the sandstones appear to show cross-bedding (Wills, 1907, fig. 1). The present sections are very limited, and the sites could not be recommended as GCR sites.

The Bromsgrove Sandstones at Bromsgrove (Figure 4.2) are approximately equivalent in age to the fossiliferous horizons at Warwick and Leamington (Walker, 1969; Paton, 1974a; Warrington *et al.*, 1980, pp. 38-9, table 4). The flora and fauna from Bromsgrove is similar to those from Warwick, and from the Otter Sandstone Formation of Devon, and they assist our understanding of these GCR sites. The fauna comprises arthropods: conchostracans (*Euestheria*), scorpionid arachnids (*Mesophonus*, *Spongiophonus*, *Bromsgroviscorpio* and *Willsiscorpio*), annelids (*Spirorbis*), and a bivalve (*?Mytilus*). The vertebrates include the shark *Acrodus*, the perleidid *Dipteronotus* and the lungfish *Ceratodus*, as well the capitosaurid amphibians *Cyclotosaurus pachygnathus* and *Mastodonsaurus* (Wills, 1916, pp. 2–7, figs 2–4, pl. 2; Paton, 1974a), and reptiles cf. *Macrocnemus, Rhynchosaurus brodiei,* rauisuchian remains (?including *'Teratosaurus', 'Cladeiodon'*), *Bromsgroveia walkeri,* a ?prosauropod tooth (Huene, 1908b, p. 242, figs 273a, b; Galton, 1985a; Benton, 1990c), a trilophosaur, a nothosaur (Walker, 1969) and other, undiagnostic, remains. The remains of *R. brodiei* from Bromsgrove are labelled as having come from 'Wilcox S. Quarry'. The Bromsgrove fauna is associated with a rich flora that includes sphenopsids (horsetails and relatives) and gymnosperms (cycads, cycadeoids, conifers).

The following Midlands Mid Triassic localities are selected as GCR sites:

1. Grinshill Quarries, Shropshire (SJ 520237). Middle Triassic (Anisian), Helsby Sandstone and Tarporley Siltstone Formations.
2. Coten End Quarry, Warwick, Warwickshire (SP 290655). Middle Triassic (Anisian), Bromsgrove Sandstone Formation.
3. Guy's Cliffe, Warwick, Warwickshire (SP 293667). Middle Triassic (Anisian), Bromsgrove Sandstone Formation.

GRINSHILL QUARRIES, SHROPSHIRE (SJ 520237)

Highlights

Grinshill Quarries have had a long history of producing skeletons and footprints of fossil reptiles. The quarries have yielded many skeletons of the small plant-eater *Rhynchosaurus*, and they are the richest site for such material in the British Isles.

Introduction

Grinshill Hill, and the adjoining Clive Hill, 300–500 m north of Grinshill village, are marked by numerous quarries, of which four large ones lie along the crest of the hill (SJ 5205 2392, SJ 5238 2387, SJ 5249 2384, SJ 5264 2380). The last of these is still operational (Figure 4.3). The quarries, exposing sections in the Tarporley Siltstone Formation and the Helsby Sandstone Formation, have yielded specimens of the reptile *Rhynchosaurus* and associated rhynchosauroid footprints. All the old quarries on Grinshill Hill are

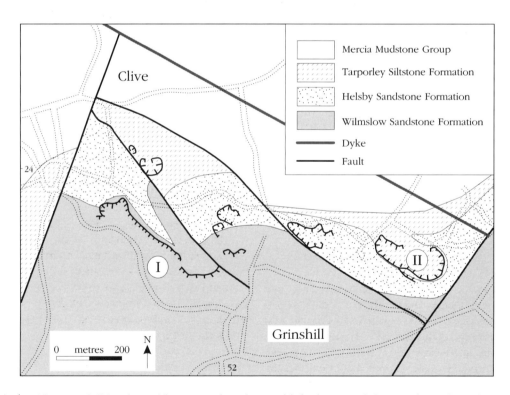

Figure 4.3 The Grinshill localities. The map is based on published maps of the British Geological Survey (BGS 1:63, 360 scale Geological Sheet 138, Wem) and on field observations by M.J.B. I. Bridge Quarries; II. working quarry (ECC Quarries Ltd).

still accessible and the currently operating quarry provides good exposure. Although excavation is slow, fresh finds of bones and footprints occur from time to time.

The buff-coloured sandstone of Grinshill was quarried in the 18th and 19th centuries and provided much of Shrewsbury and northern Shropshire with building stone. Murchison (1839, pp. 37–41) described a section taken in one of the Grinshill quarries and the Reverend Dr T. Ogier Ward of the Shrewsbury Natural History Society described (1840) vertebrate footprints, rain mark impressions and ripple marks taken from the Waterstones of Grinshill (Ward, 1840; noted in Murchison, 1839, appendix, p. 734). More footprints were reported from Grinshill by Beasley (1896, 1898, 1902, 1904, 1905, 1906), who identified most of them as of 'rhynchosauroid' type. Thompson (1985) and Benton *et al.* (1994) note also the occurrence of some cheirotherioid tracks at Grinshill.

The first bones from Grinshill, discovered during the 1840s, were noted by Ward (1840). The specimens had been found some years earlier by John Carline, quarrymaster at Grinshill, and were given to the museum of the Shropshire and North Wales Natural History Society. Between August 1840 and November 1841, Ward obtained several more bones belonging to *Rhynchosaurus* from various quarrymen, although he later claimed that these specimens were 'first discovered by myself in 1837–1838' (Ward, 1874). These he sent to Richard Owen at the Royal College of Surgeons in several packages (Owen correspondence, Coll. Sherborn, BMNH letters 110, 103, 118, 105, 109, 107, 114, 116; D.B. Thompson, pers. comm., 1988).

In a paper to the Geological Society of London on 24th February 1841, Owen referred most of the Grinshill material to a species of *Labyrinthodon* (i.e. *Mastodonsaurus*), an amphibian which Jaeger (1828) had described from the Late Triassic of Germany. Owen (1842b, 1842c) later recognized the reptilian affinities of the material and named it *Rhynchosaurus articeps*. Further specimens of *Rhynchosaurus* provided more detail for the description of the skull (Owen, 1859a, 1863a). *Rhynchosaurus* was later redescribed from Owen's specimens and from newer material by Huxley (1887), Woodward (1907a), Watson (1910a), Huene (1929a, 1938, 1939), Hughes (1968) and Benton (1990c). More details of the history may be found in Benton (1990c) and Benton *et al.* (1994).

Some *Rhynchosaurus* remains have been recovered recently from Grinshill, as well as 11 slabs bearing good vertebrate trackways, collected by Dr J. Stanley and Dr D.B. Thompson (Keele University) between 1968 and 1982. All of this material came from the single operational quarry.

Description

Grinshill Hill consists of Triassic sediments dipping north and north-west. It is bounded to the east and west by NE–SW trending faults and these are linked by two WNW–ESE trending faults.

Murchison (1839, pp. 37–41) gave a section in 'Grinshill Stone Quarries', a locality no longer identifiable exactly – it may have been generalized from several quarries (Pocock and Wray, 1925, p. 39). Hull (1869, p. 64), in his revision of the Triassic rocks of the Midlands, showed the presence of Upper Mottled Sandstone, Lower Keuper Sandstone and Waterstones at Grinshill (although he could not identify the boundary between the last two), and reproduced Murchison's section in simplified form:

	Thickness Ft in
Lower Keuper Sandstone	
1. Fee and jay (rubbly thin bedded rock)	13 0
2. Flag rock, yellowish or light brown in colour	19 0
3. Sand bed called Esk	0 9
4. Hard burr	2 6
5. Coarse freestone, mottled, of yellowish and reddish colours, best building stone	9 6
6. Grey freestone	7 6
7. Good light yellow freestone underlain by a seam of clay	11 0
8. Good white freestone	2 0
9. Strong white freestone	8 0
Upper Mottled Sandstone (Bunter)	
10. Sandy and bad freestone	2 0
11. Bad stone, sometimes used for walls, &c	9 0
12. Soft yellow sandstone, the grains of sand cemented by decomposed feldspar	4 6
13. Sandstone of deep red colour sunk through for water	222 0
TOTAL	311 7

Pocock and Wray (1925, pp. 15–16) established the 'Ruyton and Grinshill Sandstones' to include 'a group of red and yellow freestone, forming a passage-bed between the Bunter and Keuper and including at its base a small thickness of the Upper Mottled Sandstone and limited upward by the base of the Waterstones'. For the 'main Grinshill quarries, 550 yards N20 degrees E of the Elephant and Castle Hotel', Pocock and Wray (1925, pp. 39–40) offer the section:

	Thickness	
	Ft	**in**
Keuper Marl: Red marl seen to	2	0
Waterstones:		
Flag rock: grey and light-yellow		
sandstone, evenly bedded, with		
thin reddish seams; ripple marks	20	0
Esk bed: incoherent grey sandstone		
and sand, with harder patches;		
full of specks of manganese dioxide	0	9
Grinshill Sandstone:		
Hard burr: hard yellowish-white		
sandstone (coarse-grained		
sandstone)	2	6
Hard yellowish freestone	2	6
Soft yellow sand	0	2
White and pale-yellow freestone,		
with iron-stained patches		
towards the base	33	0
White freestone with iron-stained		
and speckled patches seen to	5	6

Pocock and Wray (1925, p. 40) gave another section taken in the only quarry then working (650 yards N45 degrees E of the hotel) which shows a similar succession. The Upper Mottled Sandstone of these authors (f3) has been renamed the Wilmslow Sandstone Formation, the Ruyton and Grinshill Sandstones (or the Building Stones; f4) the Helsby Sandstone Formation, and the Waterstones the Tarporley Siltstone Formation (Warrington *et al.*, 1980).

The Tarporley Siltstone Formation, typically ranging from 20 m to 250 m in thickness (Warrington *et al.*, 1980, table 4), is only about 6–10 m thick at Grinshill. The sediments are well-bedded, white, pale-green or reddish fine-grained sandstones and marls. Two facies, A and B, have been identified by Thompson (1985, pp. 119–21). Facies A, fluvial and tidal, is characterized by trough-shaped erosion channels filled with beds of ripple cross-laminated, fine- to medium-grained sandstone, which bear on their bedding surfaces ripple marks, rhynchosauroid footprints (see below), trace fossils formed by invertebrates(?), and supposed raindrop impressions, which were

reported for the first time from Grinshill by Ward (1840) and Buckland (1844). Facies B, largely intertidal and rarely hypersaline, consists of interbedded fine sandstones, siltstones and mudstones. The mud and silt horizons are generally 10–20 mm thick; the sand beds are thicker at about 100 mm. Many of these horizons show current and wave ripple marks, load casting, flute marks and prod marks. A few show adhesion ripple marks, indicating half wet, half dry conditions. Mudcracks and halite pseudomorphs have been observed occasionally, as well as rhynchosauroid footprints and poorly preserved invertebrate trace fossils.

The underlying exposure of Helsby Sandstone consists of about 30 m of buff and yellow, medium-grained, well-sorted sandstones. These are well cemented, and contain numerous small spots of manganese hydroxide. Large-scale cross-beds are sometimes visible in vertical quarry faces, bearing lamination structures which imply aeolian conditions of deposition (Thompson, 1985), relating to large transverse barchanoid dune ridges. At Grinshill, the Helsby Sandstone Formation appears to grade up into the Tarporley Siltstone Formation through a bed of loose sand, about 0.3 m thick, termed the Esk Bed (Pocock and Wray, 1925, pp. 39–40; Thompson, 1985, p. 119), of indeterminate environmental origin (Figure 4.4).

Most of the reptile specimens appear to have come from the debris associated with quarrying, but were probably derived from horizons within a thickness of about 2 m. The *R. articeps* specimens occur in two main lithologies, as noted by Owen (1842b, p. 146); a fine-grained grey sandstone and a coarser pinkish-grey sandstone (his coarse 'burr-stone').

The remains of *R. articeps* and the tetrapod trackways appear to have come from a number of quarries on Grinshill (D.B. Thompson, pers. comm., 1984). The footprints described by Ward (1840) were found on ripple-marked surfaces in a finely laminated buff-coloured sandstone beneath the rubbly red-coloured sandstone called 'Fee', presumably equivalent to part of Thompson's (1985) largely intertidal Facies B. Walker (1969, p. 470) observed that the specimens of *R. articeps* came from the siltstones and fine sandstones of the Tarporley Siltstone Formation, and possibly from the immediately underlying beds at the top of the Grinshill Sandstones (the coarser sandstone) (Walker, 1969). This was implied also in Pocock and Wray's (1925, pp. 39–40) section, in which the top of the Grinshill Sandstone is

Figure 4.4 The operational quarry on Grinshill: view of the north face, showing the massive cross-bedded Helsby Sandstone Formation at the bottom, and the softer, more thin-bedded Tarporley Siltstone Formation above. (Photo: M.J. Benton.)

described as 'Hard Burr: Hard yellowish-white sandstone, 2ft 6in'. However, Thompson (1985, p. 118) was doubtful whether any bones had been found in the aeolian Grinshill Sandstone Formation, noting (D.B. Thompson, pers. comm.) specimens only from his largely fluvial Facies A of

the Tarporley Siltstone Formation, in the operating quarry (SR 526 238).

Mr John O'Hare, the former quarry owner, is certain that the Keele University (1984) *Rhynchosaurus* specimen came from the lowest 0.2 m of the Tarporley Siltstone Formation. The

1986 and 1991 finds are in large blocks of coarse sandstone, which are most likely to have come from the hard burr stones at the top of the aeolian Grinshill Sandstone (D.B. Thompson, pers. comm., 1993).

The Grinshill specimens of *Rhynchosaurus* are largely complete (Figure 4.5) and undisturbed, but the bone material is soft and friable. The skeletons tend to lie flat in the dorso-ventral plane with the limbs stretched out to the sides: this suggests relatively rapid burial with little scavenging or transport.

The commonest tetrapod tracks at Grinshill are of the 'rhynchosauroid' type termed rhynchosauroid D1 by Beasley (1902); rarer finds include *Cheirotherium* prints, all from the Tarporley Siltstone Formation. The remainder are small prints, possibly of a different vertebrate type. These include small arcs of claw marks ('type C') and more or less complete 'hand'-shaped marks which could be of the same foot type ('type B'). The rhynchosauroid prints are generally small (15–20 mm across) and, when well preserved, they show clawed digits with an opposing associated impression. Single slabs may preserve a variety of impressions belonging to a number of overlapping trackways.

The tracks are most frequently preserved in sandstone as negative moulds on the undersurface of current and wave ripple-marked horizons. The ripples are asymmetrical and, because they are on an undersurface, their crests are well preserved. The preservation of these trackways is excellent. One slab preserves 11 distinct sets of claw prints, and in some specimens there is clear indication that the claws were twisted sideways (revealing their arcuate shape) as the foot was impressed into the sediment. Another specimen (JSW GH 3) exhibits a more or less circular area of sediment disturbance, which appears to suggest that the animal had been engaged in some activity such as eating from the ground. Other specimens exhibit regular series of impressions which appear to be teeth or jaw marks. More details of the taphonomy of the skeletons and footprints are given in Benton *et al.* (1994).

Fauna

Diapsida: Archosauromorpha: Rhynchosauridae
 Rhynchosaurus articeps Owen, 1842
  About 17 individuals: SHRBM, SHRCM, BMNH, MANCH, BATGM, Keele Univ; some specimens have been missing since the 19th century and the total could be greater

Footprints
 Rhynchosauroides sp.
  23 slabs: SHRBM, MANCH, WARMS, BGS(GSM), BUGD, others in private hands
 Cheirotherium sp.
  Two slabs: SHRBM, SHRCM.

Interpretation

On the basis of palynological evidence Warrington (1970b) dated the basal Helsby Sandstone Formation as Scythian, a view followed by Pattison *et al.* (1973), and by Warrington *et al.* (1980, p. 33, table 4) who placed the overlying Tarporley Siltstone Formation in the Anisian. However, a more recent assessment of the palynological data (Warrington, *in* Benton *et al.*, 1994) confirms an Anisian age for both formations (Figure 4.2).

The ages have also been debated on the basis of the reptiles. Walker (1969, 1970a) argued that all relevant horizons were of Mid Triassic age because of the resemblance between *Rhynchosaurus* and *Stenaulorhynchus*, and because of the purported intermediate evolutionary position of *Rhynchosaurus* between Early and Late Triassic rhynchosaurs. Its closest relative seems to be *Stenaulorhynchus* from the Manda Beds (?Anisian), of Tanzania, although it seems slightly more advanced in some respects according to Walker (1969). *Stenaulorhynchus* and *Rhynchosaurus* were grouped in the subfamily 'Rhynchosaurinae' (Chatterjee, 1974, 1980; Benton, 1983d), but this view has not been supported in more recent analyses (Benton, 1990c).

Rhynchosaurus articeps was a relatively small reptile, about 0.5 m long, and probably like a large lizard in appearance (Figure 4.5). The triangular skull (60–80 mm long) is low and broad at the back, and it shows all the typical rhynchosaur features of beak-like premaxillae, a single median naris and fused parietal. The dentition was specialized, as in other rhynchosaurs, consisting of a grooved maxillary tooth plate with several rows of teeth and a lower jaw (that slots into the groove) with teeth on the upper edge and down the inside surface. The pattern of wear, and the nature of the jaw joint, suggest that *Rhynchosaurus* had a precision shear bite, as in other rhynchosaurs, with no back and forwards

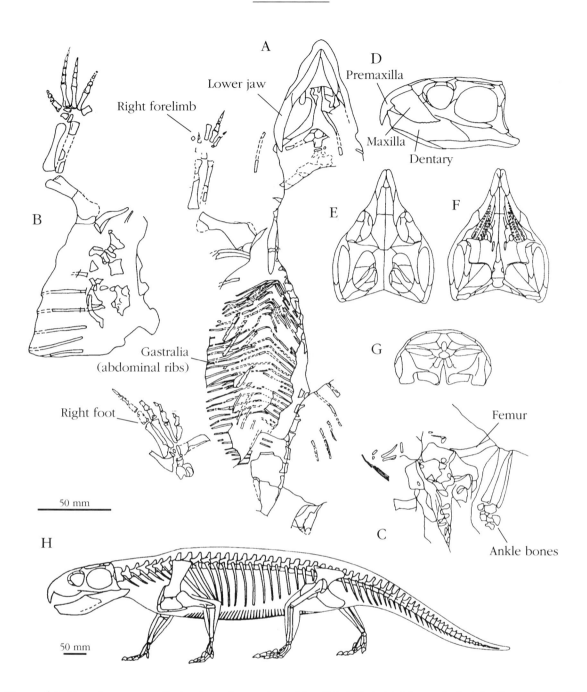

Figure 4.5 *Rhynchosaurus articeps*, the only member of the Grinshill skeletal fauna: typical fossil remains (A–C) and restorations (D–H). (A) Partial skeleton lacking the tail and the limbs of the left side, in ventral view (BMNH R1237, R1238); (B) dorsal vertebrae, ribs, and right forelimb in posteroventral view (SHRBM 6); (C) pelvic region, right leg with ankle bones, presacral vertebrae 22–25, sacral vertebrae 1 and 2, and caudal vertebrae 1–8 (BATGM M20a/b); (D)–(G) restoration of the skull, based on SHRBM G132/1982 and 3 and BMNH R1236, in lateral (D), dorsal (E), ventral (F), and occipital (G) views; (H) restoration of the skeleton in lateral view in walking pose. All based on Benton (1990a).

motion. The diet was probably tough vegetation, which was dug up by scratch digging, raked together with the fore feet or the premaxillary beak, and manipulated in the mouth by a large, fleshy tongue (Benton, 1990c).

The skeleton of *R. articeps* is relatively more slender than that of most other Mid and Late Triassic rhynchosaurs. This is probably an allometric effect resulting from its relatively smaller size. The slim body and the semi-erect limb posture deduced from the skeletons may have assisted in fast terrestrial locomotion and the strong limbs and girdles support the notion of an active lifestyle.

Comparison with other localities

Grinshill is a unique site. Its sole reptile species, *Rhynchosaurus articeps,* is most comparable with *R. brodiei* from Coten End (q.v.) and Bromsgrove, and with *R. spenceri* from Devon.

Conclusions

Grinshill is important as the site of the type species of *Rhynchosaurus*, a genus represented from other localities in England (Coten End, Bromsgrove, Sidmouth, Budleigh Salterton) and hence of some value biostratigraphically. Specimens are still coming to light at Grinshill, as are slabs with footprints. There is much potential here for new finds and for studies of the palaeobiology of rhynchosaurs. The association of excellent skeletons and footprints is unusual and gives the site considerable conservation value.

COTEN (AKA COTON) END QUARRY, WARWICK, WARWICKSHIRE (SP 29006550)

Highlights

Coten End Quarry has produced the most diverse assemblages of fossil amphibians and reptiles in the English Midlands, including more than eight species. The reptiles range from lizard-sized plant- and insect-eaters to large carnivorous forms.

Introduction

This site consists of a small quarry within the town of Warwick which is currently used as a small-bore rifle range. It displays a section in the upper part of the Bromsgrove Sandstone Formation. The list of reptiles and amphibians from Coten End is large, and the site is the most productive for Mid Triassic tetrapods in the Midlands. It is the type locality for various species of temnospondyl amphibians as well as the reptiles *Bromsgroveia walkeri* Galton, 1985 and

Rhynchosaurus brodiei Benton, 1990. Further quarrying would doubtless produce more reptile remains from the 'dirt bed', but the site is now surrounded by housing.

Coten End Quarry was worked in the early 19th century for building stone, and Murchison and Strickland (1840, p. 343) stated that it 'has been most productive in the remains of vertebrata'. Howell (1859, p. 40) described the Warwick Triassic in general terms, and Hull (1869, pp. 88-9) distinguished the Building Stones (Lower Keuper Sandstone) overlain by Waterstones containing the amphibians.

Murchison and Strickland (1840, pl. 28, figs 6-10) were the first to figure bones from Coten End which they identified as teeth of '*Megalosaurus*', and of a 'Saurian', as well as an unidentified vertebra. Owen (1841b) named one of the 'teeth' *Anisodon gracilis*, and later (Owen, 1842a, p. 535) he reinterpreted this specimen as the ungual phalanx (claw) of the temnospondyl amphibian *Labyrinthodon (Mastodonsaurus) pachygnathus*. He also examined the vertebra described by Murchison and Strickland, assigning it to the temnospondyl amphibian *L. leptognathus* (Owen, 1842a, pp. 523-4, pl. 45, figs 5-8). Both specimens have since been reidentified as rhynchosaur remains: a premaxilla (the 'tooth') and a dorsal vertebra (Benton, 1990c). Owen (1842c) described further jaw, skull and postcranial fragments from Coten End as pertaining to *L. leptognathus* and *L. pachygnathus* (now assigned to the genera *Stenotosaurus* and *?Cyclotosaurus* respectively; Milner *et al.*, 1990), and also gave an account of the microscopic anatomy of the teeth of *Mastodonsaurus* from Coten End and compared it with that of German Keuper forms. Later (1842d), he described skull fragments of various temnospondyls, including *M. leptognathus* and *M. pachygnathus*. All of Owen's descriptions were based on the extensive collections by Dr Lloyd of Leamington.

In the 1840s and 1850s the Reverend P.B. Brodie and Dr Lloyd collected jaw bones of *Rhynchosaurus* from Coten End and these were described by Huxley (1869), who mistakenly ascribed them to the related form *Hyperodapedon* from Elgin. Huxley (1870a) also described supposed dinosaur remains from Coten End and redescribed many of Owen's *Mastodonsaurus* bones as probably dinosaurian. L.C. Miall (1874) agreed with these reassignments and described further remains of *Mastodonsaurus*. Huene (1908b) redescribed

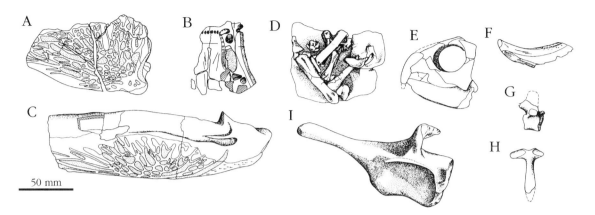

Figure 4.6 Typical elements of the Warwick fauna. (A) Left postero-external angle of the skull of '*Cyclotosaurus pachygnathus*' (Cyclotosaurinae *incertae sedis*) in lateral view (WARMS Gz13); (B) part of the snout of '*Stenotosaurus leptognathus*' (Stenotosaurinae *incertae sedis*) in palatal view (WARMS Gz38); (C) posterior portion of a left lower jaw of '*Stenotosaurus leptognathus*' in lateral view (WARMS Gz35); (D) scattered bones of cf. *Macrocnemus* (*Rhombopholis scutulata*) (WARMS Gz10); (E)–(H), assorted remains of *Rhynchosaurus brodiei*: (E) anterior part of the skull in lateral view (WARMS Gz6097/ BMNH R8495), (F) anterior part of a dentary in medial view (WARMS Gz950), (G) mid-dorsal vertebra in right lateral view (WARMS Gz17), and (H) interclavicle in ventral view (WARMS Gz34); (I) right ilium of *Bromsgroveia walkeri* in lateral view (WARMS Gz3). After various sources; from Benton *et al.* (1994).

most of the supposed dinosaur material.

More recently, in a new phase of research on British Triassic vertebrates, Walker (1969) provided reidentifications of many of the archosaurs and other reptiles from Coten End (Figure 4.6) and Paton (1974a) revised the temnospondyls. Galton (1985a) reviewed the archosaur material and named a new rauisuchian, *Bromsgroveia walkeri*, Benton (1990c) established the new species *Rhynchosaurus brodiei*, and Benton and Walker (in prep.) revised the prolacertiform *Rhombopholis*, the type specimens of all three of which came from the site.

Rhynchosauroid footprints have also been recorded (e.g., Beasley, 1906); some of these appear to be associated with large groove marks produced by the flow of water. Further details of these, and of the skeletal faunas are given in Benton *et al.* (1994).

Description

The Bromsgrove Sandstone Formation (upper part of the Sherwood Sandstone Group) is from 20–35 m thick in Warwickshire (Warrington *et al.*, 1980, pp. 38–9, table 4; Old *et al.*, 1987, p. 20), and the middle to upper portions of this formation are exposed at Coten End. These units equate with

the former 'Waterstones' and 'Building Stones' (Warrington *et al.*, 1980, p. 39). Murchison and Strickland (1840, p. 344) gave a section in the quarry:

	Thickness (ft)
a. Soft, white sandstone and thin beds of marl	8
b. Whitish sandstone, thick bedded	12
c. Very soft sandstone, coloured brown by manganese, called 'Dirt-bed' by the workmen	1
d. Hard sandstone, called 'Rag'	*c.* 2
Total	23

This section was confirmed by Hull (1869, pp. 88–9). Old *et al.* (1987, p. 23) documented 7 m of massive sandstone and flat-bedded sandstone grading up into 4 m of cross-bedded sandstone and mudstone in the quarry. This section was interpreted as lying near the middle of the thin Bromsgrove Sandstone Formation of the Warwick district and hence may lie within 10 m of the base of the overlying Mercia Mudstone Group. There are still good exposures in the quarry which show channelled and cross-bedded, water-laid, buff and red sandstone units varying in thickness from one to three metres. Laterally discontinuous marl and clay bands, 0.1–0.5 m thick, probably correspond to the fossiliferous Dirt bed.

According to Murchison and Strickland (1840, p. 344), the fossil bones were found principally in the Dirt-bed. Hull (1869, pp. 88–9) stated that the amphibian fossils occurred in the 'Waterstones', but Walker (1969, 1970a) noted that the reptiles came from the upper part of the Building Stones, a fine-grained, brown-coloured sandstone which forms a bed essentially equivalent to the 'Dirt-bed' of Murchison and Strickland.

The specimens of *Rhynchosaurus brodiei* from Coten End are preserved in a disarticulated state as far as can be determined, and no groups of elements were ever found in even moderately close association. Murchison and Strickland (1840, p. 344) described the bones as 'rolled and fragmentary', but subsequent studies have shown that they are not abraded, nor are they distorted, as Miall (1874, p. 417) noted (Benton, 1990c; Benton *et al.*, 1994).

The bone is preserved as hard, white to buff-coloured material, apparently with all of the original internal structure intact. However, Murchison and Strickland (1840, p. 344) noted that the bones were in a decomposed condition when they were freshly collected and suggested treatment with gum arabic as a useful method of curation. This description is hard to equate with the present hard and well-preserved condition of the fossil bone in the museum collections.

Fauna

The faunal list of fishes, amphibians and reptiles is derived from Huene (1908b), Allen (1908), Horwood (1909), Wills (1910), Walker (1969), Paton (1974a), Galton (1985a), Benton (1990c) and Benton *et al.* (1994).

Osteichthyes: Dipnoi: Ceratodontidae
 Ceratodus laevissimus (Miall, 1874)
  Tooth of a ceratodontid lungfish (WARMS)

'Temnospondyli': Capitosauridae
 Stenotosaurus leptognathus (Owen, 1842a)
  Jaws and other skull fragments (WARMS)
 Cyclotosaurus pachygnathus (Owen, 1842a)
  Jaws and other skull fragments (WARMS)

'Temnospondyli': Mastodonsauridae.
 Mastodonsaurus sp. indet. (Owen, 1842a)
  Jaw and skull fragments (WARMS)

Diapsida: Archosauromorpha: Prolacertiformes: Macrocnemidae
 Rhombopholis scutulata (Owen, 1842a)
  Ilium, femur (WARMS)

Diapsida: Archosauromorpha: Rhynchosauridae
 Rhynchosaurus brodiei Benton, 1990
  About 7 jaw and skull fragments and 3 skeletal elements (WARMS), 3 skull remains (BMNH),
  2 maxillary tooth-plates (BGS(GSM)).

Archosauria: Crurotarsi: Pseudosuchia: ?Poposauridae
 Bromsgroveia walkeri Galton, 1985
  Vertebrae, sacrum, ilium, ischium, ?femur (WARMS)

Archosauria: Crurotarsi: indet.
 'Large thecodontian' ilium (WARMS)
 Cladeiodon lloydi Owen, 1841
  About ten isolated teeth (WARMS, BMNH, BGS(GSM))
 'Prosauropod dinosaur' cervical vertebra (BMNH)

Interpretation

Murchison and Strickland (1840, p. 342) correlated the 'sandstones of Warwick' with those of Ombersley and Bromsgrove, which they had correlated with the Buntsandstein of Germany since they were separated from 'true Keuper sandstone' by a vast thickness of red and green marl. On the other hand, Owen (1842c, 1842d) agreed with the view of Buckland, that the Warwick sandstone was Keuper in age on the basis of identity of the temnospondyls with those of the German Keuper. The discontinuous fine-grained bands, including the bone-bearing horizon, probably represent overbank pools subsequently broken up by flood waters. The middle to upper portions of this formation, as seen at Coten End, have been interpreted as deposits of mature, meandering river channel and floodplain complexes (Warrington, 1970b).

The mistaken identification of *Hyperodapedon* from Coten End by Huxley (1869) led to correlation of the Lossiemouth Sandstone of Elgin with that termed Lower Keuper Sandstone at Warwick (e.g. Huene, 1908c). Later, Huene (1908c, 1908d) correlated the Warwick sandstone with the German Lettenkohle, of Ladinian age, on the basis of the occurrence of the temnospondyl *Mastodonsaurus giganteus* and the plant *Equisetum arenaceum* from Bromsgrove.

Walker (1969) suggested an Early to Mid Ladinian age for the upper part of the Building Stones or 'Lower Keuper Sandstone' on the basis of *Rhynchosaurus* and *Macrocnemus*. Paton (1974a) gave an Early Ladinian age and Warrington *et al.* (1980 pp. 39-40, table 4), on the basis of palynological work by Warrington, gave the age of the Bromsgrove Sandstone Formation (formerly 'Building Stones') as Late Scythian to Early Ladinian, with the reptiles occurring in the upper part. Warrington (in Benton *et al.*, 1994) reviews evidence from miospores which places the Bromsgrove Sandstone Formation in the Anisian. Indeed, north of the Warwick-Leamington area, miospores indicate an Anisian age for the lower part of the overlying Mercia Mudstone Group, hence clearly constraining the age of the Coten End site as Anisian (Figure 4.2).

The Coten End fauna (Figure 4.6) consists of fishes, up to four species of aquatic carnivorous or piscivorous temnospondyl amphibians, a moderately sized insectivore or carnivore (macrocnemid), two herbivores (*Rhynchosaurus brodiei*, ?'prosauropod dinosaur'), and two or more terrestrial carnivores ('thecodontian', *Bromsgroveia*, *Cladeiodon*) which may have fed on the herbivores. The numbers of specimens of all taxa are small, but *Rhynchosaurus*, *Bromsgroveia* and two species of *Cyclotosaurus* seem to be represented by more than five specimens each (Benton *et al.*, 1994).

The capitosaur temnospondyls, well represented here by *Mastodonsaurus, Stenotosaurus* and *Cyclotosaurus*, were heavily built moderate-sized aquatic amphibians, with heads about 200 mm long. The skull is vaguely crocodile-like, flattened, with long jaws closely lined with teeth. There were other series of teeth on the palate to assist in gripping prey and the skull was heavily ornamented and bore lateral line canals which were sensory systems for use under water. Their diet included fishes and probably small tetrapods. The deposits in which the fossils are found indicate the presence of large rivers, a probable habitat for the temnospondyls, and some fishes have been found which may have featured in their diet. Several temnospondyl species have been described from Coten End, most of which have been synonymized with the named taxa: *Mastodonsaurus jaegeri* Owen, 1842, *M. lavisi*

Seeley, 1876, *M. ventricosus* Owen (1842), *M. giganteus* Owen (1842) and *Diadetognathus varvicencis* Miall (1874). Milner *et al.* (1990) noted that the amphibians compare broadly with material from central Europe and North America. *Mastodonsaurus* is known from Anisian to Carnian units in Germany (mainly the Keuper).

The prolacertiform *Rhombopholis* turns out to be rather like *Macrocnemus*, a slender lizard-like animal, 500-800 mm long. It had large eyes and many small teeth and may have been a carnivore or piscivore. *Macrocnemus* is well known from the marine Grenzbitumenzone (Anisian/Ladinian boundary) at Tessin, in the Swiss Alps, and in neighbouring deposits in North Italy. Other species come from the Upper Buntsandstein in Germany (Scythian/Anisian) and from the Upper Muschelkalk (?Ladinian) of Catalonia, Spain. It is a prolacertiform, a largely Triassic group of reptiles closely related to archosaurs (Benton, 1985).

Rhynchosaurus brodiei was a moderate-sized rhynchosaur with a skull 90-140 mm long (estimated body length, 0.5-1.0 m) and a herbivorous diet (Benton, 1990c). It differs from *R. articeps* from Grinshill (skull length 60-85 mm) in being considerably larger and in having a broader skull. The jugal in *R. brodiei* is much deeper than that of *R. articeps*, being the largest bone in the side of the skull, the orbit in *R. brodiei* is placed relatively further forward, and the maxilla is relatively smaller than in *R. articeps*. The characteristic 'tusks', slicing dentition and tooth plate, and large eyes, are shown by these specimens.

Bromsgroveia walkeri was a moderate- to large-sized carnivorous quadruped (rauisuchid) or biped (poposaurid) (Galton, 1985a), based on an isolated ilium. It probably preyed on small terrestrial and semi-aquatic reptiles *Rhynchosaurus* and *Macrocnemus*. Other archosaurs are represented by teeth called *Cladeiodon lloydi*. These range in length from 10 to 50 mm and could belong to *Bromsgroveia*, or to some other carnivorous archosaur. The so-called 'prosauropod dinosaur' could be the oldest in the world, if it really is correctly identified, but that is uncertain.

Comparison with other localities

The nearest analogues of the Coten End fauna come from the Bromsgrove Sandstone Formation of Guy's Cliffe (see below) and Bromsgrove (see above), and the Otter Sandstone Formation of Sidmouth and Budleigh Salterton (see below).

Outside the British Isles, the fauna compares with Early to Mid Triassic faunas from France (Grès à Voltzia) and Germany (Buntsandstein) and Mid to Late Triassic faunas from Germany (Lettenkeuper).

Conclusions

The value of the Coten End fauna, and the other British examples of similar age, is linked to the difficulty in correlation. There are no mainland European terrestrial faunas of the same age, since the Muschelkalk marine transgression occupies that interval of time. Coten End preserves the richest Mid Triassic continental tetrapod fauna in Britain and probably in Western Europe. Although the potential for re-excavation is now restricted it is still possible, hence the conservation value of the site.

GUY'S CLIFFE, WARWICK, WARWICKSHIRE (SP 293667)

Highlights

Guy's Cliffe is the site of a superb specimen of a large fish-eating amphibian, *Mastodonsaurus jaegeri*, one of the best-preserved examples of this group.

Introduction

The exposures in the grounds of Guy's Cliffe House and on the banks of the River Avon below which comprise Guy's Cliffe, display good sections in the Bromsgrove Sandstone Formation which yielded a fine specimen of *Mastodonsaurus* early in the 19th century. Although not a reptile but an amphibian tetrapod, this specimen is important in correlating the Triassic Warwick sandstones with those of Bromsgrove and Devon. Guy's Cliffe House is owned by freemasons; the property is fenced off and access is difficult, but re-excavation could produce further finds.

Buckland (1837) described the 'excellent section' at Guy's Cliffe as exposing Keuper sandstone. This age assignment was based on a find made in 1823 of 'part of the jaw and other bones of a saurian . . . presented to the Oxford Museum by the late Butic Greathead, Esq.' This was probably the first find of a tetrapod to be made in the area. Buckland (1837) identified the bones as those of *Phytosaurus*, a German form. The original specimen, although now lost, is well represented by casts.

Owen (1842a, pp. 537–8, pl. 44, figs 4–6, pl. 37, figs 1–3) and Miall (1874, p. 433) described Buckland's 'saurian'; in reality fine specimens of the lower jaw of the temnospondyl *Mastodonsaurus jaegeri*. Milner *et al.* (1990, p. 878) suggest that the Warwickshire material of *Mastodonsaurus*, including *M. jaegeri*, should properly be given *nomen dubium* status and redefined as *Mastodonsaurus* sp.

Howell (1859, p. 40) and Hull (1869, pp. 88–9) reviewed the lower Keuper Sandstone at Guy's Cliffe and elsewhere, and Huene (1908c) described sections at Guy's Cliffe and proposed that the outcrops provided evidence of subaerial dunes as well as water-laid deposits.

Description

Murchison and Strickland (1840, p. 344) published the following section from a quarry in the grounds of Guy's Cliffe House:

	Thickness (ft)
Sandstone and beds of marl	8
Solid sandstone, whitish or grey, occasionally of a reddish tint	12
Red, micaceous marl, with wedges of sandstone	8
Solid, light-coloured, reddish tinted sandstone,	*c.* 20
Total	48

Huene (1908c) showed cross-bedded sandstone units that had been eroded into a channel and covered by a discontinuous breccia layer in a section at Guy's Cliffe 'below the house of Lord Algernon Percy, on the bank of the Avon'. He noted that the bedding was very irregular and that ripple marks occurred over some beds. Another section figured by Huene 'on the rocky cliff opposite Guy's Cliffe House shows contorted sandstones with laterally discontinuous marl and breccia bands'. These features he attributed to the action of moving dunes 'near the border of the sea'.

Good sections of 7–10 m of cross-bedded, buff-coloured sandstone with irregular shale lenses are still exposed in the grounds of Guy's Cliffe House.

Behind the house is a yard which is bounded to the west by the house, to the north by a chapel and to east and south by rock. The outcrop has been chiselled vertical, and stables and hermit's holes are built into the rock on the south side. On the east is the historic Guy's Cave cut into the rock.

Fauna

'Temnospondyli': Mastodonsauridae.
　Mastodonsaurus sp. (=*Mastodonsaurus jaegeri* Owen, 1842)
　　Remains of lower jaw – casts only

Interpretation

Buckland (1837) placed the Guy's Cliffe sandstone in the Keuper on the basis of the bones collected in 1823, misidentified by him as *Phytosaurus*, a form common in the German Keuper. However, Murchison and Strickland (1840, p. 346) assigned a Bunter age to the 'sandstone of Warwick, Bromsgrove and Ombersley', but were troubled by Buckland's 'saurian' which they attempted to explain away as a Bunter form.

Owen's (1842c, 1842d) recognition of the identity of the Warwick *Mastodonsaurus* with those of the German Keuper confirmed Buckland's view. Howell (1859, p. 40) and Hull (1869, pp. 88-9) confirmed the age of the sandstones of Guy's Cliffe and other Warwick localities as 'Lower Keuper'. Huene (1908c, 1908d) and Wills (1910) repeated the correlation of the Warwick and Bromsgrove sandstones and suggested their equivalence to the German Lettenkohle (Ladinian). Walker (1969) and Paton (1974a) suggested an Early Ladinian assignment on the basis of reptiles and amphibians respectively. Warrington (*in* Benton *et al.*, 1994) gave palynological evidence for an Anisian age, as at Coten End (see above), the most comparable locality.

Conclusions

Guy's Cliffe has produced a good specimen of *Mastodonsaurus jaegeri* (*Mastodonsaurus* sp.), a heavily built, fish-eating, crocodile-like amphibian (over 2 m long), the best British example of this species. The conservation value of the site relates largely to the importance of this fossil amphibian

for correlating the Warwick sandstones and the potential for future finds.

MID TRIASSIC OF DEVON

The Mid Triassic of Devon is represented by the Otter Sandstone Formation. Inland, the formation has a poorly exposed outcrop in east Devon around the districts of Budleigh Salterton and Sidmouth and further inland beyond Honiton, but on the coast, between Sidmouth and Budleigh Salterton, it is exposed in a series of fine sea cliffs and the fossil vertebrate specimens come from these coast sections (Figure 4.7). The recent discovery of a rich vertebrate fauna from several localities between Budleigh Salterton and Sidmouth has provoked interest in the Otter Sandstone as a productive source of Middle Triassic vertebrates. The locality was known to the late Victorians, who had collected among the first known remains of *Rhynchosaurus* and good material of the amphibian *Mastodonsaurus* from the same localities, but their finds were rather sparse.

Two sites, one at Otterton Point, near Budleigh Salterton, and the other covering the cliffs nearer to Sidmouth, are selected. The former is primarily of historic interest as the locality at which the remains of Triassic vertebrates were first recognized from Devon.

4. Sidmouth coast section (SY 092838–SY 131873). Middle Triassic (Anisian), Otter Sandstone Formation.
5. Otterton Point, near Budleigh Salterton (SY 07758196). Middle Triassic (Anisian), Otter Sandstone Formation.

HIGH PEAK (SIDMOUTH), EAST DEVON (SY 092838–SY 131873)

Highlights

The Otter Sandstone Formation at Sidmouth is the richest active Mid Triassic reptile site in Britain. Ten or more species of amphibians and reptiles have been found here, most of them recently, and the site represents one of the most promising terrestrial reptile localities of its age anywhere in the world.

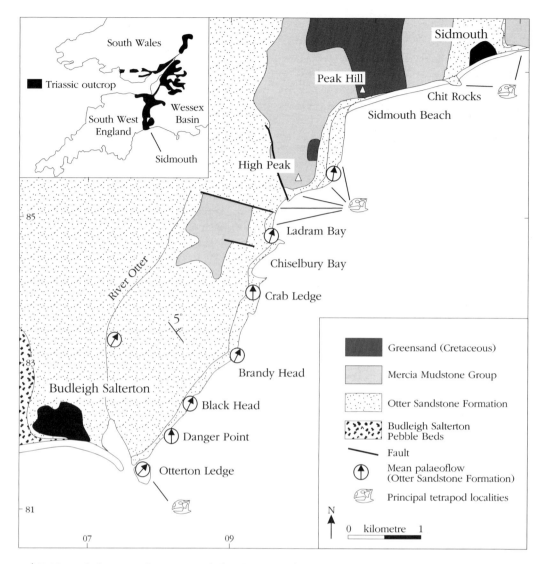

Figure 4.7 Map of the coastal outcrop of the Otter Sandstone Formation between Sidmouth and Budleigh Salterton, Devon. The major Triassic formations are indicated, together with mean fluvial palaeoflow directions, and principal tetrapod localities. From Benton *et al.* (1993).

Introduction

The fossiliferous beds are developed in the series of high cliffs to the west of Sidmouth between Chiselbury Bay (SY 092838) and Chit Rocks (SY 121869), and at Port Royal, just east of Sidmouth (SY 12978730). The whole locality (Figure 4.7) is important as one of the most productive sources of tetrapods of Mid Triassic age in Britain and fresh finds are made every year (1980–94) after cliff falls. However, it is difficult and dangerous to collect from the cliff face and most of the fossils have come from fallen blocks on the foreshore, or *in situ* from ledges at beach level (Figure 4.8).

Whitaker (1869) distinguished 'red sandstone' overlain by 'red marl' in the New Red Sandstone at High Peak (SY 144858), which is in turn overlain by Cretaceous Upper Greensand, and he reported the first finds of vertebrates from the Otter Sandstone Formation. Lavis (1876) reviewed the Sidmouth coast in more detail, and Seeley (1876a) described a fine lower jaw and other bones of *Mastodonsaurus lavisi* and a possible *Hyperodapedon* (=*Rhynchosaurus*) tooth plate which Lavis had collected. Hutchinson (1879) further reported fossil plant remains that he identified as stems of an equisete or calamite. Ussher (1876), Metcalfe (1884), Carter (1888), Irving (1888, 1892, 1893), Hull (1892) and

Figure 4.8 The Otter Sandstone Formation exposed in the cliffs west of Sidmouth, view looking east. Fish and reptile remains have been found at various horizons. (Photo: M.J. Benton.)

Woodward and Ussher (1911) discussed the stratigraphy and dating of the coastal section near Sidmouth, with particular attention to occurrences of fossil vertebrate material. Metcalfe (1884) figured remains of *Rhynchosaurus*, *Mastodonsaurus* jaws, and other bones collected from fallen blocks near High Peak, while Carter (1888) described further remains, including fish scales and coprolites.

A second phase of work on the Otter Sandstone Formation coast section began in the 1960s. Laming (1966, 1968) and Henson (1970) provided further information on the sedimentology and stratigraphy of the formation. Warrington *et al.* (1980), Laming (1982) and Warrington and Scrivener (1990) discussed the problems of correlating the Otter Sandstone with other Triassic sequences. Leonard *et al.* (1982), Selwood *et al.* (1984), Mader and Laming (1985), Lorsong *et al.* (1990), Mader (1990), Smith (1990), Purvis and Wright (1991), Smith and Edwards (1991) and Wright *et al.* (1991) carried out studies on the sedimentology of the Otter Sandstone Formation, focusing on the palaeosols and other climatic indi-

cators. Spencer and Isaac (1983), Milner *et al.* (1990), Benton (1990c) and Benton *et al.* (1994) described collections of fishes and tetrapods made between 1982 and 1994 by P.S.S. that greatly enlarged the faunal list.

The Otter Sandstone Formation has been regarded as 'sparsely fossiliferous' (Spencer and Isaac, 1983). This mistaken impression may be the result of the steepness and height of the cliffs and the fact that most fossils so far collected have come from fallen blocks on the shore. The Sidmouth to Budleigh Salterton section has yielded the largest number of remains of fossil reptiles and amphibians from the New Red Sandstone of Devon, and one of the widest ranges of fossil amphibians and reptiles from the British Middle Triassic, and it continues to produce new finds. Type specimens of *Mastodonsaurus lavisi* and *Rhynchosaurus spenceri* come from High Peak, and other unusual finds include the ?ctenosauriscid neural spine, the tanystropheid tooth and the exquisite small procolophonids. These small fossils may be of biostratigraphic value.

Description

The Otter Sandstone Formation (Sherwood Sandstone Group) is exposed in a series of fine sea cliffs along the coast from west of Ladram Bay to just east of Sidmouth. The nature of the cliffs was described by all the Victorian authors mentioned above. Whitaker (1869) noted that most of the cliff at High Peak was formed by the 'red marl', which was heavily weathered above the harder 'red sandstone'. Irving (1888, pp. 152–3) stated that the latter was underlain by 'massive, strongly current-bedded (Bunter) sandstones' which continue to the mouth of the Otter River. The succession is summarized below, with measurements estimated from Lavis (1876, fig. 1), on the assumption that High Peak is 155 m high (contour on 6-inch topographic map).

	Thickness (m)
Chalk gravel	5
Greensand	30
Upper (Keuper) Marls (unnamed formation of Mercia Mudstone Group)	60
Otter Sandstone Formation	*c.* 60

The Otter Sandstone Formation (the 'red sandstone') comprises *c.*118 m of medium- to fine-grained red sandstones which dip gently eastwards in the coast section. The formation continues northwards to Somerset and eastwards as far as Hampshire and the Isle of Wight beneath younger Triassic sediments (Holloway *et al.*, 1989). It rests unconformably on the Budleigh Salterton Pebble Beds, a 20–30 m thick unit of fluvial conglomerates (Henson, 1970; Smith, 1990; Smith and Edwards, 1991). The contact is marked by an extensive ventifact horizon (Leonard *et al.*, 1982) that represents a non-sequence of unknown duration and is interpreted by Wright *et al.* (1991) as a desert pavement associated with a shift from a semi-arid to an arid climate.

Calcretes occur abundantly at Otterton Point, Budleigh Salterton (see below), but farther east they are rarer and the formation is dominated by sandstones in large and small channels, with occasional siltstone lenses. The sandstones occur in cycles, often with conglomeratic bases, and fine upwards through cross-bedded sandstones to ripple-marked sandstones. The Otter Sandstone Formation is capped by water-laid siltstones and mudstones of the Mercia Mudstone Group.

Henson (1970), Laming (1982, pp. 165, 167, 169) and Mader and Laming (1985) interpreted the Otter Sandstone Formation as comprising fluvial and aeolian deposits. Sandstones near the base are aeolian, and middle and upper parts of the formation are of fluvial origin; sandstones were deposited by ephemeral braided streams flowing from the south and south-west (Selwood *et al.*, 1984). The comparatively thin mudstones are interpreted as the deposits of temporary lakes on the floodplain. The calcretes indicate subaerial soil and subsurface calcrete formation in semi-arid conditions (Mader and Laming, 1985; Lorsong *et al.*, 1990; Mader, 1990; Purvis and Wright, 1991). The climate was semi-arid, with long dry periods when river beds dried out, and seasonal or occasional rains leading to violent river action and flash floods.

Recent collections of amphibian and reptile bones have come from the top 40 m or so of the Otter Sandstone Formation and occur in all lithologies, but most commonly in intraformational conglomerates and breccias (Spencer and Isaac, 1983). Lower in the sequence, in breccias exposed west of Chiselbury Bay (Figure 4.7), the abundance of tetrapod finds declines significantly. The bones are generally in a fine- to medium-grained reddish sandstone that often contains clasts of pinkish, greenish or ochreous calcrete and mudflakes up to 20 mm in diameter. The more complete fish specimens are, however, preserved in dark red siltstone, sometimes in association with plants and conchostracan crustaceans. Plant remains are preserved in iron oxide in all the lower-energy deposits, and their occurrence appears to be controlled by the sedimentology.

The only specimens found *in situ* by Spencer and Isaac (1983, p. 268) came from 'the lowest of three intraformational conglomerates', but these were 'indeterminate bone fragments'. Since 1983, four rhynchosaur specimens (EXEMS 60/1985.284, 285, 292, and 7/1986.3) have been collected *in situ* from a single horizon at beach level, and a partial rhynchosaur skeleton was found at the top of the foreshore exposures in Ladram Bay in 1990 (EXEMS 79/1992). It is likely that fossils occur at numerous levels throughout the Otter Sandstone Formation, but most have been found in fallen blocks on the shore and locating the original horizons in the cliffs is difficult.

The Victorian authors believed that one or more discrete bone beds occurred at the eastern end of the outcrop. Lavis (1876) and Metcalfe

(1884) placed it 'about 10 feet from the top of the sandstone'; Hutchinson (1906) and Woodward and Ussher (1911) placed it 'about 50 feet below the base of the Keuper Marls', some 40 ft (13 m) lower in the section.

Lavis (1876) made his finds in fallen blocks from a 'fossiferous zone' consisting of up to four beds and 'characterized by lithological differences, in as much as the matrix is composed of much coarser sandstone, containing here and there masses of marl varying in size from that of a pea to that of a hen's egg. In these beds ripple-marks are very plentiful. The fragments of bone which are found in this zone seem to be very slightly water-worn'. Metcalfe (1884) gave further details of this locality at High Peak, stating that bones were found in fallen blocks of sandstone from a light-coloured band in the cliff close below the base of the 'Upper Marls' (Mercia Mudstone Group). Carter (1888) recovered bone material and coprolites from this locality.

Hutchinson (1879, p. 384) gave the most detailed account of the fossiliferous horizons. He found equisetalean plant stems in a bed at the top of the sandstone and 'about eight or ten feet above' two or three 'white bands' which appear as clear horizons in the cliff face. Then, 'one or two steps below' the White bands 'is what I venture to call the Saurian or Batrachian band, in which Mr Lavis found his Labyrinthodon; but I cannot exactly say how many feet this band is below the white bands, because the fall down of the under cliff has concealed the stratification at this place; but it may be fifty feet below and amongst the beds of red rock. Be that as it may, the Saurian band rises out of the beach somewhere under Windgate, as the hollow between the two hills is called, and ascends westwards into High Peak Hill, and having proceeded for about half-a-mile, and having attained a height of sixty or seventy feet above the sea, a fall of the cliff enabled Mr Lavis to find his specimens on the beach, and I was so fortunate as to see them soon afterwards.'

Woodward and Ussher (1911, pp. 12–13) summarized an unpublished section drawn up by Hutchinson in 1878 in which he located the bone bed '100 feet above the talus on the beach, and about 50 feet below the base of the Keuper Marls'. No trace of any tetrapod-bearing horizon in the form of a bone bed can be seen today, and there is no evidence that one existed. The Victorian geologists evidently expected to find bones at discrete levels, and had no concept of restricted lenticular deposits, such as channel lags.

The tetrapod fossils (Figures 4.9 and 4.10) are generally preserved in a fine- to medium-grained, orange to reddish sandstone that often contains clasts, including reworked rhizolith concretions, up to 20 mm in diameter, and claystone intraclasts which may have a pinkish, greenish or ochreous colour. The bones occur as generally isolated elements: jaws, teeth, partial skulls or single postcranial bones, but some occur in articulation. Exceptions are the partial articulated skull and lower jaws of *Rhynchosaurus spenceri* (EXEMS 60/1985.292), the associated humerus, radius and ulna of that species (EXEMS 60/1985.282), two sets of vertebrae (EXEMS 60/1985.15, 57), and the recently collected partial rhynchosaur skeleton (EXEMS 79/1992), which comprises much of the trunk, the pelvis and the hindlimbs, with the bones in close association, but mostly slightly disarticulated (Benton *et al.*, 1993). The tetrapod bones generally show little obvious sign of abrasion and some tiny procolophonid jaws are exquisitely well preserved. More details of taphonomy are given by Benton *et al.* (1994).

About half of the identifiable tetrapod bones found are rhynchosaur remains, and most of these are parts of the skull, especially the jaw elements, which have a high preservation potential. The amphibians are represented mainly by skull and pectoral girdle elements, all relatively dense and with characteristic sculpture. The small reptiles are represented by limited postcranial elements, a partial skull (with lower jaws articulated), teeth and small segments of jaw, and the larger archosaur(s) by teeth and vertebrae. Specimens of the fish *Dipteronotus* and fossil invertebrates (Figures 4.9J, 4.10H–J) obtained from a claystone lens, east of Windgate, are extremely well articulated and occur in association with a 'still water' fauna of branchiopod crustaceans.

Fauna

The faunal list of invertebrates, fishes, temnospondyl amphibians and reptiles is compiled from Benton (1990c), Milner *et al.* (1990) and Benton *et al.* (1994).

Arthropoda: Crustacea: Branchiopoda
  Lioestheria
    Carapaces of adults and juveniles (BRSUG)

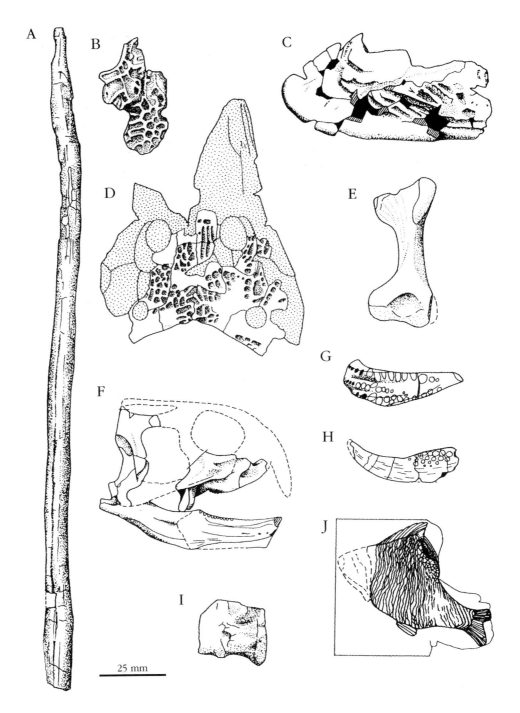

Figure 4.9 Larger elements of the Otter Sandstone Formation fauna of Devon. (A) Spine of an unknown verte-brate, possibly a dorsal neural spine of a ctenosauriscid archosaur (EXEMS 60/1985.88); (B) fragment of the skull roof of *Mastodonsaurus lavisi* in dorsal view (EXEMS 60/1985.287); (C) posterior portion of a right mandible of an unknown capitosaurid, in lateral view (EXEMS 60/1985.78); (D) incomplete skull roof of *Eocyclotosaurus* sp., in dorsal view (EXEMS 60/1985.72); (E)–(I) remains of *Rhynchosaurus spenceri*: (E) left humerus in ventral view (EXEMS 60/1985.282), (F) restored skull in right lateral view (EXEMS 60/1985.292), (G) right maxilla in ventral view (EXEMS 60/1985.292), and (H) right dentary in lingual view (BMNH R9190). (I) Vertebra of an archosaur (Bristol Univ. unnumb.); (J) the neopterygian ('palaeonisciform') fish *Dipteronotus cyphus* (EXEMS 60/1985.293). After various sources; from Benton *et al.* (1994).

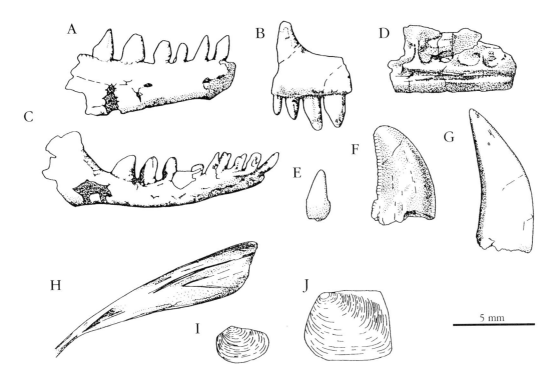

Figure 4.10 Smaller elements of the Otter Sandstone Formation fauna of Devon. Right dentaries (A, C) and a left maxilla (B) of a procolophonid, all in lateral view (EXEMS 60/1985.311, 3, and 154); (D) dentary fragment of an unknown small pleurodont reptile, showing pits for teeth, in lingual view; (E) tooth of *?Tanystropheus*, showing small accessory cusps (EXEMS 60/1985.143); (F), (G) recurved teeth of two kinds of unknown archosaurs (BRSUG unnumb.); (H) unidentified insect wing (BRSUG unnumb.); (I), (J) carapaces of the conchostracan *Euestheria* (BRSUG unnumb.). After various sources; from Benton *et al.* (1993).

Euestheria
  Carapaces of adults and juveniles (BRSUG)

Arthropoda: Crustacea: Ostracoda
  Two carapaces, apparently representing separate taxa (BRSUG)

Arthropoda: Insecta
  Insect wing (BRSUG)

Mollusca: Bivalvia
  Taxon unidentified; single valve (BRSUG)

Osteichthyes: Actinopterygii: Neopterygii: Cleithrolepididae
  Dipteronotus cyphus Egerton, 1854
    Complete specimens, pieces of flank, individual scales and spines (EXEMS)

Osteichthyes: Actinopterygii: 'Palaeonisciformes'
  Gyrolepis(?) and others
    Scales

Sarcopterygii: Dipnoi: Lepisosteidae
  Lepisosteus sp.
    Scales in coprolites

'Temnospondyli': Mastodonsauridae
  Mastodonsaurus lavisi (Seeley, 1876), *nomen dubium*
    Skull fragments and part of a lower jaw (BMNH, EXEMS)

'Temnospondyli': Benthosuchidae
  Eocyclotosaurus sp
    Remains of a skull and other fragments (EXEMS)

'Temnospondyli': Capitosauridae
  Capitosauridae *incertae sedi*
    Posterior part of mandible (EXEMS)

Anapsida: Procolophonidae
  Procolophonid *incertae sedis*
    Three small dentaries, a maxilla and an interclavicle (EXEMS, BRSUG)

Archosauromorpha: Rhynchosauridae
  Rhynchosaurus spenceri Benton, 1990
    Skull and mandible fragments, isolated maxillae and postcranial elements from about 29 individuals (BMNH, BGS(GSM), EXEMS, BRSUG)

Archosauromorpha: Prolacertiformes
  Tanystropheus sp.
    A small tricuspid tooth (EXEMS)
  ?Prolacertiform
    jaw fragments and teeth (EXEMS)

Archosauria: Crurotarsi: *indet.*
  Rauisuchids and others(?)
    Numerous teeth, a jaw, cranial and postcranial elements (BMNH, EXEMS)

Amniota *incertae sedis*
  ?Ctenosauriscid archosaur
  ?Neural spine (EXEMS)

Interpretation

Attempts to recover palynomorphs from the Otter Sandstone Formation have so far not been successful (Warrington, 1971, and pers. comm. to P.S.S., 1983). Its age is poorly constrained by occurrences of Late Permian miospores in the lower part of the Permo-Triassic succession near Exeter (Warrington and Scrivener, 1988, 1990) and Carnian taxa in the Mercia Mudstone Group, 135 m above the Otter Sandstone Formation. The only other biostratigraphic indicator, the vertebrate fauna itself, is all that is available for consideration. Walker (1969, 1970a), Paton (1974a) and Benton (1990c) favoured a Ladinian age for the fauna, but Milner *et al.* (1990) argued that an Anisian age was most likely. The association of the perleidid fish *Dipteronotus cyphus* (Anisian–earliest Ladinian), *Eocyclotosaurus* (Late Scythian–Anisian), procolophonids (Scythian–Anisian), ?ctenosauriscid (Anisian–Carnian) and ?tanystropheid (Anisian–Ladinian) identifies the Anisian as the only shared date (Figure 4.2).

The remains of three forms of temnospondyl amphibian (*M. lavisi*, *Eocylotosaurus* sp., capitosaur *incertae sedis*) are abundant in the Otter Sandstone Formation (Figure 4.9). These were all aquatic, superficially crocodile-like forms, and were probably carnivores or piscivores which fed at the waterside. The new eocyclotosaur material represents the first find of a benthosuchid from the Middle Triassic in Britain. It is similar to *Eocyclotosaurus* species from two European formations: *E. lehmani* from the *Voltzia* Sandstone of the Vosges in France and *E. woschmidti* from the Lower Röt of the Schwarzwald in Germany. There is also undescribed eocyclotosaur material from the Moenkopi Formation of Arizona (Welles and Estes, 1969; Morales, 1987). The remains of *Mastodonsaurus lavisi* show some resemblance in interorbital proportions and dermal sculpture to material from Coten End and Bromsgrove (Paton, 1974a, pp. 265–82) and these show closest resemblance to *M. cappelensis* from the Upper Buntsandstein (Anisian) of Baden-Württemburg, Germany (Milner *et al.*, 1990). *M. lavisi* is the largest temnospondyl in the Otter Sandstone herpetofauna with an estimated skull length of 500–600 mm, and a body length of 2 m or more.

Rhynchosaurus spenceri (Figure 4.9) is the largest species of the genus, with an average skull length of 140 mm, and an estimated body length of 0.9–1 m (Benton, 1990c). The maxilla had two grooves, a major and minor one, which received two matching ridges on the dentary when the lower jaw was in full occlusion. The genus *Rhynchosaurus* is also recorded from the Bromsgrove Sandstone Formation of the Midlands (*R. brodiei* at Coten End Quarry, Leamington, and Bromsgrove) and from Grinshill Quarry, Shropshire (*R. articeps*) (see above). *R. spenceri* is distinguished from these forms in having a larger skull length (140 mm), a skull that is broader than it is long (otherwise a character of Late Triassic rhynchosaurs) and a tendency for the tooth rows on the maxilla to 'meander'.

Some recently collected procolophonid remains (Figure 4.10A–C) appear to belong to a primitive form, and Fraser (in Milner *et al.*, 1990) suggested that they most closely resembled the Mid Triassic (Anisian) form *Anisodontosaurus greeri* from the Holbrook Member of the Moenkopi Formation of Northern Arizona, USA. Re-examination of the material by P.S.S. indicates that there may be up to three taxa, and the most closely related forms appear to be *Kapes, Tichvinskia* and *Phaanthosaurus* from the Lower and lower Middle Triassic Vetluga series of the Russian Platform.

A tricuspid tooth (Figure 4.10E), the sole specimen ascribed to the Tanystropheidae, is reminiscent of the teeth of *Tanystropheus* from the Anisian and Ladinian of Central Europe (Wild, 1980a).

An elongate element from the Otter Sandstone

(Figure 4.9A) may provisionally be assigned to a ctenosauriscid archosaur. Comparable occurrences of ctenosauriscids with such elongate spines are *Ctenosaurus* from the Anisian Upper Buntsandstein of Germany (Krebs, 1969), *Hypselorhachis* from the Anisian Manda Formation of Tanzania, and *Lotosaurus* from the Middle Triassic of China.

Dipteronotus cyphus, a deep-bodied perleidid fish, is represented at Sidmouth by many well-preserved partial and complete remains (Figure 4.9J). Specimens of *D. cyphus*, including the holotype, have been obtained elsewhere only at Bromsgrove, from the upper member of the Bromsgrove Sandstone Formation. The Otter Sandstone specimens are better preserved than those from Bromsgrove (Gardiner, *in* Milner *et al.*, 1990). *Dipteronotus* is known also from the Scythian of Europe and the Carnian/Norian of Morocco.

The only plants so far found in the Otter Sandstone Formation are stems and leaves of large horsetails (Hutchinson, 1879), and recent finds of fossils identified as *Schizoneura*, a form also known from the Bromsgrove Sandstone Formation (P.S.S., personal observation).

The Otter Sandstone fauna and flora (Figure 4.11) is comparable to that of Bromsgrove. It is also reminiscent of the Scythian/Anisian Upper Buntsandstein and *Voltzia* Sandstone faunas of Germany and France (Milner *et al.*, 1990), although in these assemblages, *Rhynchosaurus* is absent. The closest comparable locality is Otterton Point.

Figure 4.11 Imaginary scene during Mid Triassic times in Devon, based on specimens recovered from the Otter Sandstone Formation between Sidmouth and Budleigh Salterton. A scorpion (mid-foreground) contemplates a pair of procolophonids on the rocks. Opposite them, a hefty temnospondyl amphibian has spotted some palaeonisciform fishes, *Dipteronotus*, in the water. Two *Rhynchosaurus* stand in the middle distance and, behind them, a pair of rauisuchians lurk. The plants include *Equisetites* (horsetails) around the waterside and *Voltzia*, a coniferous tree. Drawn by Pam Baldaro, based on her colour painting.

Conclusions

The coast at Sidmouth offers vast potential for study of Mid Triassic reptiles. New finds are made all the time *in situ* and in fallen blocks, with erosion constantly supplying new specimens. This potential and the importance of past finds give the site its conservation value.

OTTERTON POINT (BUDLEIGH SALTERTON), EAST DEVON (SY 07758196)

Highlights

The Otter Sandstone Formation at Otterton Point is the source of a specimen of *Rhynchosaurus spenceri*, and offers potential for future finds of this important Mid Triassic reptile.

Introduction

The fossil-bearing site lies in a cove on the east side of the mouth of the Otter River, just north of Otterton Point (Figure 4.7), accessible over fields from South Farm. The Otter Sandstone Formation here yielded the first find of *Rhynchosaurus* from Devon. The Otterton Point locality and coast section in general is probably much as it was 100 years ago, and fresh finds are likely.

A tooth plate of *Rhynchosaurus* (i.e. *R. spenceri*) was collected by William Whitaker from a 'brecciated horizon' in the lower part of the Otter Sandstone Formation exposed in a low cliff on the east bank of the Otter River. It was described as *Hyperodapedon* by Huxley (1869) and compared with the Elgin rhynchosaur. Lavis (1876) and Ussher (1876) made general comments on the sandstone at Otterton Point, and Metcalfe (1884) reported white fragments in the harder beds of the sandstones 'at numerous points near Budleigh Salterton and Otterton Point', which were identified by him as fragmented bone. Irving (1888, 1892, 1893) and Hull (1892) further described the stratigraphy and structure of this section. Subsequent palaeontological and sedimentological work is outlined in the account of the Sidmouth section (see above).

Description

At Otterton Point hard, calcite-cemented, cross-bedded sandstone units (less than 0.5 m thick) in the Otter Sandstone Formation contain calcite-cemented rhizoliths, up to 1 m deep, and other calcrete formations (Mader, 1990; Purvis and Wright, 1991). Purvis and Wright (1991) attributed the large vertical rhizoliths to deep-rooted phreatophytic plants which colonized bars and abandoned channels on a large braidplain. The sedimentology of the Otter Sandstone Formation is more fully described in the Sidmouth account (see above).

These phenomena were noted by earlier authors. Whitaker (1869) commented that 'on the left bank of the Otter [the sandstone] has, in parts a brecciated character'. Lavis (1876) noted that the sandstones at Otterton Point 'contain curious irregular branching-shaped masses of harder texture, which withstand the weathering and give the cliff a rugged aspect' and he observed that these hard masses allowed the sandstone to resist erosion and form promontories into the sea. Ussher (1876, p. 380) observed that the sandstones here 'contain two or three conglomerate beds, and a few pebbles in false-bedded lines'. Metcalfe (1884, pp. 259–60) described these hard masses as calcareous concretions produced from the eroding debris of Devonian limestone. Irving (1888, p. 153) described 'an irregular band of breccia... intercalated with the sandstones, just above high-water mark', and containing fragments of slate, granite, sandstone and quartzite. Woodward and Ussher (1911, pp. 10–11) traced this 'brecciated horizon' as far as Ladram Bay, 3.5 km to the north-east of Otterton Point.

Fauna

Archosauromorpha: Rhynchosauridae
    Rhynchosaurus spenceri Benton, 1990
        maxillary tooth-plate BGS(GSM)

Interpretation

The original find of a *Rhynchosaurus* jaw from Otterton Point appears to have come from the zone of breccia and calcite-cemented nodules which occurs along the base of the cliff eastwards for 2 km towards Ladram Bay. The same beds occur in isolated exposures for about 1 km up the Otter River and in a small outcrop at the east end of the esplanade in Budleigh Salterton, but further remains of *R. spenceri* have not been recovered from any of these localities. Professor R.J.G. Savage of Bristol University washed and sieved loose matrix from some of these exposures along the Otter River and obtained numerous remains of fish, including isolated teeth, scales and spines, and a thecodont tooth fragment (pers. comm. to P.S.S., 1983).

Comparison with other localities
(see the account of the Sidmouth site)

Conclusions

Otterton Point yielded the first evidence of Mid Triassic reptiles from Devon, and formed a useful point of comparison with localities elsewhere in England, especially with the larger Sidmouth section. The importance of past finds and the potential for new ones gives the site its conservation value.

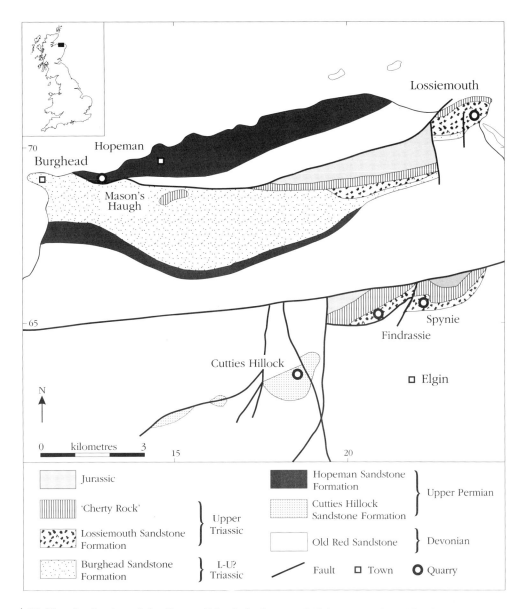

Figure 4.12 The distribution of the Permo-Triassic beds around Elgin, Morayshire. The formations are indicated by shading, and the main reptile and footprint localities are named. From Benton and Walker (1985).

LATE TRIASSIC OF SCOTLAND

The Lossiemouth Sandstone Formation (Late Carnian) of Grampian, Scotland is famous for its reptile fauna. The outcrop is distributed in three fault-bounded blocks at Lossiemouth, Findrassie and Spynie, and in an east–west strip about 2 km south-west of Lossiemouth (Figure 4.12). Finds of reptiles are restricted to several small quarry workings at Spynie and Findrassie and the coast section at Lossiemouth, where finds came from excavations at Lossiemouth East and West Quarries. The largest fauna has been obtained from the last-named sites at Lossiemouth (NJ 236707 and NJ 231704): the archosaurs *Ornithosuchus*, *Stagonolepis*, *Scleromochlus*, *Erpetosuchus* and *Saltopus* (a dinosaur), the rhynchosaur *Hyperodapedon*, the sphenodontid *Brachyrhinodon* and the procolophonid *Leptopleuron*. Spynie (quarries at NJ 223657) has yielded specimens of *Leptopleuron*, *Hyperodapedon* and *Ornithosuchus* (and doubtfully *Stagonolepis* and *Erpetosuchus*), and Findrassie (NJ 207652, NJ 204651) has produced good remains of *Ornithosuchus* and *Stagonolepis*. A glacially transported block of reptiliferous Lossiemouth Sandstone on the Hill of Meft, north-west of Urquhart (NJ 268642), 5 km ESE of Spynie, has yielded a specimen of *Leptopleuron*, and possibly *Stagonolepis* scutes (Taylor, 1920). All the reptile-bearing localities can be correlated on the basis of identical reptile assemblages and on lithology, and it is likely that the Lossiemouth Sandstone Formation at Lossiemouth East and West Quarries is the same as that at Spynie and Findrassie. Three locations are selected as GCR sites:

1. Lossiemouth East Quarry (NJ 236707), Upper Triassic (upper Carnian), Lossiemouth Sandstone Formation.

2. Spynie (NJ 223657). Upper Triassic (upper Carnian), Lossiemouth Sandstone Formation.

3. Findrassie (NJ 207652, NJ 204651). Upper Triassic (upper Carnian), Lossiemouth Sandstone Formation.

LOSSIEMOUTH EAST QUARRY (NJ 236707)

Highlights

Lossiemouth East Quarry is one of the richest Late Triassic reptile sites in Britain, the source of superb specimens of six species of archosaur, one procolophonid and a sphenodontid. Four of the reptiles have been found nowhere else. The reptiles from this site are important in making palaeobiogeographical interpretations for the Late Triassic: the most similar faunas elsewhere are in India and South America.

Introduction

The Lossiemouth Sandstone Formation is seen in a 450 m long raised cliff between Lossiemouth and Branderburgh (Figure 4.12), running from School Brae to the old railway station, bounded above by Prospect Terrace and below by Quarry Road. Lossiemouth West Quarry, which is located on the west side of School Brae (NJ 231704), was once more important for reptile finds, but is now largely filled in. Lossiemouth East Quarry, although partly filled in and surrounded by new housing, shows up to 3 m of orange or grey weathering, jointed sandstone in its eastern part (Figure 4.13) and partially to the west. Fossils could still probably be collected at the lower level of the East Quarry, which seems to have been the location of the reptile band. Neville Hollingworth obtained pieces of bone and a good skull of *Leptopleuron*, from the nearby beach in 1979.

Building stone was quarried in Lossiemouth as early as 1790, where two hills 'abound with excellent quarries in white and yellow free-stone, which is not to be found anywhere else in the Moray Firth. About 20 masons, including apprentices, and nearly double that number of labourers, are constantly employed in quarrying and dressing stones, to supply the demand for that article from this and neighbouring counties' (Lewis Gordon in Sinclair, 1793, vol. 4, p. 78). The West Quarry had ceased to work by 1912, but the larger East Quarry continued in operation until 1936 or later (according to County valuation rolls).

In 1844 a workman named Anderson collected a slab bearing casts of 31 scales, or scutes, at Lossiemouth. These he passed to the Elgin town clerk and geologist, Patrick Duff, who tried with-

Figure 4.13 Lossiemouth East Quarry: view of heavily jointed dune cross-bedded sandstones at the eastern end of the site. Reptile skeletons were found at the base of the quarry. (Photo: M.J. Benton.)

out success to have them identified in Edinburgh. Eventually, he sent drawings to the Swiss palaeontologist, Louis Agassiz, who identified them as belonging to an Old Red Sandstone (ORS) fish related to the large ganoid *Gyrolepis*. He named it *Stagonolepis robertsoni* after Alex Robertson, a local geologist (Agassiz, 1844, p. 139, pl. 31, figs 13 and 14). On the basis of this find, it seemed clear to Agassiz that the age of the Lossiemouth beds was Devonian, as had been previously assumed largely on the basis of lithological similarity with neighbouring sediments of undoubted Old Red Sandstone age.

In 1858 Mr Martin, schoolmaster in Elgin, 'detected . . . a bone, possibly the scapula of a reptile . . . It was near the same place that Mr Duff's specimen . . . of *Stagonolepis robertsoni* was found' (Gordon, 1859, p. 46). This bone was collected in sandstone beds behind the houses of Lossiemouth (Murchison, 1859, pp. 427–8), thus

the East Quarry. Murchison collected further remains of limb bones in September, 1858, in company with George Gordon, minister of Birnie and naturalist. These remains were shown (Huxley, 1859a, 1859b) to be associated with the scutes of *Stagonolepis robertsoni* and proved that it was a reptile and, in particular, an ancestral crocodile according to Huxley (1875, 1877).

Some poorly preserved remains of another reptile were found at Lossiemouth by Gordon in April and May, 1859. These were interpreted as a rhynchosaur (Huxley, 1869) and named *Hyperodapedon gordoni* (Huxley, 1859a). The presence of an animal clearly closely related to the English *Rhynchosaurus* (from rocks of known Triassic age), and of the specialized *Stagonolepis* convinced Huxley that the Elgin reptiles were Triassic in age, and not Devonian. Murchison and other geologists continued to argue for an ORS age since there was no obvious structural or lithological break between true Old Red Sandstone of the Elgin district and the reptiliferous sandstone. However, Murchison admitted that his belief was shaken by the find of *Hyperodapedon* (Murchison, 1859, p. 436).

During the 1860s several geologists claimed to have found structural evidence to separate the Elgin ORS and Triassic, and further finds of *Hyperodapedon, Rhynchosaurus* and a rhynchosaur in the Triassic Maleri Formation of India (Huxley, 1869) were enough to convert Murchison and with him most other geologists: 'To such fossil evidence as this the field geologist must bow, I willingly adopt the view established by fossil evidence, and consider that these overlying sandstones and limestones are of Upper Triassic age' (Murchison, 1867, p. 267).

Further material of *Hyperodapedon* was described by Huxley (1887), Burckhardt (1900), Boulenger (1903) and Huene (1929a, 1938, 1939). Boulenger (1903) erected a new genus, *Stenometopon*, based on a specimen that supposedly differed from *Hyperodapedon*, but which is merely a distorted example (Benton, 1983d).

A well-preserved skeleton of *Leptopleuron lacertinum* (*Telerpeton elginense*) was collected by James Grant, a Lossiemouth schoolmaster, in 1866 and described by Huxley (1867a). This was the second specimen collected, the first having been found at Spynie in 1851 (see below). Further bones of *Stagonolepis* were also collected in the 1860s (Anon., 1864) which provided materials for detailed monographs by Huxley (1875, 1877).

In the early 1890s Grant discovered a small skull

and partial skeleton built into a breakwater at Lossiemouth, probably originating from one of the Lossiemouth quarries; it was named *Erpetosuchus granti* by Newton (1894b). Between 1895 and 1920 William Taylor, a retired chemist and naturalist, collected extensively in both East and West Quarries at Lossiemouth. He supplied materials for further descriptions of *Hyperodapedon* (Boulenger, 1903), *Ornithosuchus* (originally collected at Spynie; Boulenger, 1903; Watson, 1909a; Huene, 1914), *Leptopleuron* (Boulenger, 1904a; Huene, 1912a, 1920), and type specimens of the new genera and species *Scleromochlus taylori* (Woodward, 1907b; Huene, 1914), *Saltopus elginensis* (Huene, 1910a) and *Brachyrhinodon taylori* (Huene, 1910b, 1912b). Recent redescriptions of much of this material have been published (Walker, 1961, 1964; Benton, 1983d; Benton and Walker, 1985; Fraser and Benton, 1989; Figure 4.14) and others are in preparation by M.J.B. and P.S.S.

Description

Lossiemouth West Quarry showed '30 ft of hard, white, fine-grained, laminated and even-grained sandstone . . . with about 5 ft of till on top. The rock is close jointed with a dominant west-north-west-trending set and a subordinate north-north-east set, both sets dipping nearly vertically. The former is composite (i.e. two joint directions with an angle of about 20 degrees between them) and the joints often carry fillings of barytes and brown fluorspar, such fillings being filled over an inch across in some places' (Peacock *et al.*, 1968, p. 67). There is probably a small NE–SW trending fault below School Brae since the floor of the East Quarry is rather higher than that of the West.

Lossiemouth East Quarry is located in a sea cliff which was extensively quarried, and still exposes sections showing up to 20 m of hard to friable, yellow and grey sandstone, weathering orange, and jointed in the eastern part. The sandstones may be finely laminated, but more usually they show large-scale cross-beds on well-weathered surfaces. An isolated 3 m high sea stack at the eastern end shows white dune-bedded sandstone. Further west in the quarry, near a footpath up the slope, various sections show purple ORS and mudstone at the base, surmounted by yellow or white, soft, thinly-bedded Triassic sandstone. A block of 'cherty rock' that overlies the reptile beds at Spynie and north of Lossiemouth is also seen here.

Details of cross-bedding and petrography of the Lossiemouth sandstones are given by Williams (1973). The section in the East Quarry that he gives is:

	Depositional environment	Thickness (m)
(Cherty Rock)	altered caliche	
Sago Pudding	water-lain,	
Sandstone	reworked aeolian	1-2
Lossiemouth		
Sandstone		
Formation	aeolian	18
Burghead		
Sandstone		
Formation	fluviatile	*c.* 4
Upper Old Red Sandstone		

Most of the reptiles were apparently collected in the West Quarry, but only the East Quarry is now exposed to any extent. Murchison (1859, p. 428) stated that the bones found then were collected 'in the lowest part' of the freestones being quarried at Lossiemouth which were 'underlain by red strata' (?ORS). Gordon (1859, p. 46) confirmed this, stating that the lowest beds at Lossiemouth were red clay (i.e. within the ORS?), succeeded by yellowish soft sandstone and then harder sandstone. The red clay may be equivalent to that reported by Peacock *et al.* (1968, p. 65) as 'micaceous siltstone', the yellowish soft sandstone may be the 'Burghead beds equivalent', and the harder sandstone is probably the Lossiemouth Sandstone. The bones were found 'immediately under this hard siliceous sandstone, in a quarry half-way to the new harbour from Rockhouse, and in the face of the wall of rock that overhangs the houses fronting the old harbour.' This probably refers to the east end of Lossiemouth East Quarry (NJ 237707). Judd (1873, p. 137) stated that the reptiles were found '100 ft below the top of the sandstones', which would imply at about the base of the Lossiemouth Sandstone Formation, if its complete thickness is taken into account. Judd (1886a, pp. 397, 403) added that the reptile remains all came from 'a single band of soft rock'. Further, Gordon (1892, p. 245) states that most of the fossils were found at the level of the platform made by the quarrymen in the base of the quarry where the sandstone became 'softish and rubbly'. Williams (1973, p. 130) notes that 'the quarry floor approximates to the contact of the aeolian sandstones with the floodplain deposits' (the water-laid Burghead Beds), and the reptiles seem to have been found near to this transition.

The remains of reptiles are normally well preserved in articulation and only a few show disturbance, possibly through scavenging. Individual bones, particularly the smaller ones, show few signs of crushing or compression. The larger limb bones, however, appear to have been more susceptible to crushing and distortion and may show damage even when in association with other unaffected elements. The bone material is usually in a corroded state and may be partly leached out, but in a few specimens where the original material is present, internal structure may be clear, with cavities marked by replacement minerals. These minerals, which include iron oxide (goethite) and fluorite, sometimes overgrow bone margins adhering with the surrounding matrix, and in such cases are hard to remove. Most commonly, however, the Lossiemouth reptiles are preserved as external moulds in very well cemented sandstone, and details of bone form are best obtained from casts taken from the cleared natural rock moulds. Various methods that involve use of flexible synthetic 'rubbers' (e.g. RTV silicone rubber, PVC) have been employed in order to preserve the rock moulds and produce highly detailed copies of the bone (Walker, 1961, 1964, 1973; Benton and Walker, 1981).

Fauna

Anapsida: Procolophonidae
  Leptopleuron lacertinum Owen, 1851
    (=*Telerpeton elginense* Mantell, 1852)
    c. 26 individuals: BMNH, NMS, EGNM

Lepidosauria: Sphenodontida
  Brachyrhinodon taylori Huene, 1910
    c. 10 individuals: BMNH, NMS, ELGNM

Archosauromorpha: Rhynchosauridae
  Hyperodapedon gordoni Huxley, 1859
    c. 29 individuals: BMNH, BGS(GSM), NMS, MANCH

Archosauria: Crurotarsi: Pseudosuchia:
  Stagonolepididae
  Stagonolepis robertsoni Agassiz, 1844
    6 large individuals: BMNH, AUZD, ELGNM, BGS(GSM)
    11 small individuals: BMNH, AUZD, AUGD, NMS, ELGNM, BGS(GSM)

Archosauria: Crurotarsi: Ornithosuchidae
  Ornithosuchus longidens (Huxley, 1877)
    7 individuals: BMNH

Archosauria: Crurotarsi: *incertae sedis*
  Erpetosuchus granti Newton, 1894
    1 individual: BMNH
  Scleromochlus taylori Woodward, 1907
    5 individuals: BMNH
    'Thecodontian'
    1 individual: MANCH

Archosauria: Dinosauria: *incertae sedis*
  Saltopus elginensis Huene, 1910
    1 individual: BMNH

Interpretation

The Lossiemouth Sandstone Formation at Lossiemouth, Spynie and Findrassie is of the same age since it contains the same reptiles and is lithologically similar. It is placed in the Late Triassic on the basis of its varied reptile fauna (Walker, 1961; Benton, 1983d, 1986b, 1991, 1994a, 1994b). *Hyperodapedon* is represented in the Maleri Formation of central India and *Stagonolepis* (= *Calyptosuchus*) is reported in the lower part of the Petrified Forest Member of the Chinle Formation of Arizona (Hunt and Lucas, 1991b). The latter unit is dated palynologically as uppermost Carnian and the shared phytosaur *Paleorhinus* ties the lower part of the Petrified Forest Member to the Maleri Formation, and also to the Blasensandstein in Germany, and the marine Opponitzer Schichten of Austria, which are dated by ammonoids (Hunt and Lucas, 1991a, 1991b). On the evidence of these two reptile genera, the Lossiemouth Sandstone Formation is dated firmly as latest Carnian (Late Tuvalian palynological zone; *macrolobatus* ammonoid zone).

Lossiemouth has yielded specimens of all eight Late Triassic Elgin reptiles (Figures 4.14 and 4.15), so they will be discussed here. Each of the eight reptiles is unique to the Elgin area and some, in fact, have been placed in separate monogeneric families. Some of the reptiles (rhynchosaur, ornithosuchid) compare best with Gondwanaland faunas of similar age, such as those of the Maleri Formation in India, the Santa Maria Formation of Brazil and the Ischigualasto Formation of Argentina. Other elements (aetosaur, procolophonid, sphenodontid) are shared with the lower units in the North American Chinle Formation and

Dockum Group, and with some parts of the Keuper of Germany. However, the links with these northern faunas are surprisingly weak, despite their close proximity to Elgin in the Late Triassic: Elgin lacks the temnospondyl amphibians and phytosaurs which are so important in North America and Germany.

Leptopleuron lacertinum was a specialized procolophonid, about 175 mm in length, and may have resembled the present-day North American desert horned lizard (*Phrynosoma*). It had a triangular skull, when viewed from above, which bore spines, and the eye sockets were exceptionally elongated. *Leptopleuron* may have been a herbivore, but more probably it was an omnivore capable of feeding on a variety of food items; its deep jaws and row of transversely broadened molariform teeth would have made an efficient grinding mechanism. *Leptopleuron* was identified as a lizard by Owen (1851a), a batrachian (amphibian) by Mantell (1852) and a lacertilian (lizard) by Huxley (1867a). However, Boulenger (1904a) noted the affinities of *Leptopleuron* with *Procolophon* from South Africa, and reclassified

these as Procolophonia in the Order Cotylosauria, the so-called 'stem reptiles'. *Leptopleuron* is most similar to an undescribed procolophonid from Fraser's (1985, 1988b) 'Site 1', a productive fissure fill locality at Slickstones Quarry (Cromhall Quarry), Avon, and both of these forms share affinities with the larger *Hypsognathus* from the Upper Triassic Newark Group of New Jersey, USA (Upper Triassic to Lower Jurassic: Olsen and Galton, 1977) and *Paotedon*, from Triassic rocks in Lin-Che-Yu, Pao-Te, north-western Shansi, China. P.S.S. is currently redescribing *Leptopleuron*.

The tiny sphenodontid *Brachyrhinodon* has acrodont teeth on the jaw margins and on the palate, and a very short snout (hence the name). The narial region is unusual as it overrides the premaxillary teeth. It probably lived a cryptic existence feeding on insects or fruit (Fraser and Benton, 1989). Walker (1966) suggested that *Brachyrhinodon* may be congeneric with *Polysphenodon* from the Middle Keuper of Hanover, but Fraser and Benton (1989) found that, in a cladistic analysis of sphenodontid rela-

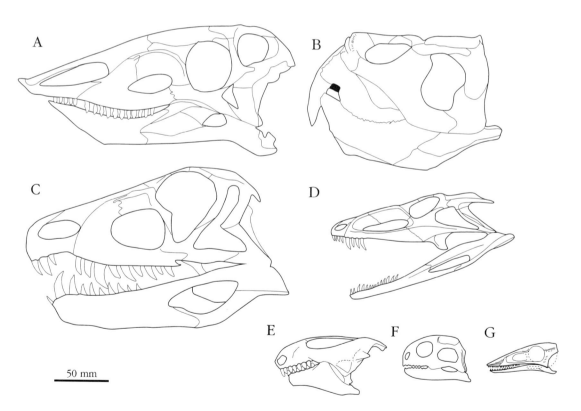

50 mm

Figure 4.14 The reptiles of the Late Triassic Lossiemouth Sandstone Formation, near Elgin, Morayshire. Skulls of (A) *Stagonolepis*; (B) *Hyperodapedon*; (C) *Ornithosuchus*; (D) *Erpetosuchus*; (E) *Leptopleuron*; (F) *Brachyrhinodon*; (G) *Scleromochlus*. From Benton and Walker (1985).

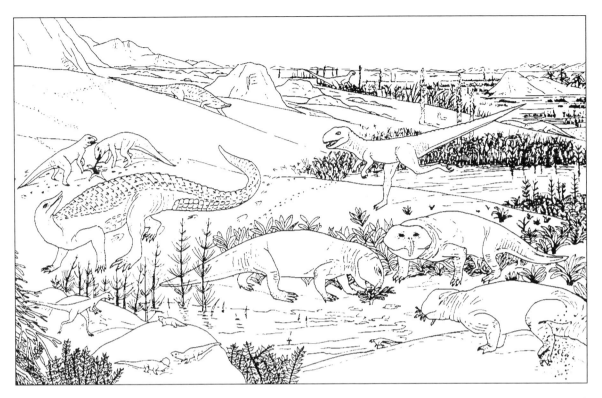

Figure 4.15 Imaginary scene at Elgin, Morayshire, in Late Triassic times, showing reconstructions of reptiles with typical Late Triassic plants. Three *Hyperodapedon* feed on seed-ferns in the foreground. Behind them, an *Ornithosuchus* runs towards the armoured *Stagonolepis* which is looking over its shoulder. Behind *Stagonolepis*, two *Erpetosuchus* feed on a small carcass. On the rocks in the left foreground are two *Leptopleuron*, a tiny *Brachyrhinodon* and a small bipedal dinosaur, *Saltopus*. To the right of the rocks is the tiny *Scleromochlus* at the side of the pond. In and around the pond there are horsetails, cycads and ferns, and there are tall lycopods in the distance. Based on a colour painting by Jenny Halstead. From Benton and Walker (1985).

tionships, the two were rather different.

Hyperodapedon was a bulky, 1.3 m long, terrestrial quadruped with strong limbs. The 100–200 mm long skull was very broad at the back – there was an anterior premaxillary 'beak' and the teeth were arranged in multiple rows on the maxilla. The dentary had a sharp edge and it cut into a groove on the maxillary tooth-plate. *Hyperodapedon* probably cut up tough plant material with a powerful precision shear bite. The massive, laterally flattened claws of the foot and the construction of the hind limb strongly suggest their use for scratch digging (Benton, 1983b, 1983d, 1984). *Hyperodapedon* was classified from the start with the English *Rhynchosaurus* and they were regarded as relatives of the living tuatara, *Sphenodon* (Huxley, 1869, 1887; Burckhardt, 1900; Boulenger, 1903; Huene, 1929a). *Hyperodapedon* is a typical Upper Triassic rhynchosaur, its closest relatives being *Hyperodapedon (Paradapedon) huxleyi* from

the Maleri Formation of India, and *Scaphonyx* from the Santa Maria Formation of Brazil and Ischigualasto Formation of Argentina, all Late Carnian in age (Benton, 1983d, 1990c).

The aetosaur *Stagonolepis* has a roughly crocodile-like skull and peg-like teeth. The snout had a curious blunt end, probably for digging. Its body was well armoured with large scutes and it had powerful digging limbs (Walker, 1961). *Stagonolepis* was thought of as a 'ganoid' fish by Agassiz (1844) and as a crocodile by Huxley (1859b, 1875, 1877). Huene (1942) classed the genus as a pseudosuchian thecodontian, and Walker (1961) recognized its close affinities with *Aetosaurus* from the Stubensandstein (middle Norian: Anderson and Cruickshank, 1978; Tucker and Benton, 1982; Benton, 1986b, 1994a, 1994b) of Germany. Murry and Long (1989) and Hunt and Lucas (1991b) report *Stagonolepis wellesi* from the lower part of the Petrified Forest Member of the Chinle Formation of Arizona. This taxon was established as a new genus (*Calyptosuchus welle-si* Long and Ballew, 1985) and was assigned to *Stagonolepis* by Murry and Long (1979, pp. 32–3), but without full justification.

The ornithosuchid *Ornithosuchus* is a small- to medium-sized carnivore with two locomotory modes – bipedal for running, quadrapedal for walking. It is one of the best known thecodontians and the fossils show a broad size range. The forelimb was adapted for grasping, and *Ornithosuchus*, when fully grown, probably preyed on *Hyperodapedon* and *Stagonolepis*, and on the smaller reptiles when younger (Walker, 1964). *Ornithosuchus* was classed provisionally as a dinosaur by Newton (1894b), a parasuchian by Boulenger (1903), a pseudosuchian by Broom (1913) and Huene (1914), and an ancestral carnosaur by Walker (1964). More recent finds in South America have suggested that *Ornithosuchus* is, in fact, a 'thecodontian' closely allied to *Riojasuchus* of the Los Colorados Formation (Norian) of Argentina (Bonaparte, 1969, 1978). The species *O. woodwardi* Newton (1894) and *O. taylori* Broom (1913) are the same as *O. longidens* (Huxley, 1877). The ornithosuchids were placed on the dinosaur/pterosaur branch of archosaur evolution by Gauthier (1986), Benton and Clark (1988) and others, but Sereno and Arcucci (1990) and Sereno (1991b) argue that they fall on the crocodilian side of the Crurotarsi.

Erpetosuchus, another crurotarsal archosaur, has a narrow 75 mm long skull with a huge antorbital fenestra and a broad 'square' posterior skull roof. The carnivorous and/or insectivorous dentition is peculiar, with long sharp recurved teeth at the front of the jaws and toothless longtitudinal ridges behind which may have been used for crushing prey or for masticating the food to an extent prior to swallowing. The need to masticate food may also connect with the presence of an incipient secondary palate. *Erpetosuchus* was classed as a parasuchian by Newton (1894b) and as a pseudosuchian by Broom (1913), Huene (1914) and Walker (1970b). The nearest relations of *Erpetosuchus* are probably *Parringtonia* from the Manda Formation (Mid Triassic) of Tanzania, and possibly *Dyoplax* from the Upper Schilfsandstein (Late Carnian) of Baden-Württemberg.

Scleromochlus, represented by five specimens, has a relatively huge skull, nearly as long as the trunk, short forelimbs, but long hindlimbs and a long tail. The long hindlimbs have been interpreted (Woodward, 1907b; Huene, 1914) as adaptations for jumping and it may have sought food on the dunes in which it is preserved using a saltating mode of locomotion. Some authors (e.g. Huene, 1914) have speculated that it could glide from tree to tree, assuming that it had membranes on the side of the body. *Scleromochlus* has been placed in the 'Dinosauria' (Woodward, 1907b) and Pseudosuchia (Broom, 1913; Huene, 1914). It is presently classified in a family on its own and its nearest relation is uncertain (Walker, 1970b, p. 361). However, its skull specializations suggest a relationship to the aetosaurs (Walker, 1970b, p. 361; Krebs, 1976, p. 90) or to the pterosaurs (Padian, 1984).

A saltating mode of locomotion was also proposed for the 'dinosaur' *Saltopus* (Huene, 1910a), represented by a partial skeleton of the hind quarters. However, *Saltopus* was probably a slender running scavenger with a very long tail for balance in rapid manoeuvring. *Saltopus* was described as a dinosaur (Huene, 1910a), but more study is required to determine whether this is correct: if it is, it would be one of the oldest known dinosaurs in the world.

The Elgin archosaurs (e.g. *Stagonolepis, Ornithosuchus, Scleromochlus, Erpetosuchus*) have forced major changes in the classification of the order (e.g. Broom, 1913; Huene, 1914; Bonaparte, 1969; Krebs, 1976; Gauthier, 1986; Benton and Clark, 1988; Sereno and Arcucci, 1990; Sereno, 1991b). *The very difficulty experienced in classifying many of these reptiles has demanded detailed redefinitions of the various families and, in particular, reappraisal of their relations to the dinosaurs and crocodiles.*

Conclusions

The site is important for its distinctive reptiles of Late Carnian age, which include four genera of archosaurs, a sphenodontid and a procolophonid; four of the genera occur nowhere else. The fauna is unusual in showing close affinities with those of southern continents (India, South America), as well as with those of the rest of western Europe and North America and, in that the remains are preserved in aeolian deposits, clearly not the natural habitat of the majority of the animals. In evolutionary terms, many of the genera are unique, or belong to rare groups (e.g. *Scleromochlus, Erpetosuchus, Brachyrhinodon*).

The conservation value of the site lies mainly in the richness and uniqueness of the fossil reptile fauna that has been obtained here and, to an extent, in its potential for future finds.

SPYNIE (NJ 223657 AND OTHERS)

Highlights

Spynie quarries were the first source for *Leptopleuron* and *Ornithosuchus*, two abundant members of the Elgin Late Triassic fauna. Excellent specimens of the herbivore *Hyperodapedon* were found at Spynie in 1947, and more material may come to light with further quarrying.

Introduction

The locality includes one main pit and up to nine smaller pits on Spynie Hill, just off the Elgin-Lossiemouth road and on the south shore of the former Loch of Spynie. The Lossiemouth Sandstone Formation here yielded the first remains of the procolophonid *Leptopleuron* and of the ornithosuchid *Ornithosuchus*, and some good material of *Hyperodapedon*. Most of the Spynie quarries are overgrown and/or filled with debris. One large pit is still clear, however (Peacock *et al.*, 1968, Quarry no. 4), and has been worked a little recently. Fossils could be found in the lower beds with further working. Neville Hollingworth collected odd bone pieces from quarry refuse in about 1980.

Spynie Hill was worked before 1790: 'Under a thin stratum of marsh soil, the whole of this ridge seems to be a mass of excellent hard free-stone; of which there is a quarry, near the summit of the hill, that supplies a large extent of the country with mill-stones, and the town of Elgin and the neighbourhood with stones for building' (A. Gordon, *in* Sinclair, 1794, vol. 10, p. 629). The various pits were worked until the 1880s and do not appear to have operated again until recently. Moray Stone Cutters lease the main pit (to the east of the others) and have blasted in the 1980s (Figure 4.16).

In October 1851 William Young, a quarryman at Spynie, showed a small reptile, preserved as part and counter-part, to Patrick Duff (Anon., 1851). This was sent to London where various people examined it, including Charles Lyell, Gideon Mantell and Richard Owen. It was immediately recognized as a tetrapod, and was thus the first identified from Elgin (*Stagonolepis* was at the time still considered to be a fish). Since all agreed that the rocks from whence it came were Old Red Sandstone (Devonian) in age, this was obviously a very important animal – the oldest tetrapod then known, and Lyell delayed publication of his *Manual of Elementary Geology* (1852) in order to include a postscript about it.

Mantell, working with Lyell, and with his old friend Lambart Brickenden, who lived in Elgin, prepared a description, but Owen became the first author on the new reptile, publishing a brief unillustrated account dated 20th December 1851 (Owen, 1851a), naming it *Leptopleuron lacertinum* and interpreting it as a lizard. Mantell's illustrated description of the same animal followed in early 1852 and he named it *Telerpeton elginense*, interpreting it as a 'batrachian' (i.e. an amphibian) largely because he had allied it with some 'frogs eggs' from the Old Red Sandstone of Forfarshire. The controversy over the description of this reptile, and the political and philosophical infighting, are described by Benton (1983c).

Specimens of the rhynchosaur *Hyperodapedon*

Figure 4.16 The main quarry at Spynie, looking northwards along the east face. Blasting has just taken place, leaving broken blocks that have, from time to time, yielded fossil reptile remains. (Photo: M.J. Benton.)

were collected at Spynie in the 1870s, and in 1891 (Anon., 1891) the first *Ornithosuchus* was discovered by George Gordon, clergyman and naturalist. This fossil, consisting of a partial skeleton and skull was described by Newton (1894b) as the new genus and species, *Ornithosuchus woodwardi*. Taylor (1920) mentions *Stagonolepis* and *Erpetosuchus* from Spynie, but there seems to be no evidence for these. Two fine skulls of *Hyperodapedon* were collected at Spynie in 1947 by Professor T.S. Westoll and are now in the NMS.

Description

In the main quarry (NJ 22256565), 20 m faces may be seen displaying grey jointed sandstone, highly siliceous at the top and more calcareous lower down, weathering orange. The joints may be filled with fluorspar, galena or sphalerite. The transition to the Cherty Rock (a sandy limestone and chert) is also exposed at the top. A 10 m deep pit, worked recently, exposes softer greyish-yellow calcareous sandstone. (This is quarry no. 4 of the memoir – Peacock *et al.*, 1968, p. 68.)

The Geological Survey borehole in Spynie Quarry no. 4 yielded the following section, in summary (Peacock *et al.*, 1968, p. 68):

	Thickness	
	Ft	**in**
Lossiemouth Sandstone Formation:		
Cherty Rock	5	0+
Sandstone, hard and siliceous at		
top, softer below	76	6
Yellowish calcareous siltstone with		
thin beds of gritty sandstone	26	10
Presumed Old Red Sandstone:		
Siltstone and sandstone with galls		
of green clay. Some reddish		
brown colouration	4	0

Early writers did not specify precisely which of the pits yielded the specimens of *Leptopleuron*, *Hyperodapedon* or *Ornithosuchus*, but the reptiles at Spynie appear to have been found low in the Lossiemouth Sandstone Formation, as at Lossiemouth East Quarry (see above). Peacock *et al.* (1968, p. 68) identify the two westernmost quarries as those that yielded the reptiles. Quarry no. 1 (NJ 21926555), now filled in, lay outside the Hill of Spynie and exposed 5 m of fine- to coarse-grained sandstone overlain by a few feet of broken rock. Large-scale dune-bedding

occurred in the top of the face. This quarry is supposed to have yielded *Hyperodapedon* (Gordon in Huxley, 1877; Linn, 1886; Peacock *et al.*, 1968, p. 68).

Quarry no. 2 (NJ 22066557) is small, but deep (12–15 m), and lies in the Spynie Hill wood. This is supposed to have been 'a much larger quarry in which specimens of *Leptopleuron (Telerpeton)* were found' (Peacock *et al.*, 1968, p. 68). It should be noted that this quarry (no. 2, Peacock *et al.*, 1968) presently contains very large trees, probably over 100 years old. The type specimen of *Leptopleuron lacertinum* 'was found . . . at the bottom of a shaft which had been sunk through 51 feet of sandstone down to a soft rubbly bed' (Duff in Murchison, 1859, p. 435). Gordon (1859, pp. 45–6), added that the type specimen of *Leptopleuron lacertinum* was found *in situ*: 'it was extracted from the living rock, deep in a quarry opened on the west end of the hill', and Martin (*c.*1860) stated that the specimen was 'found low down, in the bottom of the quarry'.

The specimens of *Ornithosuchus* collected by quarrymen in 1891 may have come from the large quarry still in operation (NJ 22256565), this being quarry no. 3 of Peacock *et al.* (1968, p. 68). This was also the site of the two skulls of *Hyperodapedon* collected in 1947.

The bone material is often powdery, or replaced by iron oxide, and casting is the best method of study, as for the material from Lossiemouth East Quarry (see above).

Fauna

Anapsida: Procolophonidae
　　Leptopleuron lacertinum Owen, 1851
　　　(=*Telerpeton elginense* Mantell,1852)
　　　2 individuals: NMS, BGS(GSM)

Archosauromorpha: Rhynchosauridae
　　Hyperodapedon gordoni Huxley, 1859
　　　3 individuals: BGS(GSM), NMS

Archosauria: Crurotarsi: Ornithosuchidae
　　Ornithosuchus woodwardi Newton, 1894
　　　3 individuals: BGS(GSM), BMNH

Interpretation

Descriptions of the reptiles *Leptopleuron*, *Hyperodapedon* and *Ornithosuchus* are given in the Lossiemouth East Quarry report (see above).

Conclusions

Spynie is important as the first recorded source of *Leptopleuron* and of *Ornithosuchus*, and the site has the best potential for future finds thereby giving it considerable conservation value. Specifically, the specimens of both *Leptopleuron lacertinum* are some of the best, and two of the best preserved skulls of *Hyperodapedon* yet known were collected in 1947, the last substantial find of reptiles from any of the Lossiemouth Sandstone Formation sites.

FINDRASSIE (NJ 207652, NJ 204651)

Highlights

Findrassie quarries produced some of the first of the Elgin reptiles to be recorded. The site is important because of the high quality of preservation of the specimens.

Introduction

The site includes a series of largely overgrown pits in a wooded area about 1 km due east of Findrassie House. The Lossiemouth sandstones here have produced good remains of *Ornithosuchus* and *Stagonolepis*. The excellent quality of preservation of fossil remains makes Findrassie worth conserving in the hope of future quarrying operations.

The Findrassie quarries were worked on a small scale in the mid 19th century: Martin (*c.* 1860) wrote that 'the quarry is now seldom worked, except occasionally for the purpose of obtaining material for road-metal'. The Findrassie quarries do not appear to have been worked after the 1860s.

The first bones of *Stagonolepis robertsoni*, which Agassiz (1844) had previously described as a fish on the basis of some scales found at Lossiemouth, were collected at Findrassie around 1857 (Gordon, 1859, p. 44; also Murchison, 1859, p. 435). At the same time a fragment of jaw with long dagger-like teeth was also collected, and ascribed tentatively to *Stagonolepis* by Huxley (1859a, pp. 434–5). He later (Huxley, 1877, pp. 43–5, pl. 4, fig. 1) described it as the new genus and species *Dasygnathus longidens*. This has since been shown to belong to *Ornithosuchus* (Walker, 1964, p. 66).

Description

The first quarry mentioned by Peacock *et al.* (1968, p. 69) (NJ 20726524) lies concealed in the southern part of Findrassie woods just beside a field. It is shallow and largely overgrown but exposes patches of hard siliceous sandstone, with occasional cavities produced by weathering.

A second set of pits (Peacock *et al.*, 1968, p. 69) (NJ 20456510) consists of three quarries, the middle one of which exposes a 7 m face of massive, hard, fine-grained sandstone, the top part being hard and siliceous and pinkish in colour with scattered larger quartz grains, and the bottom yellow to yellow-brown with rusty spots. There are several other small pits in the Findrassie woods and on the moor to the west of the wood, a large shallow quarry (NJ 20176496) shows massive, pinkish-brown sandstone with pebbles.

The East Lodge of the Findrassie Estate, where the first Findrassie specimens of *Stagonolepis* were found, is situated at NJ 20746545, and the site where the find was made might be one of the remaining Findrassie quarries lying to the south and south-west of the entrance (Peacock *et al.*, 1968, p. 69), but it could now be filled (Walker, 1961, p. 106). Linn (1886) recorded that *Stagonolepis* was found 'in the more westerly' of a line of three quarries (?NJ 20156495), but Peacock *et al.* (1968, p. 137) suggest a more easterly pit at NJ 205651 as the probable source of the reptiles.

The Findrassie specimens figured by Huxley (1877) are in the form of well-preserved moulds, but specimens in ELGNM labelled 'Findrassie' have bone preserved, which may indicate a different locality. There are occasional pebbles in the matrix of many slabs and the early specimens, at least, must have come from the base of the reptiliferous sandstone, in beds just above the ORS (Gordon, 1859; Walker, 1961).

Fauna

Archosauria: Crurotarsi: Pseudosuchia:
　Stagonolepididae
　Stagonolepis robertsoni Agassiz, 1844
　　2 large individuals: NMS, ELGNM, AUGD; 1 small individual: ELGNM

Archosauria: Crurotarsi: Ornithosuchidae
  Ornithosuchus longidens Huxley, 1877
    1 individual: ELGNM

Interpretation

The cranial remains of *Ornithosuchus* from Findrassie were originally named *Dasygnathus longidens* by Huxley (1877, p. 45): these were shown to belong to a carnivorous 'thecodontian' by Walker (1961, pp. 108–10) and synonymized with *Ornithosuchus* by Walker (1964, pp. 63–6). *O. woodwardi* from Spynie and Lossiemouth is the same as *Dasygnathus longidens*, but the better known name *Ornithosuchus* is used since *Dasygnathus* is preoccupied (a beetle named in 1819). The palaeobiology and relationships of *Stagonolepis* and *Ornithosuchus* are discussed in the Lossiemouth report.

Conclusions

This is the locality of the holotype of *Ornithosuchus longidens*, as well as three individuals of *Stagonolepis* and the first known *Stagonolepis* bones (apart from scutes). Remains are usually excellently preserved moulds that give high-fidelity casts for study, and Findrassie is a better site than Lossiemouth or Spynie in terms of the quality of preservation, which gives it special conservation value.

UPPER TRIASSIC OF SOUTH WALES AND CENTRAL AND SOUTH-WEST ENGLAND

Reptiles of Late Triassic age have been obtained in South Wales and south-west and central England from two main sources: marginal Triassic outcropping in South Glamorgan, where an assemblage of trackways of Norian age has recently been discovered, and from the 'Rhaetic Bone Bed' in the Penarth Group. A third important source of Late Triassic reptiles in Britain is from the cave and fissure fillings in the region of the Severn Estuary: these are treated separately towards the end of this chapter.

The marginal Triassic deposits of South Glamorgan have yielded rare finds of dinosaur footprints from at least two localities. Early finds of isolated dinosaur prints probably came from Scorlon, near Porthcawl. A recent discovery at Bendrick Rock, Barry, comprises numerous trackways assignable to the two dinosaur footprint ichnogenera *Anchisauripus tuberosus* and *Gigandipus*.

The 'Rhaetic Bone Bed', actually comprising several ossiferous horizons, is an unusual sequence at the base of the Westbury Beds, with a wide geographic extent. Typical localities yielding reptiles include the following: Devon: Culverhole Point (SY 275893: archosaurs); Somerset: Chilcompton railway cutting (ST 626509; *Pachystropheus*; Antia, 1979; Duffin, 1980), Hapsford Bridge (ST 755490–ST 756489; prosauropod), Blue Anchor Bay (ST 042432; ?crocodile, *Pachystropheus*; Richardson, 1911b; Huene, 1935; Sykes, 1977; Storrs and Gower, 1993); Avon: Aust Cliff, Severn Estuary (ST 566898; ichthyosaurs, plesiosaurs, dinosaurs, *Pachystropheus*; Storrs, 1994), Garden Cliff, Westbury-on-Severn (SO 718128; ichthyosaurs, plesiosaurs, *Pachystropheus*; Etheridge, 1872; Storrs, 1994), New Clifton, Redland (ST 585735; ?reptile (BMNH)), Carrefour (ST 585815; plesiosaur vertebrae); Glamorgan: Stormy Down (SS 846806; megalosaur dinosaur; Newton, 1899); Leicestershire: Wigston (SK 603991; ?phytosaur; Richardson, 1909), Spinney Hills brickpits (SP 604045; plesiosaur, ichthyosaur; Kent, 1968), Glen Parva brickworks (SP 5689; ichthyosaur, plesiosaur; Browne, 1889, 1894; Fox-Strangways, 1903; Horwood, 1916); Nottinghamshire: Bantycock Pit (SK 811502) and Staple Pit (SK 805499; dinosaurs, *Plateosaurus*, *Pachystropheus*, ichthyosaurs, plesiosaurs, crocodile; Martill and Dawn, 1986), Barnstone (Sykes *et al.*, 1970), Beacon Hill (Johnson, 1950), Stanton-on-the-Wolds (SK 637312).

A fossil reptile is also known from sediments of 'Rhaetian' age at Wedmore, Somerset (ST 4448; prosauropod *Camelotia borealis* ['*Avalonia*', '*Picrodon*'; type specimen: BMNH R2870–4, R2876–8]; Seeley, 1898; Galton, 1985c; Storrs, 1993), and Rhaetian sediments in a large glacial erratic at Linksfield, near Elgin (NJ 223641) have produced remains of fishes and reptiles, including plesiosaurs (Taylor and Cruickshank, 1993; Storrs, 1994).

Of these numerous localities, only two could be selected as GCR sites, since many of the others are no longer accessible, or offer only marginally different faunas:

1. Bendrick Rock, South Glamorgan (ST 131668).

Upper Triassic (Norian), Mercia Mudstone Group.

2. Aust Cliff, Avon (ST 565895–ST 572901). Upper Triassic ('Rhaetian'), 'Rhaetic Bone Bed', Westbury Formation.

BENDRICK ROCK, SOUTH GLAMORGAN (ST 131668)

Highlights

Bendrick Rock, Barry is the source of Britain's best dinosaur trackways. Hundreds of footprints have been recorded and collected there over the years, and the site is still extremely rich.

Introduction

Abundant dinosaur footprints have been found in Late Triassic sediments near Bendrick Rock (Figure 4.17). This is the best site in the British Isles for such early dinosaur trackways, and it may be the best in Europe. Its value is in providing clear evidence for dinosaur ichnofaunas in Europe for comparison with the well-known, and similar, ichnofaunas from the eastern United States (Newark Supergroup) and from southern Africa. Although many slabs have now been collected, natural marine erosion and excavation of the site continues to yield further material.

Three-toed dinosaur-like footprints were found by the artist T.H. Thomas in 1878 in a loose slab, possibly from a quarry at Scorlon, Newton Nottage, near Porthcawl (Thomas, 1879). These came from 'flaggy, calcareous beds with subangular pebbles of limestone' of the marginal Triassic. They were described by Sollas (1879), who named them *Brontozoum thomasi*.

The tracks at Bendrick Rock were found in 1974 on bedding surfaces. At least 450 individual prints were observed (Tucker and Burchette, 1977) and many are still *in situ* (the main slabs then exposed were removed to the NMW, and further specimens were excavated in 1990 for exhibition at the NMW).

Figure 4.17 Aerial view of a bedding plane on the foreshore at Bendrick, covered with three-toed dinosaur footprints, named *Anchisauripus*. Each small depression is a footprint. Width of field of view is about 5m. (Photo: M.J. Benton.)

Description

The sediments in the cliff and on the foreshore at Bendrick Rock form part of the marginal Triassic deposits of South Glamorgan (Marginal Triassic of Tucker, 1977), formerly called the Dolomitic Conglomerate or the Littoral Triassic (Ivimey-Cook, 1974). They include fluviatile sandstones and siltstones, and shore-zone lacustrine sediments (Tucker, 1977). The assemblage is laterally equivalent to playa-lake and aeolian deposits at Lavernock and other localities nearby.

These all form part of the Mercia Mudstone Group (Warrington *et al.*, 1980) which ranges in age from the Mid to the Late Triassic (pre-'Rhaetian'). There is no clear evidence of age, although the dinosaur footprints would strongly suggest Late Triassic and probably Norian.

Palaeogeographically, this was a low lying piedmont area, adjacent to islands or hilly upland of Carboniferous Limestone flanked by alluvial fans and talus slopes, and marginal to an inland (epicontinental) sea. The climate was warm and conditions were desert-like with only intermittent rainfall (evidence of evaporites, calcretes, sheet floods).

The sequence including footprints is logged as follows by Tucker and Burchette (1977):

	Thickness (m)
Marl	3.5+
Conglomerate (erosive base)	*c.* 1
Marl	1–1.5
Sandstone (bedded), containing footprints	*c.* 0.8
Conglomerate (channels)	0–2
Marl, with calcareous nodules	0.5–2
Sandstone, trough cross-bedded	*c.* 2
Marl	1
Sandstone, bedded and cross-bedded	1
Marl, with calcareous nodular horizon	1–4
(Resting unconformably on Carboniferous Limestone)	

The footprints occur on several bedding planes, the best being an 0.08 m thick graded sandstone, overlain by a marl parting, and then another sandstone bed, 0.03 m thick (Figure 4.18). The footprint surface also bears ripple marks. Most of the prints are reasonably clear and many are perfect in their preservation, but others are somewhat deformed or reduced to vague squelch marks, which demonstrates that the muddy sand on which the animals were walking was originally soft and damp. However, the prints appear to have been preserved by a desiccation process caused by subsequent drying out of the surface

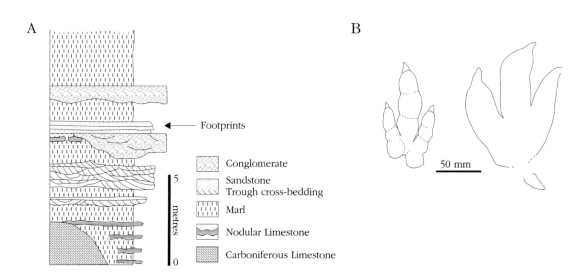

Figure 4.18 The Bendrick Rock footprints. (A) Sedimentary log showing the horizon at which footprints occur; (B) the two footprint types from the locality, each of which is ascribed to *Anchisauripus*. After Tucker and Burchette (1977).

into which they were impressed. Burial by sediments carried in by thin sheet floods of low turbulence then covered the top of the marked surface.

Fauna

Two types of print occur, a small three-toed variety (105 mm x 70 mm) and a larger four-toed variety, the latter having three toes directed forward and a fourth obliquely backwards (160 mm x 120 mm). Preservation is good, and trackways of many prints may be observed. Tucker and Burchette (1977) attributed the prints to the ichnogenus *Anchisauripus* (Lull) (synonym: *Brontozoum*), but they did not attempt an identification to specific level. Delair and Sarjeant (1985, p. 153) generally compared the smaller Bendrick Rock footprints with *A. thomasi* (Sollas, 1879) and then referred them to *A. tuberosus* (Hitchcock, 1836). They disputed referral of the larger prints to *Anchisaurus* 'since phalangeal pads are not visible' and because of their markedly smaller size, and instead attributed these prints to the ichnogenus *Gigandipus* (Hitchcock), characterizing them as *Gigandipus* sp. nov. The original specimen of *Brontozoum thomasi*, and others collected more recently, are in the NMW (Tucker and Burchette, 1977).

Interpretation

If all the prints are referrable to *Anchisauripus* sp., then it would once have been thought that they had been made by the prosauropod dinosaur *Anchisaurus*, a form known from the Early Jurassic of North America and southern Africa. However, the maker of *Anchisauripus* prints was a small theropod dinosaur (Haubold, 1971), and at least two forms may be implied by the assignment of the prints to two ichnogenera, *A. tuberosus* and *Gigandipus* sp., the latter interpreted by Haubold (1971) as a large form of *Anchisauripus*. However, there are many taxonomic problems with these kinds of trackways: many ichnospecies hitherto assigned to *Anchisauripus* have been reassigned to *Atreipus* and *Grallator* (Olsen and Baird, 1986), and these authors interpret *Atreipus* at least as the footprint of an ornithischian. Hence, these kinds of three-toed footprints have been assigned to all major dinosaur groups: sauropodomorphs, theropods and ornithischians!

Comparison with other localities

There are no other known occurrences of Late Triassic footprints in the British Isles, nor of those of a prosauropod dinosaur. *Anchisauripus* prints have been described from the Late Triassic to Early Jurassic Newark Supergroup of eastern North America (Connecticut, Massachusetts, New Jersey, Pennsylvania, New Mexico) and the Late Triassic of South America (Argentina), as well as possible *Anchisauripus* from the Mid Triassic of France (Haubold, 1971, 1986; Olsen and Baird, 1986). More detailed comparisons and interpretations of the palaeobiology and stratigraphic significance of the Bendrick Rock footprints must await a full review of the relevant ichnotaxa.

Conclusions

Constant erosion by the sea keeps this locality clear of debris, and new footprints are exposed from time to time. Recently excavated footprints from here are currently being studied and have provided material for museum exhibits, e.g. National Museum of Wales in Cardiff. This potential and the importance of the finds from here give the site its conservation value.

AUST CLIFF, AVON (ST 565895–ST 572901)

Highlights

Aust Cliff is world-famous for its superb exposure of Rhaetian marine bone beds. Abundant reptile fossils have been collected, and continue to be collected, representing a mix of mainly marine ichthyosaurs and plesiosaurs, but also rare dinosaur bones.

Introduction

Aust Cliff, at the eastern end of the Severn Road Bridge (Figure 4.19), is Britain's most prolific site for Rhaetian fossil reptiles. The cliff exposes the boundary between the Upper Triassic and Lower Jurassic, and was first described by Buckland and Conybeare (1824), and subsequent accounts have been given by many authors, including Strickland

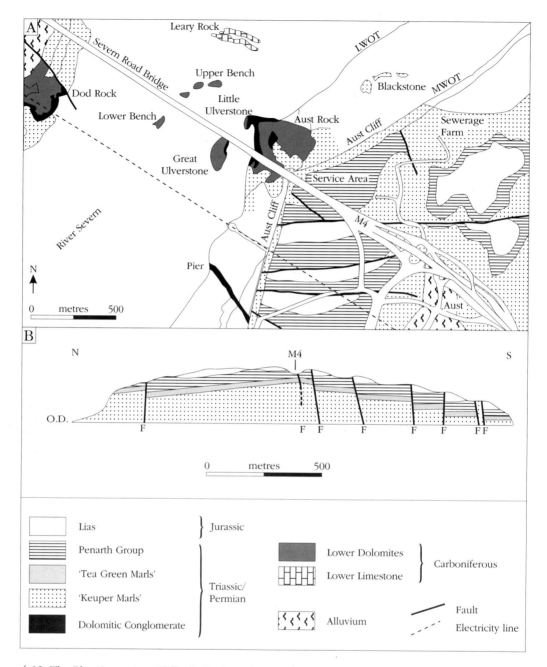

Figure 4.19 The Rhaetian at Aust Cliff. (A) Geological map of the Aust Cliff area; (B) the broad anticlinal structure of Aust Cliff, showing the Lias (1); the Rhaetian (Penarth Group) (2); the 'Tea Green Marls' (3); and the 'Keuper Marls' (4). Both after Hamilton (1977).

(1841), Etheridge (1868), Short (1904), Reynolds (1946), Hamilton (1977) and Storrs (1994). The Aust section has yielded important collections of ichthyosaurs, plesiosaurs and dinosaurs, as well as fishes. The rare dinosaur remains are generally heavily abraded and it is likely that they have been transported for some distance. The site is subject to constant erosion and occasionally produces good new specimens.

Description

Aust Cliff exposes a section through the Upper Triassic and the lower part of the Lower Jurassic (Figure 4.19). It represents the truncated face of a ridge of Triassic and Lower Jurassic rocks surrounded by alluvium. A very gentle anticlinal structure is shown, cut by five small faults with throws to the south ranging from *c.* 1 m to 4.5 m.

Both flexing and faulting have been explained by compaction of the Mercia Mudstone Group sediments. The Mesozoic succession exposed in the cliff is readily subdivided lithologically and biostratigraphically. The lower part of the cliff consists of the Mercia Mudstone Group, including the Blue Anchor Formation ('Tea Green Marls'). Macrofossils are generally absent from these beds, but occur abundantly in the overlying dark and lighter grey sediments of the Penarth Group (including the 'Rhaetic'). Limestones and shales at the top of the cliff form the lowest part of the Lias. This Mesozoic succession rests unconformably on the upturned edges of a Carboniferous Limestone ridge, exposing the Lower Dolomites which dip about 15° southwest. The section (based on Reynolds, 1946, Hamilton, 1977, and Warrington *et al.*, 1980) is:

		Thickness (m)
JURASSIC	Blue Lias	
	(Hettangian)	
	planorbis Beds	(variable)
TRIASSIC	Pre-*planorbis* Beds	(variable)
Penarth	Lilstock Formation	*c.* 3.4
Group	Westbury Formation	
	(bone beds at base)	*c.* 4.3
Mercia Mudstone	Blue Anchor	*c.* 7.0
Group	Formation	
	Red mudstones	*c.* 30.0
CARBONIFEROUS	Carboniferous	
	Limestone	(variable)

The reptile remains are found predominantly in the 'Rhaetic Bone Bed' (Figure 4.20) which occurs in places at the base of the Westbury Formation, the subdivisions of which are (Reynolds, 1946):

Figure 4.20 Aust Cliff: view on the north-eastern side of the Severn Bridge, looking south-east. The red sediments of the Mercia Mudstone Group extend about four-fifths of the way up the cliff, capped by the Penarth Group (latest Triassic). The Blue Lias of the Jurassic lies at the very top, in the vegetation line. Vertebrate remains are found in lenses of 'Rhaetic' Bone Bed, at the base of the Penarth Group. (Photo: G.W. Storrs.)

	Thickness (m)
8. Greenish-black shales	0.3
7. Hard grey limestone ('upper *Pecten* Bed')	0.13
6. Black Shales	2.4
5. Hard pyritous limestone ('lower *Pecten* Bed')	0.18
4. Black shales, hard fissile paper shale above	1.2
3. Bone Bed	0.02–0.15
'Tea-Green Marls'	

The 'Bone Bed' occurs as lenses of grit or intra-formational conglomerate (or breccia) of sedimentary rocks with a calcite-cemented sandy matrix on top of the Blue Anchor Formation, the surface of which may be ripple-marked. The conglomeratic component is made up mainly from clasts of the Blue Anchor Formation sediments, together with quartz pebbles and bone fragments. Many of the fragments of Blue Anchor Formation sediment are squeezed and plastically deformed, which suggests that they were still soft when incorporated into the bone bed. The quartz pebbles are mainly of vein quartz, are mostly well rounded, and are probably derived from older beds (although Wickes, 1904, suggested that they might represent stomach stones, or gastroliths, swallowed by plesiosaurs to aid in the digestion of food).

The vertebrate remains are mainly phosphatized bones, teeth and scales. They are

Figure 4.21 Reconstruction of the Rhaetian sea floor, based on fossils from Aust Cliff and neighbouring localities. The fishes are *Saurichthys*, at top right, *Birgeria* at top left, and a hybodont shark in front of it. The marine reptiles include ichthyosaurs (mid-top) and placodonts (lower left). After Duff, McKirdy and Harley (1985).

disarticulated and often rolled and worn, indicating some post-mortem transport. Coprolites (faecal droppings), some possibly of aquatic reptiles, are also abundant: these contain crustacean fragments and abundant fish scales and they are heavily phosphatized, containing 25–50% calcium phosphate. A reconstruction of the Rhaetian sea floor based on fossils from Aust Cliff and neighbouring localities is shown in Figure 4.21.

According to the classification of Sykes (1977), the Bone Bed had a part-primary and part-secondary origin. The indications of primary deposition include the condition and orientation of the fossils, and the poorly sorted nature of the deposit. However, most of the fossils and other clasts show signs of abrasion, which indicates that the deposit is largely reworked. This is borne out by finds of teeth of the Carboniferous fishes *Psephodus magnus*, *Psammodus porosus* and *Helodus* in the bone bed, presumably reworked from the local Carboniferous Limestone or the Coal Measures. Macquaker (1994) and Storrs (1994) conclude that the bed represents a tempestite.

Vertebrate remains are most abundant in the impersistent Bone Bed. A similar fish fauna occurs in the succeeding basal sands of the Westbury Formation and also at the base of the limestone bands and in the shales. Bone-rich debris is also sometimes present in the topmost layers of the green marls at the base of the Rhaetian sequence where the marl is intensely bioturbated. These occurrences of reptile and fish bones in Rhaetian horizons at Aust, other than in the basal bone bed itself, are classified as trace bone beds (Sykes, 1977, p. 220).

Fauna

Dozens of slabs of sediment with fish and reptile bones and teeth are preserved in BRSMG, BRSUG, BMNH, BGS(GSM), CAMSM and in most other British collections. Many of the specimens collected during the last century were identified to species level, and made the types of new species and new genera, but there is no point in listing those since there is rarely enough evidence for such precise determination (see Storrs, 1994, for discussion).

Chondrichthyes: Elasmobranchii
  Polyacrodus,   *Hybodus*,   *Nemacanthus*,
  Palaeospinax, *Pseudodalatias*, *Lissodus*

Osteichthyes: Actinopterygii: 'Palaeonisciformes'
  '*Birgeria*', *Gyrolepis*

Osteichthyes: Actinopterygii: 'Semionotiformes'
  Sargodon, *Colobodus*

Osteichthyes: Sarcopterygii: Dipnoi
  Ceratodus

Ichthyopterygia: Ichthyosauridae
  '*Ichthyosaurus*'

Sauropterygia: Plesiosauria: ?Plesiosauridae
  '*Plesiosaurus*'

Archosauromorpha: Choristodera: Pachystropheidae
  Pachystropheus rhaeticus (=*Rysosteus*)

Archosauria: Dinosauria
  ? Camelotia, megalosaur

Interpretation

The Bone Bed is assumed to have been deposited under marine coastal conditions. The conglomerate shows signs of rapid deposition and winnowing by wave action and shore-line currents. It has been suggested (Macquaker, 1994; Storrs, 1994) that the Aust bone bed represents a storm deposit: a mass of rocks and fossils picked off the shore-line and carried back down into deeper water by the ebb current of a storm sturge, or exhumed and redeposited from penecontemporaneous shallow-water sediments. Other authors suggest that reworking of strand-line deposits produced by the Rhaetian transgression might equally have produced the bone bed. This occurred with overstep across the former playa-type sediments of the Mercia Mudstone Group, a palaeoenvironment of intrinsically low relief. Although the palaeontological evidence does not give precise dating, it is probable that the marine flooding phase occurred very rapidly and would have had a strong erosive force. Kent (1970), for example, suggested that the whole of the Midlands was submerged by the transgression almost simultaneously.

Fish remains are common in the Bone Bed at Aust. These include the teeth and fin spines of sharks, teeth and scales of primitive, heavily-scaled, bony fish ('*Birgeria*', *Sargodon*, *Gyrolepis*). The most characteristic remains, however, are palatal tooth-plates of the lung fish

Ceratodus, a form close to the extant *Neoceratodus* from Australia.

Temnospondyl amphibians have been reported from Aust, based on mandibles referred to *Metopias diagnosticus* (e.g. Reynolds, 1946), but these have turned out to be the teeth and jaws of a palaeonisciform fish (identified as '*Birgeria*' *acuminata* Agassiz by Savage and Large, 1966, but probably representing a new genus; Storrs, 1994).

The most common reptile remains are ichthyosaurs and plesiosaurs, with a few possible dinosaurs. Ichthyosaurs are represented at Aust by their vertebrae, which are flat circular biconcave elements. One in the BRSUG measures 180 mm across. Other ichthyosaur remains include a humerus (BRSMG), a large lower jaw (Huene, 1912c) and numerous isolated teeth.

The vertebrae and teeth of *Plesiosaurus* are the commonest reptile remains from Aust. The vertebrae are distinguished from those of ichthyosaurs by, among other features, being thicker and having two planar surfaces on the centra. Three species have been named: *P. rugosus, P. costatus* and *P. rostratus. P. rugosus* was erected by Owen (1840a) for some vertebrae (BRSMG) which were regarded as sufficiently distinct by Seeley (1874b) to be named as a new genus, *Eretmosaurus*, together with some limbs, limb girdles and apparently similar vertebrae from Granby. Swinton (1930) doubted the validity of the genus, but Persson (1963) retained it, and classified it as a rhomaleosaurid. *P. costatus* was also erected by Owen (1840a) for certain cervical vertebrae, which Swinton (1930) regarded as a possible new genus. Other plesiosaurs reported from Aust include *P. hawkinsi* and *P. rostratus* (Reynolds, 1946). Most plesiosaur specimens from Aust are non-diagnostic teeth, vertebrae, ribs and paddle bones, which Storrs (1994) regarded as Plesiosauria *incertae sedis*. Unfortunately, most of the specimens were destroyed in the BRSMG during the Second World War.

Owen (1842b) named a small reptile vertebra, with a partial humerus and femur from Aust as *Rysosteus*. This was interpreted as a dinosaur by Reynolds (1946), but recent studies by Storrs and Gower (1993), based on material from Aust and from other British Westbury Formation localities, suggests that most *Rysosteus* specimens are the same as *Pachystropheus rhaeticus* Huene, 1935. Storrs and Gower (1993) reinterpret *Pachystropheus (Rysosteus)* as a choristodere, a

superficially crocodile-like diapsid of uncertain affinities, and a group that is best known from the Late Cretaceous and Palaeogene, but does have British Mid Jurassic representatives (*Cteniogenys*; see Bathonian site reports below).

Some very large bones, possibly dinosaurian (Reynolds, 1946), lack their articular ends, so that identification is difficult. Three large specimens, found in 1844 (370 mm long, 420 mm in circumference), 1846 (600 mm long, 125 mm in diameter at one end), and between 1846 and 1875 (200 mm long, 370 mm in circumference), were described by Stutchbury (1850) and Sanders (1876).

Some smaller bones (a vertebra, four ends of phalanges, a small rib) in the BMNH and LEICS (Reynolds, 1946) have been identified as *Zanclodon* (Lydekker, 1888a). That attribution is incorrect (*Zanclodon* is an indeterminate archosaur), but they may be termed 'megalosaur *incertae sedis*' for the present. Occasional megalosaur teeth have also been found at Aust (Reynolds, 1946), and three phalanges (BRSUG and in private collections) may also be attributed to a megalosaur. The present whereabouts of these so-called dinosaurs from Aust are uncertain.

Conclusions

Aust Cliff exposes the best section of the Rhaetian beds with the 'Rhaetic Bone Bed' in Britain thus giving the site considerable conservation value. This bone bed occurs in many other localities in England and South Wales, but it is probably best developed at Aust and here contains the most diverse fauna of reptiles. The bone beds of the Westbury Formation have been of considerable importance in Triassic vertebrate palaeontology (Storrs, in press), and have played an important role in the discussion of bone bed formation and diagenesis (e.g. Antia, 1979).

VERTEBRATE-BEARING FISSURE DEPOSITS OF SOUTH-WEST ENGLAND AND SOUTH WALES

Cave and fissure systems developed in the Carboniferous Limestones of the Mendips and Glamorgan (Figure 4.22) during the Late Triassic and earliest Jurassic contain abundant reptilian and other vertebrate remains. The Mendips and parts of South Wales appear to have comprised an archipelago of low limestone islands, and the fissures developed in these limestones preserve a detailed record of the diverse and often insular herpetofaunas of the time (Robinson, 1957a; Tarlo, 1962; Halstead and Nicoll, 1971; Kermack *et al.*, 1973; Fraser, 1986, 1988b, 1994; Savage, 1993). The nature of the palaeokarst and the geology of the caves is reviewed by Simms (1990). The fossil bones, although isolated and disarticulated, are often well preserved and lend themselves to detailed anatomical studies involving a large number of individuals, for example Whiteside (1986) on *Diphydontosaurus avonis* and Fraser (1988c) on *Clevosaurus*. The main Late Triassic (1–8) and Early Jurassic (9–13) fissures of south-west England and Wales (Figure 4.22), and their reptile faunas (see Figure 4.24), are listed below:

1. Slickstones (Cromhall) Quarry, Avon (ST 704916). Seven species of sphenodontid, including the types of *Clevosaurus hudsoni* Swinton, 1935, *C. minor* Fraser (1988c), *Planocephalosaurus robinsonae* Fraser (1982), *Sigmala sigmala* Fraser (1986) and *Pelecymala robustus* Fraser (1986), as well as two unnamed sphenodontids, a procolophonid, the gliding diapsid *Kuehneosaurus*, an ?aetosaur, a ?scleromochlid, a terrestrial crocodilomorph, a ?sphenosuchid, a rhamphorhynchoid pterosaur, the dinosaur *Thecodontosaurus*, the enigmatic diapsid/procolophonid *Variodens* and various unidentified diapsids.

2. Tytherington Quarry, Avon (ST 660890). Type of the sphenodontid *Diphydontosaurus avonis* Whiteside (1986), as well as the sphenodontids *Clevosaurus, Planocephalosaurus*, a crocodilomorph, the dinosaur *Thecodontosaurus*, a 'coelurosaur' dinosaur, and unidentified sphenodontids and archosaurs.

3. Durdham Down, Avon (ST 572747). Types of the prosauropod dinosaur *Thecodontosaurus antiquus* Morris (1843) and the (?) phytosaur *Rileya platyodon* (Riley and Stutchbury, 1840). Also *Diphydontosaurus*.

4. Batscombe Quarry, Somerset (ST 460550). Type of the gliding diapsid *Kuehneosuchus (?= Kuehneosaurus) latissimus* (Robinson, 1962).

5. Emborough Quarry, Somerset (ST 623505). Types of the gliding diapsid *Kuehneosaurus latus* Robinson (1962) and the enigmatic diapsid *Variodens inopinatus* Robinson (1957b), as well as an archosaur, a sphenodontid and the mammal *Kuehneotherium* sp.

6. Highcroft Quarry, near Gurney Slade, Somerset (ST 623499). A reptile jaw (Robinson, 1957a), *?Clevosaurus* (Fraser, 1994).

7. Pant-y-ffynon Quarry, South Glamorgan (ST 047741). Type of the terrestrial crocodilomorph *Terrestrisuchus gracilis* Crush (1984),

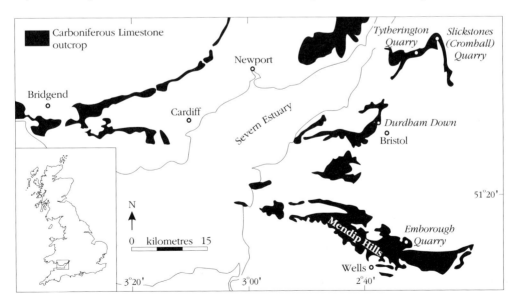

Figure 4.22 Map showing the distribution of Carboniferous Limestone and of tetrapod-bearing GCR fissure sites in south-west England. After Fraser (1985).

as well as the gliding diapsid *Kuehneosaurus*, the sphenodontid *Clevosaurus*, a sclero-mochlid, the dinosaurs *Thecodontosaurus* cf. *antiquus* (Kermack, 1984) and *Syntarsus*, and lepidosaurs.

8. Ruthin Quarry, South Glamorgan (SS 975796). Type of *Tricuspisaurus thomasi* Robinson (1957), as well as pleurodont reptiles, the sphenodontids *Clevosaurus* and *Plano-cephalosaurus* and archosaurs (Fraser, 1986, 1994).

9. Windsor Hill Quarry, near Shepton Mallet, Somerset (ST 615452). Types of the trity-lodont mammal-like reptiles *Oligokyphus major* Kühne (1956) and *O. minor* Kühne (1956).

10. Holwell Southern Quarry, near Frome, Somerset (ST 727452). Types of the early mammals *Haramiya moorei* (Owen, 1871), *H. fissurae* (Simpson, 1928), *Thomasia angli-ca* Simpson (1928), *Eozostrodon parvus* Parrington (1941) and *E. problematicus* Parrington (1941), as well as teeth of croco-dilians, and other reptiles (Robinson, 1957a), a tritylodont (Savage and Waldman, 1966) and *Clevosaurus* (Fraser, pers. comm., 1993).

11. Duchy Quarry, South Glamorgan (SS 906757). Type of *Morganucodon watsoni* Kühne (1949), as well as other triconodont teeth and a symmetrodont mammal.

12. Pont Alun Quarry, South Glamorgan (SS 899765). Types of the sphenodontian *Gephyrosaurus bridensis* (Evans, 1980, 1981) and *Kuehneosaurus praecursori* Kermack *et al.* (1968), as well as *Morganucodon/Eozostrodon*.

13. Pant Quarry, South Glamorgan (SS 896760). The sphenodontian *Gephyrosaurus*, three sphenodontids, one or more archosaurs, the tritylodont *Oligokyphus* and the mammals *Thomasia*, *Kuehneotherium* and *Morgan-ucodon watsoni*.

The dating of the fissures is difficult. Robinson (1957a) regarded all the reptile-dominated ('sauropsid') faunas as being Norian in age, and those dominated by tritylodonts and mammals ('theropsid' faunas) as Rhaetian or Early Jurassic. Fraser (1986, 1994) argued that the sauropsid/theropsid, Norian/Rhaeto-Jurassic cor-relation was not clear-cut: it could just as readily be a taphonomic division of faunas. Independent palynological evidence has established a Rhaetian age for some Tytherington fissures

(Marshall and Whiteside, 1980) and a Hettangian–Sinemurian age for Duchy, Pant and Pont Alun Quarries, based on the occurrence of *Hirmerella (Cheirolepis)* spores in the last three sites. The division into sauropsid/theropsid assemblages was further challenged by the dis-covery of a mammal tooth, *Kuehneotherium* sp., at Emborough Quarry in a fissure otherwise clearly placed in the 'sauropsid' Triassic group (Fraser *et al.*, 1985).

In the absence of further palynological evi-dence, some indication of the ages of individual fissure faunas may be obtained by comparisons of reptiles and mammals with more securely dated localities elsewhere. For example, the crocodylomorph *Terrestrisuchus* is most like *Saltoposuchus* (and may be congeneric) from the middle Norian Mittlerer Stubensandstein of south-west Germany. *Thecodontosaurus* is a basal prosauropod, like forms of Late Carnian to Norian age in North America, central Europe and southern Africa. Procolophonids, aetosaurs and scleromochlids all died out before the end of the Triassic elsewhere, and aetosaurs are exclusively Late Carnian to Rhaetian in age. *Scleromochlus* is known otherwise only from the Late Carnian of Scotland. *Kuehneosaurus* is most like *Icarosaurus* from the Late Carnian of North America. All the evidence, therefore, confirms a Late Triassic age for the fissures nos. 1–8 in the above list, and probably a range of ages from Late Carnian to Rhaetian. There is no reason why all should be regarded as contemporaneous, but all should postdate the Mid Carnian pluvial episode of Simms and Ruffell (1989, 1990), if those authors are correct that most of the fis-sures were excavated at that time.

Four fissure localities of scientific and historic importance are selected as GCR sites, while Windsor Hill Quarry has been scheduled for its tritylodont remains as a mammal site (see GCR Fossil Mammals and Birds volume):

1. Slickstones (Cromhall) Quarry, Avon (ST 704916). Late Triassic (Carnian/Norian), fis-sure fill.

2. Durdham Down, Avon (ST 572747). Late Triassic (Carnian/Norian), fissure fill.

3. Emborough Quarry, Somerset (ST 623505). Late Triassic (Carnian/Norian), fissure fill.

4. Tytherington Quarry, Avon (ST 660890). Late Triassic (Carnian/Norian), fissure fill.

SLICKSTONES (CROMHALL) QUARRY, AVON (ST 704916)

Highlights

Slickstones (Cromhall) Quarry is the site of some of the richest of the fissure deposits in the Bristol–South Wales region. The Late Triassic cave fills have produced excellent specimens of more than 20 species of small reptiles: procolophonids, sphenodontids, gliding kuehneosaurs, problematic archosauromorphs, pterosaurs, crocodilomorphs, thecodontians and rare dinosaurs.

Introduction

Slickstones Quarry contains at least seven fossiliferous cavity-fill sites, the main fissures being located at ST 70359155, ST 704916 and ST 70409165. Fissures 1 and 2 are the original sites of fossil finds described by Robinson (1957a). The quarry company (ARC Ltd) is working the Carboniferous Gully Oolite and Black Rock Limestone (Tournaisian age). The fissures noted are not in areas in which there are currently any quarrying operations; these are not likely to take place in the future, and further collecting is possible (Figure 4.23).

The vertebrate-bearing fissure deposits in Slickstones Quarry were first exposed during quarrying operations early this century and, having no commercial value, were left *in situ* as a rock promontory. They were later identified by Robinson (1957b) as representing one of the prime examples of a fossilized underground water-course fissure or cave. Robinson (1957b) recorded that F.G. Hudson discovered reptiles in the Slickstones deposits in 1938, this site being the first non-marine fissure discovery. Hudson collected specimens of a sphenodontid which was described and named as the type of *Clevosaurus hudsoni* by Swinton (1939), who did not specify the exact locality. Continued excavation and extensive collecting from the same area revealed the first remains in association, and Robinson *et al.* (1952) and Robinson (1957a) gave further details on the fauna, estimating the total number of reptile species as about nine and listing sphenodontids, squamate lizards and archosaurs. Robinson (1962) mentioned that the gliding 'lizard' *Kuehneosaurus* occurred at Slickstones. Halstead and Nicoll (1971) drew attention to dis-

Figure 4.23 Slickstones (Cromhall) Quarry: view of the north face, at the uppermost level, of this working site. A bone-bearing fissure is seen extending down the middle of the photograph. The fissure fills are late Triassic in age, and they occupy caves dissolved into uplifted Carboniferous marine limestones. (Photo: M.J. Benton.)

sociated material within recemented debris in the fissures, which formed the subject of all future work, and noted the occurrence of two further species, a procolophonid and a trilophosaur. Crush (1980) revised the number of species represented at Slickstones Quarry to 10, including *Kuehneosaurus latus*, a prosauropod dinosaur, two ?lizards and *Clevosaurus hudsoni*.

More recently, research by Fraser and Walkden (1983), Fraser (1985, 1994), and Walkden and Fraser (1993) has revealed the presence of six vertebrate-bearing deposits of Triassic age at Slickstones Quarry. Extensive collections were made from all these horizons and they identified six genera of lepidosaurs, including five sphenodontids, and at least four genera of archosaurs, comprising a theropod dinosaur, one or two 'thecodontians' and a possible terrestrial crocodile.

The sphenodontids were most abundant and diverse. Fraser (1982) and Fraser and Walkden (1984) described the dissociated remains of the most abundant form, a new small sphenodontid, *Planocephalosaurus robinsonae*. Specimens of *Clevosaurus* in the AUGD collections were found in five of the fissures and, although they too are mostly dissociated, some articulated remains were collected, including a partial skull and two lower jaws. A third sphenodontid, *Sigmala sigmala* (Fraser and Walkden, 1983; Fraser, 1986), was based on a maxilla, dentary and palatine, and a fourth, *Pelecymala robustus*, was based entirely on isolated jaw elements (Fraser, 1986). Fraser (1988c) described *Clevosaurus* in detail, and formally described the new species *Clevosaurus minor* and a third indeterminate species which he called *Clevosaurus* sp.

Fraser (1988a) described some rare and unusual skeletal elements, including jaw bones and a pro-coelous vertebra, which he suggested tentatively might represent prolacertiform, thalattosaurian and even pterosaurian remains. The remains of a rhamphorhynchoid pterosaur, the earliest known from Britain, were reported by Fraser and Unwin (1990).

Description

Fraser and Walkden (1983), Fraser *et al.* (1985) and Walkden and Fraser (1993) identified seven fissure and cavity fill sites at Slickstones, six of which yielded vertebrate remains. The seventh, a collapsed cave system, was evidently not fossiliferous and pre-dated the rest on sedimentological grounds. The fissure walls range from subhorizontal to almost vertical in relation to bedding in the Carboniferous Limestone host rock, and the fills are arranged in a linear fashion along a common axis. This alignment of the fissures is not obviously associated with any persistent joint system or with faults; it probably reflects the flow direction of an ancient water course (Fraser, 1985). The fissures have been interpreted as a series of separately filled sink-holes (dolines) and part of a small cave system. Simms (1990), however, disputes this interpretation and regards at least the western fissure as a thermal spring conduit.

The fissure material consists of Triassic land-wash laid down in the karstic fissures and cave systems in Carboniferous Limestone. A calcareous, buff-coloured matrix is lodged in Fraser's fissure 1, whereas fissures 3, 4, 5 and 7 contain marls, sandstone, bedded crinoidal limestone and soft, pale green to red mudstones. Most of the fills are nearly horizontally bedded, sometimes with sag-curvature and with some small-scale cross-bedding, and they may contain fragments of stalactites and rare cave pearls. Halstead and Nicoll (1971) mention fluting on the limestone wallrock and laminated red marls (clay grade) in the lower levels, and coarser sandy green marls higher up, the latter formed at a time when the limestone surface had been eroded to approximately the level of the water table.

In fissures 3, 4, 5 and 7 vertebrate remains occur in three principal lithologies: lenses of conglomerate with Carboniferous Limestone clasts in the mudstones enclose reptile bones and remanié fossils, paler sandy mudstones with reptile bones, and hard red calcite-cemented sandstones with the original *Clevosaurus* material and many specimens of the branchiopod crustacean *Euestheria minuta*.

The preservation of the bone is generally excellent, although few bones are complete. Some of the fossils are extremely well preserved and in a fully associated state (i.e. with bones still in position of original articulation), and complete skulls have been found; part of the skin was preserved in one specimen. However, many specimens are broken and show in addition a high degree of rounding and polishing, indicating considerable attrition during transport (Kermack *et al.*, 1973; Fraser and Walkden, 1983).

A certain degree of sedimentary sorting of elements is evident, with coarser sediments tending to contain larger bones in addition to the smallest elements. One lamina was found to contain almost exclusively dentaries of *Clevosaurus* (specimen in BRSUG). Current alignment is not uncommon in some of the beds.

The bone material is preserved as a pale yellow or white substance, but may be dark brown in colour. Unabraded elements often show fine surface detail, preserving tiny foramina and even muscle attachment points.

Fauna

Nearly twenty genera of archosaurs and lepidosaurs (Figure 4.24) have been identified from the six fissure fills, and the variety of assemblages present may represent successional ecosystems. The reptiles are mostly small, the largest, a

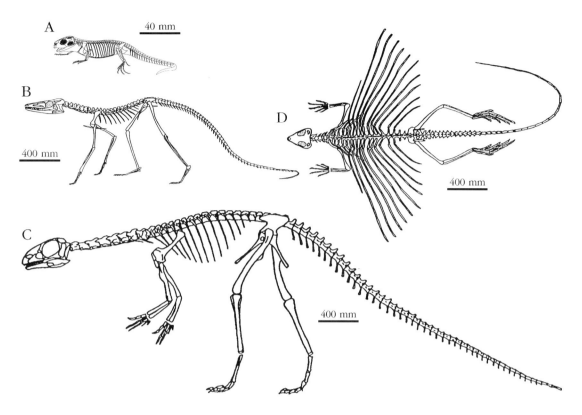

Figure 4.24 Typical reptiles from the Late Triassic fissures in South Wales and around Bristol. Skeletal reconstructions of (A) the sphenodontid *Clevosaurus*; (B) the crocodilomorph *Saltoposuchus*; (C) the prosauropod dinosaur *Thecodontosaurus*; and, (D) the gliding diapsid *Kuehneosaurus*. After various sources; in Fraser (1994).

dinosaur, being no more than one metre long. At least seven species of sphenodontids occur, and for five of these, *Clevosaurus hudsoni, C. minor, Sigmala sigmala, Pelecymala robustus* and *Planocephalosaurus robinsonae*, Slickstones is the type locality. A reconstruction of the faunas of the Bristol region in the Late Triassic, based on fossils found in fisure deposits, is shown in Figure 4.25.

Arthropoda: Crustacea: Branchiopoda
  Euestheria minuta
    Casts of carapaces (AUGD)
Arthropoda: Myriapoda: Diplopoda
  Millipedes
    Several well-preserved complete and part-enrolled specimens (AUGD)
Arthropoda: Hexapoda: Insecta
  Beetle elytra (AUGD)
Anapsida: Procolophonidae
  Procolophonid
    Abundant dissociated cranial and postcranial material (AUGD)

Lepidosauria: Sphenodontida
  Clevosaurus hudsoni Swinton, 1939
    Syntypes: BMNH R6100, R9249, R9251, R9252, R9253, R9255–R9260
  Clevosaurus minor Fraser, 1988
    Type specimen: AUGD 11377
  Clevosaurus sp.
    Various specimens (AUGD)
  Planocephalosaurus robinsonae Fraser, 1982
    Type specimen: AUGD 11061
  Sigmala sigmala Fraser, 1986
    Type specimen: AUGD 11083
  Pelecymala robustus Fraser, 1986
    Type specimen: AUGD 11140
  Diphydontosaurus avonis Whiteside, 1986
    Reptile 'B'; AUGD
Archosauromorpha: *inc. sed.*
  Kuehneosaurus latus Robinson, 1962
    Various specimens (AUGD)
Archosauromorpha: Prolacertiformes
  Prolacertiform (?)
    Premaxilla and maxilla (AUGD)

Figure 4.25 Reconstruction of the faunas of the Bristol region in the late Triassic, based on fossil remains found in several fissure deposits. The prosauropod dinosaur *Thecodontosaurus* stands in the background, while the gliding diapsid *Kuehneosaurus* passes over a sphenodontid (bottom left), the possibly lizard-like *Variodens* (bottom middle), and the early mammal *Haramiya* (bottom right). After Duff, McKirdy and Harley (1985).

Archosauromorpha: (?)Trilophosauridae
  Variodens sp.
    Various (AUGD)
Archosauria indet.
  Pterosaur or Thalattosaurian (?)
    Fused premaxilla, maxilla and vertebrae (AUGD)
  'Thecodontians'
    Numerous serrated teeth belonging to two or three genera (AUGD)
  Archosaur ('Reptile G')
    A single left dentary with three remaining teeth (AUGD)
  (?)Scleromochlid
    Various (AUGD)
  (?)Aetosaur

    Various (AUGD)
Archosauria: Crocodylomorpha
  Terrestrisuchus sp. and a sphenosuchid
    Various (AUGD)
Archosauria: Pterosauria
  Rhamphorhynchoid
    Metacarpals (AUGD)
Archosauria: Dinosauria: Saurischia: Sauropodomorpha
  Thecodontosaurus
    Various (AUGD)
Archosauria: Dinosauria: Saurischia: Theropoda
  Theropod dinosaur
    Various (AUGD)
Unidentified sauropsids

Interpretation

The recognition of pre-Rhaetian sediments at Slickstones Quarry has been based largely on the abundance of red and green marls of Mercia Mudstone lithology, since it has not been possible to establish direct stratigraphic relationships based on nearby bedded sequences, as has been achieved for other fissures (e.g. Robinson, 1957a). In their studies of the sediments, Fraser and Walkden (1983) note that there is nothing to suggest that the Slickstones deposits postdate the Rhaetian transgression. Walkden and Fraser (1993) argue further that the sediments at Cromhall overlying the fissures contain fish and tetrapod faunas typical of the Penarth Group (Rhaetian), and that this provides an upper constraint on the ages of the fissures. The fissures may have filled at different times, and it is not possible to provide a lower age limit.

Despite careful searching at Slickstones Quarry, no palynomorphs have so far been recognized to provide a reliable date for the sediments (e.g. Warrington, 1978). Only the vertebrate fauna is available for determination of the relative age, and on this basis the Slickstones deposits have been assigned a Late Triassic age (Robinson, 1957a) and specifically Late Norian for *Clevosaurus* (Robinson, 1973). Benton (1994a, 1994b) noted that certain elements of the Lossiemouth Sandstone Formation (Late Carnian), such as *Leptopleuron*, *Brachyrhinodon* and *Scleromochlus*, are very similar to elements of the Cromhall fauna, and *Terrestrisuchus* is very like *Saltoposuchus* from the Mid Norian Stubensandstein of Germany.

Marshall and Whiteside (1980) noted *Clevosaurus* (Figure 4.24A) from fissures at Tytherington, and identified a marine palynomorph assemblage which they regarded as Rhaetian. However, the Slickstone fissures contain faunal elements (e.g. *Kuehneosaurus*) not known from Tytherington, and these indicate a broad age range of Late Carnian to earliest Rhaetian, which accords with similar non-fissure faunas, such as those of the Newark Supergroup of eastern North America.

The most abundant remains at Slickstones Quarry are those of sphenodontids, of which the commonest is *Clevosaurus*. Four species of *Clevosaurus* seem to be present, based on dentitions and distinctive patterns of tooth wear (Fraser and Walkden, 1983; Fraser, 1988c): *C. minor* Fraser, 1988 is the smallest form, intermediate in size between *Planocephalosaurus* and *C. hudsoni*; *C. hudsoni* is the largest; a third and possible fourth species are as yet unnamed.

Clevosaurus hudsoni is one of the largest lepidosaurs in the deposit, averaging 150–200 mm in length, with a maximum of approximately 250 mm (Fraser, 1988c). It was superficially like a modern lizard with a slender body, long limbs and (probably) a long tail. The skull was strongly built. It has been suggested that *C. hudsoni* may have been a facultative herbivore (Fraser and Walkden, 1983), 'raking food together with its beak-like front teeth, and chopping it further back in the mouth'. However, in the adult *Clevosaurus* the jaws and worn dentition formed a powerful shearing mechanism fully capable of crushing the chitin exoskeleton of large insects, and the possibility of an omnivorous or fully insectivorous diet cannot be ruled out. Young *Clevosaurus* specimens almost certainly fed on small insects and soft-bodied invertebrates such as worms, the immature jaw bones possessing sharp pointed teeth. *Clevosaurus* is known elsewhere, at Tytherington and at Durdham Down, but other localities in the Bristol Channel area have recently also been found to contain material assignable to the genus: Pant-y-ffynon (Crush, 1980, fig. 1a), Highcroft (Fraser, 1986), and Holwell (Fraser, pers. comm., 1993).

The remains of *Diphydontosaurus* (Whiteside, 1986), a primitive sphenodontid, best known from Tytherington Quarry, are also present at Slickstones, and at Durdham Down. 'Reptile B' of Tytherington is also present in the Slickstones fauna.

Planocephalosaurus robinsonae (Fraser, 1982) usually exhibits an incomplete lower temporal bar suggesting that this character, formerly thought unique to the squamates, is more widespread in the Lepidosauria as a whole. The genus appears to have close affinities with *Clevosaurus* and *Sphenodon*. The jaw action of *P. robinsonae* is slightly propalinal (back and forth movement), and the dentition (Fraser and Walkden, 1983) indicates that it may have been primarily insectivorous, although possibly capable of taking newly hatched sphenodontids, if the opportunity arose. The genus has been found elsewhere in the fissures at Tytherington and Ruthin Quarry, South Wales, where a sphenodontid maxilla has been found very similar in shape and size to the Slickstones form (Fraser, 1982). Fraser and Walkden (1983) noted subtle differences in the adult *Planocephalosaurus* material obtained from

Tytherington and Slickstones quarries: the Slickstones variety was generally larger in size, exhibiting more robust skull elements. The range of *Planocephalosaurus* from the Norian into the Rhaetian may help explain these slight differences in morphology (Fraser and Walkden, 1983).

The rare sphenodontid *Sigmala sigmala* differs from *Planocephalosaurus* in bearing a high coronoid process and in possessing a somewhat deeper dentary bone. Well-defined facets on the teeth match precisely in opposing jaws and indicate the lack of any propalinal movement for this species. In a review of sphenodontid relationships, Fraser and Benton (1989) found that *Diphydontosaurus* and *Planocephalosaurus* were basal taxa, while *Clevosaurus* fell in a crown group containing mainly Jurassic and Cretaceous forms, as well as the living *Sphenodon*. The other Slickstones taxa are too incompletely known to be used in cladistic analysis.

The procolophonid (Halstead and Nicoll, 1971; Fraser, 1985) shows affinities with *Tricuspisaurus* of Ruthin Quarry, but appears most similar to *Leptopleuron lacertinum*, an advanced form from the Late Carnian Lossiemouth Sandstone Formation of Elgin (Fraser, pers. comm., 1991). The Slickstones procolophonid and *Leptopleuron* share certain affinities with *Hypsognathus* from the Rhaetian of the Newark Supergroup of the eastern USA.

The gliding diapsid reptile *Kuehneosaurus* (Figure 4.24D; Robinson, 1962) is comparable with *Icarosaurus* from the Late Carnian Lockatong Formation of New Jersey, USA. The other discoveries at Slickstones include several archosaurs: an articulated partial skeleton of a terrestrial crocodilomorph, a partial skull of an undescribed thecodontian and other thecodontian remains (a possible scleromochlid and a possible aetosaur) and a procolophonid jaw. Terrestrial crocodilomorphs of comparable age to the Slickstones form include *Terrestrisuchus gracilis* (Figure 4.24B) from fissures in Old Pant-y-ffynon Quarry, South Glamorgan (Crush, 1984). Ruthin Quarry also yielded a terrestrial crocodile (*?Terrestrisuchus*) (Crush, 1984), and other reptiles in the Ruthin sediments include a few jaw fragments of a small archosaur, probably a thecodontian similar to the Slickstones forms (Fraser and Walkden, 1983; Fraser, 1985). Terrestrial crocodilomorphs and aetosaurs are also known from several Late Carnian to Rhaetian localities elsewhere in the world, but scleromochlids hitherto have only been reported from the

Lossiemouth Sandstone Formation of Lossiemouth (see above). The trilophosaur/procolophonid specimen figured by Halstead and Nicoll (1971) is closest to *Variodens* from Emborough. The Slickstones fauna shares elements with many of the other fissure sites, but the greatest faunal affinity is with Tytherington, at least six reptiles being common to both localities.

The pterosaurs (Fraser and Unwin, 1990) belong to the Rhamphorhynchoidea and represent the earliest such remains from Britain. The material consists of two specimens of a metacarpal IV (from level K of site 4; Fraser, 1985). Triassic pterosaurs are rare, but well preserved material has been reported from Norian limestones of Cene, near Bergamo, Italy (*Eudimorphodon, Peteinosaurus*) and the Norian of Friuli, Italy (*Preondactylus*) (Wild, 1978a, 1983).

The only invertebrates present in the Slickstones fissures are the crustacean *Euestheria minuta*, and rare diplopod (millepede), and insect remains. An invertebrate fauna is unknown in most of the other fissure fillings of a comparable age, but carapaces of *Euestheria* are recorded from Tytherington.

Conclusions

Slickstones is the richest fissure deposit: nearly twenty reptile taxa have been recognized, and it is the type locality for five species of sphenodontids. The diversity of lepidosaurs at Slickstones is unmatched by any other Triassic locality in the world, and this is the only fissure locality to have produced articulated remains of sphenodontid reptiles. The pterosaur remains are the oldest from Britain, and the only Triassic pterosaurs known outside Italy. This palaeontological importance and the potential for future finds give the site its conservation value.

DURDHAM DOWN, AVON (ST 572747)

Highlights

Durdham Down fissure was the first of the Late Triassic Bristol fissures to be identified. It was the source of the material of the prosauropod dinosaur, *Thecodontosaurus antiquus*, the most primitive member of its group.

Durdham Down, Avon

Introduction

The fissure at Durdham Down, located in a quarry close to Quarry Steps, is important in being the site of the first discovery of a reptile-bearing fissure in the Bristol region. It was from here that the remains of the prosauropod dinosaur *Thecodontosaurus*, unique to the district around the Severn Estuary, were first described, and the remains of two other prosauropods and a phytosaur recognized. The site also produced a low-diversity fish fauna, represented by spines, scales and teeth, as well as an *Echinus* spine and reworked Carboniferous fossils (Moore, 1881). The fissure cuts through Carboniferous Limestone and was regarded as a true fissure by Tarlo (1959a) and Halstead and Nicoll (1971), in contrast to Robinson (1957a), who viewed the deposit as an infilled depression in the land surface. Although the quarry at Quarry Steps is largely built over today, fissures can be seen on the limestone faces (ST 572747). Careful excavation of the site should produce more finds of *Thecodontosaurus* and establish a detailed stratigraphy for the fissure(s) and the palaeo-environmental context of the fossils.

The first mentions of the find at Durdham Down seem to be Anon. (1834, 1835). Riley and Stutchbury (1840) described three dinosaurs, of which only two, *Palaeosaurus cylindrodon* and *P. platyodon*, were named as species, *Thecodontosaurus* being referred to only generically. Morris (1843) gave the specific name *T. antiquus*. Seeley (1895) re-examined the material, identifying two, and not three, species of dinosaur, *Thecodontosaurus antiquus* and *Palaeosaurus platyodon*. Huene (1908b) renamed the larger dinosaur *Thecodontosaurus cylindrodon*.

Huene (1902, pp. 62–3) established a new genus and species of phytosaur, *Rileya bristolensis* for a humerus and two vertebrae from Durdham Down. Huene (1908b, p. 240; 1908e) also reclassified the tooth *Palaeosaurus platyodon* as a phytosaur, allied it with the postcranial remains as the species *Rileya platyodon*, and assigned further teeth and postcranial bones to this form.

Halstead and Nicoll (1971) mention a small jaw of the sphenodontid *Clevosaurus*, which has been re-identified as *Diphydontosaurus avonis*, known also from Tytherington and Slickstones (Fraser and Walkden, 1983; Whiteside, 1986), and the articulated skeleton figured (Halstead and

Nicoll, 1971, pl. 23B) as a lizard may also belong to this species.

Description

The site of the *Thecodontosaurus* deposit is not known for certain. Pertinent information can be found in the papers of Etheridge (1870), Moore (1881) and Huene (1908a). Etheridge shows two drawings (his figs. 4 and 5) which show the reptile deposit at about 320 ft above mean sea level. Moore (1881, p. 72) mentions specifically a place known as 'The Quarry and The Quarry Steps' and states 'Looking from it [the platform of Quarry Steps], along the Down escarpment to the west, the eye takes in Bellevue Terrace [Belgrave Terrace], on the edge of the Down; and it was between these houses and the quarry, a distance of probably 200 yards, along the same face of limestone . . . that the . . . *Thecodontosaurian* remains were found . . . Unfortunately the precise spot is unknown . . . and built over.' Huene (1908e, 1908f) seemingly misunderstands Moore, naming the site of discovery as Avenue Quarry at the end of Avenue Road, but Moore mentioned this quarry as a location 680 yards away from Quarry Steps and terminating a transect of workings which produced fissures of different ages.

Fauna

Archosauria: Crurotarsi: ?Phytosauria
   Rileya platyodon (Riley and Stutchbury, 1840)
      Type tooth: BRSMG. Other putative remains:
        BRSMG, BMNH, YPM
Archosauria: Dinosauria: Saurischia:
   Sauropodomorpha
   Thecodontosaurus antiquus Morris, 1843
      Type jaw and other cranial and postcranial
        material (BRSMG, BMNH, BGS(GSM), YPM)

Interpretation

Moore (1881) regarded the deposit as of 'Rhaetic' age on the basis of a reptile vertebra from Vallis Vale, but later (according to H.H. Winwood) thought it to be 'Upper Keuper' after finding teeth of *Thecodontosaurus* at Ruishton near Somerset. Etheridge (1870) thought the deposit was equivalent to the German Lettenkohle (Ladinian). Conditions at the time of deposition of the Durdham Down fissure system and the dating of the

Tytherington *Clevosaurus* and *Diphydontosaurus* as Rhaetian suggests that the Durdham Down fissure could also be Rhaetian, and the presence of unrolled fish teeth, implying a high water-table, could appear to add support to this notion. Halstead and Nicoll (1971) mention that the matrix is virtually identical to the breccia from the Gliny sea cave; this also implies a marine influence, hence suggesting a Rhaetian age. Moore (1881) describes the fissures very near that of Durdham Down with Rhaetian and Lower Jurassic (Lias) fossils.

On the other hand, *Thecodontosaurus* (Figure 4.24C) is a basal prosauropod in cladistic terms (Gauthier, 1986; Galton, 1990), with relatives from North America and elsewhere that occur in Late Carnian and Norian deposits. *Thecodontosaurus* is known also from fissures in Slickstones, Tytherington, and Pant-y-ffynon quarries, and these are dated, on evidence of their reptile faunas, with a range of Late Carnian to Rhaetian ages, and Pant-y-ffynon even as Late Triassic to Early Jurassic (Kermack, 1984). If *Rileya* is a phytosaur, this would limit the age to Late Triassic only.

Rileya platyodon is an enigmatic form, being regarded by some as a phytosaur and by others as an aetosaur (Westphal, 1976, p. 116). The paucity of the original material (a tooth), and the later assignment of postcranial material to the same species (Huene, 1902; 1908b, p. 240; 1908e) has not helped matters. Restudy of the material is necessary.

The *Thecodontosaurus* bones from Durdham Down show similar size ranges to those from Tytherington, and it would appear that *T. cylindrodon* are adult *T. antiquus*. The new remains of *Thecodontosaurus* sp. from Pant-y-ffynon (Kermack, 1984) may belong to juvenile *T. antiquus*, based on their smaller size and on the incomplete ossification of some bones. The systematic position of *Thecodontosaurus* has been debated recently. Galton and Cluver (1976) referred it to the Anchisauridae, but cladistic analyses (Gauthier, 1986; Galton, 1990), showed that it was a basal sauropodomorph taxon, and it is assigned to its own family, Thecodontosauridae. Galton (1990) also includes *Azendohsaurus* from the Late Carnian Argana Formation of Morocco in this family, but this is based on very limited dental remains and offers little evidence for comparison. The recent analyses of the relationships of *Thecodontosaurus* have been based largely on the ?juvenile Pant-y-ffynon material (Kermack, 1984), and it has yet to be demonstrated that this is the same as the type *T. antiquus* from Durdham Down.

Conclusions

Durdham Down is the type locality of *Thecodontosaurus antiquus* and *T. cylindrodon* (if it is considered a separate species) and of the ?phytosaur *Rileya platyodon*. The only cranial material of *T. antiquus* comes from this site (an occiput held in the YPM). *Thecodontosaurus* is an important basal sauropodomorph dinosaur, seemingly unique to the British fissures.

The historical importance of the fossil finds from this site and its limited potential for re-excavation together give it significant conservation value.

EMBOROUGH QUARRY, SOMERSET (ST 623505)

Highlights

Emborough Quarry is the source of a varied fauna of Late Triassic small reptiles. It is the locality where the best specimens of the extraordinary gliding reptile *Kuehneosaurus* have been found, as well as *Kuehneotherium*, perhaps one of the oldest mammals in the world.

Introduction

Emborough is a disused quarry formerly worked for Hotwells Limestone (Carboniferous: Asbian), which dips to the north-east (Savage, 1977). The reptile-bearing cavity filling is exposed in a promontory in the south-east corner of the eastern quarry. The site is important for yielding numerous remains of the unusual gliding reptile *Kuehneosaurus* and the ?trilophosaur *Variodens*. The fissure site occupies a relatively small proportion of the quarry area, and further collecting is possible.

Reptiles were discovered in fissure sediments at Emborough by Kühne in 1946, but the first discussion on the geology of the deposit and the reptiles was that of Robinson (1957a, 1957b). Robinson (1957b) described a tricuspid reptile, *Variodens inopinatus*, from Emborough on the basis of two dentary fragments and referred it to the Tricuspisauridae. A description of the Emborough gliding diapsid *Kuehneosaurus*, the first known, was made by Robinson (1962). Fraser *et al.* (1985) reported a therian mammal

(*Kuehneotherium*) from Emborough, regarded by them as the oldest therian mammal in the world, and used this as evidence to invalidate earlier claims by Robinson (1957a, 1971) for the existence of a clear-cut distinction between sauropsid-bearing Late Triassic fissure fills and theropsid-bearing Early Jurassic ones (see above).

Description

The sediments may be divided lithostratigraphically into lower and upper units. The lower sediments are unfossiliferous, well-bedded, dark-red clays with green patches. The upper deposit consists of a conglomerate of Carboniferous Limestone clasts, up to boulder-size, set in a matrix of limestone pebbles, pale shale and silts. The silt is finely bedded and free of clasts in some places, usually red, but sometimes pale green. The reptile fossils are found in the higher part of the conglomeratic deposit in the silts. Most of the fossils are dissociated, but some *Kuehneosaurus* material is in articulation.

Fauna

Diapsida: Archosauromorpha: *inc. sed.*
  Kuehneosaurus latus Robinson, 1962
    Type material: BMNH R8172
Diapsida: Archosauromorpha: ?Trilophosauridae
  Variodens inopinatus Robinson, 1957
    Type material: BMNH
Diapsida: Archosauria
  Archosaur *incertae sedis*
Lepidosauria: Sphenodontida
  Planocephalosaurus sp.
Mammalia: 'Symmetrodonta': Kuehneotheriidae
  Kuehneotherium sp.
    Tooth AUGD 11133

Interpretation

Robinson (1957a) interpreted the sediments as the filling of a collapsed cave, the lower beds being deposited by underground streams and the upper deposit formed by a collapse of the cave roof, with fine silt and the reptile remains brought in by land-wash. Robinson (1957a) and Savage (1977) mention solution features such as water-worn faces and boulders; these and the presence of stalactite fragments in the conglomerate con-

firm the impression that the void was part of a cave system.

It has been possible to date the sediments filling the Emborough fissure as Late Triassic on the basis of direct stratigraphic evidence using the topographical relationship of the cavity to the local, normally bedded stratigraphy (Robinson, 1957a; Fraser *et al.*, 1985). This age assignment is in accord with the date given to *Icarosaurus* (a close relation of the Emborough diapsid *Kuehneosaurus*) from the Late Carnian (Lockatong Formation) of the Newark Supergroup in New Jersey.

The reptile fauna from Emborough includes the unusual gliding diapsid *Kuehneosaurus latus* (Figure 4.24D; Robinson, 1962, 1967), which is the most abundant animal present (Fraser, 1994), as well as the ?trilophosaur *Variodens inopinatus*, for which Emborough is the type locality. Robinson (1957a) mentions two other reptiles, an archosaur and a sphenodontid, and Fraser *et al.* (1985) reported two teeth of the mammal *Kuehneotherium*.

Kuehneosaurus is represented by dissociated as well as good articulated remains, and the skull including the braincase has been figured by Robinson (1962). The only other kuehneosaurs known are *Kuehneosaurus latissimus* from Batscombe Quarry and *Icarosaurus* from the Lockatong Formation of the eastern USA (Colbert, 1970). The kuehneosaurs, because of the absence of a lower temporal bar, the presence of a streptostylic quadrate and a pleurodont dentition, were formerly considered to represent primitive squamates, the true lizards and snakes. This view has been strongly doubted by Evans (1980, 1988a), Benton (1985), and others, who regard the kuehneosaurs as primitive archosauromorphs, on the basis of numerous characters that place them close to rhynchosaurs and prolacertiforms in the cladogram. *Kuehneosaurus* is a gliding reptile (Figure 4.24D), one of the earliest aerial tetrapods known, which displays a remarkable convergence with the extant gliding lizards of south-east Asia (*Draco*) and also to *Weigeltisaurus*, a gliding form from the Late Permian of Durham (q.v.), Germany and Madagascar. *Kuehneosaurus latus* has also been reported from Slickstones Quarry and the Pant-y-ffynon fissures (Fraser and Walkden, 1983). The numerous Emborough specimens, being mostly dissociated, will allow a complete anatomical study.

The ?trilophosaur *Variodens inopinatus* is known from two dentaries found at Emborough.

The trilophosaurs are a diapsid group, currently placed in the Archosauromorpha (Benton, 1985; Evans, 1988a). Trilophosaurs are known from the Triassic of Russia and North America, but *Variodens* is the only probable representative of the group described from Western Europe. Halstead and Nicoll (1971) have figured a possible trilophosaur from a fissure at Slickstones Quarry. Fraser (1986) and others have hinted that *Variodens* may be a procolophonid.

The *Kuehneotherium* tooth from Emborough was claimed to be the oldest therian mammal fossil in the world, coming as it did from probably pre-Rhaetian sediments (Fraser *et al.*, 1985). The determination of the stratigraphic age of the fissure sediments was disputed, and other equally old mammal remains may be known from France. Lucas and Hunt (1990) reported a supposed mammal skull from the Late Carnian Tecovas Formation of West Texas, USA. Nonetheless, the early record of *Kuehneotherium* points to the possibility of more substantial finds in Emborough.

The fauna is most similar to that of Slickstones (see above), but the dominance of *Kuehneosaurus* at Emborough might suggest a different depositional environment, or a different age. However, the different proportions of taxa could equally well be the result of differential sampling of the fauna, either by the nature of the fissure or of collection error.

Conclusions

Emborough is the type locality of the remarkable gliding reptile *Kuehneosaurus latus*, the only known locality of the trilophosaur/procolophonid *Variodens inopinatus*, and the site for a *Kuehneotherium* tooth, possibly the oldest record of a therian mammal in the world.

This great palaeontological importance, combined with some potential for re-excavation, give the site substantial conservation value.

TYTHERINGTON QUARRY, AVON (ST 660890)

Highlights

Tytherington Quarry has produced abundant and varied Late Triassic reptiles from fissures in the background Carboniferous limestones. This varied fauna includes small lizard-like animals as well as abundant bones of the dinosaur *Thecodontosaurus*.

Introduction

The fissure infillings and their enclosed fauna are found in the new quarry, centred on ST 660890 (Figure 4.26). The quarry company (Amey Roadstone Corporation Ltd) work the Black Rock Limestone and the Black Dolomite of the Carboniferous Limestone which dips in a south-easterly direction at about 20–30°. The fissures in the limestone have yielded a large fauna dominated by lepidosaurs and the dinosaur *Thecodontosaurus* that are dated as Rhaetian on the basis of a contained palynomorph assemblage. Realistically, the only sites that can be preserved are those of fissures which occur at the top level and at the edge of the current quarrying operations, and a palynomorph-bearing fissure on the second level. These fissures are very near the road leading to Tytherington village and it is therefore improbable that any further quarrying will take place on this face. Tytherington is a working quarry and this allows new fissures to be revealed continually, greatly adding to its potential.

The first fossils, discovered in 1975, were the postcranial bones of the prosauropod dinosaur *Thecodontosaurus*. These remains, found by two amateur geologists, Mike Curtis and Tom Ralph, were preserved in a breccia composed of clasts of limestone and dolomitized limestone of Carboniferous age set in a sandy clay matrix. The bulk of the fossil-bearing material (about 10 tonnes) was transported to the University of Bristol, where Whiteside (1983) studied the fauna, and published a description of the cranial skeleton of the most abundant sphenodontid, *Diphydontosaurus avonis* (Whiteside, 1986).

Description

The fissures exhibit a variable morphology; some are aligned vertically on joints, whereas others appear to be true caves which usually follow joints, but also cut unjointed sections of the massive crinoidal limestone and principally follow the dip. The solution fissures can be divided into two types, those formed above (vadose) or below (phreatic) the water table:

Tytherington Quarry, Avon

Figure 4.26 Tytherington Quarry: view taken in 1981. Fissures containing Triassic sediment occur in the upper levels. (Photo: R.J.G. Savage.)

1. vertical, with sub-parallel walls (vadose);
2. those with a circular cross-section (phreatic).

The sub-parallel fissures probably represent palaeo-dolines (e.g. fissure 1), and are filled with finely laminated calcareous clays and sandstones, some showing cross-bedding and some being ripple-marked, with mud cracks and water droplet impressions. Other doline-like features have an infilling of breccia composed of Carboniferous Limestone clasts in a red sandy matrix and resemble the marginal facies of the local Triassic (the Dolomitic Conglomerate); this can be seen, for example, in fissure 3. Fissure 2 appears to be a phreatic cave formed in times of very high water table, when the sea level was approximately 100 m higher than today. This fissure exhibits long, horizontal solution features and large flute marks formed on the Carboniferous Limestone wallrock. Of the other fissures currently exposed, no. 7 exhibits repeated fining-upward cycles of conglomeratic, sandy Westbury Formation facies with clasts of black shales. This sequence is best exhib-

ited on the southern side; the middle of the fissure is cut by a hydrothermal vein of baryte, galena and sphalerite. The infilling on the northern side of the hydrothermal vein is of contorted Westbury Beds sands and conglomerates which have been deformed, probably as a result of the fall of Carboniferous Limestone blocks into partly consolidated sediment. The hydrothermal vein has also been broken as a result of this fall. Simms (1990) gave further information on the karst aspects of the fissures.

Nine fossil-bearing sites have been identified. The *Thecodontosaurus*-bearing breccia formed the middle section of the infill of a large cavernous fissure (no. 6b; Whiteside, 1983), 4 m in diameter, situated at the third level of the quarry (ST 66188894). Disarticulated isolated bones of a crocodilomorph and two sphenodontids were also found in a fissure (no. 8) lying to the northwest.

A glauconitic clay, a mineral usually associated with marine conditions, recorded from the now destroyed fissure 6b is apparently unique in that it was found in a quartz-rich conglomerate which contains a predominantly terrestrial reptile fauna.

Fauna

Lepidosauria: Sphenodontida
  Diphydontosaurus avonis Whiteside, 1986
    Type specimen: BRSUG 23760 and abundant material in BRSUG
  Clevosaurus sp.
    Varied material (BRSUG)
  Planocephalosaurus robinsonae Fraser, 1982
    Varied material (BRSUG)
Archosauria: Crocodylomorpha
  Terrestrisuchus sp.
    Varied material (BRSUG)
Archosauria: Dinosauria: Saurischia: Sauropodomorpha
  Thecodontosaurus ?antiquus Riley and Stutchbury, 1840
    Varied material (BRSUG).

Interpretation

Marshall and Whiteside (1980) proposed that at least one fissure (no. 2, see below) at Tytherington was infilled in a marginal marine location, on the basis of an assemblage of Rhaetian marine and terrestrial palynomorphs. Whiteside and Robinson (1983) expanded this model, suggesting that the fissure was infilled in a fluctuating freshwater to saline environment, with evidence based upon the occurrence of a glauconitic clay. Whiteside (1983) demonstrated that some fissures were infilled in a freshwater environment, some in brackish conditions, and others in a mixed freshwater and marine regime.

An assemblage of palynomorphs from fissure no. 1 at Tytherington, consisting of 18 elements including the miospore *Rhaetipollis germanicus* and the dinoflagellate cyst *Rhaetogonyaulax rhaetica*, affords unequivocal evidence for a Rhaetian age (Marshall and Whiteside, 1980). The palynomorphs are not reworked, and the enclosing lithology and associated *Euestheria minuta* and *Pholidophorus* indicate equivalence with the Westbury Formation (Rhaetian) (Whiteside, 1983). Fissures 4, 5, 6a and 7 also have Rhaetian palynomorphs indicating equivalence with the Westbury Formation or Cotham Member. All other fissures contain infills interpreted as Rhaetian, except fissure 3, which may be older. The significance of the Tytherington find is that a

relatively precise date can be assigned to some of the fissure infillings and the presence of the reptiles *Clevosaurus* and *Diphydontosaurus* in the same matrix is a pointer to the age of the fauna as a whole.

The diversity of sphenodontids from Tytherington is only exceeded by that at Slickstones (Cromhall) Quarry, but at least one species of sphenodontid, here named 'C', is unique to the fissure 1 deposit at Tytherington. The sphenodontids *Clevosaurus* and *Planocephalosaurus* are well represented by isolated material, but *Diphydontosaurus avonis* is the best known form from the site where over 100 individuals have been recovered from fissure 6b. The entire skull (except for the auditory capsule) of this species has been reconstructed (Whiteside, 1986). This form was the smallest member of the Tytherington fauna and its numbers suggest that locally it formed high-density populations. It was probably insectivorous and had a unique dentition among sphenodontids in which pleurodont teeth and acrodont teeth occur together in the same jaws: the pleurodont teeth in the premaxilla and on the anterior margins of the dentary and maxilla, the acrodont teeth behind the pleurodont series on the maxilla and dentary. These alternate in size, an autapomorphic sphenodontid character. All other known sphenodontids have an entirely acrodont dentition, but some have successional anterior teeth as a neotenous feature. *Diphydontosaurus* thus appears to be a primitive form, and Fraser and Benton (1989, p. 440) confirmed this position when, on the basis of computer-based cladistic analyses, *Diphydontosaurus* came out as the most primitive known sphenodontid.

The prosauropod *Thecodontosaurus* is represented at Tytherington by numerous postcranial elements, some of which may be partly articulated. The bones of *Thecodontosaurus* are normally well preserved as a hard white substance with all internal structure intact, but they may be broken and abraded through transport. Such fine preservation of bone assigned to *Thecodontosaurus* is not found at any other fissure locality. See the Durdham Down account for more details of this dinosaur.

Clevosaurus hudsoni, Planocephalosaurus robinsonae and sphenodontid 'B' also occur at Slickstones Quarry (Fraser and Walkden, 1983; Fraser, 1988b), and *Clevosaurus* also occurs in the Highcroft Quarry fissure near Gurney Slade and at Pant-y-ffynon Quarries (Crush, 1980). Adult

Planocephalosaurus specimens from the Tytherington fissures appear generally larger than those recovered from Slickstones Quarry, and may represent a more derived later form (Fraser and Walkden, 1983, pp. 359–60, fig. 15). Only one maxillary fragment out of approximately 250 maxillae recovered from Slickstones Quarry is comparable in size to the Tytherington form.

Of the archosaurs, *Thecodontosaurus* has also been recorded from the Durdham Down fissure (Riley and Stutchbury, 1840) and Old Pant-y-ffynon Quarry (Kermack, 1984). The Tytherington material is better preserved than that from Durdham Down, but no cranial elements have so far been identified. The remains of terrestrial crocodilomorphs are better preserved and more numerous at other fissure localities such as Slickstones and Pant-y-ffynon (Crush, 1984).

One important, although rare, member of the Tytherington fauna is a fish whose scales resemble those of *Pholidophorus*, which is found in the Cotham Member (Lilstock Formation, Penarth Group) nearby. The only other non-marine fish found in fissure deposits is *Legnonotus,* a species also recorded from South Wales (M. Howgate, pers. comm.). Tiny reworked teeth and scales of the Rhaetian marine fishes *Gyrolepis, Hybodus* and '*Saurichthys*' have been recorded from a number of fissures at Tytherington, and have also been found at Holwell and Windsor Hill (C. Copp, pers. comm.; Savage and Waldman, 1966; Savage, 1977) and in the covering sediments at Cromhall

(Walkden and Fraser, 1994). The importance of these fish remains is that they indicate a Rhaetian (probably Westbury Formation time equivalent) age and independently confirm the probability of a saline intrusion into the fissure at the time of infilling.

Contemporaneous invertebrates are rare and, apart from the internal moulds of possible Rhaetian gastropods, the only specimens are a few individuals of the branchiopod *Euestheria minuta* var. *brodeiana* known elsewhere from the Cotham Member.

Conclusions

Tytherington provides many unique finds, particularly *Diphydontosaurus* in fissure 1, an admixture of terrestrial reptiles and non-marine fish in fissure 2, and the palynomorph assemblage of fissure 5. The Tytherington fissures as a whole provide the best evidence of infilling of an ancient subterranean cave complex. The solution-marked surface of fissure 2 is an excellent phreatic cave passage. Moreover, Tytherington is the only quarry in which fissure infillings have been dated independently of the vertebrates.

The considerable conservation value of this quarry lies in the combination of the potential for new finds from continued working and the preservation of some fissures that are marginal to the quarrying.

Chapter 5

British Early Jurassic
fossil reptile sites

INTRODUCTION: JURASSIC STRATIGRAPHY AND SEDIMENTARY SETTING

In Britain, rocks of Jurassic age occur in England in a long, almost continuous outcrop, running from Dorset to Yorkshire, also in South Wales and in scattered patches in the islands off north-west Scotland, and in north-east Scotland (Figure 5.1). The Jurassic System is represented by rocks of predominantly shallow marine origin, with mainly fine-grained sediments such as marine shales, clays and mudstones. Shallower facies, marked by greater terrestrial input, include deltaic sequences

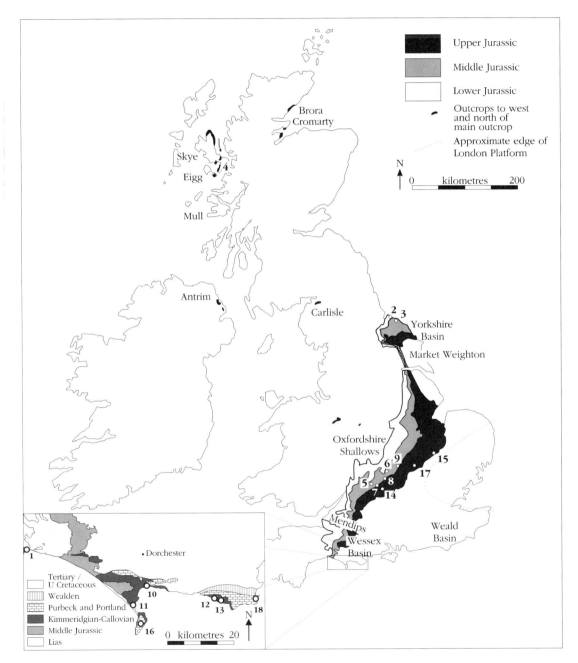

Figure 5.1 Map showing the distribution of Jurassic (Lower, Middle and Upper) rocks in Great Britain. GCR Jurassic reptile sites: (1) Lyme Regis; (2) Whitby; (3) Loftus; (4) Eigg; (5) New Park Quarry; (6) Stonesfield; (7) Huntsman's Quarry; (8) Shipton-on-Cherwell Quarry; (9) Kirtlington Old Cement Works; (10) Furzy Cliff, Overcombe; (11) Smallmouth Sands; (12) Kimmeridge Bay; (13) Encombe Bay; (14) Chawley Brickpits; (15) Roswell Pits, Ely; (16) Isle of Portland; (17) Bugle Pit, Hartwell; (18) Durlston Bay.

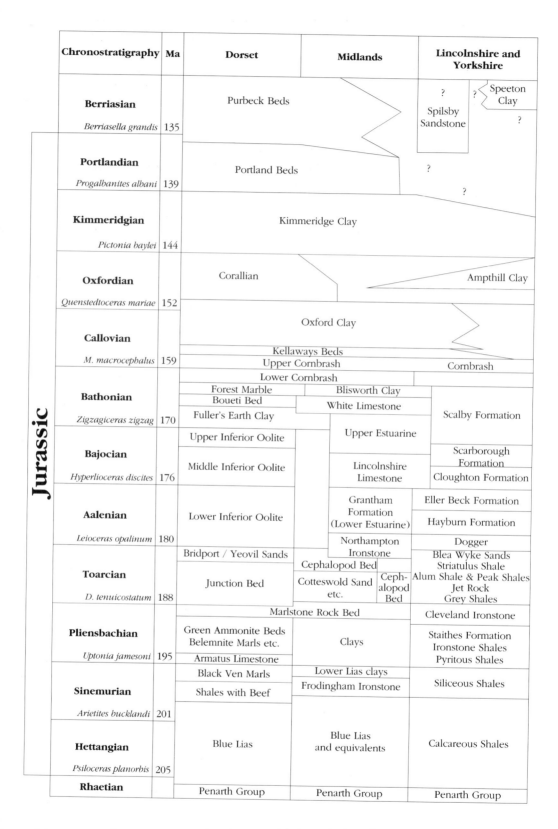

Figure 5.2 Summary of Jurassic stratigraphy, showing global standards and some major British formations (based on Harland *et al.*, 1990).

of clays and sandstones, while those with little terrestrial input include shelf carbonates. The last are characteristic of the Mid Jurassic (Aalenian–Bathonian) and Late Jurassic (Oxfordian, Portlandian) and are commonly oolitic in character.

Rich faunas of ammonites allow precise biostratigraphic correlation for most of the Jurassic (Figure 5.2). Within the Early Jurassic, for example, a time span of about 25 Ma, 20 ammonite zones are recognized, and each is further subdivided into subzones. Where ammonites are scarce or lacking, as is the case in some British Mid and latest Jurassic terrestrial sediments, correlations have been attempted using other fossils, most commonly foraminiferans, ostracods, molluscs, dinoflagellates, spores and pollen, but these give a lower resolution.

The onset of marine conditions in Britain was marked by the Rhaetian transgression in the latest Triassic (documented in the rocks of the Penarth Group) and, by the Early Jurassic, fully marine conditions had become established. A shallow epicontinental sea (or epeiric sea) flooded much of northern Europe, forming a huge shelf sea. The extensive shelf area gave protection from strong tidal or storm influences, and distinctive facies of laminated bituminous shales and rhythmic sequences of lime mud and marl accumulated. Over shallow regions (swells), oolitic ironstones and condensed cephalopod limestones developed in the relative absence of terrigenous input.

Marine conditions continued through much of Jurassic time with two major regressive intervals, one during the Mid Jurassic and one at the close of the Jurassic, when the area of epeiric seas became significantly reduced, eventually giving way to the subaerial facies of the Portland and Purbeck beds.

The lowest units of the British Jurassic (Lias) consist primarily of marine clay–shale facies which in outcrop are more calcareous in the south and more sandy in the north. Two principal shale groups, those of the Lower and Upper Lias, are developed and are separated by the shallower-water facies of sandy shales, sandstones and oolitic ironstones of the Middle Lias. Fine-grained yellow sands of Upper Lias and earliest Bajocian age (e.g. Bridport sands), outcrop from Gloucester southwards to the Dorset coast. The biota of the Lias is dominated by a variety of marine benthic forms which indicate rather harsh bottom conditions. However, at times, environmental conditions appear to have deteriorated

further so that only very low-diversity invertebrate fossil faunas occur. The sequences of unbioturbated bituminous laminated shales, characterized by the Jet Rock of Yorkshire, lack even protobranch bivalves, and represent the onset of sterile bottom conditions. Under these conditions, midwater swimmers died and sank and were buried undisturbed; consequently such sequences contain some of the best examples of the marine reptiles of the time.

Bathonian times in the Mid Jurassic were characterized by regressive facies. Fluvio-deltaic deposits were laid down in southern Britain while, in west Scotland, the Great Estuarine Group accumulated under lagoonal conditions. At the same time, lagoonal-marsh and marginal-marine conditions appear to have developed in central England, where characteristic terrigenous deposits are found. The contemporaneous rocks in southern England are rather different, being dominated by marine carbonates with a lesser component of clays (e.g. the Great Oolite and the Fullers Earth), and these appear in the Cotswolds and the south Midlands to represent nearshore deposition, with signs of subaerial exposure. Ammonites there are consequently rare and correlation is difficult.

The succeeding rocks demonstrate a resumption of marine clastic sedimentation following commencement of the second major transgressive phase during the Callovian. The facies are predominantly monotonous, laterally extensive, dark bituminous clays which, in essence, mark a return to restricted muddy marine environments like those of the Early Jurassic (Duff, 1975). In southern Britain, these beds are represented by the Lower Oxford Clay. The deeper-water Kimmeridge Clay (clays, mudstones and shales) is comparable, being rich in preserved organic material (including kerogen), and in containing a restricted marine benthos.

The Portland Group shows evidence of shallowing and renewed regression, and preserves a range of facies. The Cherty Beds are rich in sponge spicules and seem to have been deposited in calm marine water. The upper parts of the succession include shallow-water oolites, micrites and eventually evaporites (represented by halite and anhydrite) and soils, which document the progress of the regression. Marine incursions, including the Cinder Bed 'event', occur in the mid to Late Purbeck beds which are predominantly non-marine, and which span the Jurassic–Cretaceous boundary.

REPTILE EVOLUTION DURING THE JURASSIC

Marine reptile evolution during the Jurassic is essentially the story of the radiations of the ichthyosaurs and the sauropterygians (Benton 1990a, 1990d). Ichthyosaurs had arisen in the Triassic, and they are known abundantly from the Muschelkalk of central Europe and, further afield, from Japan, Spitsbergen and Canada. Late Triassic ichthyosaurs are known from all parts of the world, but they are represented in Britain only in the Rhaetian 'Bone beds'. However, by Early Jurassic times, ichthyosaurs are found abundantly in Britain, and all phases of their evolution may be followed. Ichthyosaurs in the Jurassic were dolphin-like animals that for the most part fed on cephalopods, fishes and other reptiles, judging from the evidence of stomach contents and coprolites, and they show relatively little morphological diversification. Elsewhere, Jurassic ichthyosaurs are well known from the Early Jurassic of Germany (especially from Holzmaden in Baden-Württemberg) and from the Late Jurassic of Germany and France, but they are rare elsewhere.

Plesiosaur evolution in the Jurassic is also well documented in Britain (Brown, 1981), with complete specimens known from the marine formations. An apparent split into long-necked plesiosauroids and short-necked pliosauroids may be traced back into the British Lias, and comparative materials are known only from Germany.

Other marine niches were occupied during Jurassic times by pleurosaurs, relatives of the sphenodontids (lepidosaurs), which are known mainly from Germany, with no British representatives, and by crocodiles. The steneosaurs and teleosaurs of the British Early and Mid Jurassic are excellently preserved and compare very well with the Early Jurassic German and the Mid Jurassic French material respectively. These crocodiles were slender gavial-like fish-eating animals with long slender snouts and evidently marine habits. The Late Jurassic metriorhynchids (geosaurs), known from the Late Jurassic of Britain and Germany, were even more aquatically adapted, having fully formed paddles for limbs and a tail fin.

On land, the Jurassic Period heralded the rise of the dinosaurs which came to dominate all terrestrial tetrapod faunas (Benton 1989, 1990a, 1990d). Early Jurassic faunas worldwide were still dominated by Triassic hold-over groups, such as the prosauropods and the basal ornithischians.

However, new groups appeared in the Mid Jurassic, such as the sauropods, large theropods, avialan theropods (bird-relatives), stegosaurs and ankylosaurs. During the Late Jurassic, the huge sauropods dominated as top herbivores and the theropods occupied a range of niches as carnivores. Among the ornithischian dinosaurs, the thyreophorans (armoured dinosaurs) were the most important.

Other diapsid reptiles diversified during the Jurassic into a wide range of new forms, and many Triassic groups continued to radiate (Benton 1990a, 1990d). Archosaurs were the most abundant diapsids on land. Apart from the dinosaurs, crocodilians radiated extensively, and a range of carnivorous and piscivorous forms evolved: these are represented in Britain only in the Mid Jurassic. Lepidosaurs such as sphenodontids and squamates (lizards) remained generally small. Choristoderes, an enigmatic diapsid group hitherto known mainly from the Late Cretaceous and Palaeogene, appeared by Mid Jurassic times, and possibly in the Rhaetian (see Aust Cliff report, above).

BRITISH JURASSIC REPTILE SITES

Most fossil reptiles obtained from the Jurassic of Britain are marine forms, but these are supplemented by important terrestrial reptiles (dinosaurs and others) collected from the subaerial facies of the Mid Jurassic (e.g. Forest Marble, Stonesfield Slate), but also from all the representative marine units (Figures 5.1 and 5.2). The most spectacular remains, including those of plesiosaurs, ichthyosaurs and marine crocodilians, derive from bituminous shale units, and important collections have come from the Early and Late Lias (Hettangian–Sinemurian), the Oxford Clay (Callovian) and the Kimmeridge Clay (Kimmeridgian). These remains are commonly complete, or nearly complete, articulated skeletons, the result of their original deposition on undisturbed stagnant bottom waters unique to the northern European Jurassic shelf sea. The marine reptiles from the Oxford Clay (Callovian) are uniquely well preserved and form a centre-point of all international taxonomic studies.

Well articulated Early Jurassic (Hettangian-Sinemurian) plesiosaurs and ichthyosaurs from Lyme Regis are unique, including forms apparently intermediate between the long-necked elasmosaurids and the shorter-necked pliosauroids typical of the later Jurassic and Cretaceous. Plesiosauroids and pliosauroids from the Mid and Late Jurassic are also well represented in British Jurassic rocks. The ichthyosaurs from Britain are among the best in the world and contain many unusual forms marking the wide diversity of a group otherwise adapted for fast marine locomotion (e.g. the swordfish-like forms *Eurhinosaurus* and *Excalibosaurus*).

British Jurassic sites also provide good coverage of terrestrial reptiles, particularly the Early Jurassic dinosaur *Scelidosaurus* from Lyme Regis, the oldest thyreophoran and the unique Mid Jurassic dinosaurs: these are matched only in China. The Mid Jurassic sauropods (*Cetiosaurus*), theropods (*Megalosaurus*), thyreophorans (*Lexovisaurus, Dacentrurus*) and other less well-known forms fill an important gap in terrestrial records of Europe and North America.

The Mid Jurassic sites of central England and north-west Scotland contain the earliest members of several groups including choristoderes (unless the Rhaetian *Pachystropheus* is a choristodere; see Aust Cliff report), possible squamates, as well as some of the youngest known mammal-like reptiles (tritylodontids). British Jurassic squamates are particularly important, with the oldest in the world having been recognized recently in the Middle Jurassic rocks of the Cotswolds (Evans and Milner, 1991, 1994). In addition, the fauna of Late Jurassic/Early Cretaceous lizards from the Purbeck of Dorset is the most diverse of this age in the world. Comparable forms are known from Portugal, Germany and North America (Estes, 1983). Jurassic sphenodontids are less well represented in Britain than in Germany. Turtles are also reported from the British Mid and Late Jurassic, and the latter are important (especially those from Portland) as some of the best preserved of their age. Comparable material is known from the Late Jurassic of Switzerland and North America.

EARLY JURASSIC

The Early Jurassic (Lias) of Britain is famous for its faunas of marine reptiles. Hundreds of good specimens have been obtained from localities along the entire length of the outcrop which stretches in a continuous belt between Dorset and the Yorkshire coast. Sites, other than Lyme Regis, that have yielded Early Jurassic reptiles are listed below. The listings are based on material in BATGM, BMNH, BRSMG, CAMSM, LEICS, OUM,

SDM, YORYM, and Hawkins (1840), Woodward and Sherborn (1890), Fox-Strangways (1892), H.B. Woodward (1893, 1894, 1895), Arkell (1947a), Delair (1958, 1959, 1960, 1968, 1973) and Macfadyen (1970). Reptile-bearing fissures of Early Jurassic age from the areas of Bristol and South Wales are listed in the Triassic chapter. Note that the use of the names *Ichthyosaurus* and *Plesiosaurus* is based on old documentation: all specimens require revision.

Lower Lias

The British Lower Lias has yielded remains of ichthyosaurs and plesiosaurs from dozens of localities from Dorset to Yorkshire. Many of these finds are only isolated bones, so that the majority of sites may be classed as not significant. Other reptiles include two dinosaurs, a possible sphenodontid and a pterosaur. Ichthyosaurs and plesiosaurs have been collected from at least 40 localities in the Lower Lias of England, along the entire length of its outcrop from Dorset to the Yorkshire coast. Abundant remains have come from the quarries around Street, Somerset (ST 4836) and Barrow-upon-Soar, Leicestershire (SK 5818), but there is very little chance of more finds unless excavations are resumed. All other sites have produced only sparse remains and those that still offer exposure can be said to have only low potential for future finds. These other sites, listed by county from the south-west to the north-east, are:

DEVON: Axminster, Tolcis Quarry (SY 280010; *Ichthyosaurus*, shale between half foot and foot limestone).

SOMERSET: Street – 18 or more quarries (ST 4836; *planorbis* Zone; Thomas Hawkins' 'Sea-Dragons'; two species of ichthyosaur, including neotype of *Leptopterygius tenuirostris* (McGowan, 1989a), type of *Protichthyosaurus protaxalis* and five species of plesiosaur, including types of *Plesiosaurus arcuatus*, *P. eleutheraxon* and *P. hawkinsi*); Street on the Fosse, south-east of Glastonbury (type of *Plesiosaurus megacephalus*); Walton, near Street (ST 4636; *Ichthyosaurus, Plesiosaurus*); Somerton, near Street (ST 4828; *Plesiosaurus*); Glastonbury (ST 5039; *Ichthyosaurus*); West Pennard (ST 5438; *Ichthyosaurus*); Keinton Mandeville (ST 5530; *Ichthyosaurus*, from *planorbis* Zone); Watchet (ST 0743; *Ichthyosaurus, Plesiosaurus* from Blue Lias on shore); Kilve, St Audrie's Bay (ST 144447; *Ichthyosaurus*; Deeming *et al.*, 1993; Lilstock foreshore (ST 196463; *Excalibosaurus*; McGowan, 1986).

SOUTH GLAMORGAN: Penarth (ST 1871; *Plesiosaurus*).

GWENT: Sedbury Cliff (ST 559930; possible sphenodontid (M.J. Simms, pers. comm.).

AVON: Bath (ST 4765; *Ichthyosaurus, Plesiosaurus*); Weston, near Bath (ST 7267; *Ichthyosaurus, Plesiosaurus*); Saltford, near Bath (ST 6867; *Ichthyosaurus* from railway cutting; donated to BRSMG by Brunel); Keynsham (ST 6568; *Ichthyosaurus, Plesiosaurus* near station); Bitton, Keynsham (ST 6869; *Plesiosaurus*); Nempnett (ST 5360; *Ichthyosaurus*); Barrow Gurney (ST 5367; *Plesiosaurus*); Banwell (ST 5959; *Ichthyosaurus*); Willsbridge, near Bitton (ST 6670; *Ichthyosaurus, Plesiosaurus*); Westfield, Radstock (ST 6854; *Ichthyosaurus*); Stoke Gifford (ST 6279; *Ichthyosaurus*); Bristol (exact locality uncertain; *Ichthyosaurus, Plesiosaurus*); Ashley Hill, Bristol (ST 6069; *Plesiosaurus*); Hengrove, Bristol (ST 6069; *Plesiosaurus*).

GLOUCESTERSHIRE: Gloucester (SO 8518; *Ichthyosaurus, Plesiosaurus*); Cheltenham: Battledown Brickworks (SO 967225; *Plesiosaurus* from *ibex* Zone); Hock Cliff, Saul (SP 7310; *Ichthyosaurus, Plesiosaurus*); Stenehouse, Strand (SO 8005; *Plesiosaurus*); Westbury-on-Severn (SO 8505; *Ichthyosaurus, Plesiosaurus*); Eastington (SO 7705; *Ichthyosaurus* from *bucklandi* Zone); Tewkesbury (various localities: Woolbridge (SO 8023), Brockridge Common (SO 8938), Hill Croome (SO 8940), Defford Common (SO 9043): *Ichthyosaurus, Plesiosaurus* in the 'saurian beds' (J. Buckman)); Bredon (SO 9237; *Ichthyosaurus* from *semicostatum/obtusum* Zone); Blockley (SP 1635; *Plesiosaurus*).

HEREFORD AND WORCESTER: Bengeworth (SO 9443; *Ichthyosaurus*); Himbleton (SO 9458; *Ichthyosaurus, Plesiosaurus*); Grafton (SO 9837; *Ichthyosaurus*); Bickmarsh (SP 1049; *Ichthyosaurus*); Honeybourne (SP 1143; *Ichthyosaurus, Plesiosaurus* from *turneri* Zone).

WARWICKSHIRE: Stratford-upon-Avon (exact locality uncertain, around SP 1559; *Ichthyosaurus*); Wilmcote (SP 168583; *Ichthyosaurus, Plesiosaurus*, ?exact locality; *Megalosaurus tibia–angulata* Zone, near railway station (Woodward, 1908b)); Harbury, Portland Cement Co. Quarry (SP 3959; *Ichthyosaurus, Plesiosaurus*, type of *Macroplata tenuiceps* (plesiosaur) from *angulata* Zone (Swinton, 1930)); Temple Graften Quarry (SP 121539; 'reptiles'); Shipston-on-Stour (SP 2540; ichthyosaur, dinosaur); Southern (SP 4161; *Ichthyosaurus, Plesiosaurus*); Little Lawford (SP 4677; *Ichthyosaurus*); Rugby, Victoria Quarry (SP 4976; *Ichthyosaurus, Plesiosaurus*); Newbold (SP 4977; *Ichthyosaurus, Plesiosaurus*); Stockton, Nelson's Quarry and others (SP 4363; *Ichthyosaurus, Plesiosaurus*).

LEICESTERSHIRE: Barrow-upon-Soar, quarries around and in the town (SK 595163, 598161, and many others; *Ichthyosaurus, Stenopterygius, Temnodontosaurus, 'Plesiosaurus', Rhomaleosaurus* and type of the megalosaur *Sarcosaurus woodi* (Andrews, 1921a; Martin *et al.*, 1986)); Normanton Hills (SK 539245; ichthyosaur remains; LEICS).

LINCOLNSHIRE: Long Bennington (SK 8445; *Plesiosaurus*).

NOTTINGHAMSHIRE: Elston (SK 7748; ?*Plesiosaurus*); Barnstone Quarry (SK 736356; *Ichthyosaurus, Plesiosaurus* from bed 3S in the pre-*planorbis* beds); Cropwell Bishop, near Barnstone (?pliosauroid).

NORTH YORKSHIRE: Robin Hood's Bay (NZ 9604; *Ichthyosaurus, 'Teleosaurus'*, plesiosaurs, bed 18 of *bucklandi* Zone).

Middle Lias

Ichthyosaurus has been reported from the Middle Lias of Golden Cap, near Charmouth, in Dorset; and from Ilminster and Dundas, in Somerset, but remains are so poor that the sites are not worth tracing. A recent find from the Middle Lias of Dorset probably comes from the Eype Clay at Thorncombe Beacon (SY 436912; Ensom, 1989b). Three other sites include Houston Quarry, Ilminster, Somerset (?ST 362153; *Stenopterygius hauffianus*, in upper *margaritatus* Zone

(McGowan, 1978)); Wotton-under-Edge, Avon (ST 7593; *Ichthyosaurus* in Middle Lias); Bugbrooke, Northamptonshire (SP 6757; *Ichthyosaurus, Plesiosaurus*); Isle of Raasay, Inner Hebrides (Scalpa Sandstone Formation; articulated plesiosaur remains; Martill 1985a).

Upper Lias

A few sites in the Upper Lias of Somerset, Northamptonshire and North Yorkshire have yielded good specimens of ichthyosaurs, plesiosaurs, marine crocodiles and one pterosaur.

SOMERSET: Strawberry Bank, Ilminster (ST 361148; 30 specimens of the marine crocodile *Pelagosaurus typus* from the 'Fish and Saurian Bed' (*exaratum* Zone, *falciferum* Subzone), quarry now filled (Duffin, 1979b)).

NORTHAMPTONSHIRE: Bugbrook(e) (SP 6757; *Ichthyosaurus, Rhomaleosaurus*; Middle-Upper Lias); Greens Norton (?SP 664492; *Steneosaurus*); Blisworth (SP 7253; *Ichthyosaurus*); Crick (SP 5872; *Thaumatosaurus*, from railway cutting); Market Harborough bypass, near Dingley (SP 753882; ichthyosaur; LEICS); Kingsthorpe, Northampton (SP 7662; *Ichthyosaurus, Thaumatosaurus, Steneosaurus*, type of *Rhomaleosaurus thorntoni* Andrews, 122 (BMNH R4853) – quarries at SP 758643 and SP 765653); Wellingborough (SP 8969; *Microcleidus* in *bifrons* Zone).

LEICESTERSHIRE: Rutland Water Dam excavations (SK 9307; ichthyosaurs in LEICS).

LINCOLNSHIRE: Stibbington (TL 092991; *Ichthyosaurus, Plesiosaurus, Steneosaurus* in *thouarsense* and *bifrons* Zones).

NORTH YORKSHIRE: Kettleness alum-works (NZ 8316; types of *Thaumatosaurus cramptoni* and *Plesiosaurus propinquus*, and *Stenopterygyius, Steneosaurus* in *communis* Zone); Saltburn (NZ 6621; *Plesiosaurus* in *capricornis* Beds); Staithes–Runswick Bay coast section (NZ 7919-NZ 8116, including Port Mulgrave; *Steneosaurus* (Walkden *et al.*, 1987)).

Three Lias localities, one from Dorset and two from Yorkshire, are selected for protection as GCR sites for their unusually prolific faunas of marine reptiles, as well as important terrestrial reptiles including pterosaurs and dinosaurs, some of which are not known outside Britain:

1. Lyme Regis coast (Pinhay Bay-Charmouth), Dorset (SY 3291-SY 3793). Early Jurassic (Hettangian-Pliensbachian), Lower Lias (*Ostrea* Beds-Green Ammonite Beds).
2. Whitby Coast (East Pier-Whitestone Point), Yorkshire (NZ 901115-NZ 928104). Early Jurassic (Toarcian), Upper Lias (Grey Shales Formation, Jet Rock Formation, Alum Shale Formation).
3. Loftus, Yorkshire (NZ 736200-NZ 757193). Early Jurassic (Toarcian), Upper Lias (Grey Shales Formation, Jet Rock Formation, Alum Shale Formation).

LYME REGIS (PINHAY BAY–CHARMOUTH), DORSET (SY 3291–SY 3793)

Highlights

Lyme Regis is the most famous British Early Jurassic marine reptile site, and one of the best in the world. For over 200 years abundant skeletons of ichthyosaurs and plesiosaurs have been found in the cliffs near the town, and the value of the site is enhanced by additional finds of rare terrestrial animals, such as the pterosaur *Dimorphodon* and the armoured dinosaur *Scelidosaurus*.

Introduction

The Lias exposures on the coast around Lyme Regis, Dorset (Figures 5.2 and 5.3A, B), are world-famous for their fossil reptiles. Specimens have been collected since at least 1790 (Delair, 1969a), and from 1810 to 1840 the younger Mary Anning found many fine ichthyosaurs and plesiosaurs. These were offered for sale and formed the basis of the earliest detailed descriptions of Mesozoic fossil reptiles (Home, 1814, 1816, 1818, 1819a, 1819b; De la Beche, 1820; De la Beche and Conybeare, 1821; Conybeare, 1822, 1824). Since then many hundreds of specimens, including pterosaurs, have been collected and finds are still being made.

Lyme Regis is historically important as the place where the first unarguably complete skeletons of ichthyosaurs and plesiosaurs were found which, because of the collecting and selling efforts of

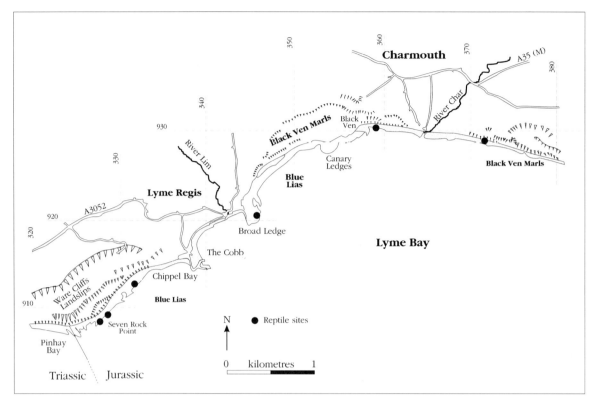

Figure 5.3 The reptile-bearing Lower Jurassic of Lyme Regis. (A) Map of the coastal section from Pinhay Bay to Charmouth, showing the major units, and indicating areas that have yielded fossil reptiles in the past.

Mary Anning, formed the basis for the study of Mesozoic marine reptiles during most of the 19th century (Taylor and Torrens, 1987).

Description

There are numerous detailed accounts of the stratigraphy of the Lyme Regis section (e.g. Lang, 1914, 1924, 1932; Lang *et al.*, 1923, 1928; Lang and Spath, 1926; Palmer, 1972). The general succession (Getty, *in* Cope *et al.*, 1980a) is:

	Lang's Bed Numbers	Thickness (m)
----- unconformity -----		
Green Ammonite Beds	122–130	32
Belemnite Stone	121	0.15
Belemnite Marls	106–120	23
Armatus Limestone	1050.4	
Black Ven Marls	76–104	43
Shales with Beef Beds	54–75	23
Blue Lias	25–53	27
Ostrea Beds	1–24	2.5
(=pre-*planorbis* Beds)		

The Blue Lias is a sequence of laterally extensive, alternating thin-bedded (and nodular) limestones and shales exposed in cliffs and on the foreshore west of the Cobb, and in Church Cliffs, just east of Lyme Regis (Figure 5.4). Large ammonites and bivalves are abundant in certain limestone beds. The Shales with Beef Beds between Lyme Regis and Charmouth consist of thin papery shales, marls and limestone nodule beds with much fibrous calcite ('beef'), pyrite and selenite. Fossils include ammonites, poorly preserved bivalves, belemnites and fishes. The Black Ven Marls, in the cliff and foreshore west and east of Charmouth, consist of blue-black mudstones and paper shales with occasional limestones. Many species of ammonites, bivalves, brachiopods, foraminifers and insects occur. Deposition of all units was marine and, although not marginal, was probably close to shore because of the presence of insect, plant and dinosaur remains.

Reptiles have been collected from the *bucklandi* Zone (McGowan, 1989a, p. 424), the Saurian Shales at the top of the Blue Lias (Lang's Bed 52: *scipionianum* Subzone, *semicostatum* Zone, Early Sinemurian), from the Shales with

Lyme Regis (Pinhay Bay–Charmouth)

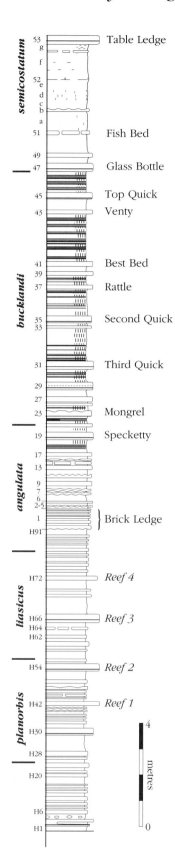

Formation	Lang Bed numbers	Stage	System
Black Ven Marls	90-104 gap 76-89	Sinemurian	Jurassic
Shales with Beef	54-75		
Blue Lias	1-53		
	H25-84	Hettangian	
Ostrea Beds	H1-24	Rhaetian	Triassic
Penarth Group			

Figure 5.3B The reptile-bearing Lower Jurassic of Lyme Regis. The rock succession of the Blue Lias, based on the work of W.D. Lang. From House (1990).

Figure 5.4 The Lower Lias sediments at Lyme Regis: view of the succession below Ware Cliffs, showing interbedded limestones and shales. (Photo: G.W. Storrs.)

Beef Beds (*semicostatum-turneri* Zones, Early Sinemurian) (Macfadyen, 1970, p. 97), Bed 85 of the Black Ven Marls (McGowan, 1993), and rarely from the 'Obtusum Shale' of the Black Ven Marls (*obtusum* Zone; Late Sinemurian) (Delair, 1960, p. 75; Martill, 1991) and the lower Belemnite Marls (Ensom, 1987a, 1989a). A partial ichthyosaur in the Philpot Museum, Lyme Regis, from Charton Bay apparently came from the pre-*planorbis* Beds, well below the usual reptile-bearing beds (Taylor, 1986, p. 312).

Specific localities include the eastern end of Pinhay Bay (Seven Rock Point) where the Saurian Shales crop out twice (SY 32629277 and SY 32779285: Lang, 1924; Pollard, 1968; McGowan, 1989a, p. 424), Devonshire Point (SY 332913: Delair, 1966, p. 62), Broad Ledge, Church Cliffs (which used to be quarried; SY 346921; McGowan, 1974b, p. 20), Black Ven Rocks (SY 358930; Delair 1960, p. 75; SY 360931; Ensom, 1989a), Stonebarrow (SY 370929; McGowan, 1993), Seaton (SY 371917; Ensom, 1987a), Stonebarrow Beach (SY 372928: Delair, 1960, p. 75), and further west (SY 376927). Recent collect-

ing has focused mainly on the Charmouth end of the section, and the Charmouth ichthyosaur (BRSMG) apparently came from the same horizon, as did Owen's original *Scelidosaurus* specimen, as well as the more recent discoveries of the latter taxon (M.A. Taylor, pers. comm., 1993).

The reptile remains generally occur in the darker shale interbeds, and they may be associated with ammonites and bivalves. The skeletons, usually extremely well articulated, stand out clearly in the soft dark shale, but are rapidly broken up by wave action. Some skeletons have been obtained from impure limestone beds (Sollas, 1881). Fossilized skin of the dinosaur *Scelidosaurus* has been preserved, showing scales and internal structure, in the Black Ven Marls (Martill, 1988), and the marine reptiles may show stomach contents within the rib cage region (e.g. Pollard, 1968).

Fauna

Delair (1958–60) reviewed the fossil reptiles of Dorset and gave an extended list of 21 species and three forms ascribed only to genera from the

Lower Lias. However, this list should be much reduced to give a truer impression of the diversity of the reptiles. Ichthyosaur taxonomy is based on McGowan (1974a, 1974b), who reduced about 50 species to seven. Delair (1986) lists a number of additional ichthyosaur specimens. The plesiosaurs have not been revised recently, but the list given here is also reduced from 40–50 species. The estimates of numbers of specimens are based on collections in the BMNH, BGS(GSM) and OUM. They are intended to give an impression of the relative abundance of each species.

	Numbers
Sauropterygia: Plesiosauria	
Plesiosaurus conybeari Sollas, 1881	5
Type: BRSMG Cb 2479	
Plesiosaurus dolichodeirus Conybeare, 1824	20
Type: BMNH 22656	
Plesiosaurus eleutheraxon Seeley, 1865	3
Types: BMNH 39851, R227	
Plesiosaurus (?)hawkinsi Owen, 1840	1
Plesiosaurus macrocephalus Buckland, 1837	10
Type: BMNH R1336	
Plesiosaurus rostratus Owen, 1865	8
Type: BMNH 38525	
Eurycleidus arcuatus (Owen, 1840)	3
Plesiosaurus sp.	*c.* 100
Ichthyopterygia: Ichthyosauridae	
Ichthyosaurus breviceps Owen, 1881	7
Type: BMNH 43006	
Ichthyosaurus communis Conybeare, 1822	45
Neotype: BMNH R1162	
Ichthyosaurus conybeari Lydekker, 1888	2
Type: BMNH 38523	
Leptopterygius tenuirostris (Conybeare, 1822)	9
Leptopterygius solei McGowan, 1993	1
Holotype: MRSMG Ce 9856	
Temnodontosaurus eurycephalus McGowan, 1974	1
Type: BMNH R1157	
Temnodontosaurus platyodon (Conybeare, 1822)	10
Type: BMNH 2003	
Temnodontosaurus risor McGowan, 1974	3
Type: BMNH 43971	
Ichthyosaurus sp.	*c.* 300

Archosauria: Pterosauria: 'Rhamphorhynchoidea'	
Dimorphodon macronyx Owen, 1859	50
Type: BMNH R1034	
'rhamphorhynchoid'	1
Archosauria: Dinosauria: Saurischia: Theropoda	
?megalosaurid	2
Archosauria: Dinosauria: Ornithischia: Thyreophora	
Scelidosaurus harrisoni Owen, 1863	3
Type: BMNH R1111	

Interpretation

About 100 'new species' were described from Lyme Regis in the 19th century, when every specimen was given a name. According to our present taxonomic list, Lyme Regis has yielded type specimens of 14 species, and nine of these species only occur at Lyme Regis (*Plesiosaurus conybeari*, *P. rostratus*; *Ichthyosaurus breviceps*; *Leptopterygius solei*; *Temnodontosaurus eurycephalus*, *T. platyodon*, *T. risor*; *Scelidosaurus harrisoni* and *Dimorphodon macronyx*).

The plesiosaurs from the Lower Lias of England are the earliest well-preserved specimens known (Figure 5.5B). Specimens of comparable age consist of a few poorly preserved remains from the Schwarzjura alpha and beta of Germany. In all, only about 10 species of Lower Lias plesiosaurs are known, and the Lyme Regis material is the most abundant and varied in the world. The species of plesiosaurs are identified on characters of the pelvis and limbs, and on the relative length of the neck and size of the head. The Lyme Regis animals show a range of neck lengths from rather short (*P. rostratus*) to rather long (*P. conybeari*) and these foreshadow the later pliosaurs and elasmosaurids, respectively. The animals vary from about 2 to 6 m in total length, and the relative size of the skull and shape of the jaw indicates diets of cephalopods, fishes and marine reptiles.

The ichthyosaurs likewise are the earliest good specimens and the most abundant and well preserved from the Lias (Figure 5.5A). Material from Lyme Regis has formed the basis of recent revisions of ichthyosaur relationships and evolution (McGowan, 1973a, 1973b, 1974a, 1974b, 1989a, 1989b). Ichthyosaurs have been classified on characters of the skull and forefin, and on this basis at least eight of the Lyme Regis taxa are presently regarded as valid. The Lyme Regis species vary in

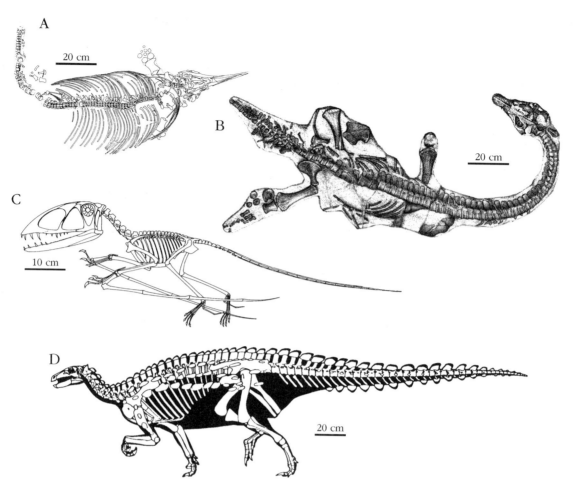

Figure 5.5 Typical reptiles from the Lyme Regis section. Skeletons of (A) *Ichthyosaurus*; (B) *Plesiosaurus*; (C) *Dimorphodon*; (D) *Scelidosaurus*. (A) and (B) from various sources; (C) from Padian (1984); (D) from Coombs *et al.* (1990).

length from 0.8 m to 9 m and they clearly fed on a wide range of sizes of fishes and invertebrates, as indicated by studies of coprolites (Buckland, 1829a) and stomach contents (Pollard, 1968). Several species of *Ichthyosaurus* were common enough for studies of growth series to be carried out in *I. communis* and *I. breviceps* (McGowan, 1973b).

I. communis is the most abundant species of ichthyosaur found at Lyme Regis, accounting for about half of the determinate skeletons. It was a moderate-sized form, reaching a maximum total length (measured from the tip of the snout to the tip of the tail) of about 2.5 m (McGowan, 1974b). The ichthyosaur *I. breviceps* is characterized by having a short snout, whereas *Leptopterygius tenuirostris* and *I. conybeari* have longer and more slender snouts. Although *L. tenuirostris* is much less common in terms of complete skele-

tons, it is abundantly represented by isolated remains of humeri, partial fins and rostral segments. This form is somewhat longer than *I. communis*, reaching lengths in excess of 2.5 m, while *L. solei* was over 7 m long (McGowan, 1993). The larger species of *Temnodontosaurus* are rarer. *T. eurycephalus* has a short snout and massive skull and it may have fed on other ichthyosaurs. *T. platyodon* is the second largest ichthyosaur of all time (length up to 9 m), and it occurs only at Lyme Regis. The species *T. risor* has a curved jaw-line (hence the name), but may represent immature *T. platyodon* (C. McGowan, pers. comm., 1993).

Dimorphodon is one of the oldest known pterosaurs, and it is represented by much skull and skeletal material (Figure 5.5C). Its anatomy is well known (Buckland, 1829b; Owen, 1870, 1874a; Wellnhofer, 1978, p. 33; Padian, 1983;

Unwin, 1988b). The skull is relatively large and high-vaulted, rather than long and pointed as in later pterosaurs. The limbs and girdles are strongly built. All of these features are primitive and *Dimorphodon* provides unique information on early pterosaur evolution. It appears to be a relative of the Late Triassic *Peteinosaurus* from Italy (Unwin, 1991).

The dinosaur *Scelidosaurus* (Figure 5.5D) is represented by one skull and skeleton, a juvenile collected recently (BRSMG) and other isolated remains (BMNH, DORCM, Philpot Museum, Lyme Regis; Ensom, 1987a, 1989a). It is the oldest known armoured ornithischian dinosaur. Its taxonomic position is uncertain, and it has been variously ascribed to the Stegosauria, Ankylosauria and the Ornithopoda (Owen, 1861a, 1863b; Newman, 1968; Rixon, 1968; Charig, 1979; Galton, 1975; Thulborn, 1977; Norman, 1985). Recent cladistic analyses define *Scelidosaurus* as the sister group of the Ankylosauria and Stegosauria, and these taxa together comprise the Thyreophora (Norman, 1984; Sereno, 1986). Coombs *et al.* (1990) identify a motley assemblage of basal thyreophorans, including *Scelidosaurus* and *Scutellosaurus* from the Hettangian of North America, as well as other poorly represented taxa. The type skeleton is fairly complete and shows a 4 m long animal with a small skull, strong hind limbs and dermal armour. The recently found juvenile specimen preserves the forelimbs and most elements of the skull and lower jaws, including some skin (Martill, 1988), thus complementing the previously known remains, and permitting an almost complete reconstruction of the skeleton (Norman, 1985). *Scelidosaurus* is currently of great interest because of its controversial taxonomic position close to the origin of the ornithischian dinosaurs, and a redescription is underway (Charig and Norman, in prep.). Other bones once ascribed to *Scelidosaurus* include limb bones of a ?megalosaurid.

Conclusions

For studies of fossil reptiles, the Lyme Regis coast section is one of the most important sites in Britain. It has yielded many type specimens, the remains are extremely well preserved, it still yields skeletons, and there is no comparable site of the same, earliest Jurassic, age outside Britain. The faunas of ichthyosaurs and plesiosaurs are the most diverse and abundant from the Early Jurassic of the world. The dinosaur *Scelidosaurus* and the pterosaur *Dimorphodon* are unique animals of great interest in studies on the early evolution of their respective groups. Historically, Lyme Regis is unique, its potential for future finds is excellent and so its conservation value is extremely high, even on an international level.

WHITBY–SALTWICK (EAST PIER–WHITESTONE POINT), YORKSHIRE (NZ 901115–NZ 928104)

Highlights

The Whitby coast has produced some of the best Upper Lias fossil reptiles in the world. Specimens of more than 10 species of plesiosaur, marine crocodile and ichthyosaur have been found there, some of them unique to Yorkshire.

Introduction

The Whitby coast section comprises a series of sea cliffs and ledges of Upper Lias mudstones and alum shales which rise from the east of Whitby harbour and extend to Whitestone Point (Figure 5.6A,B). The site is of historic interest in being one of the earliest localities in Britain to be exploited for its fossil reptiles. It has produced many important finds of marine crocodiles, ichthyosaurs and plesiosaurs which form part of a distinct marine fauna, and which are similar to those known from the famous localities at Holzmaden in Germany. The cliffs at Whitby are subject to continuing erosion, and the site has produced many good recent finds.

The wave-cut platform and cliffs east of Whitby harbour have been famous for their marine reptiles since the middle of the 18th century. In 1758, Mr Wooller described 'the fossil skeleton of an animal found in the alum rock . . . buried . . . by the force of the waters of the universal deluge.' In the same year, William Chapman described the same specimen as 'the fossile bones of an allegator', and the figures show that it clearly was a fine specimen of an early crocodile. The first recorded ichthyosaur from the Yorkshire coast was collected in 1819, and another one in 1821 was described by Young (1820). Further crocodiles were collected soon after from the same area in

1824 (Young, 1825; Charlesworth, 1837). Plesiosaur remains had been found by 1822 (Young and Bird, 1822), and the first plesiosaur skeleton was collected before 1842, but described somewhat later (Owen, 1865). Further crocodiles, ichthyosaurs, plesiosaurs and remains of a ?thero-pod dinosaur have been collected and described since then. The history is reviewed in detail by Benton and Taylor (1984).

Description

The stratigraphy of the Upper Lias (Toarcian, Early Jurassic; Figure 5.6C; see also Figure 5.7) of the Yorkshire coast has been described in detail for the sections between Port Mulgrave and Kettleness, Whitby harbour mouth and Whitestone Point, and at Ravenscar (Dean, 1954; Howarth, 1955, 1962, 1973). The general succession at Whitby, summarized by Howarth (*in* Cope *et al.*, 1980a), and with revised nomenclature from Powell (1984), is:

	Thickness (m)
------- unconformity ---------	
Whitby Mudstone Formation	
Alum Shale Member (lower part of *Hildoceras bifrons* Zone)	
Cement Shales	5.8
Main Alum Shales	15.2
Hard Shales	6.3
Jet Rock Member (*Harpoceras falciferum* Zone)	
ovatum Band	0.25
Bituminous Shales	23.0
Jet Rock	7.1
Grey Shales Member (upper and middle parts of *Dactylioceras tenuicostatum* Zone)	13.3
Cleveland Ironstone Formation (upper part)	0.6

The beds are nearly flat-lying in the sections to the east of Whitby (Figure 5.6). The Jet Rock Member occurs in the seaward portions of the wave-cut platforms at Saltwick Nab and Black Nab just to the east of Saltwick Bay. Behind these, the Bituminous Shales, *ovatum* Band and Hard Shales outcrop on the platform. The Main Alum Shales and Cement Shales occur mainly in the lower part of the cliff, and the upper part consists of the Mid Jurassic rocks above the unconformity. The Main Alum Shales and the Cement Shales were formerly quarried for the manufacture of alum at Saltwick Nab and at Black Nab.

The Jet Rock Member is a sequence of well-cemented, finely-laminated, grey or brown shales. The shales are frequently bituminous, and contain bands of small to large calcareous concretions known as 'doggers', up to 5 m in diameter. The shale unit is 1–3 m thick and the concretion bearing horizons vary between 0.1 and 1.0 m in thickness. Typical ammonites belong to the genera *Harpoceras, Hildaites* and *Eleganticeras* in the lower five metres of the Jet Rock Member and the bivalve *Inoceramus dubius* occurs above (Howarth, 1962; Hemingway, 1974).

The Bituminous Shales, like the Jet Rock, contain soft jet, but this is considerably less abundant. Likewise, there are fewer calcareous concretions. The shales are less well laminated than the Jet Rock and contain less bitumen. The shale units are 3–8 m thick, and there are three or four 0.15 m bands of pyrite-coated concretions. The most common ammonites belong to the genus *Harpoceras* and the bivalve *Inoceramus* also occurs. Fossils are often pyritized (Howarth, 1962; Hemingway, 1974).

The *ovatum* Band consists of a 2 m thick bed with two dominating bands of large sideritic doggers, which weather to a dark reddish brown. The ammonite *Ovaticeras ovatum* occurs commonly and belemnites are found in associated aggregations.

The Hard Shales are a non-bituminous grey shale unit characterized by scattered calcareous concretions. A thin bed of siderite mudstones is present. The typical ammonite is *Dactylioceras commune*.

The Main Alum Shales are a sequence of alternating soft, grey, flaggy shales (0.25–5.00 m thick) and irregular bands containing scattered calcareous concretions and sideritic mudstone horizons. The shales typically weather to distinctive brittle flakes (Hemingway, 1974, p. 176). *Dactylioceras commune* is the typical ammonite in the lower 12 m of the unit, *Peronoceras fibulatum* in the upper 3 m, with the latter form occuring in association with *Hildoceras, Phylloceras, Dactylioceras, Zugodactylites, Pseudolioceras* and *Peronoceras* (Howarth, 1962; Hemingway, 1974).

The Cement Shales (0.25–4 m thick) consist of grey shales which contain the ammonite *Hildoceras bifrons* and species of *Porpoceras, Catacoeloceras* and *Phylloceras*. The bivalves *Nuculana* and *Gresslya* and belemnites, occur abundantly (Howarth, 1962). At Whitby this unit is unconformably overlain by the Dogger Formation (Aalenian, Mid Jurassic).

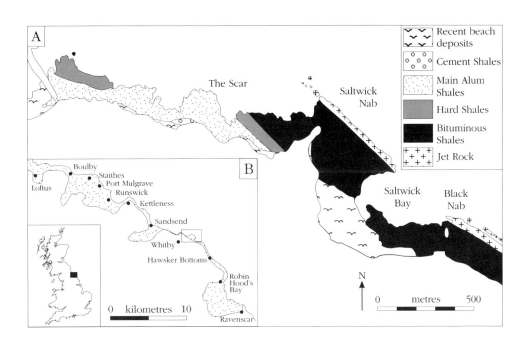

Zones	Subzones	Lithostratigraphy			Howarth's Bed Numbers	Thickness	Tate & Blake (1876)	Buckman (1910, 1915)
Hildoceras bifrons	*Catacoeloceras crassum*	Whitby Mudstone Formation	Alum Shale Member	Cement Shales	65-72	5.8 m	*communis* Zone	*braunianum* Z.
	Peronoceras fibulatum			Main Alum Shales	51-64	15.2 m		*fibulatum* Zone
	Dactylioceras commune							*subcarinatum* Z.
				Hard Shales	49,50	6.3 m		*pseudovatum* Z.
Harpoceras falciferum	*Harpoceras falciferum*		Jet Rock Member	*Ovatum* Band	48	0.25 m	*serpentinus* Zone	*exaratum* Zone
				Bituminous Shales	41-47	23.0 m		
	Harpoceras exaratum			Jet Rock	33-40	7.1 m		
Dactylioceras tenuicostatum	*D. semicelatum* *D. tenuicostatum* *D. clevelandicum* *Protogrammoceras paltum*		Grey Shales Member		1-32	13.3 m	*annulatus* Zone	*tenuicostatum* Zone
		Cleveland Ironstone Formation (part) 0.65 m						

Figure 5.6 The reptile-bearing Lower Jurassic (Toarcian) of Whitby. (A) Map of the Upper Lias (Jet Rock Member and Alum Shale Member) exposed on the foreshore between Whitby Harbour and Saltwick Bay. (B) North-east Yorkshire with fossil reptile localities marked. The coastal outcrop of Lias rocks is stippled and the area shown in (A) is outlined. (C) The Lower Toarcian sequence at Whitby, showing ammonite zones and subzones, formations, bed numbers from Howarth (1962), and thicknesses for sections near Whitby (after Cope *et al.*, 1980a). The terminology used by earlier authors is also indicated. From Benton and Taylor (1984), after Howarth (1962).

The reptiles appear to have been obtained from various horizons, but since most of the material has remained unstudied until recently there has been much confusion over the precise provenances. This difficulty has been brought about by a combination of reasons, but principally through poor collection data and contradictory statements by the early authors. Recent changes in the nomenclature of ammonite zones have created further problems. Benton and Taylor (1984) reviewed the provenance of specimens on the basis of early collectors' reports and on a study of

Figure 5.7 The Upper Lias sediments east of Whitby at Saltwick Nab, showing the fossiliferous rocks on the wave-cut platform. (Photo: C. Little.)

the matrix and ammonites associated with specimens, and a clearer picture of the sources for most of the more important specimens has emerged.

The 'allegator' collected in 1758 (BMNH R1088) was originally described as coming from 'the sea-shore, about half a mile from Whitby. The ground that they lay in is what we call alum-rock, a kind of black slate that may be taken up in flakes. The bones were covered five or six feet with water every full sea' (Chapman, 1758, p. 688). Wooller (1758, p. 790) noted that 'this skeleton lay about six yards from the foot of the cliff, which is about sixty yards in perpendicular height' and that the fossil was found 'about 10 or 12 feet deep in . . . the black slate or alum rock.' Thus, the locality was most probably The Scar, a small promontory in the Alum Shales about 700 m ('half a mile') east of Whitby harbour mouth (NZ 909115). The cliff here is 50–55 m (sixty yards) high, exactly as Wooller (1758) described, and the wave-cut platform is easily accessible from Whitby. Westphal (1962, p. 106), however, contradicted this account and, in following Simpson (1884, p. XI), stated that the skeleton was found in the *Hildoceras serpentinum* Zone (= *H. falciferum* Zone), which occurs 12 m–18 m above the Jet Rock 'Series', therefore in the Bituminous Shales of the Jet Rock Formation. There is some confusion in earlier writings on the Jet Rock

'Series' and the Alum Shales, and where the intervening beds are to be placed, i.e. whether early writers ascribed the Bituminous Shales to the Alum Shales. However, none of the components of the Jet Rock Formation occurs 'half a mile' from Whitby and all the evidence points to an assignment of this crocodile to the Main Alum Shales, contrary to Westphal's statement (Benton and Taylor, 1984).

The crocodile collected in 1824 (WHIMS 770S) was found 'in the face of a steep cliff, not far from the town (Whitby)' (Young, 1825, p. 76), and Westphal (1962, p. 106) stated that it came from an alum pit within the Main Alum Shales. This would restrict the locality to the old alum works at Saltwick Nab (NZ 914112) or at Black Nab (NZ 921107).

The specimen named *Steneosaurus brevior* by Blake (*in* Tate and Blake, 1876, pp. 244–6) came from the old 'Zone of *Ammonites serpentinus*' (= *Hildoceras falciferum* Zone). This places it in the Jet Rock Formation 'immediately below . . . the Alum Shale', according to Westphal (1962, p. 106).

The first Yorkshire ichthyosaur to be reported 'was imbedded in the alum-rock, where it is washed by the tide, and covered at high water, about half a mile east from the entrance of Whitby harbour, and ten yards from the face of the steep cliff. . . . The cliff . . . is about sixty yards in height. . . . The skeleton lay in the upper part of the great aluminous bed, which here descends below high-water mark' (Young, 1820, p. 451). This leaves little doubt that the locality and horizon were the same as for the first 'allegator'. A second, more complete, ichthyosaur skeleton was 'found in the compact shale . . . on the scar' in October, 1821 (Young and Bird, 1828, p. 282). These two specimens apparently came from The Scar (NZ 909115), the source of the first crocodiles and possibly also from the Main Alum Shales there. The specimens have not been traced, but the figures indicate that they may be examples of *Leptopterygius acutirostris* (Owen, 1840a).

Forty or so specimens of ichthyosaurs were collected from the Whitby area from about 1820 (Young and Bird, 1828, pp. 283–6), but most of these were purchased by private collectors and cannot at present be traced. The bulk of these 'were found at or near Saltwick, in the main bed of the alum shale'. The Scar is mentioned again for some of the specimens, but others may have come from excavations in the alum shale cliff at Saltwick Nab (NZ 914112).

The first important plesiosaur was found 'by Mr Marshall of Whitby, imbedded in a hard rock belonging to the upper lias beds, situated between Scarborough and Whitby, near the place where that gentleman had formerly discovered the remains of a crocodile' (Dunn, 1831). If the 'crocodile' is WHIMS 770S, this plesiosaur came from the vicinity of Saltwick Bay, and probably from a nodule in the Alum Shale Formation. It was a partial postcranial skeleton, lacking much of the neck, apparently of a large plesiosauroid with a body about 3 m long. The best documented find of a plesiosaur was an almost complete articulated skeleton of a plesiosauroid about 4.5 m long with a 0.2 m long skull (CAMSM J35182). It was referred to the Lower Lias species *Plesiosaurus dolichodeirus,* or alternatively to the Owen MS species *P. grandipennis* (Phillips, 1853), but was renamed by Seeley (1865a, 1865b) who described it as the type of *P. macropterus.* Watson (1911a) redescribed it as the neotype of *Eretmosaurus* Seeley, 1874, a genus that had been erected on the basis of undiagnosable material. It was found in the early summer of 1841 by Matthew Green and two other jet collectors of Whitby in the Lias cliffs at Saltwick (Browne, 1946, p. 57).

All later finds to be described from Whitby can only be localized on the basis of crude zonal data which are the only clues as to the provenance of specimens provided by the later authors. Blake (*in* Tate and Blake, 1876) listed the following reptiles from the 'Zone of *A. communis*' (i.e. Alum Shale Formation): *Plesiosaurus homalospondylus*, *P. coelospondylus* (from Saltwick Alum Pit; Simpson, 1884, p. 9), *Ichthyosaurus acutirostris* and *I. longirostris*. Blake (*in* Tate and Blake, 1876, pp. 250–2) stated that '*Plesiosaurus*' *longirostris* came from the 'Zone of *A. serpentinus*' (i.e. Jet Rock Formation). White (1940, p. 452) notes the old zonal assignment of *Macroplata (P.) longirostris*, but mistakenly listed the specimen as coming from near the bottom of the Alum Shale.

A few specimens in collections offer some additional information on the typical occurrence of the Whitby reptiles. A recently collected ichthyosaur in the British Museum (BMNH R8309) carries the label 'Bituminous Shale, Black Nab', and a pair of ichthyosaur jaws collected in 1981 came from below the High Lighthouse (NZ 929103), most probably from the Bituminous Shales. A second skeleton of *Macroplata longirostris* was found in 1960 in the *bifrons* Zone 'between Old Peak and Blea Wyke Point, southeast of Robin Hood's Bay' (Broadhurst and Duffy, 1970). This specimen (MANCH unnumb.) is about 4 m long. A *Steneosaurus* lower jaw (BMNH R12011) was collected in 1989 in the Bituminous Shales just south of Black Nab (NZ 926104).

In conclusion, the bulk of the reptiles from Whitby appear to have come initially from the Main Alum Shales of The Scar, and later from the alum workings in the cliff at Saltwick Nab and Black Nab. A few specimens appear to have been found in the Jet Rock Formation (?Bituminous Shales), probably on the foreshore between Saltwick Nab and Black Nab.

Taphonomic study of the Whitby marine reptile remains has been hampered by the lack of suitable collection data and in addition by the incompleteness of some specimens, the result of collection failure and through artificial 'improvements' made to certain specimens. An examination of museum specimens shows most skeletons to be well preserved in an articulated state with only slight damage, probably as a result of scavenging. This was presumably minimized by the prevailing anoxic conditions in the bottom sediment, as suggested by their bituminous nature. Other partial skeletons may have been broken up prior to burial or by recent wave action before the specimens were collected from the foreshore.

Fauna

About 20 species of marine reptile have been described from the Whitby area (Benton and Taylor, 1984), of which seven may be valid, but further revision might alter the figure. Of these seven, four (*Steneosaurus brevior, S. gracilirostris, Rhomaleosaurus longirostris* and *Sthenarosaurus dawkinsi*) occur only at Whitby, and one (*Stenopterygius acutirostris*) probably occurs only in Yorkshire. The taxonomy of the Upper Lias crocodiles from Whitby has been reviewed by Westphal (1961, 1962) and Duffin (1979a, 1979b), the ichthyosaurs by McGowan (1974b, 1976, 1978, 1979), and the plesiosaurs by Watson (1909c, 1910b), White (1940), Persson (1963) and Taylor (1992b). Approximate numbers of specimens in the BMNH, CAMSM, WHIMS and YORYM are given.

Numbers

Sauropterygia: Plesiosauria
Eretmosaurus macropterus
(Seeley, 1865a) 1
Macroplata longirostris (Blake, 1876) 1+
Type: MCZ 1033
Microcleidus homalospondylus
(Seeley, 1865) 6
Type: YORYM G502
Sthenarosaurus dawkinsi Watson,
1909 2
Type: MANCH L8023
Thaumatosaurus propinquus
(Blake, 1876) 2
'*Plesiosaurus*' sp. 4
Ichthyopterygia: Ichthyosauridae
Stenopterygius acutirostris
(Owen, 1840) 8
Type: BMNH 14553
Eurhinosaurus longirostris
(Mantell, 1851) 1
Type: BMNH 14566
'*Ichthyosaurus*' sp. 12
Archosauria: Crocodylia: Thalattosuchia:
Teleosauridae
Steneosaurus bollensis (Jaeger, 1828) 9
Steneosaurus brevior Blake, 1876 6
Type: BMNH 14781
Steneosaurus gracilirostris
Westphal, 1961 4
Type: BMNH 14792
Pelagosaurus brongniarti
(Kaup, 1835) 8
(incl. ?*Teleosaurus chapmani*)

Pelagosaurus typus Bronn, 1841 1
Steneosaurus sp. 6
Archosauria: Dinosauria: Saurischia
?theropod 1

Interpretation

The Whitby plesiosaurs divide up into forms with long necks and small skulls, others with relatively large skulls and one with long pointed jaws. They also range in total body length from 2 m to 6 m, and clearly used a range of hunting and feeding strategies. Their principal diet was probably cephalopods and fishes, and the larger species might also have eaten other marine reptiles. Their range of forms indicates four qualitative lineages and they provide the best information on plesiosaur evolution in the Upper Lias. Holzmaden and other German localities have also yielded good specimens of the same age, but these localities lack the variety of forms found at Whitby. At Whitby, there are at least two possible pliosauroids (plesiosaurs with short necks and large skulls), *Macroplata longirostris* which has a gracile snout and the *R. cramptoni–R. zetlandicus–R. propinquus* group with robust snouts (Figure 5.8C). There are also two or three plesiosauroids (plesiosaurs with long necks and small heads), namely *Microcleidus macropterus, M. homalospondylus* and *Sthenarosaurus dawkinsi*. *Macroplata longirostris* was about 5 m long with a head about 0.7 m long. It had a remarkably slender head and elongate rostrum, a character unknown in any other Jurassic plesiosaur. *Microcleidus homalospondylus* is represented by nearly complete skeletons which show an animal about 6 m long with an extremely long neck (2.5 m) and a relatively small skull. It had large paddles and is distinguished by characters of the vertebrae and limb girdles. *Sthenarosaurus dawkinsi* based on a partial skeleton collected at Saltwick, is another long-necked form with strong limbs. It is currently regarded as a plesiosauroid (Brown, 1981, p. 339). *Rhomaleosaurus propinquus* was about 2.5 m long and had a 0.6 m skull – relatively large.

The ichthyosaurs from Whitby are the best Upper Lias forms from Britain. However, those from Holzmaden, of approximately the same age, are more abundant, better preserved and show greater variety. Some of the ichthyosaurs from Whitby may also occur at Holzmaden, although most of the German specimens belong to different

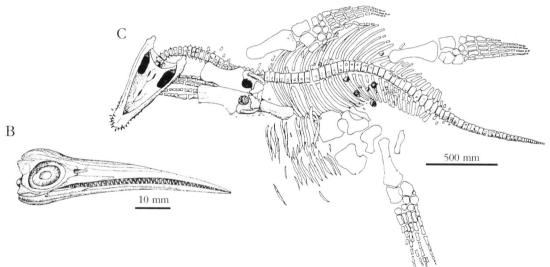

Figure 5.8 Marine reptiles from the Lower Jurassic Alum Shale Member of Whitby. (A) The crocodile *Steneosaurus gracilirostris* Westphal, 1961, type specimen (BMNH 14792); (B) the ichthyosaur *Temnodontosaurus longirostris* (Mantell, 1851), type specimen (BMNH 14566); (C) the pliosauroid plesiosaur *Rhomaleosaurus cramptoni* (Carte and Baily, 1863), type specimen (NMI F8785), skull and skeleton.

species (McGowan, 1974b, 1989b). The two Whitby ichthyosaur species recognized as valid by McGowan (1974b), *Eurhinosaurus longirostris* and *Stenopterygius acutirostris*, are distinguished largely by the relative proportions of parts of the skull (Figure 5.8B). For example, *S. acutirostris* has a larger orbit and nasal opening than *E. longirostris* in relation to the overall skull length. *S. acutirostris* is generally larger than *E. longirostris*, with skull lengths from 0.6 m to 1.50 m compared with skull lengths of less than 1 m. *Eurhinosaurus* is a swordfish-like form showing a remarkable disparity in the lengths between the upper and lower jaws, the mandible being only about half the length of the skull (McGowan, 1986, 1989b, 1989c). *Stenopterygius acutirostris* was larger than *Eurhinosaurus*, with a skull up to 1 m long, and had a long pointed snout and large orbit.

The crocodiles from Whitby represent the best Jurassic marine forms in Britain, but the preservation is not as good as in material from localities in Germany, such as Ohmden, Holzmaden, Boll, Holzheim (Baden-Württemberg) and Neumarkt (Oberpfalz). The age of these German sites is similar to that of the Whitby sediments (Posidonienschiefer, Lias epsilon 1, 2, 3, Early Toarcian). *Steneosaurus* (Figure 5.8A) and *Pelagosaurus* are teleosaurs which differ in size and in certain features of the skull and skeleton. *Steneosaurus* was 2.5–5.0 m long, whereas *Pelagosaurus* was under 1.75 m. *S. brevior* has a shorter snout (64% of skull length) than *S. bollensis* (72%) or *S. gracilirostris* (77%) (Westphal, 1961, 1962). However, Steel (1973) synonymized all three species as *S. bollensis*. In *P. typus* the snout is not sharply demarcated from the skull.

The Early Jurassic teleosaurs were specialized water-dwellers with elongate snouts and numerous teeth that suggest a diet of fish. The hind legs were twice as long as the forelegs and were doubtless powerful organs of propulsion. In general, teleosaurs are found in estuarine, shallow marine sediments, and they probably lived partly on land and fed in brackish and salt water. Teleosaurs had a strong bony armour. The group has a long history, the later forms evolving into several narrow- and broad-snouted forms in the Late Jurassic and Early Cretaceous of Europe in particular. The Early Jurassic remains of Whitby and Baden-Württemberg are the best preserved and most useful for an assessment of the relationships and biology of early marine crocodilians. These Early Jurassic teleosaurs represent the first radiation of crocodilians into the sea, after their origin in the latest Triassic as small terrestrial insectivorous animals.

Huene (1926, pp. 36–71) and Wild (1978b, p. 2) cite an undescribed specimen (WHIMS) of a 'middle sized femur' of a carnivorous theropod dinosaur from Whitby. Huene noted that he had not himself seen the specimen and cited a personal letter from 'Dr (D.M.S.) Watson'. The specimen had the fourth trochanter placed above the midpoint of the femur. This specimen has not been traced; if it is found, it will be of great interest as the only find of a theropod dinosaur from the Upper Lias of any locality. Indeed, only two other dinosaurs are known from the Upper Lias: the hind limb of the sauropod *Ohmdenosaurus* from Ohmden, near Holzmaden (Wild, 1978b), and a nearly complete skull and skeleton of an early thyreophoran, *Emausaurus*, from Klein-Lemahagen, near Rostock (Haubold, 1990).

Two probable pseudofossils from the Whitby Lias have been interpreted as reptilian: a possible teleosaur egg (YORYM 505; Melmore, 1931) and a supposed group of embryos or juveniles of four plesiosaurs (BMNH R3585; Seeley, 1887a, 1888a, 1888b, 1896). The former is certainly egg-shaped, but it consists of a mudstone and calcite 'core' surrounded by a pyrite skin, and is probably a concretion (Benton and Taylor, 1984, p. 418). The latter was reinterpreted by Thulborn (1982) as infilled *Thalassinoides* burrows surrounding a concretion, whereas Benton and Taylor (1984, pp. 418–19) suggested that the nodule was wholly inorganic in origin (?a septarian concretion). Such calcareous and pyritic mudstone 'doggers' occur abundantly in the Lias, and in the Jet Rock Formation in particular (Howarth, 1962).

Comparison with other localities

Other comparable Upper Lias localities occurring along the Yorkshire coast that have yielded a similar marine reptile fauna include Saltwick and the old alum quarries at Kettleness (NZ 8316) and Loftus (NZ 7420). Further reptile localities, including Runswick Bay, Robin Hood's Bay, Port Mulgrave, Staithes, Sandsend, Hawsker Bottoms, Boulby and Ravenscar (Old Peak–Blea Wyke Point), have also produced a comparable fauna, although the remains of marine reptiles from these localities are less abundant. The Upper Lias of England is not as rich in marine reptile fossils as the Lower Lias. Various localities in Somerset, Northamptonshire, Leicestershire, Lincolnshire and North Yorkshire have yielded isolated ichthyosaurs, plesiosaurs and steneosaurs (see above). The localities at Blisworth (SP 7354) and Wellingborough (SP 9868) are still accessible but most of the other sites are now inaccessible and have little potential for future finds.

The reptile faunas most similar to those from Yorkshire are those recorded from various localities in the Upper Lias of south-west Germany (e.g.

Holzmaden, Ohmden, Boll, Banz, Altdorf) and France (e.g. Normandy, Franche-Comté). Most of these sites cannot be compared readily with the Whitby section since the recorded finds are too sparse to constitute a 'fauna'. The exception is Holzmaden, Baden-Württemberg, where the bituminous laminated shales and grey mudstones of the Posidonienschiefer, a subdivision of the Schwarzjura ε (*tenuicostatum* to *bifrons* Zones of the Early Toarcian; Urlichs 1977), have produced hundreds of specimens. Hauff (1921) noted that the bulk of these came from his subdivisions II 2 to II 13 (middle ε, upper *tenuicostatum* Zone to upper *falciferum* Zone), thus rather older on average than the reptiles from the Yorkshire coast. Hauff (1921) records ten specimens of plesiosaurs, including four almost complete skeletons, about 350 specimens of ichthyosaurs, many of which are relatively complete, about 70 specimens of crocodiles many of which are also complete, and about 10 skeletons and bones of pterosaurs. Thus plesiosaurs and crocodiles are relatively less abundant, and ichthyosaurs are much more common at Holzmaden than around Whitby.

Several species of reptile are shared between Whitby and Germany. Among the crocodiles, *Steneosaurus bollensis*, *Pelagosaurus brongniarti* and *P. typus* occur in both areas. Among the plesiosaurs, the only Holzmaden pliosauroid is specifically different from the Yorkshire forms, but it is not clear whether any of the plesiosauroids are shared. McGowan (1979) ascribes German '*L. acutirostris*' to *L. burgundiae* (Gaudry, 1892).

Conclusions

The Yorkshire coast sites are undoubtedly the best for British Upper Lias reptiles. The coast between Whitby and Whitestone Point has yielded more specimens, and type specimens, than any other Upper Lias marine reptile site in Britain, and many of these are articulated. The fauna differs from the Upper Lias faunas of south-west Germany (e.g. Holzmaden) and France. It has produced the best collections of fossil crocodiles from the Early Jurassic of Britain. The ichthyosaurs and plesiosaurs from Whitby are the most numerous and varied of British Upper Lias sites, and the plesiosaurs in particular show a broad range of separate lineages.

The great importance and conservation value of the Whitby–Saltwick section lies, like that of Lyme Regis, in the combination of the richness of historical finds and the potential for future discoveries.

LOFTUS, YORKSHIRE (NZ 736200–NZ 757193)

Highlights

Loftus Alum Quarries have produced a diverse assemblage of marine fossil reptiles, plesiosaurs, ichthyosaurs and crocodiles. They are especially notable as the site where the pterosaur *Parapsicephalus* was found, a remarkable specimen that preserves the outline of the brain.

Introduction

The former alum workings in the Upper Lias Alum Shale at Loftus have yielded many important fossil reptile remains. These appear to form a fauna distinct from that found at Whitby, and Loftus is thus an important companion site. The quarried platform at Loftus is partly grassed over. It has a hummocky appearance (?quarry spoil) and there are several tracks still visible. The lower parts of the cliff behind (i.e. in the Upper Lias) are still largely exposed. Thus, much of the site is still available for further examination and additional finds could be made. However, the site is isolated from the sea above a cliff of Lower to Middle Lias and erosion is probably less than at Whitby. The geology has been described by Fox-Strangways (1892) and the reptiles by Carte and Baily (1863), Seeley (1865a), Tate and Blake (1876), Newton (1888), Watson (1911a), Melmore (1930), Wellnhofer (1978) and Taylor (1992a, 1992b).

Description

The extensive alum quarries on the Yorkshire coast between Loftus and Boulby yielded many reptile remains when they were in operation. Fox-Strangways (1892, p. 134) notes that 'the saurian remains were so numerous that one of the walks at Boulby House is edged with the vertebrae of these reptiles'. Although the two quarries are now linked and the former boundary cannot be detected, they operated separately throughout the

19th century. Loftus Alum Quarry (known as Lofthouse or Lingberry in the past) was operated by the Earl of Zetland and was closed in 1863, whereas Boulby Alum Quarry was closed in 1861 (Fox-Strangways, 1892, p. 453).

The sequence of the Upper Lias at Loftus is approximately the same as in the Staithes and Whitby sections, consisting of an ascending sequence through the upper part of the Cleveland Ironstone Formation and the Whitby Mudstone Formation (Grey Shales, Jet Rock Alum Shale members; Howarth, *in* Cope *et al.*, 1980b; see Whitby report above).

The Lower and Middle Lias are exposed on Hummersea Scar, west of the Alum Quarries, and on the foreshore below the quarry (*jamesoni* Zone, Early Pliensbachian, on the wave-cut platform; *margaritatus* and *spinatum* Zones, Late Pliensbachian, Cleveland Ironstone Formation on the 80–90 m cliff). The Alum Quarries have been dug back from this lower cliff line, forming a broad shelf well above sea-level, and a high cliff rises behind to a total height of 200 m. The upper part of the cliff consists of three Early Toarcian (Upper Lias) members, capped by Mid Jurassic sediments (Dogger Formation (1.5 m), Hayburn Formation (25 m), Aalenian; Fox-Strangways, 1892).

The reptiles appear to have been found in the Loftus Alum Quarries rather than in the Boulby Alum Quarries, since the specimens are labelled 'Lofthouse'. They are all recorded as 'Zone of *A. communis*' by Blake (in Tate and Blake, 1876, pp. 246, 250, 253–4) (=Early Toarcian, *Hildoceras bifrons* Zone), and they probably came from the Main Alum Shales. The provenances of some specimens can be traced from the early literature and also from examination of matrix associated with the remains, and these confirm Blake's statement. Young and Bird (1828, p. 287) noted a plesiosaur vertebra from Loftus, while Seeley (1880) described an ichthyosaur specimen (CAMSM 35176), presumably from Loftus, as *Ichthyosaurus zetlandicus*. The type specimen of the pliosaur *Rhomaleosaurus zetlandicus* (YORYM G503) also came from Loftus (Phillips, 1854; Tate and Blake, 1876, p. 250; Taylor, 1992a, 1992b), and presumably from the Cement Shales or the upper part of the Main Alum Shales, as confirmed also by the matrix of the specimen, a flaky, grey, pyritous shale containing concretions around the bones. The histories of the specimens from Loftus are detailed in Benton and Taylor (1984, pp. 410–14, 416).

Fauna

Loftus Alum Quarries have yielded many specimens according to past records, but only six may still be traced. However, these are rather important. They are preserved in the BMNH, BGS(GSM), CAMSM, WHIMS, and YORYM.

Numbers

Sauropterygia: Plesiosauria:
   Eretmosaurus macropterus
     (Seeley, 1865a)
      Type: CAMSM J35182   1
   Rhomaleosaurus zetlandicus
     (Phillips, 1853)
      Type: YORYM G503   1
   '*Plesiosaurus* sp.'   1
Ichthyopterygia: Ichthyosauridae
   Stenopterygius acutirostris
     (Owen, 1840)
      Type of *Ichthyosaurus crassimanus*
      Blake, 1876: YORYM G497   1
   Temnodontosaurus platyodon
     (Conybeare, 1822)   1
Archosauria: Pterosauria: 'Rhamphorhynchoidea'
   Parapsicephalus purdoni (Newton, 1888)
      Type: BGS(GSM) 3166   1

Interpretation

Loftus Alum Quarries have yielded type specimens of four species, of which at least two are apparently unique to this locality. *Eretmosaurus macropterus* has been recorded from Whitby.

The plesiosaur *Eretmosaurus macropterus* is represented by a fine articulated skeleton with a total length of 5 m. The skull is relatively short (0.25 m) and the teeth are long and curved. The neck is long (2 m) and consists of 39 vertebrae, and the tail is 1.3 m long. The limbs are very large: they all measure over 1 m in length. The only descriptions (Seeley, 1865a; Blake, *in* Tate and Blake, 1876, p. 246; Watson, 1911a) are brief and the specimen has never been figured.

Rhomaleosaurus zetlandicus is about 6 m long, with a long skull (1.1 m), short neck (1.5 m) and long tail (2 m). The limb bones are large. The specimen was collected in about 1850 (Phillips, 1853, 1854; Carte and Baily, 1863; Blake, *in* Tate and Blake, 1876, pp. 249–50) and has recently been redescribed in detail (Taylor, 1992a, 1992b).

Ichthyosaurus crassimanus Blake (1876, pp. 253–4) is 10 m long and has a 2 m skull. The front paddles (0.8 x 0.3 m) are larger than the hind paddles (0.6 x 0.25 m). It was described in some detail by Melmore (1930). McGowan (1974a, pp. 31-2) ascribed it to *Stenopterygius acutirostris*, but later (McGowan, 1976, p. 675, footnote; 1979, pp. 120-1) provisionally placed it in *Leptopterygius*. There is a problem regarding the locality of this specimen. Blake (*in* Tate and Blake, 1876, p. 254) noted its provenance as 'Lofthouse', but Simpson (1884, p. 12) stated that it came from Kettleness. Later authors have been non-committal: 'Alum Shale Quarries north of Whitby' (Melmore, 1930, p. 615); 'Alum Shales north of Whitby' (McGowan, 1974a, p. 32); 'near Whitby' (Pyrah, 1979, p. 423). We assume that it came from Loftus since that is the locality quoted by its original describer. A large specimen of *Temnodontosaurus platyodon* (5 m long), with a 1.25 m head, has also been recorded (Simpson, 1884, p. 12).

The pterosaur *Parapsicephalus purdoni* (Figure 5.9), originally ascribed to the genus *Scaphognathus* (from the Late Jurassic Solnhofen Beds of Germany) by Newton (1888), is represented by a partial skull, which lacks dentition and the snout tip. The skull is long and low and has large openings, especially the antorbital fenestra. The preservation of this specimen is such that a fine brain cast is displayed, which shows how the relatively large brain fits obliquely in the skull behind the eye. The forebrain is large and the olfactory lobes are short. The large optic lobes are a reptilian feature, but they are not quite dorsal in position, which is a bird-like feature. The cerebellum is small and there are large flocculi (Newton, 1888; Wellnhofer, 1978, pp. 30, 39). *Parapsicephalus* is one of the earliest true rhamphorhynchids (Unwin, 1991) and it falls in a time interval when relatively few pterosaurs are known. Other late Early Jurassic pterosaurs include *Campylognathoides* and *Dorygnathus* from Germany and India (Wellnhofer, 1978, pp. 73-4).

Comparison with other localities

Loftus Alum Quarries are most immediately comparable with Kettleness Alum Quarries to the south (NZ 8316) and the coast at Port Mulgrave (NZ 8018). The coast east of Whitby (NZ 9012-NZ 9310) has produced more species of reptiles and more specimens, but those from Loftus are broadly different taxa. Only the plesiosaur *Eretmosaurus macropterus* is shared with

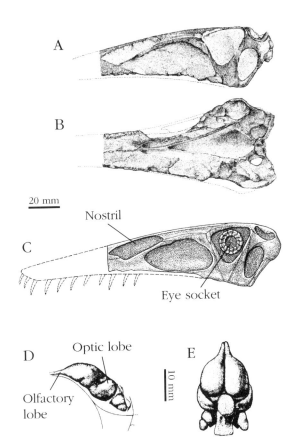

Figure 5.9 The pterosaur *Parapsicephalus purdoni* (Newton, 1888) from the Lower Jurassic Alum Shales Member of Loftus, Yorkshire. (A), (B) and (C) skull in lateral and ventral views; (D) and (E) brain cast in left lateral and dorsal views. From Westphal (1976).

Whitby. The pterosaur *Parapsicephalus* is unique to Loftus.

Conclusions

Loftus Alum Quarries have never been as rich in reptile remains as the Whitby coast, but the taxa are different. The ichthyosaurs are much larger than those from any other British Upper Lias locality. The two species of plesiosaur are also large and probably unique to Loftus. Of particular importance is the unique specimen of *Parapsicephalus*, the only British Upper Lias pterosaur described, and of great significance in general because of the fine brain cast that is preserved.

The combination of this historic importance and some potential for future finds gives the site considerable conservation value.

Chapter 6

British Mid Jurassic fossil reptile sites

INTRODUCTION: BRITISH MID JURASSIC REPTILE SITES

Fossil reptiles have been found in numerous localities in the Mid Jurassic (Aalenian–Callovian) of southern England and west Scotland, but the most productive sources for reptiles are mainly in rocks of Bathonian and Callovian age. The typically shallow-water lagoonal and littoral marine facies of the Bathonian (e.g. Forest Marble) have produced many important finds of dinosaurs, pterosaurs and mammal-like reptiles (some of the last of this group in the world), in addition to marine reptiles, while the Callovian Oxford Clay is famous for its plesiosaur remains, which occur throughout the outcrop. Fuller details of British Jurassic geology, reptile evolution worldwide and British Jurassic sites are given in the introduction to Chapter 5.

British Mid Jurassic reptile sites are listed below, grouped roughly in stratigraphic order, and excluding the selected GCR sites, which are listed at the end. Details of these sites were obtained from Fox-Strangways (1892), H.B. Woodward (1894, 1895) and Waldman (1974), as well as from museum records and other unpublished sources.

Aalenian–Bajocian (Inferior Oolite)

There are relatively few reptile sites in the Inferior Oolite (Aalenian–Bajocian) of Britain. The remains are mainly teeth and jaws of the dinosaur *Megalosaurus* and odd pieces of the crocodiles *Steneosaurus* and *Teleosaurus*.

DORSET: Eype, near Bridport (SY 4592; *Teleosaurus*); Bradford Abbas (SY 5915; 'Stegosaurus' spines); Nethercombe Quarry, Sherborne (ST 636175; type of *Megalosaurus nethercombensis* from *humphriesianum* Zone); Lower Eastham Farm, Crewkerne (ST 458104; *Ichthyosaurus* in Yeovil Sands); Cold Harbour Road Quarry, Sherborne (?ST 642173; type of *Megalosaurus hesperis* from *parkinsoni* Zone in quarry behind the houses on the north side of Cold Harbour Road, now built over).

SOMERSET: Doulting Quarries, Shepton Mallet (ST 6543; *Megalosaurus*).

GLOUCESTERSHIRE: Stroud – ?exact locality (SO 8505; *Megalosaurus*); Frith Quarry, Stroud (SO 868083; plesiosaur tooth in Lower *Trigonia* Grit [*discites* Zone, Lower Bajocian]); Rodborough Hill, Stroud (SO 8404; *Teleosaurus*); Leckhampton Quarries (SO 950185; *Steneosaurus, Teleosaurus* from the Gryphite Grit [*laeviuscula* Zone, Lower Bajocian], also fragments of *Ichthyosaurus, Pliosaurus*); Crickley Hill Quarry (SO 928164; ?*Megalosaurus*).

NORTH YORKSHIRE: White Nab, Scarborough Bay (TA 058864; ?*Cetiosaurus* from *humphriesianum* Zone; ?also *Ichthyosaurus, Plesiosaurus*).

Bathonian (including White Limestone, Great Oolite, Forest Marble, Lower Cornbrash, etc.)

Reptile remains are common in the 'Great Oolite' and Forest Marble of Gloucestershire and Oxfordshire in particular, but dozens of localities are known throughout the British Bathonian (Figures 6.1 and 6.2). The commonest forms are crocodilians (*Steneosaurus, Teleosaurus*), dinosaurs (*Megalosaurus, Cetiosaurus, 'Stegosaurus', Lexovisaurus*), pterosaurs (*Rhamphocephalus*), plesiosaurs ('Cimoliasaurus'), and turtles (*Protochelys*). References include Phillips (1871), H.B. Woodward (1894), A.S. Woodward (1910), Huene (1926), Arkell (1933, 1947a, 1947b), Torrens (1968, 1969a, 1969b), Palmer (1973), Sellwood and McKerrow (1974); Metcalf *et al*. (1992) and Evans and Milner (1994).

DORSET: Long Burton, near Sherborne (ST 6513; ?*Megalosaurus* from ?Forest Marble); Yetminster (ophthalmosaur; from a nodular Cornbrash limestone immediately underlying the Oxford Clay; Delair, 1986); Watton Cliff (West Cliff) (SY 451901–SY 453907; microvertebrate remains, including frogs, salamanders, lizards, crocodiles, dinosaurs from the Forest Marble: Evans, 1991, 1992b); Swyre (SY 525868; amphibian and reptile microvertebrate bones and teeth from the Forset Marble: Evans, 1991b, 1992b).

SOMERSET: Closworth (ST 5610; type of *Steneosaurus stephani* from the Cornbrash).

AVON: Bath – ?exact locality (ST 7565; type of *Steneosaurus temporalis* from the Great Oolite [*aspidoides* Zone, Upper Bathonian]).

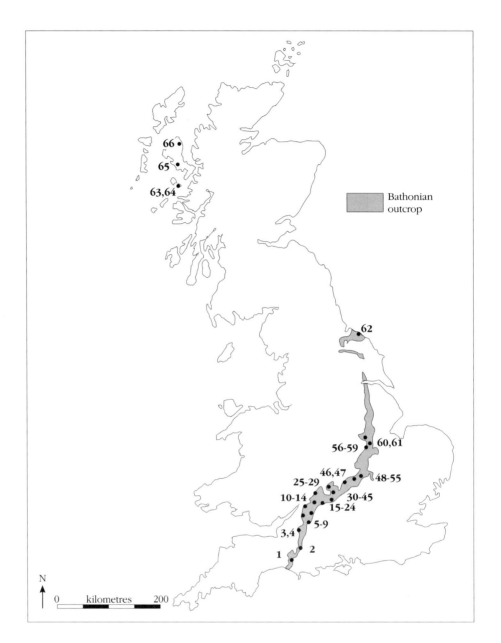

Figure 6.1 Distribution of British Bathonian tetrapod localities. Dorset: Long Burton (1), Watton Cliff (1a), Swyre (1b); Somerset: Closworth (2); Avon: Bath (3); Wiltshire: Avoncliff (4), Bradford-on-Avon (5), Frankley (6), Box Tunnel (7), Atford (8), Malmesbury (9), Leigh Delamere (9a); Gloucestershire: Minchinhampton (10), Sapperton Tunnel (11), Avening (12), Cirencester (13), Tarlton Clay Pit (13a), Sevenhampton (14), Chedworth (15), Stanton (16), Bibury (17), Naunton (18), Kyneton Thorns (19), Huntsman's Quarry (20), Eyeford (21), New Park Quarry (22), Oakham (23), Longborough Road Quarry (24); Oxfordshire: Chipping Norton (25), Sarsden (26), Over Norton (27), Sharp's Hill (28), Temple Mills Quarry (29), Enstone (30), Stonesfield (31), Slape Hill (32), Glympton (33), Bladon (34), Hanborough (35), Enslow Bridge (36), Bletchingdon Station (37), Shipton Quarry (38), Kirtlington (39), Hampton Common (40), Fritwell (41), Littlemore (42), Woodeaton (43), Ardley (44), Stratton Audley (45); Buckinghamshire: Stony Stratford (46), Olney (47); Northamptonshire: Blisworth (48), Cogenhoe (49), Northampton (50), Kingsthorpe (51), Rushden (52), Thrapston (53), Ilchester (54), Oundle (55); Leicestershire: Essendine (56), Belmesthorpe (57); Cambridgeshire: Peterborough (58), Botolph's Bridge (59), Orton Longueville, Peterborough (60), Stilton (61); Yorkshire: Scarborough (62); Hebrides: Eigg (63), Muck (64), Elgol, Skye (65a, b), Bearreraig, Skye (66). Based on information in Evans and Milner (1994), Metcalf *et al.* (1992), and original.

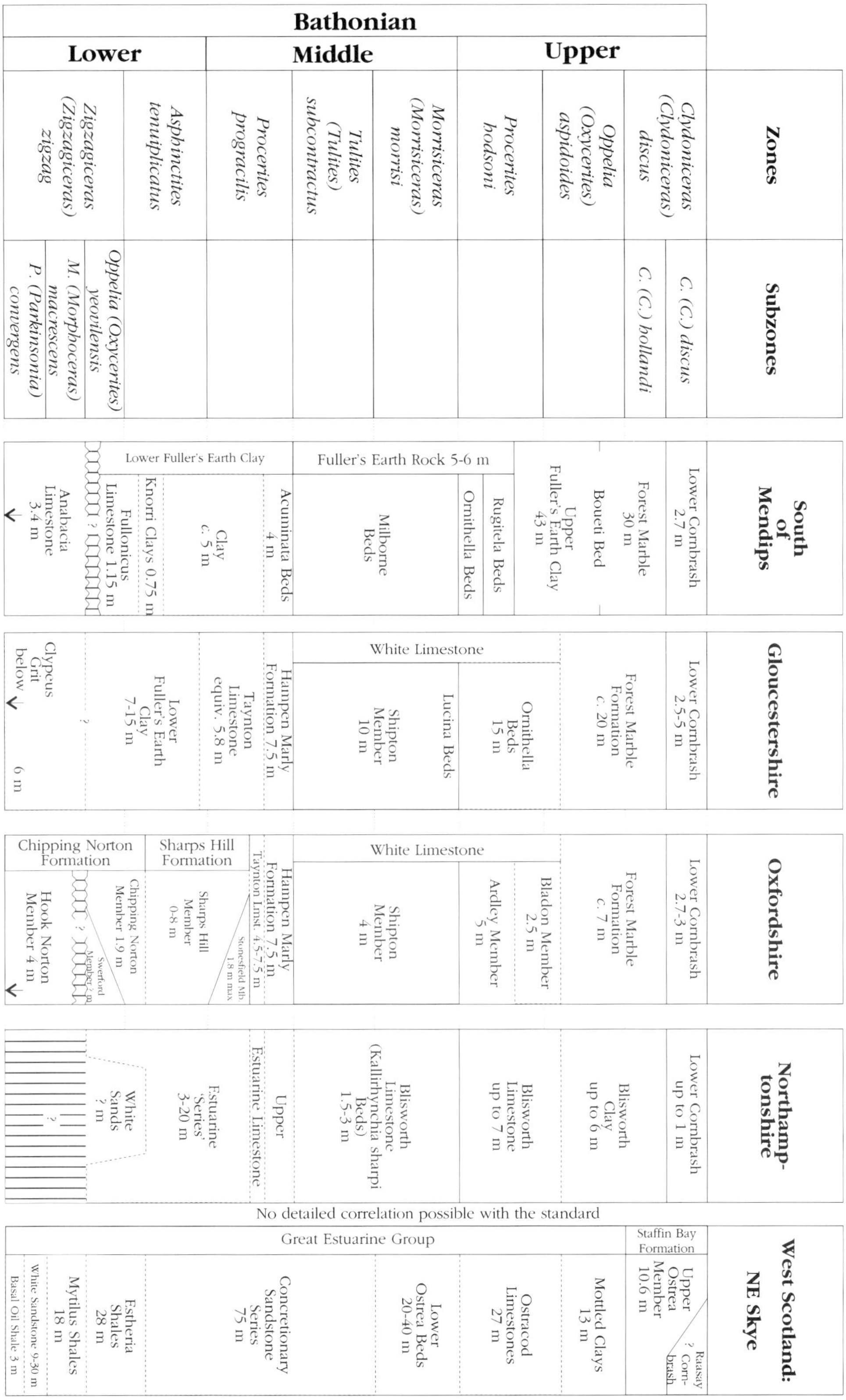

British Mid Jurassic reptile sites

Bathonian										
Lower		**Middle**			**Upper**					
Zigzagiceras (Zigzagiceras) zigzag	*Asphinctites tenuiplicatus*	*Procerites progracilis*	*Tulites (Tulites) subcontractus*	*Morrisiceras (Morrisiceras) morrisi*	*Procerites hodsoni*	*Oppelia (Oxycerites) aspidoides*	*Clydoniceras (Clydoniceras) discus*			**Zones**
Oppelia (Oxycerites) yeovilensis / *M. (Morphoceras) macrescens* / *P. (Parkinsonia) convergens*							*C. (C.) hollandi*	*C. (C.) discus*		**Subzones**

South of Mendips

- Anabacia Limestone 3-4 m
- Knorri Clays 0.75 m
- Fullonicus Limestone 1.15 m
- Clay c. 5 m
- Acuminata Beds 4 m
- Lower Fuller's Earth Clay
- Millborne Beds
- Fuller's Earth Rock 5-6 m
- Rugitela Beds
- Ornithella Beds
- Upper Fuller's Earth Clay 43 m
- Boueti Bed
- Forest Marble 30 m
- Lower Cornbrash 2.7 m

Gloucestershire

- Clypeus Grit below — 6 m
- Lower Fuller's Earth Clay 7-15 m
- Taynton Limestone equiv. 5.8 m
- Hampen Marly Formation 7.5 m
- White Limestone
- Shipton Member 10 m
- Lucina Beds
- Ornithella Beds 15 m
- Forest Marble Formation c. 20 m
- Lower Cornbrash 2.5-5 m

Oxfordshire

- Chipping Norton Formation: Hook Norton Member 4 m; Chipping Norton Member 1.9 m
- Sharps Hill Formation: Sharps Hill Member 0-8 m; Stonesfield Mb. 18 m max.; Swerford Mb.
- Taynton Lmst. 4.5-5 m
- Hampen Marly Formation 7.5 m
- White Limestone
- Shipton Member 4 m
- Ardley Member 5 m
- Bladon Member 2.5 m
- Forest Marble Formation c. 7 m
- Lower Cornbrash 2-3 m

Northamptonshire

- White Sands ? 7 m
- Estuarine 'Series' 3-20 m
- Upper Estuarine Limestone
- Blisworth Limestone (Kallirhynchia sharpi Beds) 1.5-3 m
- Blisworth Limestone up to 7 m
- Blisworth Clay up to 6 m
- Lower Cornbrash up to 1 m

No detailed correlation possible with the standard

West Scotland: NE Skye

- Great Estuarine Group
- Basal Oil Shale 3 m
- White Sandstone 9-30 m
- Mytilus Shales 18 m
- Estheria Shales 28 m
- Concretionary Sandstone Series 75 m
- Lower Ostrea Beds 20-40 m
- Ostracod Limestones 27 m
- Mottled Clays 13 m
- Staffin Bay Formation: Upper Ostrea Member 10.6 m; ? Raasay ? Cornbrash

Figure 6.2 Stratigraphy of the British Bathonian (after Cope *et al.*, 1980b), with ages of the localities listed in Figure 6.1 indicated.

British Mid Jurassic fossil reptile sites

WILTSHIRE: Avoncliff (ST 8059; *Teleosaurus* from the Fuller's Earth); Bradford-on-Avon (ST 8260; *Cetiosaurus*, 'Cardiodon' from the Forest Marble [discus Zone, Late Bathonian]); Atford (ST 8666, ?Atworth; teeth of 'Cardiodon', *Megalosaurus*, 'Hylaeosaurus', *Plesiosaurus* from the Forest Marble); Frankley, near Bradford (?Frankleigh, ST 822622; *Cetiosaurus*); Box Tunnel (ST 8469; *Megalosaurus*); Leigh Delamere (ST 890790; microvertebrates from the Forest Marble; Evans and Milner, 1994); Malmesbury (ST 9387; *Cetiosaurus*).

GLOUCESTERSHIRE: Minchinhampton Reservoir (?SO 855113; type of *Megalosaurus bradleyi* from the White Limestone); Avening (SO 8897; *Teleosaurus* from Forest Marble); Sapperton Tunnel, Hayley Farm (SO 949018; *Megalosaurus*, ?*Cetiosaurus*, from Kemble Beds; *Plesiosaurus*, *Steneosaurus* – ?same locality); Tarlton Clay Pit, near Cirencester (SO 970001; assorted microvertebrates from the Forest Marble; Evans and Milner, 1994); Cirencester, Jarvis' Old Quarry (SO 995999; *Cetiosaurus*, 'Goniopholis', from the Kemble Beds); Ready Token, near Cirencester (SP 100050; microvertebrates; Evans and Milner, 1994); Bibury (SP 1106; *Cetiosaurus* from Forest Marble); Chedworth (SP 0511; *Steneosaurus* from lower White Limestone); Sevenhampton (SP 0321; 'Pterodactylus' from Cotswold Slate); Naunton (SP 1123; *Teleosaurus* from Cotswold Slate [?Taynton Stone or Hampen Marly Formation, *progracilis* Zone, Middle Bathonian]); Kineton Thorns Quarry (SP 123263; *Megalosaurus*; type of *Rhamphocephalus prestwichi* from Cotswold Slate); Oakham Quarry, Little Compton (SP 279306; *Megalosaurus, Cetiosaurus*, ?*Lexovisaurus* from the Chipping Norton Limestone Formation [*zigzag* Zone, Lower Bathonian]); Stanton (SP 0734; *Megalosaurus* from the Forest Marble); Hornsleasow (Snowshill) Quarry (SP 131322; *Cetiosaurus, Megalosaurus*, small carnivorous dinosaurs, pterosaurs, crocodiles, tritylodontid, chelonians, 'lizard' from Chipping Norton Limestone Formation [*zigzag* Zone, Lower Bathonian]; (Vaughan, 1989; Metcalf *et al.*, 1992).

OXFORDSHIRE: Smith's Quarry, Sarsden (SP 300266; type of *Rhamphocephalus depressirostris, Megalosaurus, Cetiosaurus* from the 'basement bed of Great Oolite Series' [?Sharp's Hill Beds, Lower Bathonian]); Padley's Quarry, Chapelhouse, Chipping Norton (SP 329281;

Cetiosaurus from Sharps Hill Member [*tenuiplicatus/progracilis* Zone, Lower-Middle Bathonian]); Workhouse Quarry, Chipping Norton (SP 318276; *Cetiosaurus*, *Megalosaurus* in Sharps Hill Formation); Temple Mills Quarry, Sibford Ferris (SP 3537; *Steneosaurus*); Enstone (SP 3724; *Cetiosaurus*); Over Norton (SP 3128; *Cetiosaurus*); Sharps Hill Quarry (SP 338358; *Lexovisaurus* and microvertebrates in Sharps Hill Formation [*progracilis* Zone, Lower Bathonian]); Woodeaton (SP 533123; *Cetiosaurus* and microvertebrates from Hampen Marly Formation [*progracilis* Zone, Middle Bathonian]); Enslow Bridge (SP 475178; type of *Steneosaurus meretrix*; *Megalosaurus* from the Stonesfield Slate (*progracilis* Zone, Middle Bathonian); types of *Cetiosaurus oxoniensis* and *Lexovisaurus? vetustus* from the Forest Marble [Upper Bathonian]); Gibraltar (Bletchington Station) Quarry (SP 483183; *Steneosaurus*, *Megalosaurus*, *Cetiosaurus* from *fimbriatus–waltoni* Beds (top White Limestone Formation [*aspidoides* Zone, Upper Bathonian]); Slape Hill, Woodstock (SP 423196; *Steneosaurus, Teleosaurus, Cetiosaurus* from White Limestone Formation); Glympton Quarry (SP 427217; type of *Cetiosaurus glymptonensis* from the Forest Marble [Upper Bathonian]); Tolley's Quarry, Bladon (SP 4414; *Cetiosaurus, Ichthyosaurus*); Hanborough Railway Station (SP 4415; crocodilian from the Cornbrash); Hampton Common (SP ?5015/?4816; *Steneosaurus, Rhamphorhynchus, Plesiosaurus*); Fritwell (SP 5229; *Teleosaurus*); Littlemore (SP 5302; *Megalosaurus* from the Corallian); Stratton Audley (SP 6026; *Cetiosaurus* from the Forest Marble (Upper Bathonian); Ardley Quarry (SP 539272; *Teleosaurus* from Ardley Member of White Limestone [*hodsoni* Zone, Upper Bathonian]).

NORTHAMPTONSHIRE: Blisworth railway cutting (SP 725543; *Cetiosaurus, Steneosaurus* from 'Great Oolite' [Blisworth Limestone, *hodsoni* Zone, Upper Bathonian]); Cogenhoe (SP 8360; *Cetiosaurus* from Cornbrash); Northampton, Buttock's Booth (SP 7864; *Steneosaurus* from 'Great Oolite' [Blisworth Limestone, *hodsoni* Zone, Upper Bathonian]); Kingsthorpe, Northampton (SP 7563; *Steneosaurus* from 'Great Oolite'); Rushden Quarry (SP 951661; 'Cimoliasaurus' from Cornbrash); Irchester (SP 8968; *Steneosaurus*); Islip Ironstone Quarry, Thrapston (SP 975782 etc.; *Steneosaurus, Megalosaurus, Muraenosaurus* from Cornbrash); Oundle (TL 0388; crocodile from 'Great Oolite').

BUCKINGHAMSHIRE: Olney (SP 8851; *Plesiosaurus, Cetiosaurus* from the top of the Cornbrash).

CAMBRIDGESHIRE: Peterborough – ?exact locality (TL 1998; *Steneosaurus, Teleosaurus,* from Cornbrash); Orton, Peterborough (TL 1796; crocodile); Norman Cross Brickworks, Stilton (TL 170912; *'Cimoliasaurus', Ichthyosaurus, Teleosaurus* from Cornbrash).

LEICESTERSHIRE: Essendine/Banthorpe railway cutting (TF 0412; *'Cimoliasaurus', Cetiosaurus* from Blisworth Clay [Upper Bathonian]); Belmesthorpe (TF 0410; *Steneosaurus, Teleosaurus, Rhamphocephalus* from 'Great Oolite' [Upper Bathonian] and Upper Estuarine 'Series' [Lower-Middle Bathonian]); Great Casterton, Rutland (*Cetiosaurus*, LEICS).

NORTH YORKSHIRE: Scarborough (?SE 8606; *Plesiosaurus* from the Cornbrash).

INNER HEBRIDES: Elgol, Skye (NG 531164, NG 519154, NG 518168; plesiosaur, crocodilian, tritylodont remains from Lealt Shale Formation and Kilmaluag Formation (Waldman and Savage, 1972; Harris and Hudson, 1980; Savage, 1984; Waldman and Evans, 1994).

Callovian

Reptiles are known from the Upper Cornbrash (*macrocephalus* Zone) of Stilton, Cambridgeshire (Martill, 1986), but it is not clear whether the overlying Kellaways Clay has produced any finds. The overlying Kellaways Sand (*calloviense* Zone) of Lincolnshire has recently produced numerous plesiosaur and marine crocodile remains from a scattering of temporary exposures around Lincoln (Brown, 1990; Brown and Keen, 1991) and in the Peterborough district (Martill, 1985b), and the Kellaways Rock (also *calloviense* Zone) has yielded a plesiosaur (*Cryptoclidus*) in Yorkshire. The Lower Oxford Clay (particularly the *jason* Zone) has produced abundant plesiosaurs (*Cryptoclidus, Liopleurodon, Muraenosaurus, Peloneustes, Pliosaurus, Simolestes, Tricleidus*), ichthyosaurs (*Ophthalmosaurus*), crocodilians (*Metriorhynchus, Steneosaurus*), pterosaurs (*'Rhamphorhynchus'*) and dinosaurs (*Callovosaurus, Cetiosauriscus, Dryosaurus, Eustreptospondylus, Lexovisaurus, Metriacanthosaurus,*

Ornithopsis, and *Sarcolestes*) from localities from Dorset to Peterborough, with the richest localities lying in and around Peterborough (Martill and Hudson, 1991). Martill (1986) notes reptile finds in nearly all Lower Oxford Clay horizons, particularly Beds 7, 8, 10, 11, 13 and 17. References include Seeley (1869a), Phillips (1871), H.B. Woodward (1895), Andrews (1910), Arkell (1933), Leeds (1956), Tarlo (1960), Galton (1980a), Brown (1981), and Martill (1985b, 1986, 1988, 1990, 1992), Adams-Tresman (1987a, 1987b), Martill and Hudson (1991) and Martill *et al.* (1994).

DORSET: Backwater, Weymouth (SY 677790; *'Cimoliasaurus', Ichthyosaurus, Lexovisaurus durobrivensis, Pliosaurus* from the Lower Oxford Clay); Putton Lane Brick Pit, Chickerell (SY 650801; type specimen of *Cryptoclidus richardsoni, Muraenosaurus, Pliosaurus, Steneosaurus, ?Dacentrurus* from Lower Oxford Clay [*calloviense* Zone]); Radipole (SY 8167; *'Plesiosaurus'*); Rodwell (SY 6778; ichthyosaur limb; Delair 1987); Shore of Fleet (?Tidman Point and bay to the west; *Steneosaurus, 'Plesiosaurus'*); Bowleaze Cove (SY 19702; *Steneosaurus, Muraenosaurus*).

WILTSHIRE: Melksham (ST 9063; ?cryptoclidid, *Metriorhynchus* from Oxford Clay); Devizes (SU 0061; *'Cimoliasaurus'* from Oxford Clay); Chippenham (ST 9173; *'Cimoliasaurus', Metriorhynchus, Muraenosaurus, Pliosaurus,* from Oxford Clay); Christian Malford (ST 957774; *'Cimoliasaurus', Peloneustes, Pliosaurus, Steneosaurus, Ophthalmosaurus* from Lower Oxford Clay [*jason* Zone]); Wootton–Bassett (?Old Park brickpit, SU 0582; *'Cimoliasaurus'* from (?Upper) Oxford Clay [*cordatum* Zone]).

OXFORDSHIRE: Long Marston [?= Marston] (SP 5309; *'Cimoliasaurus', Steneosaurus* from Oxford Clay); Shotover Hill (SP 5706; *'Cimoliasaurus', Pliosaurus* from Oxford Clay); Cowley Field, Oxford (SP 5703, pit filled; *'Cimoliasaurus', Ophthalmosaurus, Cryptoclidus,* type of *'Plesiosaurus' hexagonalis* [*nomen dubium*] from Oxford Clay); St Clements, Oxford (SP 5306, pits filled; *Ophthalmosaurus, Rhamphorhynchus* from Oxford Clay); Summertown Brick Pit, near Oxford (SP 5109; type of *Eustreptospondylus oxoniensis, Muraenosaurus plicatus* [*nomen dubium*], *'Cimoliasaurus' oxoniensis,* as well as crocodiles,

Ophthalmosaurus from Middle Oxford Clay [Upper Callovian, *athleta* Zone]); Wolvercote Brick Pit (SP 494105; *'Megalosaurus'*, *Cryptoclidus, Ophthalmosaurus* from Oxford Clay); Shellingford Crossroads Quarry (SU 326942; *Pliosaurus, ?Goniopholis* from Oxford Clay); St Edmund Hall, Oxford (SP 518063; crocodile from Oxford Clay [*lamberti* Zone]); Iffley Road Sports Ground, Oxford (SP 524054; *'Plesiosaurus'* from Oxford Clay); Cumnor (?SP 4704; *'Pliosaurus'* from Oxford Clay).

BUCKINGHAMSHIRE: Calvert Brick Pit, and other localities (?), Buckingham (SP 6933; *Ophthalmosaurus* from Oxford Clay); Bletchley Brick Works (SP 868327; *Ophthalmosaurus* from Oxford Clay); Newton Longville Brickworks (SP 853322; *Pliosaurus* from Oxford Clay); Fenny Stratford (SP 8834; *Steneosaurus* from Oxford Clay); Caldecotte Reservoir, Milton Keynes (SP 892352; *Ophthalmosaurus* from Lower Oxford Clay [*coronatum* Zone]; Martill, 1986).

BEDFORDSHIRE: Marston Moretaine (SP 9941; *Peloneustes* from Oxford Clay); Stewartby Clay Pit (TL 0142; *Ophthalmosaurus, Liopleurodon* from Oxford Clay); Kempston (?Clay Pits around TL 0345, or Green-End Old Pit, TL 007475; *Peloneustes* from Oxford Clay); Bedford (TL 0449; *'Cimoliasaurus', ?Lexovisaurus, Steneosaurus* from Oxford Clay); Ravensden, Bedford (TL 0754; *Pliosaurus* in Oxford Clay).

CAMBRIDGESHIRE: Orton Brick Pit (TL 165937; *Cryptoclidus, Ophthalmosaurus* from Lower Oxford Clay); Norman Cross Brick Pit (TL 173916; unidentifiable reptile from Lower Oxford Clay); Yaxley Brick Pit (TL 178932; *Cryptoclidus, Pliosaurus, Steneosaurus, Metriorhynchus* from Lower Oxford Clay); Eyebury Brick and Tile Works (TL 1859; type of *Pliosaurus evansi* from Lower Oxford Clay); St Neots Brickyard (TL 1860; *Pliosaurus* from Oxford Clay [*coronatum* Zone]); London Road, Peterborough (?TL1896; *Muraenosaurus, Cryptoclidus, Steneosaurus, Metriorhynchus, Cetiosauriscus* from Lower Oxford Clay); Woodston Lodge, Peterborough (TL 1897; *Ophthalmosaurus, Liopleurodon* from Oxford Clay); Fletton Brick Works (various pits around TL 1995; have probably yielded the majority of 'Peterborough' reptiles; types of *Neopterygius entheciodon, Ophthalmosaurus icenicus, Apractocleidus teretipes, Cryptoclidus eurymerus, Muraenosaurus durobrivensis, M.*

leedsi, Peloneustes philarchus, Simolestes vorax, Tricleidus seeleyi, Metriorhynchus cultridens, M. durobrivensis, Mycterosuchus nasutus, Steneosaurus depressus, S.durobrivensis, S. hulkei, S. leedsi, S. obtusidens, Cetiosauriscus leedsi, Lexovisaurus durobrivensis, 'Stegosaurus' priscus, Sarcolestes leedsi, amongst others from Lower Oxford Clay [*jason* Zone]); Peterborough Gas Works (?TL 199991; type of *Ornithopsis leedsi* probably from the junction of the Kellaways Clay with the overlying Kellaways Sand; Martill, 1986; Brown and Keen, 1991); Barrow Pit, Farcet (TL 200958; *Cryptoclidus, Liopleurodon, Metriorhynchus* from Lower Oxford Clay); Stanground (TL 2097; *Ophthalmosaurus, Steneosaurus?, Muraenosaurus* from Lower Oxford Clay [*jason* Zone]); Dogsthorpe Brick Pit, Peterborough (TF 219019; *Metriorhynchus?* from Lower Oxford Clay [*jason* Zone]; *Liopleurodon*; Dawn, 1991); Eye (TL 2202; *Steneosaurus* from Lower Oxford Clay); Whittlesey Clay Pits (TL 252976 and/or TL 250976; *'Cimoliasaurus', Lexovisaurus durobrivensis, Muraenosaurus, Ichthyosaurus, Peloneustes, Pliosaurus, Steneosaurus, Metriorhynchus* from Lower Oxford Clay [*jason* Zone]); St Ives Brickyard (?TL 304718; type of *Rhamphorhynchus jessoni, Pliosaurus* from Middle Oxford Clay).

LINCOLNSHIRE: Reepham (TF 046747; *Cryptoclidus eurymerus, Muraenosaurus leedsii, Steneosaurus* sp., *Metriorhynchus* sp., *Liopleurodon ferox* from the Kellaways Sand [*calloviense* Zone]; Brown, 1990; Brown and Keen, 1991).

HUMBERSIDE: Mill Hill, Elloughton, near Brough (SE 942278; *Cryptoclidus, Muraenosaurus*, etc. from Kellaways Sand [*calloviense* Zone]; Brown and Keen, 1991).

YORKSHIRE: Hackness, Scarborough (SE 9690; *Ichthyosaurus, Plesiosaurus*); Gristhorpe (TA 1283; *Steneosaurus* from Kellaways Beds).

Six reptile-bearing sites have been selected as GCR sites from the huge numbers that have been noted in the literature, as those representing the greatest range of faunas and preservation types, and as having the greatest potential for future collecting. These are all Bathonian; none of the Aalenian or Bajocian sites was strong enough for inclusion. In addition, none of the important Callovian localities could be selected because they have either been lost to infill or degradation, or they are currently worked in a way that prevents the conservation of fossiliferous horizons. In addition, it is not possible to say that any one or two Oxford Clay sites is likely to be more or less productive than any other. The Mid Jurassic sites are:

1. Kildonnan and Eilean Thuilm, Eigg (NM 495870, NM 483913). Mid Jurassic (?Lower Bathonian). Kildonnan Member, Great Estuarine Group.
2. New Park Quarry, Longborough, Gloucestershire (SP 171296). Mid Jurassic (Lower Bathonian), Chipping Norton Member, Chipping Norton Formation.
3. Stonesfield, Oxfordshire (SP 387171). Mid Jurassic (Middle Bathonian), Stonesfield Member, Sharps Hill Formation.
4. Huntsman's Quarry, Naunton, Gloucestershire (SP 126253). Mid Jurassic (Middle Bathonian), Eyford Member ('Cotswold Slates').
5. Shipton-on-Cherwell Quarry, north-west corner, Oxfordshire (SP 475178). Mid Jurassic (Upper Bathonian), Ardley and Bladon members, White Limestone Formation, Forest Marble Formation and Lower Cornbrash.
6. Kirtlington Old Cement Works, Kirtlington, Oxfordshire (SP 494199). Mid Jurassic (Upper Bathonian), White Limestone Formation to Lower Cornbrash.

MID JURASSIC (BATHONIAN) OF SCOTLAND

The lagoonal facies of the Great Estuarine Group (Bathonian) of the Inner Hebrides in west Scotland have been known as sites for reptiles since the mid-19th century when Hugh Miller noted reptile material, mainly of plesiosaurs, in the Kildonnan Member of Eigg ('Hugh Miller's Bone Bed'). Further remains of reptiles have been found recently throughout the Group from several locations (Harris and Hudson, 1980; Martill, 1985a). Productive Bathonian sites in the Great Estuarine Group are restricted to the Kilmaluag Formation (Figure 6.2), in which some of the youngest tritylodontid mammal-like reptiles, and other specimens of reptiles and mammals, have been found (Waldman and Savage, 1972; Savage, 1984; Waldman and Evans, 1994), and to the Kildonnan Member of the Lealt Shale Formation.

KILDONNAN AND EILEAN THUILM, EIGG (NM 495870, NM 483913)

Highlights

Kildonnan and Eilean Thuilm, Eigg are the site of Hugh Miller's Bone Bed, a famous and extraordinary occurrence in the Middle Jurassic of the Hebrides (Figure 6.3). Reptile bones were first found here in 1844 and 1845 and since then, bones of marine turtles, crocodiles and plesiosaurs have been found. This is unusual, as most other British Bathonian sites represent fully terrestrial situations.

Introduction

Hugh Miller first found reptile bones in the Great Estuarine Group on the northern and eastern coasts of the island of Eigg in 1844 and 1845. The bone bed, now known as Hugh Miller's Reptile Bed, was relocated early this century by the Geological Survey fossil collector Tait (Barrow, 1908) and again in the 1950s, and several small collections of bones have been made since then. The bone-bearing horizons may be seen *in situ* at Kildonnan, while only isolated blocks containing bones have been found on the reworked raised beach opposite the small island Eilean Thuilm.

Hudson (1966) described the location, exposure and sedimentology of the Reptile Bed in great detail. The locality on the north coast, opposite Eilean Thuilm, is readily accessible from the settlement of Cleadale on the western side of the island, by crossing a shoulder of the main raised plateau between Guala Mhor and Leit an Aonaich. The eastern site is now reached in a rather different way from that described by Hudson (1966): it is best to descend the cliff further north from Kildonnan, at the field boundary near NM 492858 named Bealach Clith, where a well-marked path leads diagonally down the cliff northwards. The dolerite sill and the shelter rocks are still to be seen, as Hudson (1966) describes.

Hugh Miller visited Eigg in the Free Church yacht *Betsey* in 1844 and 1845. In 1844 he found reptile bones in loose blocks opposite the island of Eilean Thuilm at the northern tip of the island; he called this locality Ru-Stoir, a name which does not occur on any map. In 1845 he found the bed *in situ* on the eastern coast of the island about midway between the headland Rudha nan

Figure 6.3 Hugh Miller's Bone Bed on Eigg, showing collecting operations in 1972. (Photo: R.J.G. Savage.)

Tri Clach and the settlement of Kildonnan. Miller described the locality in his book *The Cruise of the Betsey* (1858). Miller's collections went to the Royal Scottish Museum, Edinburgh, where they remained unrecognized for nearly a century.

The Reptile Bed was referred to the Lower Shales of the Great Estuarine Series by Barrow (1908), who listed some of the fossils. Hudson (1962) established that the Great Estuarine Group was Late Bajocian and Bathonian in age and that the 'Lower Shales' were equivalent to the *Estheria* Shales of Skye. He designated the outcrop north of Kildonnan, which includes the Reptile Bed, the type locality of the *Mytilus* Shales, a lower subdivision of the Lower Bathonian *Estheria* Shales.

Hudson (1966) rediscovered the precise locations of the Reptile Bed and collected some reptile bones. Barney Newman also made collections in 1961 (Newman, *in* Persson, 1963), but these have never been described. Further collections were made by D.S. Brown (The University, Newcastle upon Tyne) in 1974–7, and these may be described in the future (D.S. Brown, pers. comm., 1993).

Description

Harris and Hudson (1980) revised the stratigraphy of the Great Estuarine Group of the Inner Hebrides, and named the unit with the Reptile Bed the Kildonnan Member of the Lealt Shale Formation. The new name, Kildonnan Member, is directly equivalent to the older term *Mytilus* Shales. It falls in the lower portion of the Great Estuarine Group, near the base of the Bathonian (Hudson, *in* Cope *et al.*, 1980b). Hudson (1966) and Harris and Hudson (1980, p. 239; fig. 6, p. 237) identify the occurrence of two bone beds, the lower horizon being Hugh Miller's Bone Bed ('Reptile Bed') and the upper, occurring 12.5 m above, containing the remains of fish only ('Fish Bed').

The section of the Kildonnan Member exposed in patches on the wave-swept bench and beach is based on Hudson (1966), Harris and Hudson

(1980, p. 237) and J.D. Hudson (pers. comm., 1993):

	Thickness (m)
9. Algal Bed	0.40
8. Limestones with *Placunopsis*	c. 2.00
7. *Unio* Bed	0.22
6. Shales with *Neomiodon*, etc.; includes Bivalve–Septarian Bed and Fish Bed	1.80
5. Shales with *Praemytilus*	c. 12.00
4. Complex Bed (sandstone with abundant phosphatic debris)	1.00
3. Shales with small *Praemytilus*	2.50
2. Reptile Bed, limestone	0.15
1. Shales with fish scales and *Praemytilus* (base not exposed)	3.00
Total	c. 23.07

The Reptile Bed is a very hard, dark grey, shelly sideritic limestone, only a few centimetres thick, which weathers to a deep red on the surface. It contains shells (abundant gastropods and rarer bivalves), as well as black, phosphatic fish scales, teeth and fin spines and black reptile bones. Some layers contain *Unio* shells which often have a nacreous appearance. The fish remains are noted as scales of *Lepidotus*? and teeth of *Hybodus*, *Acrodus* and ?*Saurichthys apicalis* Agassiz (Miller, 1858; Barrow, 1908). Patterson (*in* Hudson, 1966, p. 275) confirmed one or two species of *Hybodus* from Newman's more recent collections. Some of Newman's non-plesiosaurian specimens appear to have come from a greenish marl, which is clearly not the Reptile Bed, but may be from a neighbouring horizon (D.S. Brown, pers. comm., 1993).

Fauna

The reptiles from Hugh Miller's Reptile Bed have never been described fully or figured, but various details may be gleaned from the literature and other sources. Miller (1858) recorded crocodilian ribs and many plesiosaur bones from the northern locality and plesiosaur bones from the eastern locality. Barrow (1908) listed a ?crocodilian tooth, a ?dinosaur vertebra and other reptilian remains from the northern shore, and a pterosaur bone, reptilian vertebra and other bones from the eastern shore.

Newman (*in* Persson, 1963, p. 22; *in* Hudson, 1966, p. 275) noted turtles, crocodilians, turtles

and plesiosaurs in his collections, a fauna confirmed by D.S. Brown (pers. comm., 1993), but Brown stresses that the Reptile Bed itself apparently yields only plesiosaur remains and fish remains. The plesiosaur material includes skull elements, vertebrae, ribs, pelvic and limb elements. None of these remains was found associated, although Martill (1985a) noted an articulated plesiosaur specimen from Eigg (noted as 'Mull' in his account). Newman considers that two kinds of plesiosaurs are present. The reptile collections are located as follows: Miller Collection (NMS); Newman Collection (BMNH R8159-61, and unnumb.); Brown Collection, Newcastle upon Tyne, Dental School (temporary). Brown (pers. comm., 1983), lists the known reptile remains as:

1. ?crocodilian elements (Newman Coll.; Brown Coll.);
2. ?turtle bones (Newman Coll.; Brown Coll.);
3. plesiosaur: vertebrae, ribs, limb and girdle elements, several teeth and disarticulated skull bones of an elasmosaur, a long-necked form, rather like the Late Jurassic *Muraenosaurus*.

Interpretation

Hudson (1962, 1966) suggested that the Great Estuarine Group was deposited in shallow lagoons with variable, but generally low, salinity. Harris and Hudson (1980) noted desiccation cracks and algal stromatolites at several horizons which demonstrated that the lagoons at times dried out, or had extensive marginal mud flats. The rarity of *in situ* plant remains further suggested that, during Great Estuarine Group times, the Inner Hebrides area never became a fully vegetated land surface. The Reptile Bed could have originated as a winnowed shell and bone concentrate on the lagoon floor. The abundant and well-preserved *Unio* shells cannot have travelled far, and they suggest that the water could have been almost fresh at times. Dispersed reptile bones and fish remains are to be accounted for by slow rates of deposition rather than by long-distance transport.

Vertebrate remains, consisting mainly fish teeth and scales, are common throughout the Great Estuarine Group, but only in the Kildonnan Member of the Lealt Shales on Eigg are they concentrated to form bone beds, and it is only at this location that the bones of marine reptiles are found in any abundance.

Comparison with other localities

The most closely comparable localities to Hugh Miller's Reptile Bed occur on Skye in units of the Great Estuarine Group:

1. 'Vertebrate Beds' with reptile bones in the upper part of the Kilmaluag Formation (formerly called Ostracod Limestones) near the top of the Great Estuarine Group, on the northern side of Glen Scaladal, Elgol (NG 519165; Harris and Hudson, 1980, pp. 244–6, who give the map reference in reverse). Waldman and Savage (1972) noted true mammals and therapsids (*Stereognathus hebridicus*) from 'marlstone bands' at their unspecified locality UB 7111, which is essentially the same site (NG 520157, fide Evans and Milner, 1994). Also the site of Waldman and Evans's (1994) choristodere skull, from a location on the shore just north of Glen Scaladal.

2. Kilmaluag Formation (?) on the shore and cliff south of Glen Scaladal 'at a small promontory of limestone about 100 m South of Carn Mor' (J.D. Hudson, pers. comm., 1982) – ?the promontory at NG 519154 (Andrews, 1985, p. 1135).

3. Hudson and Morton (1969, p. D29) noted a plesiosaur from the Bearreraig Sandstone (Late Bajocian) near Rigg, Trotternish (NG 521566).

4. Andrews and Hudson (1984) reported a single large cast of a dinosaur footprint from the Lonfearn Member of the Lealt Shale Formation, beneath the cliffs south of Rudha nam Braithairean (NG 526625), Trotternish, Skye.

5. Martill (1985a, p. 162) notes six ichthyosaurs from the Great Estuarine Group of the Isle of Skye. No more locality information given.

Elsewhere in Britain, plesiosaur remains are rare in the Bathonian. A plesiosaur humerus (CAMSM J5736) has been found in the Cornbrash (Callovian) of Scarborough (?exact locality), and several plesiosaur teeth, vertebrae and limb bones are known from the Stonesfield Slate (Mid Bathonian) of Stonesfield, Oxfordshire (BMNH, CAMSM – see Stonesfield report). Occasional teeth and isolated bones ascribed to plesiosaurs have been noted from several other localities in the Bathonian of Oxfordshire, Buckinghamshire and Cambridgeshire, but none of these matches the more extensive remains from Eigg.

Conclusions

The two localities of Hugh Miller's Reptile Bed on Eigg are historically important, and they have yielded small collections of identifiable reptile bones. The sites are still accessible, depending on how storms have moved the loose boulders on the beach, and the Reptile Bed is readily identifiable, but thin. The sites are better documented than their approximate equivalents on Skye. There are no comparable formations outside the Hebrides. These are the best Bathonian sites for marine reptiles and an important intermediate between the well-known Early and Late Jurassic marine faunas, hence their national importance and considerable conservation value.

MID JURASSIC (BATHONIAN) OF SOUTHERN ENGLAND

The remains of dinosaurs, crocodilians, pterosaurs, plesiosaurs and turtles are known from a large number of localities in the Bathonian of Gloucestershire and Oxfordshire (Figure 6.1). Many of these are important from an historical point of view as well as being productive sources of reptiles. The majority of the remains come from rocks that are interpreted as subaerial lagoonal and lacustrine deposits, and these document a number of major faunal changes. Most notable was the disappearance of the tritylodontids, the last surviving members of the essentially Permo-Triassic mammal-like reptiles. A number of new groups appeared, including the choristoderes, the avialan dinosaurs, and possibly the lizards (see Figure 6.5). These British records are international 'firsts' and 'lasts' (more detail in the introduction to Chapter 4).

The GCR sites, New Park Quarry (SP 171296), Stonesfield (SP 387171), Huntsman's Quarry (SP 126253), Shipton-on-Cherwell (SP 475178) and Kirtlington (SP 494199), provide a coverage of faunas which range in age from the Early to the Late Bathonian.

NEW PARK QUARRY, LONGBOROUGH, GLOUCESTERSHIRE (SP 176282)

Highlights

New Park Quarry is the source of specimens of fossil crocodiles and dinosaurs, including the theropod *Megalosaurus*, the sauropod *Cetiosaurus* and the stegosaur *Lexovisaurus*. The site is important as one of the oldest Bathonian sites in the world.

Introduction

New Park Quarry, Longborough, 3 km NNW of Stow-on-the-Wold, has yielded the finest fauna of Mid Jurassic (Bathonian) dinosaurs in Britain this century. The quarry was in operation in the 1920s when the remains of several crocodiles came to light (Richardson, 1929). Around 1935, a collection of well-preserved dinosaur bones, representing several genera, was obtained (Gardiner, 1935), and the British Association for the Advancement of Science sponsored an excavation that produced many more specimens (Gardiner, 1937, 1938).

Figure 6.4 Exposure of the Chipping Norton Limestone Member in New Park Quarry. Reptile bones were recovered from the top of the underlying Hook Norton Limestone Member, in the floor of the quarry. (Photo: M.J. Benton.)

Reynolds (1939), Galton and Powell (1983) and Galton (1985b) described the stegosaur remains, some of the oldest of that group in the world, but the other material is yet to be studied. The quarry still offers good exposures in the Chipping Norton Limestone Member and could be re-excavated for further reptile finds (Figure 6.4).

Description

H.B. Woodward (1894, pp. 143-4) briefly described the geology of New Park Quarry, and Richardson (1929, p. 89) added a section. Arkell and Donovan (1952, p. 249) gave a fuller account with stratigraphic information:

	Thickness	
	ft	**in**
Chipping Norton Limestone		
6. Flaggy oyster limestone. *Exogyra* sp.	2	0
5. Marl with oysters, chiefly *Exogyra* sp.		10
4. False-bedded, white, shelly oolite	*c.* 18	0
?Roundhill Clay		
3. Impersistent seam of clay and locally sandy marl with small irony claystone pellets	nil to	8
Hook Norton Limestone		
2. Nodular buff limestones, with dark fossil bones 1-2 feet down. *Parkinsonia neuffensis* auct. from this bed, *teste* P.J. Channon	5	0
1. Brownish and buff limestones with sandy marl partings	seen to 5	0

The Chipping Norton Limestone (5-6 m thick) consists of wavy-bedded white or cream coloured sandy limestone which weathers to a reddish colour in places. Low-angle cross-beds are present. The rock takes on a more nodular appearance in the lower portions. Richardson (1929, p. 89) reported that the 'black pebble-like bodies' (in bed 3 of Arkell and Donovan 1952) consisted of manganese, limonite, calcium carbonate and phosphate. The lower unit (1-2 m visible; bed 1 of Arkell and Donovan, 1952) is similar to the main limestone, but appears to be less nodular.

Richardson (1929, p. 89) noted crocodiles (9 fragments of a mandible of *Steneosaurus* and a scute of *Teleosaurus subulidens*) in the Chipping Norton Limestone (Arkell's bed 4), and the bones collected in the 1930s were found in a hard cream-coloured limestone which was worked for

road metal (Reynolds, 1939, p. 193). Arkell and Donovan (1952, p. 249) clearly identify their bed 2 as the source of bones of *Steneosaurus* and *Megalosaurus*, and they indicate that the other bones in the British Museum and the Stroud Museum came from this horizon.

The sequence at New Park Quarry spans most of the Chipping Norton Formation, which is dated as belonging to the *zigzag* Zone, the basal zone of the Early Bathonian (Torrens, *in* Cope *et al.*, 1980b). The Chipping Norton Limestone (=Chipping Norton Member) is ascribed to the *yeovilensis* Subzone of the *zigzag* Zone on the basis of specimens of the ammonite *Oppelia* (Torrens, 1969b, p. 74). The Hook Norton Limestone (=Hook Norton Member) at New Park Quarry has yielded specimens of the ammonites *Parkinsonia neuffensis* Oppel and *P. subgalatea* Buckman (Reynolds, 1939; Arkell, 1951-8, p. 160; Arkell and Donovan, 1952, p. 249), which indicate a zonal assignment to the *convergens* Subzone (Torrens, 1969b), the lowest Subzone of the *zigzag* Zone, hence earliest Bathonian (Torrens, 1969b; Cope *et al.*, 1980b, fig. 6a). Thus, some of the crocodiles come from the top of the *zigzag* Zone, and the dinosaurs and some crocodiles from the base of the *zigzag* Zone.

The reptile remains from New Park Quarry in museum collections have been prepared so that no matrix remains; hence, nothing can be said of the relationship of the bones to the sediment nor of the relative association of the bones. Nevertheless, it seems evident that all elements were disarticulated and must have been transported at least a short distance. There is no sign of major abrasion to the bones, although some delicate processes have been lost.

Fauna

The reptilian fauna from New Park Quarry was described by Reynolds (1939), and the species are listed together with a note of major specimens in the BMNH, BGS(GSM) and SDM. The Stroud Museum display of these specimens is described by Walrond (1976):

Archosauria: Crocodylia: Thalattosuchia: Steneosauridae
  Steneosaurus cf. *subulidens* (Phillips, 1871)
    BMNH R.6307; BGS(GSM) 37520-2 (mentioned, Richardson 1929, p. 89); SDM 44.42, 44-8, 51-7, 59-68

Dinosauria: Saurischia: Theropoda:
  Megalosauridae
  Megalosaurus sp.
    BMNH R.9666-76, R.9678-9, R.9681-7,
      R.9689-700; SDM 44. 1, 4-10, 12-5, 18,
      21, 23-6

Dinosauria: Saurischia: Sauropoda: Cetiosauridae
  Cetiosaurus sp.
    SDM 44. 30-40

Dinosauria: Ornithischia: Stegosauria:
  Stegosauridae
  Lexovisaurus? vetustus (Huene, 1910)
    SDM 44. 41; BMNH R.5838

Interpretation

The remains of the marine crocodile *Steneosaurus* include skull bones, a braincase, several partial snouts, isolated teeth, scutes, vertebrae and a scapula. The New Park crocodile evidently had a skull 0.3-0.6 m long and probably a total body length of 1.5-3.0 m, thus a fairly large animal. Comparisons are difficult because of the confused taxonomy of Jurassic marine long-snouted crocodiles and, in particular, the distinction between the genera *Pelagosaurus, Steneosaurus* and *Teleosaurus* and the multiplicity of their included species (Westphal, 1962; Steel, 1973). Bathonian species include *S. boutilieri, T. cadomensis, T. geoffroyi* and *T. gladius* from the 'Fullers Earth' and 'Great Oolite' of Normandy; *S. brevidens, S. latifrons, T. cadomensis* and *T. subulidens* from the 'Great Oolite' of Oxfordshire and Northamptonshire; and *S. stephani* from the Cornbrash of Closworth, Somerset. All of these specimens come from the Mid or Late Bathonian, younger than the New Park material.

The remains of *Megalosaurus* include a good sacrum, some caudal vertebrae, a rib, two coracoids, a scapula, a humerus, three ischia, a femur, a metatarsal and a partial lower jaw. The postcranial bones are well preserved, and the lower jaw shows seven teeth in various stages of growth. Other Mid Jurassic material of *Megalosaurus* is known from the Upper Inferior Oolite of Sherborne, Dorset (*M. hesperis* Waldman, 1974; *M. nethercombensis* Huene, 1923), the Stonesfield Slate (*sensu lato*) of Oxfordshire, Gloucestershire and Dorset (*M. bucklandi* Meyer, 1835; *M. bradleyi* Woodward, 1910; *M. incogni-*

tus (Huene, 1932)), the Oxford Clay of Dorset (*M. parkeri* Huene, 1926) and the Bathonian of Morocco (*M. mersensis* Lapparent, 1955). The New Park material could belong to any of these species, or to some other.

The bones of the sauropod *Cetiosaurus* from New Park Quarry include ribs, a coracoid and a pair of ischia. Because of their size (the ischia are 1.0 m long), these specimens undoubtedly belong to *Cetiosaurus*, but they cannot be assigned to a species since the taxonomy is in need of revision. At least eleven species have been described from the Mid and Late Jurassic of England, France and Morocco, and the Early Cretaceous of England. The Mid Jurassic forms are *C. rugulosus* (Owen, 1841), *C. oxoniensis* Phillips, 1871 and *C. glymptonensis* Phillips, 1871 from the 'Great Oolite' of Wiltshire, Gloucestershire, Oxfordshire and Northamptonshire, and *C. mogrebiensis* Lapparent, 1955 from the Bathonian of Morocco. It has never been made clear what the diagnostic characters of each species are supposed to be. The New Park sauropod is slightly older than all of these forms.

The 'dermal plates of *Stegosaurus*' from New Park represent some of the earliest records of stegosaurid dinosaurs. Galton (1983c) and Galton and Powell (1983) described a dorsal vertebra (OUM J29770) and a cervical centrum (OUM J29827) from Sharps Hill, Oxfordshire as the oldest British stegosaurid, and this claim was repeated by Boneham and Forsey (1991) who reported further stegosaur remains from Sharps Hill. However, the Sharps Hill Member lies partly in the *progracilis* Zone, the earliest zone of the Mid Bathonian, and partly in the *tenuiplicatus* Zone, the topmost zone of the Early Bathonian (Torrens, 1968; Torrens *in* Cope *et al.*, 1980b), hence above the Chipping Norton Formation. Therefore, the New Park stegosaur finds are one or two ammonite zones older than those from Sharps Hill, and in any case, *Tatisaurus oehleri*, from the Lower Lufeng Formation (Sinemurian-Hettangian) of Yunnan Province, China, hitherto regarded as an unidentifiable ornithischian, may be a true stegosaur (Dong, 1990).

Reynolds (1939) referred the two large dermal plates to the North American genus *Stegosaurus*, but Galton *et al.* (1980, p. 41) disputed this referral and suggested that the plates were probably the sacral ribs of a sauropod. However, later, Galton and Powell (1983) and Galton (1985b), in their reviews of Bathonian stegosaurs from

England, provisionally reassigned the specimens to the Stegosauridae. They noted that the plates were reasonably massive when compared with the complete series of plates of *Stegosaurus* and the few known plates of *Lexovisaurus*, and tentatively referred them to *Lexovisaurus ?vetustus* (Huene, 1910). The New Park stegosaur must have been large, since the plates are 0.25 m high.

Stegosaurs have been reported from higher in the British Mid Jurassic, from the Lower Oxford Clay (Callovian) of the Peterborough area. Hulke (1887) noted remains of a stegosaur which he termed *Omosaurus* (=*Dacentrurus*), believing it to come from the Kimmeridge Clay; these represent *Lexovisaurus durobrivensis* (Galton, 1985b). Hulke also reported armour plates, but these turn out to be from the giant teleost fish *Leedsichthys* (Martill, 1988). Stegosaur armour

plates were later found in the area. A second Oxford Clay stegosaur was '*Stegosaurus*' *priscus* from Fletton (Nopsca, 1911), synonymized by Galton (1985b) with *L. durobrivensis*. Other isolated remains of this species have been reported from the Lower Oxford Clay of Whittlesey and Weymouth.

Comparison with other sites

The fossil reptile sites nearest in age to New Park are Oakham Quarry (SP 279306), which has produced remains of *Megalosaurus* and *Cetiosaurus*, Longborough Road Quarry (SP 171296), which has yielded *Steneosaurus*, Sharps Hill Quarry (SP 338358), which has yielded *Lexovisaurus* and Hornsleasow Quarry (SP 131322), which has yielded a variety of crocodilian and dinosaur remains. Oakham Quarry lies in the Chipping

Figure 6.5 Scene in Early Bathonian times, showing a small lake in Gloucestershire surrounded by seed ferns and conifers. Fishes (*Lepidotus*) live in the water, and frogs (*Eodiscoglossus*) disport themselves around the sides. Dinosaurs include some of the earliest stegosaurs and maniraptorans (?), plated and small carnivorous dinosaurs respectively. A carcass of the large sauropod, *Cetiosaurus*, is rotting in the water, and *Megalosaurus* scavenges. Lizard-like animals, crocodiles, pterosaurs, mammals and tritylodont mammal-like reptiles complete the scene. Based on a restoration painting by Pam Baldaro, showing the scene at Hornsleasow Quarry, Gloucestershire. Reproduced with permission of the University of Bristol.

Norton Formation (*zigzag* Zone) and Longborough Road Quarry is in the Inferior Oolite (Aalenian, Bajocian) (Richardson, 1911a, pp. 227-8; Arkell and Donovan, 1952, pp. 248-9). Sharps Hill Quarry is dated as *tenuiplicatus* Zone and Hornsleasow as *zigzag* Zone, both Early Bathonian.

Richardson (1929, p. 88) recorded a vertebra of *Megalosaurus* (BGS(GSM) 37523) from a quarry in the Chipping Norton Formation in a field near the Fosse Way (?SP 193271). Richardson (1929, p. 95) also mentioned that bones had been found in the Chipping Norton Formation of a quarry ENE of Swell Buildings (?SP 164264). Only Oakham Quarry and Hornsleasow Quarry come near to New Park for the abundance and diversity of their reptile remains, but better comparisons may be made with the richer faunas of the White Limestone Formation (e.g. Eyford, Chipping Norton, Stonesfield, Slape Hill, Enslow Bridge, Kirtlington). Hornsleasow has proved extremely productive as a result of recent studies (Metcalf *et al.*, 1992), and has yielded thousands of microvertebrate remains, scales and teeth of fishes, and teeth and bones of a diverse array of amphibians (frogs, salamanders), reptiles (turtles, 'lizards', choristoderes, crocodilians, dinosaurs, pterosaurs, tritylodontids) and rarer mammals (Figure 6.5). However, that site has been essentially worked out, since the bulk of the fossiliferous clay containing fossils has been removed from the site.

The New Park stegosaur (*Lexovisaurus*? *vetustus*) is also known from isolated remains in the Lower Cornbrash (*discus* Subzone, Late Bathonian of Oxfordshire) and from the Sharps Hill Formation, Oxfordshire (*tenuiplicatus* Zone; latest Early Bathonian) (Galton, 1985b).

Elsewhere, the crocodiles are comparable with those from the Fuller's Earth of Caen and Calvados, Normandy (?Early/Mid Bathonian), and the megalosaur and cetiosaur with remains of these forms from the Bathonian of El Mers in the Moyen Atlas of Morocco (Steel, 1970, pp. 36, 64-5).

Conclusions

New Park Quarry is a very important site for terrestrial fossil reptiles, especially in view of its age. Several other sites in the Chipping Norton Limestone have yielded isolated reptile remains, but the fauna from New Park Quarry is by far the richest for large-sized reptile remains. It is comparable to many of the younger and better known sites in the Mid and Late Bathonian north of Oxford and at Eyford, Gloucestershire. The preservation of the reptiles from New Park Quarry is excellent. They will be of great value in assessing the relationships of early theropods and sauropods. The dermal plates from New Park may be the oldest remains of true stegosaurids in the world. The fact that the fossils were discovered this century suggests that more may be forthcoming if the site is excavated and enhances the conservation value.

STONESFIELD, OXFORDSHIRE (SP 387171)

Highlights

Stonesfield is the most important of the British Bathonian localities in the Cotswolds, and arguably the best Middle Jurassic terrestrial reptile site in the world. It is the source of over 15 species of fossil reptiles, including turtles, crocodilians, pterosaurs, dinosaurs and rare marine forms (ichthyosaurs, plesiosaurs), as well as mammal-like reptiles and mammals.

Introduction

The series of quarries and mines which formerly worked the Stonesfield Slate at Stonesfield (Figure 6.6) are famous for yielding one of the most diverse reptile faunas of Mid Jurassic time known. The remains are all isolated elements, but include those of terrestrial animals such as saurischian and theropod dinosaurs, pterosaurs and a tritylodont mammal-like reptile, as well as aquatic turtles and marine crocodilians, ichthyosaurs and plesiosaurs. Specimens include the type materials of *Megalosaurus bucklandi*, the first dinosaur to be described (Buckland, 1824). Although the quarrying industry at Stonesfield is now extinct, re-excavation could produce many more finds.

The Stonesfield 'slates', or 'tilestones' (Richardson *et al.*, 1946, p. 33) of Stonesfield have been well known for their fossils, which contain an unusual mixture of marine, freshwater and terrestrial forms (Arkell, 1947a, pp. 40-1). In Roman times local country houses were roofed with squared slabs of limestone 'slate'. In the 16th or 17th century it was discovered that when the

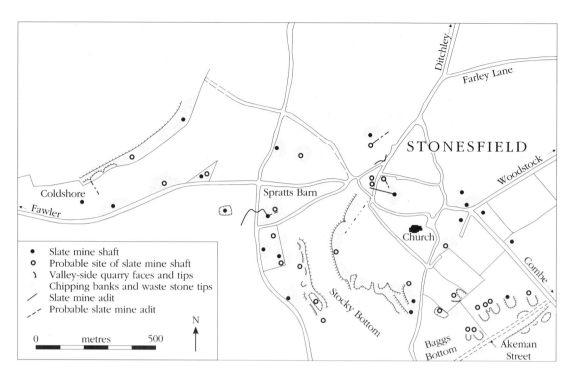

Figure 6.6 The Stonesfield Slate mines. Map based on ground surveys and studies of historical records by Aston (1974).

freshly dug stone was exposed to the frost, it would split into thinner sheets. The quarries continued and expanded production, providing roofing materials for local houses and building material for more important buildings further afield (Arkell, 1947b), and they remained productive until the late 19th century. The stone was reached by vertical shafts, usually about 6 m deep, and horizontal galleries which were driven through the bed. During the 18th and 19th centuries slate-digging was a major industry (Figure 6.6) and employed numerous craftsman. The slate-makers examined each slab and put aside fossils for sale to tourists and dons at Oxford and other universities. The last mine closed in 1911 (Arkell, 1947b; Aston, 1974, p. 35).

Stonesfield has figured prominently in the history of the study of dinosaurs (Swinton, 1970; Delair and Sarjeant, 1975). The first reptile fossil to be described from Stonesfield was a megalosaur tooth figured by Edward Lhuyd (1699, pl. 16, opposite p. 63). Joshua Platt (1758) described three large vertebrae and a left femur as the 'fossile thigh-bone of a large animal', now identified as probably that of *Megalosaurus* or *Cetiosaurus*. The femur was 29 inches (0.81 m) long and was found surrounded by fossil shells of 'sickle oysters' (*Liostrea acuminata*). A partial scapula of *Megalosaurus* from Stonesfield was acquired by

the Woodwardian Museum, Cambridge in 1784, but it has never been described (Delair and Sarjeant, 1975). An anonymous article published in the *Gentleman's Magazine* for 1757 (A.B., 1757) records some finds thought to be of significant note, and suggests that the site was well known to the wider public for its large fossil bones well before the 1820s, contrary to what is normally assumed: 'But I think we can boast of as great a variety [of fossils] . . . at a small village called Stonesfield, near Woodstock in this country'. This records sharks' teeth, fish palates, echinoids, oysters, belemnites, 'nautilites', ferns and 'bones of quadrupedes, ribs, vertebrae &c. some of birds . . .'. Also mentioned is Joshua Platt's 'thigh bone', as that of 'the *Hippopotamus*, or sea-horse'. The anonymous author goes on to say 'I formerly met with two pieces of bone and some vertebrae of the same kind, and of a proportional bulk, at the same place. . . . Those I have been speaking of must [by analogy] be the remains of some animal of greater bulk than the largest ox'.

William Buckland (1784–1856) is well known as the describer of *Megalosaurus* from Stonesfield, although it would be truer to say that he was the first person to realize its reptilian nature. He did not record details of when he acquired his specimens, as did Mantell who col-

Figure 6.7 View of the entrance to an adit at Stonesfield, re-opened in 1980 in an operation funded by the Nature Conservancy Council. (Photo: W.A. Wimbledon.)

lected material of *Iguanodon* in 1822 and earlier. Archive evidence suggests that Buckland had good specimens of *Megalosaurus* from Stonesfield around 1818 (Delair and Sarjeant, 1975), and the 'huge lizard' was well known to Cuvier, Parkinson, Conybeare, Mantell and others before its eventual description in 1824 (Buckland, 1824). Further abundant remains of reptiles were collected during the 19th century, but finds ceased with the extinction of the quarrying industry. Figure 6.7 shows a cleaning exercise when an adit at Stonesfield was re-opened in 1980.

Description

Stratigraphic sections through the type Stonesfield Slates have been given by Fitton (1836, p. 412), Phillips (1871, pp. 148–9), H.B. Woodward (1894, pp. 29–33, 312), Walford (1895, 1896, 1897), Richardson *et al.* (1946, pp. 29–33), McKerrow

and Baker (1988, pp. 63, 64) and Boneham and Wyatt (1993). Richardson *et al.* (1946, p. 30) gave the following section from the sides of a shaft at Stonesfield:

	Thickness (ft)
[White Limestone]	
Rubbly limestone	
[Hampen Marly Beds]	
Clay with *Terebratulites*	
Limestone	
Blue clay	
Oolite	
Blue clay	in total 32
[Taynton Stone]	
'Rag', consisting of shelly oolite, with casts of bivalves and univalves	*c.* 25
[Stonesfield Slate Beds]	
'Soft stuff', yellowish sandy clay, with thin courses of fibrous transparent gypsum	0.5
'Upper Head', sand enveloping a course of spheroidal laminated calcareous gritstones which produce the slate. These are called 'Potlids', from their figure, and receive with the other slaty bed the name of 'Pendle' as characteristic of workable stone. The stone is partially oolitic and shelly, sometimes full of small fragmentary masses	1.5
'Manure' or 'Race', slaty friable rock	1
'Lower Head', sand and grit, including a course of spheroidal concretions of slate, as above	1.5–2
'Bottom stuff', sandy and calcareous grit, with admixture of oolitic grains	1
[Chipping Norton Limestone]	

The Stonesfield Slate (=Stonesfield Member) consists of quartz sands and siltstones with fine laminae (0.1–0.3 m apart) of ooliths. A 20 mm thick conglomerate, containing clasts of limestones from the underlying Chipping Norton Formation and bored pebbles of other oolites, occurs in the middle of the unit (Sellwood and McKerrow, 1974). The Stonesfield Slate is no more than 1.8 m thick at its type locality and it is confined to an elliptical area within 1.5 km around Stonesfield (Aston, 1974).

The fauna of the Stonesfield Slate consists of marine invertebrates (rare ammonites, belemnites, large numbers of bivalves and gastropods [80 species altogether], rarer brachiopods, crustaceans, annelids and corals), land-derived plants (13 species), insects (seven species), about 40 species of fish, including sharks, 'holosteans' and a species of *Ceratodus*, as well as reptiles, and mammals (*Amphilestes, Phascolotherium, Amphitherium*) (Phillips, 1871, pp. 167–237; H.B. Woodward, 1894, pp. 314–17; Richardson *et al.*, 1946, pp. 28–9).

The bone in the Stonesfield Slate is well preserved and rarely abraded, although delicate processes may be broken off. The remains range from small elements (e.g. teeth, scutes, pterosaur limb bones) to complete vertebrae and partial skulls. Skeletons are disarticulated. Thus, there is evidence of short-term transport and sometimes violent breakage, and the bones may be associated with other coarse clasts (pebbles, shells, etc.).

Fauna

Major collections may be seen today in the BMNH, BGS(GSM), CAMSM and OUM. Most older university, city and private fossil collections in Britain have some teeth or bone scraps from Stonesfield, but it is clearly pointless to record all of these. The type specimen numbers are noted and an estimate of the numbers of specimens of each species in major collections is appended:

	Numbers
Testudines	
Protochelys stricklandi	
(Phillips, 1871)	
Type specimen: OUM	1
Archosauria: Crocodylia: Thalattosuchia:	
Steneosauridae	
Steneosaurus boutilieri	
Deslongchamps, 1869	
Steneosaurus brevidens	
(Phillips, 1871)	1
Teleosaurus ?geoffroyi	
Deslongchamps, 1867	4
Teleosaurus subulidens	
Phillips, 1871	
Type specimen: OUM J1419	1
Steneosaurus/Teleosaurus sp.	125+

	Numbers
Archosauria: Pterosauria:	
'Rhamphorhynchoidea'	
Rhamphocephalus bucklandi	
(Meyer, 1832)	
Type specimens: ?OUM J23043,	
23047–8, 28266, 28297,	
28311, 2831	250+
Rhamphocephalus depressirostris	
(Huxley, 1859)	
Type specimen: BMNH 47991	
[?Stonesfield or Sarsden]	2
?Rhamphocephalus sp.	125+
Dinosauria: Saurischia: Theropoda	
Iliosuchus incognitus Huene, 1932	
Type specimen: BMNH R83	2
Megalosaurus bucklandi Meyer, 1832	
Type specimen: OUM J12142	
(=OUM J13505)	110+
Dinosauria: Saurischia: Sauropoda:	
Cetiosauridae	
Cetiosaurus oxoniensis Phillips, 1871	3
Dinosauria: Ornithischia: Ornithopoda	
'hypsilophodontid'	1
Sauropterygia: Plesiosauria	
'*Cimoliasaurus*'/*Plesiosaurus* sp.	17
Ichthyopterygia: Ichthyosauridae	
Ichthyosaurus ?advena Phillips, 1871	
(*nom. dub.*)	2
Synapsida: Therapsida: Cynodontia:	
Tritylodontidae	
Stereognathus ooliticus	
Charlesworth, 1855	
Type specimen: BGS(GSM) 113834	1

Interpretation

The highly localized distribution of the Stonesfield Slate around the village of Stonesfield (Figure 6.6) is best explained by the deposition of clastic sediments during a transgressive event within isolated hollows above an intermittent hardground which occurs at the top of the Chipping Norton Formation (Sellwood and McKerrow, 1974, p. 206). Sellwood and McKerrow (1974, pp. 204–5) note sedimentary structures indicative of deposition of the Stonesfield Slate in upper-flow regime conditions. Storm-produced scours filled with shell-lags occur.

The fossil content points to a shallow-marine environment with a large input of terrestrial material. The bones, plants and insects may have been concentrated and preserved by rapid burial in sands brought offshore by storm-induced rip-currents. The features of bone preservation in a disarticulated state and, in coarse clastic units, point to sorting and rapid deposition, possibly during storm-events.

As in much of the Great Oolite Group of Oxfordshire, the clastic sediments and the land-derived plants and animals reflect the influence of the nearby London landmass, but the ammonites indicate that the Stonesfield Member is one of the few beds in the Bathonian of Oxfordshire to be deposited in proximity to open marine conditions.

The dating and precise stratigraphic position of the Stonesfield Slate has been problematic because of its limited exposure and outcrop, and because of the small number of ammonites. Fitton (1836) placed it below the Taynton Stone (Taynton Limestone Formation) by correlating it with the Cotswold Slate. Hull (1860) and H.B. Woodward (1895) recorded information that suggested that the Slate Beds lay at or near the top of the Taynton Stone. At this time, most authors assumed that the Stonesfield Slate was laterally equivalent to the 'Stonesfield Slate' of Eyford, Gloucestershire (the Cotswold Slate or Eyford Member), and even with units near Bath and in Northamptonshire; it was regarded as a handy marker bed for the base of the Great Oolite. However, as Arkell (1947a, pp. 37–41) pointed out, the Sharps Hill Clays and thin limestones (the Sharps Hill Member of the Sharps Hill Formation; Torrens, *in* Cope *et al.*, 1980b) are present over much of north Oxfordshire while, in the vicinity of Stonesfield, the lithologically similar Stonesfield Slate (the Stonesfield Member of the Sharps Hill Formation) comprises a separate unit.

Arkell (1931; 1933, pp. 294–7) and Richardson *et al.* (1946, pp. 25–33) placed the Stonesfield Slate between the Sharps Hill Clays and the Taynton Stone on the basis of correlations with the Ashford Mill railway cutting section (SP 387159). The 'Coral and *Rhynchonella* Bed' there was ascribed to the 'Stonesfield Slate Beds' and the Stonesfield Slate itself, which was absent there, was supposed to occur between this and the Sharps Hill Beds below. Arkell (1947a, pp. 38–41) suggested that the Stonesfield Slate occurred immediately below the Taynton Stone and that its lateral equivalent was the Upper Sharps Hill Beds (Torrens, 1968, p. 231). Subsequently, Sellwood and McKerrow (1974) suggested that the sandy beds of the Stonesfield

Member could represent the reworked top of the underlying sandy beds of the Chipping Norton Formation. They would then be expected to rest directly on sandy limestones or calcareous sandstones rather than on clays, and should thus lie below the clays of the Sharps Hill Member in those localities where both members are present. However, Torrens (*in* Cope *et al.*, 1980b, fig. 6a, column B12) was not convinced by these arguments and continued to assume that the Stonesfield Slate was above the Sharps Hill Clays.

McKerrow and Baker (1988) examined the stratigraphy of two newly opened shafts at Stonesfield (as part of the GCR programme) (Hillside, SP 39181730; Home Close, SP 39181724) and in an adit to the west of the village, and compared these with the greatly extended workings in Town Quarry, Charlbury (SP 365189). They corroborated the view of Sellwood and McKerrow (1974), demonstrating that when the clay-rich Sharps Hill Member and flaggy limestones of the Stonesfield Member occur together at the same locality, the Stonesfield Member underlies the Sharps Hill Member and rests directly on the Chipping Norton Formation (both in the Sharps Hill Formation), whereas at Home Close and Hillside the Stonesfield Member is overlain directly by oolitic limestones (Taynton Limestone Formation). Boneham and Wyatt (1993) suggest that the Stonesfield Slate was formerly worked at three levels within the Taynton Limestone Formation, and that the 'Stonesfield Member' can no longer be regarded as a valid subunit of the Sharp's Hill Formation.

The ammonite fauna of the Stonesfield Member consists of 10 species belonging to the genera *Clydoniceras, Microcephalites, Oppelia (Oxycerites), Paroecotraustes, Turrelites* and *Procerites (P. progracilis, P. mirabilis, P. magnificus)* (Arkell, 1951–8, p. 240). There are problems of identification of some of these, and of comparison with similar specimens elsewhere in Britain and abroad (Torrens, 1969b, pp. 71–3; *in* Cope *et al.*, 1980b, p. 38). This fauna has been assigned to the *progracilis* Zone (early Mid Bathonian) with the Stonesfield Member at Stonesfield as the stratotype (Torrens, 1974, pp. 586–7).

The chelonian species *Testudo stricklandi* was established by Phillips (1871, p. 182) for isolated scale-like elements found in the Stonesfield Slate (Blake, 1863). Mackie (1863) had also announced the discovery of a turtle coracoid from Stonesfield, and had named it *Chelys* (?)*blakii.*

Lydekker (1889b, pp. 220-2) combined these in his new genus *Protochelys*. The scales are unusual in that they lack bony material, and in that they are supposedly from the crest of the back of a turtle carapace. The coracoid is probably indeed chelonian, but the scales may be remains of invertebrates, though the possibility remains that they might represent shed turtle scutes (certain extant turtles shed their carapace scutes in order to facilitate growth). We have found no recent discussion of the Stonesfield 'chelonians', although Romer (1956, 1966) classes *Protochelys* tentatively as a pleurosternid turtle, a group typical of the Jurassic. The most recent monograph on turtles (Młynarski, 1976) makes no mention of them, and quotes *Protochelys* Williston (1901) for a Late Cretaceous American animal: clearly a preoccupied name. Turtles are rare in the Mid Jurassic but, until the exact nature of the Stonesfield specimens is reassessed, nothing can be said of their significance.

The long-snouted crocodilians from Stonesfield include four species of *Teleosaurus* and *Steneosaurus*. They had long curved teeth and were clearly fish-eaters in fresh and marine water. These genera are well known from the Bathonian of Britain and France, but the Stonesfield specimens are extremely well preserved and formed the basis of a revision (Phizackerley, 1951). *T. subulidens* Phillips (1871) may be a synonym of *T. cadomensis* Geoffroy St-Hilaire (1825). The type specimens of this species, and of the others from Stonesfield, came from the Fuller's Earth and Great Oolite of Normandy, but they were all destroyed at Caen in 1944. Thus, the Stonesfield specimens take on an added significance. Although *Steneosaurus* and *Teleosaurus* are common in the English Bathonian from localities in Somerset, Avon, Wiltshire, Gloucester, Oxfordshire and Northamptonshire, only Enslow Bridge, Oxfordshire has more than one or two species. Stonesfield represents the best site in Britain for Mid Jurassic crocodiles in terms of abundance and variety.

The pterosaur *Rhamphocephalus* is known from the Stonesfield Slate (Figure 6.8A) and the Eyford Member (=Cotswold Slate) of the Eyford area, Gloucestershire (Kyneton Thorns Quarry (SP 122264), ?Sevenhampton (SP 0321), ?Huntsman's Quarry (SP 125254)). By far the best and most abundant material comes from Stonesfield, and it may include type specimens of two or three species. The remains show a 120 mm long mandible with long pointed teeth. The restored wing is about 0.75 m long, and individual vertebrae 25 mm long. Details of the bone microstructure were also observed (Huxley 1859d; Phillips, 1871, pp. 219-29). There is a problem over the original locality of the type specimen of *R. depressirostris*: Huxley (1859d) and Wellnhofer (1978, p. 41) give the locality as Sarsden ('Smith's Quarry': SP 300226), whereas the label on BMNH 47991 and Lydekker's Catalogue (1888a, p. 36) give the locality as Stonesfield. *Rhamphocephalus* is important as the only Mid Jurassic pterosaur known, apart from fragmentary remains of a single pterosaur from the Oxford Clay of St Ives, Cambridgeshire, and isolated teeth from many Cotswolds sites. The Pterosauria radiated during the Late Triassic and four main 'rhamphorhynchoid' genera are known from the Early Jurassic (*Dimorphodon* from Lyme Regis, *Parapsicephalus* from Loftus, Yorkshire, *Campylognathoides* from Holzmaden and India, and *Dorygnathus* from Germany). There is a gap in the Mid Jurassic before abundant fine remains of *Rhamphorhynchus*, *Scaphognathus* and *Anurognathus* are found in the Late Jurassic of Solnhofen, in particular. The crown-group pterosaurs, the Pterodactyloidea, are known from skeletons first from the Late Jurassic of Solnhofen (Bavaria), England and France (*Pterodactylus, Germanodactylus, Ctenochasma*), and by teeth from the Cotswolds Bathonian. It is not clear whether *Rhamphocephalus* is a pterodactyloid or not (Wellnhofer, 1978, p. 41).

Stonesfield is perhaps best known as the source of the original and most extensive material of *Megalosaurus* (Figure 6.8B; Buckland, 1824). Remains are incomplete, but most of the skull and skeleton are known. The skull was up to 1 m long, with sharp recurved teeth, each 50-150 mm long. The femur was 1 m long and the hind feet bore large recurved claws. *Megalosaurus* reached total lengths of 3.5-7.0 m, and was clearly a major and fearsome predator (Buckland, 1824; Phillips, 1871, pp. 196-219; Huene, 1906, 1923, 1926; Walker, 1964; Steel, 1970, pp. 33-4). Many more species of *Megalosaurus* have been established and *M. bucklandi* is now largely restricted to the Mid Jurassic specimens from Oxfordshire, and Stonesfield in particular. The preservation of the Stonesfield specimens is good enough to allow detailed studies of the braincase (Huene, 1906). Other species of carnivorous dinosaur from the Late Triassic of Wales, Early, Mid and Late Jurassic and Early Cretaceous of England, France, Austria, Morocco and Arizona have been ascribed to 20 or

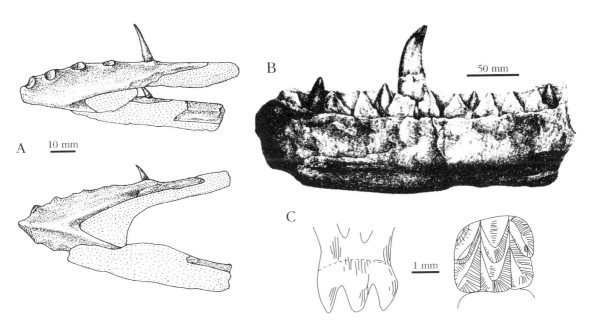

Figure 6.8 Bathonian reptiles from Stonesfield. (A) The rhamphorhynchoid pterosaur *Rhamphocephalus buck-landi* (Meyer, 1832), anterior part of the lower jaw; (B) the type specimen of *Megalosaurus bucklandi* Meyer, 1832, a partial lower jaw, seen from the inside; (C) *Stereognathus ooliticus* Charlesworth, 1855, reconstructed right upper molar, showing posterior and crown views. (A) After Wellnhofer (1978); (B) after Buckland (1824); (C) after Simpson, 1928).

more species of *Megalosaurus*. Several of these have now been placed in other genera (Huene, 1923, 1926, 1932; Walker, 1964; Waldman, 1974; Molnar, 1990; Molnar *et al.*, 1990), but more work is needed before a true picture of the morphological variation and distribution in time and space of *Megalosaurus* (*sensu stricto*) can be established.

A related species is *Iliosuchus incognitus* Huene (1932), often placed in *Megalosaurus* (e.g. Romer, 1966, p. 369), based on a small ilium from Stonesfield. Galton (1976b) has argued that it is a distinct genus with a close relative in the Morrison Formation (Late Jurassic) of Utah, which would be evidence for faunal ties between Europe and North America in the Jurassic. However, Molnar *et al.* (1990, p. 202) do not confirm this assignation.

The sauropod *Cetiosaurus* is represented by only a few elements at Stonesfield, and it was either not a major part of the fauna or the elements have not been preserved. It is much better known from the younger Bathonian White Limestone and Forest Marble of other places in Oxfordshire, Gloucestershire, Cambridgeshire and Northamptonshire.

The only ornithischian dinosaur from Stonesfield is a tooth (YPM 7367) identified as that of an ornithopod, and possibly a hypsilophodontid (Galton, 1975, pp. 742, 745, 747; Galton 1980b, pp. 74-5). If it belongs to a hypsilophodontid, it is the oldest known member of that group, of an age comparable with *Yandusaurus* He (1979) from the Xiashaximiao Formation (Bathonian-Callovian) of China (Sues and Norman, 1990).

The remains of plesiosaurs and ichthyosaurs from Stonesfield have been mentioned by Phillips (1871, p. 183) and Lydekker (1889a, p. 245). They consist of teeth, vertebrae and limb bones, but they have never been adequately described. In any case, these marine forms are rare in the British Bathonian, the only other specimens being vertebrae and teeth identified as '*Ichthyosaurus* sp.' or '*Plesiosaurus* sp.' from the White Limestone, Forest Marble and Cornbrash in a few sites in southern and midland England. Stonesfield has produced the best range of specimens of these forms, even if they are not very impressive.

Finally, the tritylodont *Stereognathus* (Figure 6.8C), represented by a jaw fragment (BGS(GSM) 113834) and one other specimen, was initially interpreted as a mammal (Charlesworth, 1855; Owen, 1857; 1871, pp. 18-21; Phillips, 1871, pp. 236-7; Goodrich, 1894, p. 424; Simpson, 1928, pp. 22-6). It is in fact one of the last surviving tritylodontids (e.g. Romer, 1956, 1966), a group which is known mainly from the Late Triassic and Early Jurassic of China, South Africa, Germany and some of the Mendip fissures (Savage, 1971; Kühne, 1956). Waldman and Savage (1972, pp. 121–2) reported several molars of a new species of *Stereognathus* from the Ostracod Limestones of the Great Estuarine Group of Skye (Late Bathonian). In addition, Ensom (1977) recorded a therapsid tooth, tentatively ascribed to *Stereognathus*, from the Forest Marble (Late Bathonian) of Bridport, Dorset, and Evans and Milner (1994) note further finds from Watton, Swyre and Kirtlington. Although fragmentary, the Stonesfield *Stereognathus* remains are important evidence of one of the last tritylodontids, and hence of one of the last 'mammal-like reptiles'. The only younger taxon is *Bienotheroides wanhsiensis* Young (1982) from the upper Xiashaximiao Formation (Bathonian–Callovian) of China (Benton, 1993).

Comparison with other localities

The fauna of the Stonesfield Slate (Stonesfield Member) is unique. However, comparisons may be made with other Early and Mid Bathonian faunas which contain some of the same species, and in particular with the Cotswold Slate (Eyford Member) of the west of Stow-on-the-Wold (*progracilis* Zone, Mid Bathonian). Localities such as Huntsman's Quarry (SP 125254), Eyford Quarries (SP 135255, etc.) and Kyneton Thorns Quarry (SP 122264) have yielded *Steneosaurus, Teleosaurus, Megalosaurus, Rhamphocephalus* and other genera. The conditions of deposition and faunal composition of the Cotswold Slate are very like those of the Stonesfield Slate, and the two units were formerly regarded as identical. The rich tetrapod fauna from the earliest Bathonian Hook Norton Member at Hornsleasow Quarry (SY 131322) (Vaughan, 1989; Metcalf *et al.*, 1992) is comparable.

Some quarries in Oxfordshire have also produced similar animals. The quarry at Sarsden, near Chipping Norton (SP 300226) produced teeth of *Megalosaurus*, a *Cetiosaurus* limb bone and bones of *Rhamphocephalus*, probably from the Taynton Stone (*progracilis* Zone). Padley's Quarry (SP 317269) at Chipping Norton yielded specimens of *Megalosaurus, Cetiosaurus* and *Teleosaurus* from the Sharps Hill Formation (*progracilis* or *tenuiplicatus* Zones). Further similar faunas are known from the White Limestone and

Forest Marble (Mid–Late Bathonian) of Oxfordshire (e.g. Slape Hill, Glympton, Enslow Bridge (Shipton-on-Cherwell; Bletchington Station), Kirtlington, Ardley and Stratton Audley), but these are often dominated by the sauropod *Cetiosaurus* and are younger in age.

The most clearly comparable sites to Stonesfield outside the British Isles are in the Bathonian of Caen, Calvados and other sites of Normandy ('Great Oolite', 'Fuller's Earth', 'Dogger'), which are well known for several species of crocodiles *Steneosaurus, Teleidosaurus and Teleosaurus*, as well as a dinosaur *Megalosaurus*. Similar crocodilians have been found as isolated remains elsewhere in the Bathonian of France (Steel, 1973). Elsewhere, Mid Jurassic reptiles are extremely rare, with small dinosaur faunas known from the Bathonian of El Mers in the Moyen Atlas of Morocco (*Megalosaurus, Cetiosaurus*) and in the Bathonian of north-western Madagascar (*Bothriospondylus*).

Conclusions

Stonesfield is arguably the most important Mid Jurassic fossil reptile site in the world, and its fauna is diverse and abundant. The four species of *Steneosaurus* and *Teleosaurus* are represented by good material and are probably the best Mid Jurassic crocodilians in the world since most of the Normandy specimens were destroyed in the war. *Rhamphocephalus*, best represented at Stonesfield, is important for studies of pterosaur evolution since it is one of the few members of this group known from the Mid Jurassic. Stonesfield is the most important site for remains of *Megalosaurus* in the world: it yielded the first and type material in the early 19th century, and continued to produce hundreds of specimens while the mines were in operation. The ornithopod tooth, interpreted tentatively as that of a hypsilophodontid, could be the oldest representative of this predominantly Cretaceous group. Apart from these terrestrial forms, Stonesfield has also produced one of the best collections of Mid Jurassic plesiosaur and ichthyosaur remains in the world, although they are very fragmentary. Finally, the tritylodont *Stereognathus* is one of the last surviving members of its group and one of the youngest 'mammal-like reptiles' in the world. Stonesfield is important for the study of fossil reptiles for two reasons: firstly, its fauna is abundant, diverse and well preserved, and secondly, because of the rarity of Mid Jurassic fossil reptiles outside Britain.

The site's historic international importance and potential for future finds from re-excavation give it a very high conservation value.

HUNTSMAN'S QUARRY, NAUNTON, GLOUCESTERSHIRE (SP 126253)

Highlights

Huntsman's Quarry, Naunton, is the best Cotswolds Slate locality, and source of six or seven species of reptiles. The fauna is comparable with that from Stonesfield, but occurs much further west, on the other side of a major palaeogeographic barrier.

Introduction

Huntsman's Quarry is the only major quarry still working in the Cotswolds Slate (=Eyford Member), formerly equated with the Stonesfield Slate (see Stonesfield report). It lies 2 km northeast of Naunton on an unclassified road that crosses Eyford Hill. The quarry has yielded a wide range of fossil reptiles in the past and is currently operated by Huntsman Quarries Ltd for road-metal and gravel and has been extended much towards the north. The exact location of the older finds is unknown, but they were probably made in the older portions of the quarry, several of which still offer good exposures (e.g. SP 123254, where a 5 m section is to be seen on the long north face of an old pit, and at SP 123252, where a 3–4 m face is still visible). The quarry has potential for future finds of reptiles since the old pits are still accessible and new quarrying operations have exposed further large areas. Some *Megalosaurus* specimens have been found recently (OUM).

The Cotswolds Slate was extensively worked for roofing slates in the 19th and early 20th century around Sevenhampton, Kineton, Naunton and Eyford (H.B. Woodward, 1894, pp. 294–6, 484–5; Richardson, 1929, pp. 102–16, 144–6; Arkell, 1933, p. 278). The fossil reptiles were examined by various Victorian authors and the fauna was reviewed by Richardson (1929), but the older, and more recent, finds are in need of redescription.

Description

H.B. Woodward (1894, p. 295) provided a section at Summerhill, Eyford (?SP 129246, one of the Old Huntsman's Quarry pits), but Huntsman's Quarry itself was first described by Richardson (1929), who gave a section (p. 114) as follows:

	Thickness	
	ft	in
Limestone, grey, rather hard seen	2	6
Marl, grey and yellow; shell fragments–mostly of oysters; one specimen of *Rhynchonella ?concinna*	0	6
Limestone, yellowish, sparsely oolitic; *Lima cardiiformis*	2	9
'The Crop'. Oolitic, grey, obliquely-laminated: makes 'Presents'	2	8
[?*Rhynchonella* Bed, and *Ostrea acuminata* Limestones remanié]. Marl, brown, clayey, with oysters and a few pebbles bored by *Lithophaga*	0	2
[? Sevenhampton Marl] Marly limestone, crowded with oysters; *Lima cardiiformis, Modiola* sp., *Pholadomya solitaria, Pinna ampla*. and occasional plant remains, locally passing into a hard limestone	2	0
?1–5. Flagstones. Limestones, massive, flaggy, in the main blue-hearted, but locally yellow and oolitic. Contain locally pebbles of oolite, and, in cavities in the lower part of the bottom stratum (5), green clay	6	6
6. Green Clay, locally sandy average thickness	0	2
7–8. Pendle. (a) Sandstone, hard, grey (weathering brown), rarely passing into a brown, irregularly laminated fissile sandstone	2	8
(b) Limestone, hard, grey, sandy [Non-sequence: no evidence for the equivalent to the Planking]	1	6
9. Sandy limestone, hard, grey at the top, brown and shaly, containing flat marl pellets. Surface undulating and water-worn. *Placunopsis socialis* common in the top layer. Said to be a very bad hard bed and not worked		

Richardson (1929) equated the bulk of the exposure in Huntsman's Quarry with the 'Stonesfield

Slate Series' low in the Great Oolite Group in the classic Hampen Railway cutting section. Arkell (1933, p. 278) and McKerrow and Baden-Powell (1953, p. 92) followed this assignment. Sellwood and McKerrow (1974, p. 193) assigned beds 1–9, and the three units above these in Richardson's section, to the Cotswold Slate, which they renamed the Eyford Member (Figure 6.9). They assign the succeeding 2 m of oolitic limestone to the Taynton Limestone Formation (?upper *progracilis* Zone). The fossils came from the 'Slate Bed' (Richardson, 1929), probably equivalent to bed 7 in the log ('Pendle').

The fauna from Huntsman's Quarry consists of reptiles, fishes, arthropods, molluscs (including rare ammonites), annelids, starfish and plants (Richardson, 1929). The invertebrate fauna is distinctive, differing in many ways from that of the beds above and below, and also from stratigraphically equivalent units elsewhere. The bivalve *Myophorella impressa* (formerly *Trigonia*) is the most abundant fossil. Other bivalves include *Liostrea, Gervillella* and *Chlamys*. Rhynchonellid brachiopods and gastropods are also a common component of the fauna, but other marine invertebrates, including ammonites, echinoderms (crinoids and starfish) and barnacles (represented by plates), are rare. Insects are also present, the commonest being beetles, which are represented in the deposit by their resistant elytra. There is a substantial flora which includes ferns and early conifers (leaves, seeds and fruit), ginkgo leaves and 'carpolithes' seeds.

The reptile remains apparently occurred as isolated bones. Most of the specimens are relatively small (crocodile and dinosaur teeth, pterosaur limb bones), which suggests some degree of sorting. The association with marine invertebrates (e.g. barnacles, starfish, ammonites, belemnites) indicates transport or reworking of the terrestrial forms at least. Nothing is known of the taphonomy of the vertebrates since the matrix has been removed from the museum specimens.

Fauna

The reptilian fauna from Huntsman's Quarry includes dinosaurs, crocodilians, pterosaurs and a turtle. One of the best collections was made by the Rev. E.F. Witts (1813–86) and it is now preserved in GLCRM (Savage, 1963). Unfortunately, many museum specimens are labelled simply 'Eyeford', which could refer to any of the quarries on the east side of Eyford Hill (e.g. SP 125252, SP 126253, SP 126254, SP 128251, SP 128253, SP 128254, SP 130251, all of which may still be seen, at least in part). Specimens labelled 'Naunton' may come from Huntsman's Quarry (SP 131322), from Summerhill Quarries (?SP 129246, SP 113245), or from New Buildings Quarries (SP 135237, SP 134239). In the following list, all specimens are labelled 'Eyford' (often spelt Eyeford), unless otherwise stated.

Testudines
  'Chelonian indet.'
    Carapace; BMNH R2634
Archosauria: Crocodylia: Thalattosuchia:
  Steneosauridae
  Steneosaurus brevidens (Phillips, 1871)
    Teeth: BMNH 28611, R2631; GLCRM
      G.53–58

Figure 6.9 Exposure of the Eyford Member, or 'Cotswolds Slates', at Huntsman's Quarry. Reptiles occur as isolated bones at various levels in the succession. (Photo: M.J. Benton.)

Teleosaurus sublidens Phillips, 1871
> Teeth: BMNH 28611, R2632-3; BGS(GSM) G.1-51; jaws: BGS(GSM) G.52, 77; BGS(GSM) (various: 'Naunton')

Steneosaurus/Teleosaurus sp.
> Teeth: BMNH R6777, R6778 ('Huntsman's Quarry'), R6779-81; BGS(GSM) 113735-6, 113759, 113764-8; rib: BGS(GSM) 11838; scute: BGS(GSM) 72280 ('Naunton'), BGS(GSM) GLCRM 59-60, various bones: GLCRM 558-62

Dinosauria: Saurischia: Theropoda:
Megalosauridae

Megalosaurus bucklandi Meyer, 1832
> Teeth: BMNH 28608, R2635; limb bones and ribs GLCRM G.70-1, G.72-3, G.74-6

Megalosaurus sp.
> Tibia: OUM J.29759

Archosauria: Pterosauria: 'Rhamphorhynchoidea'

Rhamphocephalus sp.
> Limb-bones, a ?proximal phalanx and pectoral girdle elements: BMNH R6782, Munchen 1976 I.41-4 ('Huntsman's Quarry'); BGS(GSM) 113728-31, 113733, 113738, 113747, 113753, 113758, 113670, un-numb.; GLCRM G.61-2

Interpretation

The Cotswold Slates (Eyford Member) are dated to the *progracilis* Zone (early Mid Bathonian) on the basis of sparse ammonites. Richardson (1929, p. 114) noted the occurrence of *Perisphinctes gracilis* (Buckman), reidentified as *Procerites progracilis* Cox and Arkell, in the Cotswold Slates of the Eyford area. *Procerites mirabilis* Arkell was found at Eyford and Huntsman's Quarry (Arkell 1951-8, pp. 199-201; Torrens 1969b, pp. 71-2). These isolated finds place the Cotswold Slates within the *progracilis* Zone (Torrens, 1969b, pp. 71-3; *in* Cope *et al.*, 1980b, p. 35, fig. 6a), although correlation of all the beds in Richardson's (1929) section is unclear.

The turtle carapace has not been identified, but it may belong to the genus *Protochelys* which occurs at Stonesfield in rocks of similar age.

The long-snouted marine crocodilians *Steneosaurus* and *Teleosaurus* from Huntsman's Quarry are represented largely by teeth. These have been identified by comparison with other Bathonian specimens from the Stonesfield Slate (Stonesfield Member), White Limestone and Forest Marble of Oxfordshire and Northamptonshire. These croco-

dilians were evidently relatively rather abundant in the Cotswold Slate, rather more so than in the Stonesfield Slate.

The teeth and tibia of *Megalosaurus* have, again, been identified by comparison with material from the White Limestone of Oxfordshire, Dorset and other localities in Gloucestershire.

The pterosaur bones have been hard to identify since they consist mainly of wing elements. They may belong to the genus *Rhamphocephalus*, but the significance of that genus is unclear. A fine pterosaur skull impression from the Cotswold Slate of Kyneton Thorns Quarry nearby (SP 122264) was named as the type specimen of *Rhamphocephalus prestwichi* Seeley (1880).

In a manuscript catalogue of the Chaning Pearce collection in BRSMG, teeth of *Ichthyosaurus* and *Plesiosaurus* from Eyford are mentioned, but the specimens have not been located. Some probable plesiosaur teeth are preserved in Gloucester (GLCRM G.63-9, 78-80).

Comparison with other localities

The reptile fauna from Huntsman's Quarry may be compared with that of other quarries in the Cotswold Slate, some of which have already been mentioned. The cluster of ten or more quarries on the east side of Eyford Hill (scattered around SP 135255) have probably yielded a similar range of fossil reptiles. Specimens labelled 'Naunton' include crocodile teeth and a scute. These may have come from various sites to the north and east of Naunton (see above). Pterosaur limb bones have been obtained from the quarries on Sevenhampton Common (SP 012232) (H.B. Woodward, 1894, p. 294). *Megalosaurus* teeth and the fine skull cast of *Rhamphocephalus prestwichi* were found in Kyneton Thorns Quarry (SP 122264) (H.B. Woodward, 1894, p. 295). Comparable sites further afield in Gloucestershire and Oxfordshire are detailed in the Stonesfield report (see above).

Conclusions

Huntsman's Quarry contains the largest extant exposure of the Cotswold Slate (Eyford Member), and it has yielded the best reptile fauna of that unit. The fauna is placed temporally between that of New Park Quarry and other quarries in the Chipping Norton Formation (*zigzag* Zone), and sites in the White Limestone and Forest Marble (*subcontractus–discus* Zones) of Oxfordshire. The fauna is far less abundant and diverse than that of the Stonesfield Slate, but it is of considerable importance in view of the great local variations in deposition in the Cotswolds and Oxfordshire at the time. Its conservation value lies in this importance and its potential for future finds.

SHIPTON-ON-CHERWELL QUARRY, NORTH-WEST CORNER, OXFORDSHIRE (SP 475178)

Highlights

Shipton-on-Cherwell Quarry is the best source of Mid Jurassic crocodilians in Britain, and perhaps the best source of freshwater forms in the world. The site has also produced a range of bones of other species, including turtles and dinosaurs.

Introduction

From the older portions of Shipton-on-Cherwell Quarry the earliest recorded find was a partial *Steneosaurus* skull collected in the early 19th century (Conybeare, 1821, p. 591), and this specimen (OXMFS J1401) was the first crocodile known from the British 'oolites'. Later finds of crocodilian and dinosaurian remains were described by Phillips (1871). The crocodiles were later redescribed by Phizackerley (1951). A stegosaur femur was made type of *Lexovisaurus vetustus* by Huene (1910c), and it was redescribed by Galton and Powell (1983) and by Galton (1985b). The quarry at about SP 477177 was known as Gibraltar Quarry or 'Enslow Bridge', until about 1920,when it was engulfed by the huge Shipton Cement Quarry (Arkell, 1931, pp. 577-8). Parts of the old faces are still extant at SP 475178, and they display sections in the Upper White Limestone, Forest Marble and Cornbrash. The potential for further finds, particularly through careful sampling in the *fimbriatus-waltoni* Beds, and other clay units, adds to its value.

There is a problem over the nomenclature of the seven or so quarries around the bend in the River Cherwell in this area (Figure 6.10). Hull (1859, pp. 20-1) refers to sections at Kirtlington Station (the railway cutting, SP 482181) and 'on

the right bank of the river, in a semi-circular cliff' (probably Quarry Bank, Enslow Bridge, SP 475183). Phillips (1860, pp. 117-18) describes these same sections. H.B. Woodward (1894, p. 322) gives a more detailed log for Quarry Bank: 'the section south-west of Enslow Bridge'. Further, he describes a series of quarries 'south of Bletchington station, and on the western side of the railway'. These are, no doubt, the Greenhill Quarries (SP 483179, SP 485177) on the eastern side of the railway, and an old portion of the Shipton Cement quarry (?SP 480175) on the western side of the railway. The 'Old quarry a little east of Bletchington railway-station' (H.B. Woodward, 1894, pp. 323, 373-4) is probably Phillips' *Cetiosaurus* Quarry (SP 484182). We may summarize these and later references:

1. Bletchington Station Quarry /Kirtlington Station Quarry/*Cetiosaurus* Quarry (SP 484182): Phillips (1871, p. 251); H.B. Woodward (1894, p. 323); Pringle (1926, p. 25); Richardson *et al.*, 1946, p. 67); Arkell (1947a, pp. 57-8); Huene (1910c, p. 75).
2. Quarry Bank, Enslow Bridge (SP 475183): Hull (1859, p. 21); Phillips (1871, p. 239); H.B. Woodward (1894, p. 321); Arkell (1947a, p. 58 (?): 'Bunker's Hill quarries in the woods west of Enslow Bridge').
3. North-west corner of Shipton-on-Cherwell Quarry/Gibraltar Quarry (SP 475178): Phillips (1871, pp. 151-2, 247); Odling (1913, pp. 496-8); Arkell (1931, pp. 579-80; 1947a, p. 58); Richardson *et al.* (1946, pp. 66-7); Palmer (1979, pp. 191, 202-3, 205, 208, 210: 'Shipton Quarry').
4. Lower Greenhill Quarry (SP 483179): H.B. Woodward (1894, p. 322); Odling (1913, pp. 495-6); Pringle (1926, p. 24); Arkell (1931, pp. 580-2; 1947a, pp. 58-9); Richardson *et al.* (1946, p. 68).
5. Upper Greenhill Quarry (SP 485177): H.B. Woodward (1894, p. 322); Odling (1913, pp. 494-5); Pringle (1926, p. 24); Douglas and Arkell (1928, pp. 129-30; 1932; 1935, p. 319); Richardson *et al.* (1946, pp. 67, 77-8); Arkell (1947a, pp. 59-60); Torrens (1968, p. 248); Palmer (1979, pp. 190, 208: 'Greenhill Quarry').
6. Oxford Portland Cement Works (old pit, ?SP 480175): H.B. Woodward (1894, p. 322), see no. 3.
7. Whitehill Quarry, Gibraltar. (SP 477186): Palmer (1979, pp. 190, 206, 209).

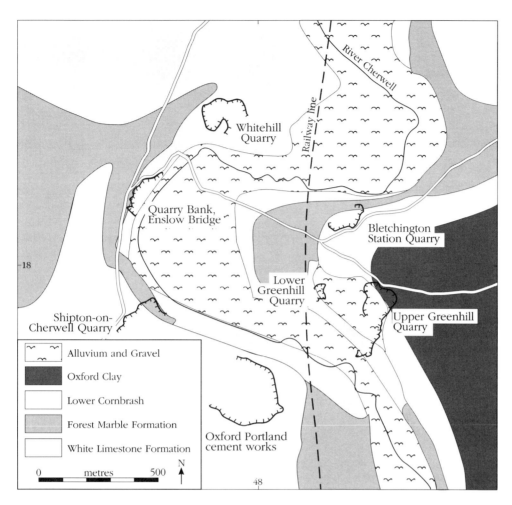

Figure 6.10 The quarries around Shipton-on-Cherwell. Up to seven quarries (detailed in the text) appear to have yielded fossil reptiles from the White Limestone and the Forest Marble formations. Based on old Ordnance Survey maps.

Richardson *et al.* (1946, p. 67) noted that Bletchington Station quarry was 'probably the original 'Gibraltar Quarry' and they refer to Phillips (1871, p. 247). However, Phillips (1871, p. 247) does not make clear whether the quarries at Enslow Bridge and Gibraltar were separate localities or part of one quarry, stating that the bones of *Cetiosaurus* came from 'the quarries at Gibraltar, near Enslow Bridge, and close to the railway station for Kirtlington and Bletchingdon'. Huene (1910c, p. 75) referred to two reptile quarries, that at Bletchington Station and one '300 m west of Bletchington Station on the other side of the river in the western valley wall'. This could be Quarry Bank or Gibraltar, about 300 m west of the station in the middle of the flood plain in the bend in the River Cherwell. McKerrow *et al.* (1969, p. 61) further complicated the issue by giving a map reference for a 'Gibraltar Quarry (477185)'. The quarry at SP 477185 (Whitehill Quarry) is a rela-

tively recent excavation, and has nothing to do with the old Gibraltar quarry. Galton and Powell (1983, p. 220) failed to mention Gibraltar Quarry, referring to the locality simply as a 'series of quarries, about 9.6 km north of Oxford, in the west bank of the River Cherwell, which extend about 0.8 km southwards downstream from Enslow Bridge itself', but provided the grid reference SP 42477177. The grid reference appears to coincide with the assumed location of Gibraltar Quarry as is indicated in the accounts of Odling (1913, pp. 486, 496–8) and Arkell (1931, pp. 577–80).

Description

The sequence at Shipton-on-Cherwell Quarry (SP 475178) is based on Arkell's (1931, pp. 579–80) description, with modifications from Richardson *et al.* (1946, pp. 36–8).

Thickness (m)   **Thickness (m)**

Lower Cornbrash
  In field immediately above the
  quarry (also at top of section to
  the south; Arkell 1947a, p. 58)

Forest Marble Formation
9. ('Wychwood Beds'). Clays,
   greenish grey dominant at N
   end; thinly laminated cross-
   bedded sand gradually replaces
   clays elsewhere (?=beds a–h of
   Odling 1913, p. 496)   1.8–3
8. ('Upper Kemble Beds') Limestone,
   cross-bedded, hard, white, blue-
   hearted, coarsely oolitic, locally
   split up by thick lens of dark-blue,
   shaly clay. (?=beds i, j of Odling,
   1913, p. 496)   *c.* 3
White Limestone Formation
Bladon Member
7. Upper *Epithyris* Bed (Coral - *Epithyris*
   Limestone) (='Fossiliferous
   Cream Cheese' Bed of Odling
   1913, p. 496, and Arkell, 1931,
   p. 579;'Cream Cheese Bed' of
   Barrow, 1908; Upper *Epithyris*
   Bed of Richardson *et al.*, 1946,
   p. 66). Limestone, cross-bedded,
   hard, similar to 8. Abundant
   Epithyris, *Modiola*, etc.   0.6–0.9
   ---- plane of erosion with a few pebbles ----
Beds 6–3: *fimbriatus–waltoni* Beds
6. (=Beds 2–4 of Odling, 1913, p.497).
   Marl, green, lignitiferous, black
   and shelly at base; 90 cm thick at
   N end, reducing to 15 cm band
   of white pellets at S end   0.1–0.9
5. (=Beds 5 and 6 of Odling, 1913,
   p. 497). Limestone, greenish grey,
   argillaceous, weathering soft.
   Abundant *Gervillia, Astarte*.
   Expands N   0.15–0.3
4. (=Beds 7 and 8 of Odling, 1913,
   p. 497). Clay, dark green with
   white pellets. Thins N   0.23–0.45
3. Limestone, hard, unfossiliferous,
   thins N   0–0.25
2. Oyster-*Epithyris* Marl (=Bed 9, first
   Terebratula bed of Odling, 1913,
   p. 497; Middle *Epithyris* Bed of
   Arkell 1931, p. 580, and
   Richardson *et al.*, 1946, p. 66).

Marl, brown, ferruginous. In
places, rolled fragments of
limestone and corals at base. Abundant
Epithyris.   0.23
   ------ plane of erosion ------

Ardley Member
1. (=Beds 10–12 of Odling, 1913,
   p. 498). Limestones, creamy
   white, compact. In places the
   Epithyris Limestone (=Lower
   Epithyris Bed) is typically
   represented, while elsewhere
   it is absent.   seen *c.* 2.0

Odling (1913, p. 498) recorded a further 3 m of section in the old Gibraltar Quarry, down to a compact limestone below the *Nerinea eudesii* Beds (lower Ardley Member).

Problems of lithostratigraphy and disagreement over the placing of the Forest Marble/White Limestone boundary are discussed by Richardson *et al.* (1946), Arkell (1947a), Palmer (1973, 1979), Torrens (*in* Cope *et al.*, 1980b, pp. 36–8) and Sumbler (1984). The main difficulty concerns the classification of beds which are, to an extent, transitional between the White Limestone and Forest Marble formations, and the confusion arising from different usage of the two formation names has been increased by the introduction of various subdivisions which have been used in different senses by different workers. The stratigraphic position of the transitional beds has recently been standardized by their inclusion in the Bladon Member of the White Limestone Formation (Sumbler, 1984).

Reptiles were found at three or four levels within the sequence: in the ?Ardley Member (?bed 1, or lower of Arkell, 1931; Richardson *et al.*, 1946), in the *fimbriatus–waltoni* Beds, in the lowest unit and the top of the Forest Marble (Arkell, 1931; Richardson *et al.*, 1946: beds 8 and 9), and in the Lower Cornbrash.

The most abundant remains appear to have come from the *fimbriatus–waltoni* Beds here and nearby. The bones excavated at Bletchington Station quarry between 1868 and 1870 lay 'on a freshly-bared surface of the Great Oolite . . . and covered by the laminated clay and thin oolitic bands which occupy the place assigned to the Bradford Clay of Wiltshire' (Phillips, 1871, p. 248). The bones clearly lay within clay bands above an oolitic limestone and, because of their size, they passed up into different thin clay and marl units. Phillips' detailed section (1871, p. 251) shows seven thin argillaceous and calcareous units that cannot be matched with sections recorded nearby in the railway cutting (Hull, 1859, p. 20; Phillips, 1860, p. 117, 1871, p. 154), nor at Shipton. Nevertheless, it may be concluded that all of these belong to the *fimbriatus-waltoni* Beds. This is confirmed by H.B. Woodward (1894, p. 154) and Arkell (1931, pp. 565, 566). Richardson *et al.* (1946, pp. 39, 65, 67, 70) and Palmer (1979, p. 221) note the occurrence of *Cetiosaurus* bones in these beds at several quarries.

De la Beche and Conybeare (1821, p. 591) recorded 'an undoubted species of crocodile, somewhat resembling the Gavial . . . in the upper beds of the Great Oolite, or in the Cornbrash . . . at Gibraltar, eight miles north of Oxford'. Phillips (1871, p. 251) later noted that 'heads of teleosaurs are not infrequent at Enslow Bridge, and in beds of Great Oolite below the strata containing ceteosaurus', and in summarizing an excursion to Enslow Bridge (Phillips, 1871, p. 239) he noted that 'the members were highly gratified to learn that during the morning a very fine skeleton of *Teleosaurus* had been found and the head was exposed to view. This quarry is in the Great Oolite, the lower and uppermost strata of which in Oxfordshire yield remains of *Megalosaurus* while in the middle beds we find *Teleosaurus*'. These probably refer to finds in the old Gibraltar/Shipton Quarry in or below the *fimbriatus-waltoni* Beds. Further, 'remains of *Teleosaurus* were obtained at Enslow Bridge (south of Kirtlington) a little below the *Terebratula*-bed' (H.B. Woodward 1894, p. 323), thus below bed 2 of Arkell's (1931) section, probably at Shipton. *Teleosaurus* occurs as low as the Upper Shipton Member at Lower Greenhill Quarry (Odling, 1913, p. 496, Bed 18).

The large *Cetiosaurus* femur obtained in 1848 from the railway cutting south of Bletchington Station was apparently 'assigned to the base of the Forest Marble' by Prestwich and another specimen was found 'within two feet of the Cornbrash' (H.B. Woodward, 1894, p. 323). These indicate horizons equivalent to beds 8 and 9 respectively in the section given above.

Galton and Powell (1983) and Galton (1985b) suggest that the holotype of the stegosaur *Lexovisaurus? vetustus* (a right femur) probably derived from the top of the Forest Marble, because of its eroded nature and the fact that bones are virtually unknown in the Cornbrash, but do occur in the lagoonal facies of the Forest Marble Formation. The bone, however, bears an example of the bivalve *Meleagrinella echinata* on its surface (P. Powell, pers. comm.), firm evidence of a Lower Cornbrash derivation. This relocation of the find is of little significance, since both the Lower Cornbrash and the top of the Forest Marble are within the *Clydoniceras discus* Subzone of the *C. discus* Zone (Late Bathonian, Mid Jurassic; Torrens, *in* Cope *et al.*, 1980b, fig. 6a).

The bones from all levels at Shipton are generally well preserved, but disarticulated. Taphonomic information is only available for the *Cetiosaurus* find of 1868-70 at nearby Bletchington Station quarry. Phillips (1871, pp. 248-51) noted that the large bones were largely shattered (?by compression from the weight of the superincumbent sediment). The separate elements were disarticulated, but associated, remains of three individuals of different size being preserved within an area measuring 6 m by 6 m. Vertebrae and ribs were much broken and mixed in 'confused' groups. No cranial remains were found. Thus, as Phillips (1871, p. 249) realized, the animals had died elsewhere, 'the parts separated by decay; the massive limbs disjointed, and the bones displaced'. The bones were washed in, but have not suffered much wear.

Fauna

The majority of the specimens labelled 'Enslow Bridge' or 'Gibraltar' appear to have come from the old Gibraltar Quarry which is now part of the Shipton-on-Cherwell cement works. Specimens from Bletchington Station Quarry are normally noted as 'Bletchington Station' or 'Kirtlington Station', and they are not listed here. Most specimens are labelled 'Great Oolite', and they probably came from beds 3-6 (*fimbriatus-waltoni* Beds) or lower.

Testudines
  'turtle scute'
    OUM J17567.
Archosauria: Crocodylia: Thalattosuchia:
  Steneosauridae
  Steneosaurus boutilieri J.A. Deslongchamps, 1869
    OUM J1401-4, J1412, J1416-7
  Steneosaurus brevidens (Phillips, 1871)
    BMNH R78-79, 44821

Steneosaurus aff. *larteti* (J.A. Deslongchamps, 1866)
  OUM J.1408-10
Steneosaurus megistorhynchus J.A. Deslongchamps, 1866
  OUM J.1414-5
Steneosaurus meretrix Phizackerley, 1951
  Type specimen: OUM J.29850. Also, OUM J.29851, J.1407
Steneosaurus sp.
  BGS(GSM) (old no.); OUM J.10590-1, J.29495; CAMSM J.21952-3
Teleosaurus subulidens Phillips, 1871
  OUM J.13599-600
Archosauria: Crocodylia: Thalattosuchia: Metriorhynchidae
Metriorhynchus cf. *geoffroyi* Meyer, 1832
  OUM J1418
Dinosauria: Saurischia: Theropoda: Megalosauridae
Megalosaurus bucklandi Meyer, 1832
  OUM J.13598, J.29773, J.13882, J.29765
Dinosauria: Saurischia: Sauropoda: Cetiosauridae
?Cetiosaurus
  OUM J.29806
Dinosauria: Ornithischia: Ornithopoda
?Iguanodon
  OUM J.29805
Dinosauria: Ornithischia: Stegosauria: Stegosauridae
Lexovisaurus? vetustus (Huene, 1910)
  Type specimen: OUM J.14000

Interpretation

The biostratigraphy is difficult to establish because of the general absence of ammonites here and in comparable units nearby. Three or four specimens of *Tulites* and *Procerites* have been recorded from Bletchington Station and Enslow Bridge, but they are either lost or stratigraphically unlocalized (Torrens, 1969b, p. 69, *in* Cope *et al.*, 1980b, pp. 37-8). The stratigraphic units are assigned within the Late Bathonian as follows (Palmer, 1979; Torrens, *in* Cope *et al.*, 1980b): Ardley Member (?lower *hodsoni* Zone), Bladon Member (?upper *hodsoni*-lower *aspidoides* Zones), Forest Marble Formation (?*aspidoides*-lower *discus* Zone), Lower Cornbrash (?upper *discus* Zone).

The reptiles from Shipton are generally large, with none of the 'lizards', pterosaurs, amphibians or mammals that are known from deposits of the same age at Kirtlington. This is probably the result of different collecting techniques, rather than any major habitat distinction. Nevertheless, the dominance by crocodilians is typical of most British Bathonian sites.

The long-snouted crocodilians *Steneosaurus, Teleosaurus* and *Metriorhynchus* are well known from several British Mid Jurassic sites from the Early Bathonian (e.g. New Park Quarry), the Mid Bathonian (Huntsman's Quarry; Stonesfield) and the Late Bathonian (e.g. Kirtlington, Oxfordshire). These forms were revised by Phizackerley (1951), but Steel (1973), in a recent review, was unable to clarify their complex taxonomy. The distinctions between the species, and the assignment of valid species to different genera, have yet to be assessed in an overview. In other words, the total of seven species from Shipton is almost certainly an overestimate and it is not clear to which genus each should be ascribed. Nevertheless, the Shipton specimens are largely skulls and lower jaws, which are taxonomically and functionally important elements, and these should be of extreme importance when a review is undertaken. The importance is heightened by the fact that the type specimens of most of the species erected by Deslongchamps and French authors, from the 'Fullers Earth' and 'Great Oolite' of Normandy, were destroyed in 1944.

Several of the crocodile skulls from Shipton are significant. OUM J.1401 (*S. boutilieri*) was the first recorded British Mid Jurassic crocodile (Figure 6.11A), and casts of it were used by E.E. Deslongchamps to supplement his studies of the Normandy crocodiles. OUM J.1403 (*S. boutilieri*) unusually preserves posterior parts of the skull and palate very well. OUM J.1416, part of the type material of *Teleosaurus brevidens* Phillips (1871), is a remarkably complete lower jaw, ascribed to *S. boutilieri* by Phizackerley (1951, pp. 1177-85). OUM J.1414 (*S. megistorhynchus*) is part of the type material of *Teleosaurus subulidens* Phillips (1871) and is a fine lower jaw. OUM J.29850 and J.29851 (formerly Oxford Zool. Dept. 1639/1 and 1639/2) are holotype and paratype respectively of *S. meretrix* Phizackerley (1951). They, and OUM J.1407, show an animal with a 1 m long, very low skull, a depressed snout and little anterior rostral expansion.

Other reptiles are less well represented. There is one plate from a turtle carapace, about which little can be said. The carnivorous dinosaur

Megalosaurus is very rare here, being represented by a scapula and isolated dagger-shaped teeth. This compares with its rarity at Kirtlington and other Late Bathonian sites also, but in the Mid Bathonian it is one of the commonest finds (e.g. Huntsman's Quarry and Stonesfield, see above). A bone referred to *Cetiosaurus* was reported in a footnote in Phillips (1871, p. 213). *Cetiosaurus* bones are relatively abundant nearby in the Late Bathonian (e.g. Bletchington Station quarry, Kirtlington Cement Quarry, Glympton (SP 427217), Stratton Audley (SP 6026) and Blisworth (SP 7253)). In the same footnote, Phillips (1871) records a specimen that he tentatively refers to *Iguanodon*, but this is questionable since *Iguanodon* comes mainly from the Early Cretaceous of Europe (Norman, 1980, 1986), with only a single occurrence from the Late Jurassic (a referred mandible from the Portlandian; see below).

Finally, the 700 mm long right femur of the stegosaur *Lexovisaurus* (type specimen of *Lexovisaurus? vetustus* (Huene, 1910c)), probably from the Lower Cornbrash of Shipton (see above) is of great importance (Figure 6.11B). The femur exhibits certain juvenile features such as a gentle curve between the head and shaft in anterior view, the persistence of the cleft between the lesser and greater trochanter, and the lack of prominent longitudinal ossified cords proximally and on the shaft. It is proportionately more massive when compared with those of other stegosaurs from England, such as *Dacentrurus armatus* (Owen) from the Kimmeridgian and *D. ?phillipsi* (Seeley) from the Oxfordian (both Late Jurassic) and *Lexovisaurus durobrivensis* (Hulke) from the Callovian (Mid Jurassic). *Lexovisaurus? vetustus* is one of the oldest stegosaurs known (see New Park Quarry report), and it is similar to *Kentrosaurus* (Hennig) from the Tendaguru Shale (Kimmeridgian) of East Africa and shows similarities to the Chinese Bathonian *Huayangosaurus*.

Comparison with other localities

The reptiles from Shipton-on-Cherwell Cement Works/Gibraltar Quarry (SP 475178) must first be compared with the other 'Enslow Bridge' localities. The best known is Bletchington Station Quarry (SP 484182), which has yielded remains of *Cetiosaurus*, *Megalosaurus* and *Steneosaurus* from the *fimbriatus–waltoni* Beds (Phillips,

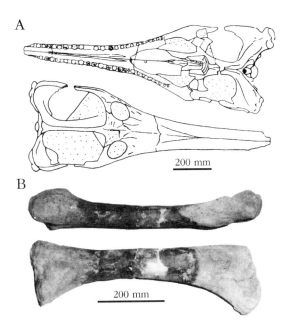

Figure 6.11 Bathonian reptiles from Shipton-on-Cherwell. (A) The crocodile *Steneosaurus boutilieri* Deslongchamps, 1869, skull in dorsal and ventral views; (B) the stegosaur *Lexovisaurus vetustus* Huene, 1910, right femur in lateral and anterior views.

1871, pp. 151, 247-94; Arkell, 1933, p. 289). The productive layer here was quarried up to the road (the A4095, also known as Lince Lane), but cannot be worked any further. A few *Cetiosaurus* bones were also found in the railway cutting south of Bletchington Station (SP 482181), possibly in the *fimbriatus–waltoni* Beds and Forest Marble. Some *Steneosaurus* teeth and bones are recorded from Lower Greenhill Quarry (SP 483179). A more useful comparison may be made with the better known faunas from Kirtlington Cement Works (SP 494199). Here, steneosaurs are relatively common in the *fimbriatus–waltoni* Beds, with some remains of *Cetiosaurus* and *Megalosaurus*. The Kirtlington Mammal Bed, at the Forest Marble/White Limestone boundary is dominated by fishes, crocodilians (*?Goniopholis*), turtles and 'lepidosaurs', with rarer frogs, salamanders, choristoderes, pterosaurs, ornithischian dinosaurs, theropods, *Cetiosaurus*, tritylodontids and mammals (Evans and Milner, 1994). The fauna of the latter unit is biased towards small fossils because of sedimentological and palaeontological factors, as well as by the means of collection. Such remains may occur at Shipton and detailed sampling of the *fimbriatus–waltoni*

Beds and other argillaceous units would be useful.

The stegosaur *L.? vetustus* is known elsewhere from the Early Bathonian (Bed 18, Perna bed, *tenuiplicatus* Zone) of the Sharps Hill Formation at Sharps Hill Quarry, near Hook Norton, and the Chipping Norton Member (*convergens* Subzone of *zigzag* Zone) of the Chipping Norton Formation of New Park Quarry, Longborough. The Mid Callovian form (*L. durobrivensis* Hulke) is known from the Oxford Clay of brick pits around Fletton, Peterborough (Galton, 1985b, p. 236).

Other comparable Late Bathonian localities in Oxfordshire and Northamptonshire include Slape Hill, Woodstock (SP 425196), which has yielded crocodile bones and teeth from the White Limestone, Glympton Quarry (SP 427217; *Cetiosaurus* vertebrae, Forest Marble), Tolley's Quarry, Bladon (?SP 449150: ?*Cetiosaurus* scapula and other bones, *fimbriatus-waltoni* Beds: Richardson *et al.*, 1946, p. 65), Ardley quarries (SP 539272, SP 541265; crocodile teeth and bones from the White Limestone), Stratton Audley (SP 6026; *Cetiosaurus* and other ?dinosaur bones, Forest Marble), Blisworth railway cuttings (SP 725543; *Cetiosaurus, Steneosaurus* bones in Blisworth Limestone or Clay), Kingsthorpe (SP 7563; *Steneosaurus* bones) and Thrapston LMS station quarry (SP 998777; *Steneosaurus, Cetiosaurus, Megalosaurus, 'Plesiosaurus'* from 'Great Oolite' or Cornbrash). Clearly, only Kirtlington, and possibly Thrapston, are of comparable stature to Shipton for Late Bathonian reptiles.

Conclusions

Shipton/Gibraltar quarry has yielded the largest variety of British Mid Jurassic crocodiles. The specimens consist of skulls and jaws which are of prime importance for classification and ecological studies. More specimens have been found at Stonesfield (early Mid Bathonian), but these are largely isolated teeth, scutes and bones. In view of the fact that the Normandy type specimens have been destroyed, the Shipton steneosaurs are the best in the world for studies of Mid Jurassic crocodiles. The stegosaur *Lexovisaurus? vetustus* is the oldest member of its genus, and one of the oldest members of Stegosauria, a group which radiated in the Late

Jurassic and Early Cretaceous of Europe, Africa and North America. Shipton's crocodiles and its stegosaur make it a Mid Jurassic site of international importance, and this importance combined with a potential for future finds give its considerable conservation value.

KIRTLINGTON OLD CEMENT WORKS QUARRY, KIRTLINGTON, OXFORDSHIRE (SP 494199)

Highlights

Kirtlington Old Cement Works is the richest site in the world for small terrestrial vertebrates from the Bathonian. The diverse tiny bones of 30 frogs, salamanders, turtles, lizards, crocodilians, pterosaurs, dinosaurs, mammal-like reptiles and mammals have been found there, many of them representing the oldest occurrences of their groups in the world.

Introduction

Kirtlington Old Cement Works Quarry has produced good faunas of fossil reptiles from the White Limestone and Forest Marble (Late Bathonian). The quarry was formerly worked for the manufacture of cement, and it closed about 1930. Although exposures were excellent (Odling, 1913; Arkell, 1931), some of the faces became obscured more recently (McKerrow *et al.*, 1969; Palmer, 1973; Freeman, 1979). Fossil amphibians, reptiles and mammals have been collected in recent years from the *fimbriatus-waltoni* Beds and from the Kirtlington Mammal Bed, a microvertebrate locality near the base of the Forest Marble (Freeman, 1976, 1979; Evans *et al.*, 1988, 1990; Evans, 1989, 1990, 1991, 1992a; Evans and Milner, 1991, 1994).

Description

The succession in the quarry has been described by Odling (1913, pp. 493, 494), Arkell (1931, pp. 570-2), Douglas and Arkell (1932, pp. 123-4) and Richardson (1946, pp. 69-71, 78-9). Additional information has been provided by McKerrow *et al.* (1969) and Freeman (1979). The following composite section is based on these authors, and

Richardson *et al.* (1946), in particular, with additions from Palmer (1973, 1979) and Torrens (*in* Cope *et al.*, 1980b, p. 36):

Thickness (m)

Lower Cornbrash
1. Limestone, rubbly and marly   1.07
2. Limestone, tough   0.76
3. Marl and rubbly limestone, in places nodular   0.23
4. *Astarte-Trigonia* Bed. Limestone, very hard, grey   0.61
5. Clay, brown, marly   0.30

Forest Marble Formation
1. Clay, grey and buff, with some thin, irregular hard bands   1.53
2. Clay, dark grey (=beds 3w-z of Freeman, 1979)   0.69
3. Limestone, yellowish, flaggy, locally marly and 'shaly', oolitic, with occasional inclusions of white lithographic limestone; ripple marks, rain pits (?=bed 3v of McKerrow *et al.*, 1969; Freeman, 1979)   0.61-0.92

(White Limestone Formation)
4. Clay, grey-blue, with three pale mudstone layers, one at the bottom (=beds 3p-u of McKerrow *et al.*, 1969; Freeman 1979; = 'Unfossiliferous Cream Cheese Bed' of Odling, 1913 and Arkell, 1931). The basal unconsolidated 0.04-0.25 m brown marl unit (Bed 3p) is the Kirtlington Mammal Bed of Freeman (1979)   2

(White Limestone Formation)
5. Coral-*Epithyris* Limestone (Upper *Epithyris* Bed or 'Fossiliferous Cream Cheese Bed' of Odling, 1913 and Arkell, 1931; ? Beds 3n-o of McKerrow *et al.*, 1969). Limestone; at northern end an extremely hard white blue-hearted lithographic rock. Passes locally into unfossiliferous oolite   1.22-2.21
6. *fimbriatus-waltoni* Beds (=Bed 10 of Arkell 1931; Beds 3k, l of McKerrow *et al.*, 1969). Clay, grey-green to greenish black, with some white pellets

Thickness (m)

at top; bed largely made up of bivalves; when bed 7 is absent, there is a lignite at the base   1.07
7. Oyster-*Epithyris* Marl (=Bed 9; Middle *Epithyris* Bed of Arkell, 1931; Bed 3k of McKerrow *et al.*, 1969). Marl, brown. Locally, a thin layer of corals occurs below   0-0.75
8. Limestone, hard, blue-hearted (?=Beds 3i,j of McKerrow *et al.*, 1969)   0.92m
9. Marl (?=Bed 3h of McKerrow *et al.*, 1969)   0.23m
10. Limestone, similar to 8 (?=Bed 3g of McKerrow *et al.*, 1969)   0.84-0.92 m
11. *Epithyris* Limestone (=Lower *Epithyris* Bed of Arkell, 1931; =Bed 3a-f, Bed 1e of McKerrow *et al.*, 1969). Limestones, white, at west end of pit a mass of *Epithyris*. Thins out eastwards and replaced from beneath by lenticular limestones   2.44
12. *Aphanoptyxis ardleyensis* Bed. Limestones, well bedded   0.46-0.61
13. *Nerinea eudesii* Beds. Limestones in three courses   1.68

This section was recorded by Arkell (1931) in various parts of the quarry, which means that it is not a true log because of the large amount of lateral facies variation. The lower parts (beds 8-13 in particular) are hard to match with the logs given by McKerrow *et al.* (1969, p. 58) because certain units, such as the *Epithyris* Limestone (Bed 11; Bed 1e of McKerrow *et al.*, 1969), are laterally impersistent.

There are considerable problems with the lithostratigraphy of the units in this quarry and these particularly concern the placing of the boundary between the White Limestone and the Forest Marble. Odling (1913, pp. 493-4) placed it above his 'Bed 1. Fossiliferous Cream-Cheese Bed', thus between beds 4 and 5 of the section of Richardson *et al.* (1946). Arkell (1931) renamed and subdivided the Forest Marble into the Wychwood Beds (beds 1-3 of the section of Richardson *et al.*, 1946) and the Kemble Beds (beds 4-7). Thus, he moved the Forest Marble/White Limestone boundary to between beds 7 and 8 on the basis of correlations with supposedly similar lithologies and fossils in

Oxfordshire and Wiltshire. Richardson *et al.* (1946, pp. 69-71) changed the Wychwood Beds/Kemble Beds boundary to lie between their beds 2 and 3, and moved the Forest Marble/White Limestones boundary to lie between their beds 5 and 6. Arkell (1947a, p. 57) interpreted the sequence as follows: Wychwood Beds (beds 1-3), Kemble Beds (beds 4-5), Bladon Beds (beds 6-7), ?Bladon Beds (beds 8-10), Ardley Beds (beds 11-13), the division of the White Limestone being based on gastropods.

More recently, McKerrow *et al.* (1969) attempted a definition based largely on the occurrence of oysters and took the basal bed of the Forest Marble to be the base of the Oyster-*Epithyris* Marl (bed 7), as Arkell (1931) had initially. Palmer (1973, p. 61) points out that at Kirtlington the Coral-*Epithyris* Limestone (bed 5) contains oysters, but otherwise shows a typical White Limestone fauna and lithology, and he proposed that the Forest Marble/White Limestone boundary should be moved to between beds 4 and 5. This view was also expressed by Barker (1976) on the basis of a study of the gastropods. Palmer (1979) further argued this point and divided the White Limestone Formation into three members, of which the Ardley Member (beds 8-13) and the Bladon Member (beds 5-7) are seen at Kirtlington. Palmer (1979, p. 208, fig. 5) makes it clear that his Bladon Member is intended to include both the *fimbriatus-waltoni* and Upper *Epithyris* Beds of the Cherwell valley which rest on the *A. bladonensis* Bed. In general, Torrens (*in* Cope *et al.*, 1980b, p. 36) recommends that the base of the Forest Marble be taken as 'the base of the clay overlying the Coral-*Epithyris* bed, or of the bed above at Kirtlington' (i.e. the base of bed 3 or 4).

Reptiles occur in the *fimbriatus-waltoni* Beds (beds 2o, 3i, 4e, 6f of McKerrow *et al.*, 1969; base of the Bladon Member, Palmer, 1979) and the Kirtlington Mammal Bed. Arkell (1931, p. 572) noted that he saw the bones of *Cetiosaurus oxoniensis* Phillips (1871) associated with lignite at the base of the *fimbriatus-waltoni* Beds where they rest on the eroded surface of the underlying limestone. Richardson *et al.* (1946, p. 70) repeated this observation, but noted that the bones and lignite occurred when the Oyster-*Epithyris* Marl (bed 7) was absent and lay on the eroded top of bed 8. However (p. 71) they say that 'the main horizon for Ceteosauran [sic] remains appears to be between the clay and the Middle *Epithyris* Bed, although here at Kirtlington

and elsewhere the remains are often enclosed by the clay'. *Cetiosaurus* has been found elsewhere in Oxfordshire in the *fimbriatus-waltoni* Beds (Phillips, 1871; Arkell, 1931; Richardson *et al.*, 1946). The bones in this unit are usually disarticulated, but appear to have been associated (Phillips, 1871, p. 250).

The Kirtlington Mammal Bed (bed 3p of McKerrow *et al.*, 1969) is an impersistent lens, 21.5 m long and 0.04-0.25 m thick in the northeastern corner of the quarry (Freeman, 1979, p. 136). The contacts of this bed with the Coral-*Epithyris* Limestone below (bed 3o of McKerrow *et al.*, 1969) and another limestone above (bed 3q) are extremely sharp and probably erosional. Associated fossils (Evans and Milner, 1994) include microscopic freshwater charophytes, indeterminate plant fragments, and ostracods, as well as the dissociated remains of a variety of bony fishes (cf. *Lepidotes*, pycnodontoid, ?amioid) and sharks (*Asteracanthus, Hybodus, Lissodus,* batoid). The tetrapod remains include a variety of amphibians, reptiles and mammals (Evans and Milner, 1991, 1994). Most of these animals are represented only by their more durable parts - teeth, scutes, jaws and vertebral fragments. By contrast, a few genera (possibly those which have been least transported) have most of their skeletal elements preserved.

Fauna

The older reptile specimens labelled 'Kirtlington' in collections are assumed to come from the *fimbriatus-waltoni* Beds, since the Mammal Bed was not exploited before the work of Freeman (1976, 1979) and its fossils are generally small.

1. *fimbriatus-waltoni* Beds
Archosauria: Crocodylia: Thalattosuchia:
   Steneosauridae
  Steneosaurus brevidens (Phillips, 1871)
    BMNH R5149
  '*Steneosaurus*' aff. *larteti* (J.A. Deslongchamps, 1866)
    OUM J.1413
  Steneosaurus sp.
    BMNH R4809, R6323; OUM J.10597, J.12007;
     CAMSM J.21949-51, J.21954
  Teleosaurus sp.
    BRSMG Cb1271 (specimen destroyed in
     World War 2)

Dinosauria: Saurischia: Theropoda:
  Megalosauridae
  Megalosaurus sp.
    BMNH R5797
Dinosauria: Saurischia: Sauropoda
  Cetiosaurus sp.
    BMNH R5152-3, R5156-7; OUM J.13526-57,
      J.13596
  Bothriospondylus sp.
    BMNH R5150-1
Sauropterygia: Plesiosauria
  ?Plesiosaurus sp.
    BMNH R2986, R5154

2. Kirtlington Mammal Bed (data from Freeman,
  1979 and Evans and Milner, 1991, 1994)
Anura: Discoglossidae
  Eodiscoglossus oxoniensis Evans, Milner and
    Mussett, 1990
    Holotype: BMNH R11700
Caudata: Albanerpetontidae
  Albanerpeton sp.
Caudata: *inc. sed.*
  Marmorerpeton freemani Evans, Milner and
    Mussett, 1988
    Holotype: BMNH R11364
  Marmorerpeton kermacki Evans, Milner and
    Mussett, 1988
    Holotype: BMNH R11361
  Salamander A
  Salamander B
Testudines: Cryptodira
  cf. Pleurosternidae
Lepidosauromorpha: *inc. sed.*
  Marmoretta oxonienesis Evans, 1991
    Holotype: BMNH R12020
Lepidosauria: Sphenodontida
  Sphenodontian (Evans, 1992a)
Lepidosauria: Squamata: Sauria
  Saurillodon sp.
  Scincomorphs
  Anguimorph
  ?Gekkotan
Archosauromorpha: Choristodera
  Cteniogenys sp.
Archosauria: Crocodylia: Neosuchia
  ?Goniopholis/Nannosuchus sp.
  atoposaurid
Archosauria: Pterosauria
  Rhamphorhynchoid
  Pterodactyloid
Archosauria: Saurischia
  Megalosaurus sp.
  'maniraptoran' and other small theropods
  ?Cetiosaurus

Archosauria: Ornithischia: Ornithopoda
  Fabrosaurid, cf. *Alocodon*
Synapsida: Therapsida: Cynodontia:
  Tritylodontidae
  Stereognathus ooliticus Charlesworth, 1855
Mammalia: Triconodonta: Morganucodontidae
  Wareolestes rex Freeman, 1979
Mammalia: Docodonta: Docodontidae
  Simpsonodon oxfordiensis Kermack, Lee, Lees
    and Mussett, 1987
Mammalia: Symmetrodonta: Kuehneotheriidae
  Cyrtlatherium canei Freeman, 1979
Mammalia: Eupantotheria: Peramuridae
  Palaeoxonodon ooliticus Freeman, 1979
Mammalia: Eupantotheria: Dryolestidae
  ?Dryolestid

Interpretation

The biostratigraphy of the Bathonian at
Kirtlington is difficult since no ammonites have
been found locally, and very few elsewhere in
comparable rocks (Torrens, 1969a; *in* Cope *et al.*,
1980b). Finds of ammonites in the White
Limestone of the Oxford area have permitted correlation of this unit with the *subcontractus* and
morrisi Zones (Mid Bathonian), and the *hodsoni*
and lower *aspidoides* Zones (Late Bathonian),
while the Forest Marble Formation is largely *aspidoides* and basal *discus* Zones (Late Bathonian),
on the basis of correlation of beds above and
below.

The approximate zonal assignments of the
three members of the White Limestone Formation
are: Shipton Member, *?subcontractus, morrisi*
Zones, Ardley Member, ?lower *hodsoni* Zone, and
Bladon Member, ?upper *hodsoni*-lower *aspidoides* Zones (Palmer, 1979; Torrens, *in* Cope *et
al.*, 1980b). However, the evidence for zonation
of these members is 'not compelling' (Torrens, *in*
Cope *et al.*, 1980b, p. 37). Ostracod zonation
(Bate, 1978) places the White Limestone of the
Oxford area in ostracod zones 5-8, the Forest
Marble and Cornbrash resolving to the top of
zone 8 and above (=upper *discus* Zone).

The reptile-bearing *fimbriatus-waltoni* Beds
(base of the Bladon Member) are dated as ?upper
hodsoni Zone (basal Late Bathonian) (Torrens, *in*
Cope *et al.*, 1980). However, the occurrence of
the ostracod *Glyptocythere penni* in the *fimbriatus-waltoni* Beds led Bate (1978) to suggest that
this unit belongs to the *discus* Zone. The
Kirtlington Mammal Bed falls within the

aspidoides or *discus* Zone (Freeman, 1979, p. 136).

Environmental interpretations have been made on the basis of the sedimentology of the *fimbriatus-waltoni* Beds. McKerrow *et al.* (1969, pp. 61-4, 80) noted the abundance of lignite and occasional caliche-like nodules which they interpreted as indicating shallow water with occasional subaerial exposure. The nodules appear to be distinct from the small pellets of 'race' common in many calcareous clays close to the ground surface, which are produced by recent weathering. Klein (1965, p. 173) considered that similar nodules from other Great Oolite clays represent caliche, indicating emergence, although Palmer (1979, p. 210) regarded them as pebbles formed by erosion of an incompletely cemented limestone bed. Palmer (1979, pp. 210-11) noted the complex channelled interdigitations of this unit at Shipton (SP 4717), and suggested that deposition of some of the clays was local and catastrophic, and that the nodules were derived from elsewhere. There is a non-sequence at the top of the *fimbriatus-waltoni* Beds, and localized emergence at this level is probable, which may be related to nodule formation. Palmer (1979) supposed a quiet-water lagoonal environment subject to periodic current activity and influx of new sediment, perhaps during storms.

The marl sediment of the Kirtlington Mammal Bed contains subangular pebbles of oolitic limestones, comminuted shell debris, individual ooliths and rare silica sand grains, all of which suggest a temporary freshwater pool that received periodic influxes of poorly sorted sediment derived from local erosion of earlier Mid Jurassic limestones (Freeman 1979, p. 139). The ostracods, charophytes and fishes lived in the pool, and the plants, amphibians, reptiles and mammals presumably lived nearby. Freeman (1979) noted that the mammal and theropod teeth were distributed in clumps, and that this might indicate their concentration in the faeces of larger animals, such as carnivorous dinosaurs ('coprocoenoses').

As outlined by Evans (1990, p. 234), in Bathonian times Kirtlington lay on or near the south-west shore of a small island barrier some 30 km from the coast of the Anglo-Belgian landmass at a subtropical latitude of about 30°N (Palmer, 1979). Lignite, charophytes and freshwater ostracods and gastropods in the marly sediments suggests a coastal environment, which had low relief, with creeks, lagoons and freshwater lakes, rather like the Florida Everglades (Palmer, 1979).

The vertebrate fauna of the Kirtlington Mammal Bed, with its amphibians and aquatic reptiles (choristoderes, crocodilians and turtles), agrees well with such a palaeoenvironmental scenario. Terrestrial elements are rather rare, being largely represented by reptile jaw fragments and teeth; these components may have been reworked from localities further inland.

The faunas of the two reptile-bearing beds at Kirtlington are rather different, which probably relates to preservational and environmental conditions rather than to the very slight age difference. They will be discussed separately.

The *fimbriatus-waltoni* Beds fauna is dominated by crocodilians and sauropod dinosaurs. The long-snouted crocodilians *Steneosaurus* and *Teleosaurus* are represented by vertebrae, teeth and jaws. Their long recurved teeth, strong jaws and adaptations for swimming suggest that they were fish-eaters in fresh or marine water. The taxonomy of these forms is complex (Steel, 1973), so that the species assignments may be incorrect. These crocodiles are relatively common in the Bathonian of England and France (see above). In the Late Bathonian of England specimens are known from the upper White Limestone of a few localities elsewhere in Oxfordshire and in the Blisworth Limestone (=White Limestone) and Blisworth Clay of Northamptonshire.

The carnivorous dinosaur *Megalosaurus* is represented only by a tooth. However, a variety of vertebrae, limb bones and skull elements (including the brain case) of the large sauropod *Cetiosaurus* have been found. More than 10 species of this genus have been erected for Jurassic and Cretaceous material (Steel, 1970, p. 64). The Mid Jurassic forms are *C. rugulosus* (Owen, 1845) from Wiltshire, *C. oxoniensis* Phillips (1871) and *C. glymptonensis* Phillips (1871) from Oxfordshire and Northamptonshire and *C. mogrebiensis* Lapperant (1955) from the Moyen Atlas of Morocco. The morphological distinctions between these species have not been elucidated (*C. rugulosus* is based on a tooth, *C. glymptonensis* on a caudal vertebra and the other two on incomplete postcranial skeletons). Further, many other generic names have been applied to large sauropod bones, and the differences have often not been made clear. Nevertheless, most of the Mid Jurassic English material may be placed in *C. oxoniensis*, the best-known species. This animal had a 1.65 m femur, and was about 15 m long overall. The braincase resembles that of the Triassic *Plateosaurus*, the neck was relatively short and the vertebrae showed primitive features (almost solid construction, and no bifurcation of the neural spines).

Two vertebrae of another sauropod, *Bothriospondylus*, have also been found at Kirtlington. This genus is known mainly from the Late Jurassic and Early Cretaceous, but two forms occur in the Bathonian, *B. robustus* Owen (1875) from Wiltshire and *B. madagascariensis* Lydekker (1895) from Madagascar. The vertebrae of *Bothriospondylus* are deeply excavated, presumably to reduce their weight. Its total body length was 15–20 m.

Two vertebrae have been named as those of a plesiosaur. If correctly identified, these may belong to the genera '*Cimoliasaurus*' of *Muraenosaurus*, known from the Bathonian of sites in Northamptonshire, Cambridgeshire, Leicestershire and Eigg, western Scotland. Assuming their correct identification, the presence of plesiosaurs in the *fimbriatus–waltoni* Beds would indicate marine conditions, but their rarity here may connect with a predominantly lagoonal/coastal situation.

The amphibians, reptiles and mammals from the Kirtlington Mammal Bed have been summarized by Freeman (1979) and Evans and Milner (1991, 1994). Details of the collecting and preparation techniques are given in Freeman (1976, 1979), Kermack *et al.* (1987) and Evans (1989). The amphibians and reptiles (Figure 6.12) are described here (the mammals will be detailed in the GCR Fossil Mammals and Birds volume).

The amphibians include a frog referrable to the family Discoglossidae (*Eodiscoglossus oxoniensis*) and five species of salamander (*Albanerpeton, Marmorerpeton kermacki, M. freemani* and two unnamed forms). *Eodiscoglossus oxoniensis* (Figure 6.12A) is the earliest identifiable discoglossid frog known, and one of the oldest frogs of any sort (Evans *et al.*, 1990). The specimens of *E. oxoniensis* from Kirtlington are comparable with *E. santonjae* from the Early Cretaceous of Montsech, Lérida, Spain, but they may be clearly distinguished by characters of the ilium and premaxilla. The only older frogs are the primitive *Triadobatrachus* from the Early Triassic of Madagascar and *Vieraella* from the Early Jurassic of Argentina.

The record of *Albanerpeton* is one of the oldest of this enigmatic family, the oldest being from the Bajocian of Aveyron, France (Evans and Milner, 1994). The albanerpetontids are also known from the Cretaceous of North America and the Miocene

of France. *Marmorerpeton kermacki* (Figure 6.12B) and *M. freemani* are the earliest known salamanders (i.e. true Caudata; Evans *et al.*, 1988), more primitive than any other known forms by the absence of intravertebral spinal nerve foramina in the atlantal centrum. However, in other features these taxa resemble members of the family Scapherpetonidae, which comprises neotenous forms otherwise known only from the Late Cretaceous and Palaeocene. Salamanders A and B are yet to be described.

Turtles are represented by many specimens (Freeman, 1979; Evans and Milner, 1994) which augment the sparse Mid Jurassic record of that group. The oldest turtles, *Proterochersis* and *Proganochelys*, come from the Late Triassic of Germany and the oldest cryptodire, the main modern group is *Kayentachelys* from the Early Jurassic of North America. Turtle fragments are known from Stonesfield and other British Bathonian sites, but the Kirtlington material includes more diagnostic skull and carapace fragments of a pleurosternid (Evans and Milner, 1994).

Lepidosauromorphs are represented at Kirtlington by a variety of forms. The Lepidosauromorpha (Benton, 1985; Evans, 1988a; Gauthier *et al.*, 1988c) include the Lepidosauria (sphenodontids plus squamates) and a number of basal Permo-Jurassic groups. *Marmoretta oxoniensis* (Figure 6.12C) is a small probably insectivorous form, apparently a common component of the fauna (Evans, 1991). True lepidosaurs are represented by some sphenodontids (Evans, 1992a) and squamates (two scincomorph lizards, one of which is *Saurillodon*, an anguimorph and a possible gekkotan). These are important since lepidosaurs are not well known in the Mid Jurassic: the nearest well-documented faunas are the sphenodontids from the Late Triassic and Early Jurassic fissures of the Bristol area and South Wales (see above) and from the Early Cretaceous of Durlston Bay (Purbeck). Sphenodontids are known also from the Late Jurassic of Germany (Solnhofen), France (Cerin) and North America (Morrison Formation) (Fraser and Benton, 1989). The first true lizards are known otherwise only from the Late Jurassic (Oxfordian of Guimarota, Leiria, Portugal; Kimmeridgian of Cerin, Ain, France; Portlandian of Solnhofen, Bavaria, Germany; Benton, 1993).

Cteniogenys (Figure 6.12D) is represented by many isolated skull and postcranial elements. The genus was named on the basis of some isolated

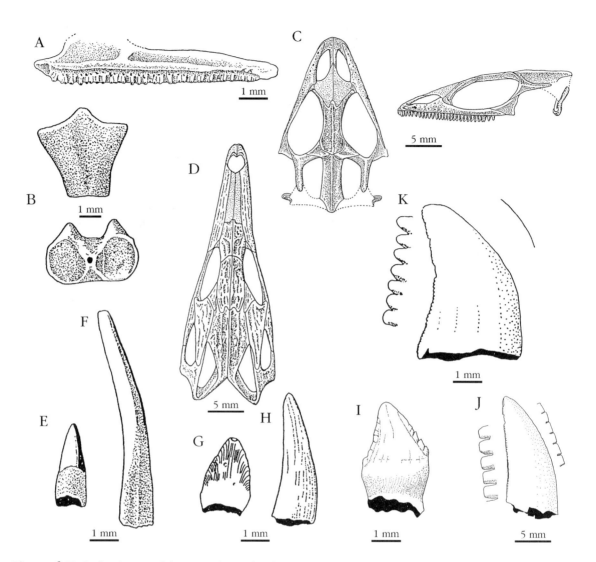

Figure 6.12 Bathonian amphibians and reptiles from Kirtlington Old Cement Works Quarry. (A) The frog *Eodiscoglossus oxoniensis* Evans, Milner and Mussett, 1990, right maxilla in medial view; (B) the salamander *Marmorerpeton kermacki* Evans *et al.*, 1988, atlantal centrum in ventral and anterior views; (C) the lepidosauro-morph *Marmoretta oxoniensis* Evans, 1991, reconstructed skull in lateral and dorsal views; (D) the choristodere *Cteniogenys oxoniensis* Evans, 1990, reconstructed skull in dorsal view; teeth of: (E) rhamphorhynchoid pterosaur; (F) pterodactyloid pterosaur; (G) goniopholidid crocodile; (H) atoposaurid crocodile; (I) fabrosaurid dinosaur; (J) megalosaurid dinosaur; (K) maniraptoran dinosaur. All after Evans and Milner (1994).

dentaries from the Late Jurassic of Wyoming by Gilmore (1928), who identified the bones as representing a lizard. This interpretation was also given for specimens from Guimarota in Portugal by Seiffert (1973) and Estes (1983). The material from the Kirtlington Mammal Bed, however, demonstrates that *Cteniogenys* is a choristodere, an archosauromorph diapsid (Evans, 1989, 1990, 1991). It appears to be the smallest choristodere known but, allowing for its size, it appears to be related to the Rhaetian *Pachystropheus* (see Aust

Cliff report), the gavial-like aquatic Late Cretaceous and Palaeocene *Champsosaurus* and *Simoedosaurus*, and a new form from the Oligocene of France. Comparison with known choristoderes suggests that *Cteniogenys* is the most primitive of the known genera (Evans, 1989). The available skull and postcranial material indicate that the Kirtlington form is represented by animals of more than one age class; the largest specimens are well ossified and can be assumed to be the adults.

The abundant crocodile teeth are nearly all shed crowns; Freeman (1979, p. 140) reports only three with roots. They apparently show little resemblance to *Teleosaurus* and Freeman (1979, p.140) compares them with the small Late Jurassic goniopholid *Nannosuchus* from the Purbeck (?juvenile *Goniopholis*). Evans and Milner (1993) note also some rare *Theriosuchus*-like atoposaurid teeth, a form otherwise known from the Wealden. If the identifications are correct, these would be the oldest records in the world of goniopholidids and atoposaurids (Figure 6.12 G,H).

The pterosaur teeth have been identified as of rhamphorhynchoid and pterodactyloid types (Evans and Milner, 1991, 1994), and similar identifications have been made from Hornsleasow. The long slender rhamphorhynchoid teeth (Figure 6.12E) could correspond to *Rhamphocephalus*, a form better known from the Mid Bathonian of Stonesfield and the Early Bathonian of sites around Eyford (see above). If the shorter blunter pterosaur teeth (Figure 6.12F) are truly pterodactyloid, this would be another oldest record, since pterodactyloid skeletons are reported first from the ?Oxfordian (Guimarota, Portugal) and the Kimmeridgian (Morrison Formation, Wyoming; Kimmeridge Clay, Dorset; Tendaguru Beds, Tanzania; Benton, 1993).

The theropod teeth are described by Freeman (1979, p. 142) as 'smaller than those of ... *Megalosaurus bucklandi*, ranging in height from 1.6 to 7.4 mm. They may be the teeth of either juvenile *M. bucklandi* or of coelurosaurs'. It is important to note their relative rarity here, especially since *Megalosaurus* teeth are among the commonest reptile finds at Stonesfield. Many of the smaller theropod teeth (Figure 6.12K) most closely resemble those of maniraptoran dinosaurs such as *Deinonychus* and *Troodon*, typically Cretaceous forms (Evans and Milner, 1991, 1994).

Ornithischian dinosaurs are represented by teeth similar to those of the ornithopod *Alocodon*, possibly a fabrosaurid (Evans and Milner, 1991, 1994, Figure 6.12I). Freeman (1979, p. 142) compared his ornithischian teeth with those of *Scelidosaurus* (Sinemurian, Charmouth, Dorset) or *Echinodon* (Berriasian, Durlston Bay, Dorset).

Finally, the rare tritylodont teeth (*Stereognathus*) described by Freeman (1979) and Evans and Milner (1994) are of considerable importance. The tritylodonts are best known from the Late Triassic and Early Jurassic of South Africa, China, Germany and some of the British fissures (Kühne, 1956, Savage, 1971). Mid Jurassic forms are known from the Stonesfield Slate of Stonesfield (*progracilis* Zone, Mid Bathonian; Simpson, 1928, pp. 22-6), the Ostracod Limestones of the Great Estuarine Group of Skye (Late Bathonian; Waldman and Savage, 1972) and the Forest Marble of Bridport, Dorset (Late Bathonian; Ensom, 1977). The Kirtlington specimens are the youngest known tritylodonts, and the last surviving mammal-like reptiles from Britain, and are superseded in age only by *Bienotheroides* from the Mid or Late Jurassic of China.

Comparison with other localities

The reptiles from Kirtlington Cement works compare best with faunas collected nearby in the Mid and Late Bathonian. Sites around Shipton-on-Cherwell Quarry (see above) have yielded remains of turtles, the crocodilian *Steneosaurus*, and the dinosaurs *Megalosaurus, Lexovisaurus* and *Cetiosaurus* from the *fimbriatus-waltoni* Beds (upper White Limestone) and from the Forest Marble and Lower Cornbrash. However, none of these sites has yielded *Bothriospondylus*, plesiosaurs, choristoderes, lepidosaurs, tritylodontids or the other small vertebrates known from Kirtlington. This is probably because careful washing and sorting of large amounts of sediment has not yet been carried out. Other comparable, but less abundant, faunas have been collected from the Forest Marble of Wiltshire, the upper White Limestone Formation and Forest Marble of Gloucestershire and Oxfordshire, and the Blisworth Limestone and Blisworth Clay of Northamptonshire (see above).

Some older localities in the British Bathonian may be better for comparison because they have yielded rich faunas: Hornsleasow (earliest Bathonian), New Park Quarry (Early Bathonian), Stonesfield (early Mid Bathonian) and Huntsman's Quarry (early Mid Bathonian). None of these has yet turned up such an array of well-preserved microtetrapod material.

The Kirtlington Mammal Bed fauna bears a significant resemblance to later Mesozoic freshwater assemblages, rather than earlier ones (Evans *et al.*, 1988; Evans and Milner, 1994). The salamanders cannot with certainty be referred to later families, but elements of the salamander-discoglossid-albanerpetontid-turtle-crocodile-choristodere association are found in later assemblages, such as those of Late Jurassic age at Guimarota (Oxfordian) and Solnhofen (Portlandian), the Late Jurassic/Early Cretaceous Purbeck in Dorset (q.v.), the Early Cretaceous at Una, Spain, and the Late Cretaceous of the Judith River (=Lance Formation) of North America.

Conclusions

Kirtlington Quarry represents the best Late Bathonian site for a variety of amphibian and reptile groups, and it is the source of numerous new forms. The *fimbriatus-waltoni* Beds reptiles are comparable with those from the same unit at several other sites in Oxfordshire, but the variety of material is greater than elsewhere, and the site is still readily accessible for further excavation. The fauna of the Kirtlington Mammal Bed is without rival for its age; the selection of large and small reptiles has still to be studied fully, but they could rival the older Stonesfield fauna in their importance. The Mammal Bed fauna includes a unique freshwater assemblage of small reptiles and amphibians, several of which are the earliest known occurrences of their respective groups (the first discoglossid frog, salamanders, pleurosternid turtle, true lizards of several groups, goniopholidid and atoposaurid crocodilians, pterodactyloid pterosaur and ?maniraptoran dinosaur). The amphibian and reptile fauna is extensive, including frogs, salamanders, turtles, lepidosauromorphs, sphenodontids, lizards, choristoderes, crocodilians, pterosaurs and small dinosaurs.

The diversity and importance of the fossil vertebrates and potential for future finds give the site its high conservation value.

Chapter 7

British Late Jurassic fossil reptile sites

INTRODUCTION: BRITISH LATE JURASSIC REPTILE SITES

Late Jurassic reptiles have come from many localities along the length of the English outcrop, between the Dorset coast and Yorkshire, and are represented in rocks ranging in age from Late Oxfordian to the Portlandian. The recorded British Late Jurassic reptile sites are detailed below stage by stage. The listings are based on sources as noted, together with examinations of major museum collections.

OXFORDIAN

Reptile remains are rare in the Upper Oxford Clay and Corallian Beds. A few sites have yielded plesiosaurs and marine crocodilians, but the most important remains are isolated finds of dinosaurs from Dorset, Cambridgeshire and Yorkshire. References include Newton (1878), Fox-Strangways (1892), H.B. Woodward (1895), Strahan (1898), Galton (1980a, 1980b), Martill (1986, 1988) and Martill and Hudson (1991).

DORSET: Sandsfoot Castle, Weymouth (SY 676774; *Pliosaurus, Teleosaurus* jaw from Sandsfoot Grit (*regulare* Zone; Upper Oxfordian)); Hill Crest Road, Weymouth (?SY 673775; *Muraenosaurus*); Nothe, Weymouth (SY 688788; *Teleosaurus* from Lower Calcareous Grit); Osmington (?SY 7252; *Ophthalmosaurus* from the Corallian); Preston (?SY 6983; *Muraenosaurus, Macropterygius*).

WILTSHIRE: Steeple Ashton (ST 9157; *Plesiosaurus* from the 'Coral Rag'); Rood Ashton (ST 887560; '*Plesiosaurus*' from the 'Coral Rag'); Heddington (SU 0067; *Pliosaurus* teeth).

BERKSHIRE: Hatford, Faringdon (SU 3394; '*Plesiosaurus*', dinosaur from the Corallian).

OXFORDSHIRE: Stanford-in-the-Vale (SU 3493; *Megalosaurus* from the 'Coral Rag'); Marcham (SU 4596; *Teleosaurus* from the Corallian); Cothill (SU 4699; crocodile from the 'Lower Calcareous Grit'); Bladon Fields (SP 4414; *Teleosaurus* from the Corallian); Littlemore, Oxford (SP 5302; *Megalosaurus*); Wheatley (SP 5907; *Pliosaurus* teeth); Headington 'Quarry Field' (?SP 555071; *Cetiosaurus, Metriorhynchus* from the 'Coral

Rag' and 'Lower Calcareous Grit'); Garsington (SP 5802; '*Plesiosaurus*').

BEDFORDSHIRE: Ampthill (TL 0438; *Steneosaurus, Ophthalmosaurus* in the Ampthill Clay).

CAMBRIDGESHIRE: Great Gransden (TL 260561 or TL 252564; type of dinosaur *Cryptodraco eumerus, Pliosaurus* from Ampthill Clay [?*serratum* Zone)]; Warboys Brick Pit (TL 308818; *Ophthalmosaurus* from Upper Oxford Clay [*mariae* Zone]; Martill, 1986); Gamlingay (TL 2352; *Ophthalmosaurus*); Mepal (TL 4480; '*Plesiosaurus*' in the Ampthill Clay).

YORKSHIRE: Slingsby (SE 708744; *Pliosaurus, Steneosaurus* teeth; type of *Dacentrurus? phillipsi* from Malton Oolite Member [*vertebrale* Subzone; *densiplicatum* Zone]); Malton (SE 7871; *Pliosaurus, Steneosaurus, Metriorhynchus*); Appleton-le-Street (?SE 734737; *Pliosaurus* from 'Birdsall Calcareous Grit'); Scarborough (TA 0488; *Pliosaurus*); North Grimstone (SE 8467; *Pliosaurus, Teleosaurus*); locality unknown (type of *Priodontognathus phillipsii* ?from the Calcareous Grit).

KIMMERIDGIAN

Overall the Kimmeridge Clay has yielded abundant turtles, crocodiles, pterosaurs, plesiosaurs, pliosaurs, ichthyosaurs and dinosaurs, and includes some of the best Late Jurassic marine reptile faunas in the world. There are many localities, although most sites have yielded only remains of one or two marine reptiles. However, abundant remains have been found in Dorset, Wiltshire, Oxfordshire, Cambridgeshire and Yorkshire. References include Seeley (1869a), Phillips (1871), H.B. Woodward (1895), Strahan (1898), Arkell (1933, 1947a, 1947c), Tarlo (1958, 1959b, 1959c, 1960), Delair (1959, 1960, 1966), Brookfield (1978b), Cope (1967, 1978) and Taylor and Benton (1986).

DORSET: Ringstead Bay (SY 751813; *Steneosaurus, Pliosaurus, Ophthalmosaurus*); Osmington Mills (SY 7382; *Pliosaurus; Ophthalmosaurus*, some at least from the *mutabilis* Zone, one plesiosaur limb bone from the *eudoxus* Zone; Clarke and Etches, 1992); Isle of Portland coast (SY 6872–SY 7072; *Ophthalmosaurus, Cimoliasaurus, Pliosaurus, Thaumatosaurus,*

Colymbosaurus; recent *Colymbosaurus* specimen from Grove Cliff (SY 706722) in the *pavlovi* Zone, Brown, 1984; Palmer, 1988); Upwey (SY 6684; '*Plesiosaurus*'); Hazelbury Bryan (*Pliosaurus*); Motcombe (ST 8425; *Liopleurodon*); Gillingham Brick Pit (ST 809258; *Liopleurodon, Dacentrurus, Ophthalmosaurus*, type of *Ophthalmosaurus pleydelli* from the *mutabilis* and *cymodoce* Zones (Lower Kimmeridgian)); Fiddleford, near Sturminster Newton (ST 8013; *Ophthalmosaurus, Liopleurodon*).

SOMERSET: ?Ilminster (ST 3614; *Ophthalmosaurus, Liopleurodon*).

WILTSHIRE: Stour cutting (ST 779305; *Ophthalmosaurus*; Bristow *et al.*, 1992, p. 141); Broughton-Gifford (ST 8763; *Peloneustes*); Westbury Clay Pit (ST 880527; '*Plesiosaurus*'; recent finds of thalassemyid turtle and *Metriorhynchus* skull from *mutabilis* Zone, and *Pliosaurus* skull from *eudoxus* Zone; plesiosaur limb bone from the *eudoxus* Zone, Clarke and Etches, 1992); Chippenham (ST 9173; *Pliosaurus*); Foxhangers (ST 937615; *Ophthalmosaurus, Cimoliasaurus, Pliosaurus, Liopleurodon, Metriorhynchus, Megalosaurus*; Delair, 1973); Rodborne, near Swindon (ST 9383; ?nodosaurid ankylosaur; Galton, 1980b (?Lower Kimmeridgian)); Westbrook(e), Bromham (ST 9666; type of *Ophthalmosaurus trigonus*); Potterne (ST 9958; *Cimoliasaurus, Ophthalmosaurus*); Devizes (SU 0161; *Thalassemys, Ophthalmosaurus, Cimoliasaurus, Pliosaurus, Liopleurodon, Metriorhynchus*); Wootton Bassett (SU 0682; *Ophthalmosaurus, Cimoliasaurus, Pliosaurus, Liopleurodon, Metriorhynchus*; type of *Dacentrurus hastiger* from the Early Kimmeridgian); Swindon Brick and Tile Pits (SU 156834; *Ophthalmosaurus, Megalosaurus, Ornithocheirus, Cimoliasaurus, Pliosaurus, Liopleurodon, Thaumatosaurus*; types of *Plesiochelys passmorei* (turtle), *Bothriospondylus suffossus* and *Omosaurus armatus* (dinosaurs), *Pliosaurus brachydeirus* and *P. macromerus* from *baylei* and *cymodoce* Zones (Lower Kimmeridgian); Delair, 1982a); Stanton-Fitzwarren (SU 1790; *Ophthalmosaurus, Liopleurodon*).

BERKSHIRE: Stanford-in-the-Vale (SU 3493; *Ophthalmosaurus*, '*Plesiosaurus*'); Oday Common, south of Abingdon (SU 492949; pliosaur; Delair, 1982a; ichthyosaur in Abingdon Museum; turtle).

OXFORDSHIRE: Hardwick (SP 3706; *Machimosaurus*); Marcham (SU 4596; *Pliosaurus*, '*Plesiosaurus*'); Drayton (SU 4794; *Ophthalmosaurus*, '*Plesiosaurus*' from Lower Kimmeridge Clay); Foxcombe Hill (SP 4901; *Cimoliasaurus, Pliosaurus, Ophthalmosaurus*); Culham (SU 5095; *Dakosaurus*); Radley Sand-pits (SU 5199; *Pliosaurus*); Oxford (SP 5106 – ?exact locality; *Cimoliasaurus, Pliosaurus*); Sandford-on-Thames (SP 5301; *Pliosaurus*); Headington Pits, Oxford (SP 555072; *Cimoliasaurus, Ophthalmosaurus*, type of *Pliosaurus brachyspondylus*); Nuneham Courtenay (SU 5599; *Ophthalmosaurus, Plesiosaurus*); Baldon (SP 5600; *Cimoliasaurus, Macropterygius*); Horspath (SP 5704; *Pliosaurus*); Shotover Hill (SP 588065; *Ophthalmosaurus, Cimoliasaurus, Pliosaurus, Dakosaurus, Metriorhynchus*, type of *Liopleurodon macromerus, Pliosaurus grandis, P. nitidus, Cimoliasaurus trochantencus, Metriorhynchus palpebrosum*); Garsington (SP 5802; *Metriorhynchus*, '*Plesiosaurus*', *Pliosaurus, Ophthalmosaurus*); Wheatley (SP 5905; *Cimoliasaurus, Ophthalmosaurus, Macropterygius*).

BUCKINGHAMSHIRE: Hartwell (SP 7916; *Liopleurodon*); Hardwick (SP 8019; *Pliosaurus*); Winslow (SP 7627; *Ophthalmosaurus*); Denbigh Hill, Fenny Stratford (SP 8834; *Pliosaurus*); Newport Pagnell (SP 8743; *Cimoliasaurus*); Aylesbury (SP 821134; 'marine reptiles', including a pliosaur recently, from Lower and Upper Kimmeridge Clay, mainly from the Homan's Bridge Shale Member; Upper Kimmeridge Clay, *wheatleyensis* Zone).

NORTHAMPTONSHIRE: Higham Ferrers (SP 9668; *Pliosaurus*).

CAMBRIDGESHIRE: Cottenham (TL 452684; *Dakosaurus*, '*Plesiosaurus*', *Pliosaurus, Ophthalmosaurus*); Haddenham (TL 4674; *Cimoliasaurus, Pliosaurus*); Wood Walton (TL 2180; *Pliosaurus*); Witcham (TL 4680; plesiosaur); Downham (TL 5283; *Pliosaurus*); Great Ouse River Board Pit, Stretham (TL 516743; *Liopleurodon, Ornithopsis*); Chettisham, Ely (TL 5583; *Ophthalmosaurus, Cimoliasaurus, Teleosaurus*); Peterborough (TL 1998; *Pliosaurus*); Oakley Cutting (TL 0254; *Steneosaurus*); Bourn (TL 3256; *Dakosaurus*).

NORFOLK: Stow Pumping Station (TL 589057; ?pliosaur); Downham Bridge (TF 601033;

?pliosaur); Downham Market Brickyard (TR 608039, TR 610031; *Ophthalmosaurus, 'Plesiosaurus', Teleosaurus, Dakosaurus*); Southery Pumping Station (TL 612932; ?pliosaur); Ten Mile Bottom (TL 612966; ?pliosaur); Denver Sluice (TF 591013; *Pliosaurus, ?Ornithopsis*); Setchey (TF 6313; *Pliosaurus*); Stowbridge (TF 604069; type of ichthyosaur *Grendelius mordax* from *wheatleyensis* Zone).

LINCOLNSHIRE: Market Rasen (TF 1089; *Ophthalmosaurus*, type of *Pliosaurus brachydeirus?*); Sweaton (*Ophthalmosaurus*).

YORKSHIRE: Speeton, Filey Bay (TA 1576; *Ophthalmosaurus*).

PORTLANDIAN: PORTLAND BEDS

The term 'Portlandian' is used here to refer to the last stratigraphic stage of the Jurassic, in preference to 'Tithonian', the primary international reference standard. This is because a basal boundary stratotype for the Tithonian has not been selected, and because the Kimmeridgian stage as used by British workers is much longer than that used elsewhere. Stratigraphic equivalents are:

UK northern France	Tethys	Russia, Poland
Portlandian Upper Kimmeridgian	Tithonian	Volgian
Lower Kimmeridgian	Kimmeridgian	Kimmeridgian

Cope (1993) has attempted to resolve this problem by reintroduction of the Bolonian Stage for the Upper Kimmeridgian *sensu anglico*, thus allowing the standard use of the Portlandian and Volgian stage names. For the present work, we use the traditional British 'long' Kimmeridgian stage name.

Reptiles have only been found on the Isle of Portland (turtles, plesiosaurs, ichthyosaurs). Isolated specimens of marine reptiles have also been collected in Wiltshire, Oxfordshire and Buckinghamshire, with a few dinosaur teeth from near Aylesbury. References include Phillips (1871), H.B. Woodward (1895), Strahan (1898) and Arkell (1933, 1947a, 1947c). Sites not included in the reptiles GCR coverage are listed here.

DORSET: Winspit Quarry, Seacombe (SY 985767; type of '*Plesiosaurus winspitensis*'); Haysoms' Quarry, St Aldhelm's Head (SY 964761; turtle); Preston, Weymouth (SZ 7038; turtle).

WILTSHIRE: Tisbury (ST 9429; *Cimoliasaurus*); Chicksgrove (sauropod, ?stegosaur, and megalosaur teeth, lizards, pterosaurs, *Goniopholis*); Town Gardens Quarry, Swindon (SU 152834; *Ichthyosaurus*).

OXFORDSHIRE: Oxford (SP 5106 – ?exact locality; type of *Metriorhynchus gracile*); Shotover Hill (SP 5906; *Cimoliasaurus*); Garsington (SP 5802; *Cetiosaurus*).

BUCKINGHAMSHIRE: Brill Brick Yard (SP 655144; *Plesiosaurus, Teleosaurus*); Quainton Hill (SP 7420; *Cimoliasaurus*).

LATE PORTLANDIAN TO EARLY BERRIASIAN: PURBECK BEDS

The Purbeck Limestone Formation is split between the Jurassic and Cretaceous. Late Jurassic forms include the lizards, 'dwarf' crocodilians and some turtles. Early Cretaceous forms are most of the turtles, pterosaurs, crocodilians and footprints. Extremely abundant remains have been obtained from Durlston Bay and quarries west of Swanage, and the smaller freshwater and terrestrial animals are of particular importance (lizards, turtles, small crocodilians, dinosaurs and pterosaurs). References include H.B. Woodward (1895), Delair (1958, 1966, 1982b), Charig and Newman (1962) and Anderson and Bazley (1971).

DORSET: Swanage Quarries (SZ 021781, etc.; many of forms found at Durlston Bay – exact information not available; type of *Pholidosaurus laevis* (crocodilian) from Keat's Quarry); Townsend Road, Swanage (SZ 02657835; dinosaur footprints; Ensom, 1982a); Herston Quarries (SZ 020784, etc.; *Goniopholis, Pleurosternum*, dinosaur footprints in the Pink Bed (Bristow's Bed 50) of the Upper Building Stones; Charig and Newman, 1962); Acton Quarries, Langton Matravers (SY 990783; *Pleurosternum*, type of pterosaur *Ornithocheirus validus*, dinosaur footprints); opposite lane leading to Old Court Pound (former underground workings with access at SY 99197872; footprints *Purbeckopus pentadactylus*; Delair, 1963; Ensom,

1984a, 1986a); Reynold's Quarry, Dancing Ledge (SY 997773 or SY 998769; *Goniopholis*, dinosaur footprints); Seacombe (SY 984767; *Goniopholis*); Sunnydown Farm Quarry, Langton Matravers (SY 98227880; sauropod and other dinosaur tracks and microvertebrate remains; Ensom, 1987a, 1987b, 1988, 1989c, 1990; West, 1988; Ensom *et al.*, 1991); Encombe (?SY 944793; '*Plesiosaurus*'); Nelson Burt Quarry, Worth Matravers (?SY 9777; '*Iguanodon*' in the 'Pinkstone Bed'); Woodyhyde Farm (SY 97507978; footprint; Ensom, 1986b); Corfe (?SY 9782; *Pleurosternum, Goniopholis*); Preston (SZ 7083; *Pleurosternum*); Portland (SZ 6771; *Cimoliasaurus* in the 'Cinder Bed'); Weymouth (exact locality?; ankylosaur); Lulworth Cove (SZ 8380; nodosaurid, *Goniopholis* from Dirt Bed); Worbarrow Tout (SY 869796; dinosaur footprints from several horizons; West *et al.*, 1969; Delair, 1982b; Ensom, 1982b, 1984b, 1985a, 1985b, 1987c; West and El-Shahat, 1985).

WILTSHIRE: Chicksgrove Quarry; Town Gardens Quarry, Swindon (SU 152835; 'reptiles', including megalosaur tooth, in Purbeck; Delair, 1973).

EAST SUSSEX: Darwell Beech Farm (?TQ 7119; *Goniopholis* and 'other reptiles' from a temporary exposure); Poundsford Farm (TQ 637226; *Megalosaurus*); Archer Wood (TQ 741819; 'reptile bones'). All from Cretaceous part of Purbeck.

Most of the listed sites could not serve as candidates for inclusion in the GCR since they have been lost to landfill and building. For the most part, only coastal sites could be considered, and the richness and accessibility of the Dorset coast is reflected in the choice of nine proposed GCR sites:

1. Furzy Cliff, Overcombe, Dorset (SY 698818). Late Jurassic (Early Oxfordian), Upper Oxford Clay.
2. Smallmouth Sands, Weymouth, Dorset (SY 669764–SY 672771). Late Jurassic (Early Kimmeridgian), Lower Kimmeridge Clay.
3. Roswell Pits, Ely, Cambridgeshire (TL 555808, TL 551805). Late Jurassic (Early Kimmeridgian), Lower Kimmeridge Clay.
4. Chawley Brick Pits, Cumnor Hurst, Oxfordshire (SP 475043). Late Jurassic (Early Kimmeridgian), Lower Kimmeridge Clay.
5. Kimmeridge Bay (Gaulter Gap–Broad Bench), Dorset (SY 9179). Late Jurassic (Early Kimmeridgian), Kimmeridge Clay.
6. Encombe Bay, Swyre Head–Chapman's Pool, Dorset (SY 937773–SY 955771). Late Jurassic (Late Kimmeridgian), Upper Kimmeridge Clay.
7. Isle of Portland, Dorset (SY 6478). Late Jurassic (Portlandian), Portland Sand–Purbeck Beds.
8. Bugle Pit, Hartwell, Buckinghamshire (SP 793121). Late Jurassic (Portlandian), Portland Stone and lowest Purbeck Beds.
9. Durlston Bay, Dorset (SZ 034780). Late Jurassic–Early Cretaceous (Portlandian–Berriasian), Portland Stone–Upper Purbeck Beds.

LATE JURASSIC (OXFORDIAN)

Oxfordian reptiles are rare, especially those of Early Oxfordian age, the only recorded sites being at Cothill and Headington in Oxfordshire and Furzy Cliff, Overcombe Dorset. This may be because of a lack of suitable exposure in the Upper Oxford Clay, but is more probably related to a genuine scarcity of reptilian remains. The rarity of reptiles from British Early Oxfordian rocks is also matched abroad, and the few British sites are of great importance in bridging the faunal gap between the underlying Callovian Middle Oxford Clay and the overlying beds of the Mid Oxfordian (*densiplicatum* Zone), both of which preserve better reptile remains. The best-known Early Oxfordian site is at Furzy Cliff and this is selected as a GCR site.

British Mid and Late Oxfordian localities have produced more reptile remains. The Coral Rag (Mid Oxfordian) of various sites in Wiltshire, Berkshire, Oxfordshire and Yorkshire have yielded teeth of plesiosaurs, pliosaurs, crocodiles, vertebrae of plesiosaurs and dinosaurs, and isolated dinosaur limb bones. The Mid Oxfordian of Yorkshire is important as the source of rare dinosaurs; a nodosaur ankylosaur (*Priodonto-gnathus*) from the Calcareous Grit, and a stegosaur femur, the type of *Dacentrurus? phillip-sii*, from the Malton Oolite Member of the Coralline Oolite Formation of Slingsby, are the only evidence for the Stegosauridae so far from the Oxfordian anywhere in the world (Galton, 1985b). Late Oxfordian reptiles are known from the Sandsfoot Grits of Weymouth and the Ampthill Clay of Cambridgeshire and Bedfordshire.

Comparable Mid and Late Oxfordian sites abroad include Vaches Noires, near Dives, Normandy (*Eustreptospondylus, Steneosaurus*), Calvados, Bourgogne and La Vendée

(*Steneosaurus*), Boulogne-sur-Mer (*Steneo-saurus*), the La Turbie-Cap d'Aggio region, Monaco ('Megalosaurus', ?crocodile) and Baden-Württemberg, Germany (*Teleosaurus*).

FURZY CLIFF, OVERCOMBE, DORSET (SY 697817–SY 703819)

Highlights

Furzy Cliff is Britain's best Oxfordian age reptile locality. It is the source of the unique specimen of the meat-eating dinosaur *Metriacanthosaurus*, as well as an ichthyosaur and a plesiosaur.

Introduction

The Upper Oxford Clay exposed at Furzy Cliff, or Jordan (Jordon) Cliff, Overcombe, has yielded sparse fossil reptile remains, but these are of considerable importance because of their age. A recent cliff fall has re-exposed large portions of the site, and the prospects for future finds are good (Figure 7.1). The fossil reptiles have been described by Huene (1923, 1926), Walker (1964) and Cope (1974).

Description

Buckman (1925) was the first to examine the stratigraphy of Furzy Cliff in detail, and he used it to distinguish one component of his three-fold division of the Upper Oxford Clay (Early Oxfordian) in south Dorset. This included: (1) clays with *Quenstedtoceras*; (2) clays with large *Gryphaea dilatata*, named the Jordan Cliff Beds; and (3) clays with reddish-brown nodules and large perisphinctids, named the Red Beds. Arkell (1947c) revised and amended the nomenclature, naming unit (1) Furzedown Clays, (2) Jordan Cliff Clays and (3) Red Nodule Beds. Buckman did not give thicknesses for the units and those of Arkell (1947c) and of Torrens (1969a) have since proved

Figure 7.1 The rapidly eroding exposure of Oxford Clay at Furzy Cliff, Overcombe. Fossil reptile bones came from the Jordan Cliff Clays, at the bottom of the sequence. (Photo: M.J. Benton.)

to be overestimates, particularly in the case of the Red Nodule Beds (Wright, 1986). In the summer of 1983 a major landslip at Furzy Cliff permitted the first accurate measured sections to be made (Wright, 1986). The Red Nodule Bed was found to be a thin unit in the middle of a distinctive clay sequence named by Wright (1986) the Bowleaze Clay Member (formerly the Red Beds). The revised section at Furzy Cliff (Wright, 1986) is:

Thickness (m)

Nothe Grit
  11. Argillaceous sandstone   seen 1.5
Bowleaze Clays
  10. Pale and medium grey clay   *c.* 6
  9. Red Nodule Bed   0.35
  8. Pale and medium grey clay   0.40
  7. Dark, carbonaceous clay with intraformational bored surfaces   1.40
  6. Pale and medium grey clay with bored surfaces, *Gryphaea* and perisphinctids   *c.* 1
  5. Interbedded dark, carbonaceous and pale and medium grey clay with scattered nodules occurring toward the middle of the unit. *Gryphaea* and perisphinctids   *c.* 2
  4. Pale and medium grey clay   *c.*3
  3. Pale and medium grey clay containing perisphinctids   *c.* 0.15
  2. Pale and medium grey clay with nodule-bearing horizon at the base   1.10
Jordan Cliff Clays
  1. Pale and medium grey clays with *Gryphaea*   seen 10.5

The Jordan Cliff Clays (*c.* 10 m) are unbioturbated, extremely fine-grained, fissile, dark slaty-blue clays that weather to a lighter, greenish-grey or reddish-brown colour. Although the lowest clays are now obscured by new sea defences, they were described briefly by H.B. Woodward (1895) and Damon (1884). H.B. Woodward (1895, p. 16) noted the unit as 'beds of bluish-grey clay, with small hard cement-stones' from below the coastguard station (i.e. western end of the section at about SY 698818). Damon (1884, p. 29) noted 'thin stony layers' and a serpulid bed, 50–90 mm thick, within the clay unit. The clays exposed today contain numerous *Isognomon promytiloides*. They are succeeded by 8 m of a tough, blocky, silty clay with numerous *Gryphaea dilatata* encrusted with *Serpula* sp. In this unit, encrusted *Modiolus bipartitus* and cardioceratids are common.

The Bowleaze Clays (*c.* 14.5 m) are predominantly very fine-grained, but with frequent incursions of sandy clay. The base is marked by a persistent band of white, elliptical limestone nodules which are associated with coarse silt and fine sand. There are about 4 m of these very fine, pale-grey clays which contain 'nests' of *Lopha gregaria* and *Liostrea* sp. Above these, the lithology changes with the first of several inputs of dark, sandy clay. The Red Nodule Beds consist of two bands of nodules which occur in the cliff about 8 m above the base of the Bowleaze Clays. Their colour is an artefact of weathering in a zone 0.5 m below the soil surface. The nodules are typically found coated with a layer of iron oxide and were formerly known as 'kidney stones' (H.B. Woodward, 1895, p. 16), but when fresh they are pale buff-coloured, dense and sideritic, frequently with septarian cracks infilled with calcite and zinc blende. The nodules commonly enclose the remains of *Modiolus bipartitus*, *Modiola*, *Astarte* and *Pleuromya alduini*. The predominant lithology of the Red Nodule Beds, however, consists of fine grey clays with layers containing very large *Gryphaea dilatata* and some aggregations of *Lopha gregarea* (Arkell, 1947c, p. 34).

The Red Nodule Beds are succeeded by some 6 m of pale grey, very fine clay. An upwards coarsening trend is reversed at the top with the sands of the Nothe Grit resting on the clay with a very sharp junction. This is exposed only at Ham Cliff (SY 712817). The highest clay is markedly calcareous, containing micrite nodules and numerous Foraminifera and small bryozoa.

Ammonites occur as flattened white impressions in the Jordan Cliff Clays and within the nodules of the Red Nodule Beds. Arkell (1933, p. 343) assigned all of these to the '*praecordatum* Zone' (?*mariae* Zone in part). Later, Arkell (1941) revised the zonation of the Early Oxfordian and placed the Jordan Cliff Clays in the *mariae-cordatum* Zones (*scarburgense-bukowskii* Subzones) and the Red Nodule Beds (Bowleaze Clays) in the *costicardia* Subzone.

The reptile remains at Furzy Cliff appear to have come from the Jordan Cliff Clays. Cope (1974) specifies this unit as the source of some ichthyosaur and plesiosaur remains collected in 1972–3. There is some doubt about the stratigraphic position of *Megalosaurus parkeri* (i.e. *Metriacanthosaurus parkeri*), but it may also have come from the Jordan Cliff Clay. It is certainly from the Oxford Clay, as an oyster, *Gryphaea dilatata,* found adhering to one of the vertebrae, has been taken to indicate an Upper Oxford Clay (Early Oxfordian) age for the specimen (Walker, 1964, p. 117).

The ichthyosaur was preserved in a semi-articulated state. A series of 39 vertebrae, with associated neural spines and ribs, was excavated. However, the limbs, neck region and skull seem to have been lost. Several other ichthyosaur and plesiosaur vertebrae and teeth were found associated. The evidence suggests limited transport and winnowing, but the skeleton was clearly not excessively disturbed or the neural spines and ribs would have been lost. Huene (1926) gave no taphonomic information on the *Megalosaurus* specimen. The remains consist of elements of the pelvis and hind-limb region, and they are rather distorted and cracked.

Fauna

Archosauria: Dinosauria: Saurischia: Theropoda: Carnosauria
 Metriacanthosaurus parkeri (Huene, 1923)
  Type specimen: OUM J.12144
Ichthyopterygia
 Ophthalmosaurus sp. repository?
Sauropterygia: Plesiosauria
 'plesiosaur' repository?

Interpretation

The remains of the carnivorous dinosaur (three dorsal vertebrae, four proximal caudal vertebrae, right ilium, fragments of right and left ischium, right and left pubis, right femur, upper part of right tibia) were collected together, probably in the 19th century. They were described as a new species of *Megalosaurus, M. parkeri* Huene (1923), characterized by the high, elongate neural spines on the dorsal vertebrae, the shape of the ilium and ischium, and the expansion of the pubic 'foot' (Huene, 1923, 1926). The femur was a slender bone with the lesser trochanter toward the top and the cnemial process bearing a strong upward projection. In these respects, *M. parkeri* differs from the typical Bathonian *Megalosaurus bucklandi*. Because of these differences, Walker (1964, pp. 109, 116–17) named

M. parkeri as the type species of the new genus, *Metriacanthosaurus*. The relationship of *Metriacanthosaurus* with other theropods has been problematic. Most workers (e.g. Walker, 1964; Steel, 1970) assign it to the *Megalosauridae*, but no other megalosaur bears the same enlarged neural spines. On the basis of the height of the neural spines, Huene (1926) suggested that *Metriacanthosaurus* could represent an early member of the Spinosauridae, and this assignment was discussed further in Walker (1964). It has since been realized that the development of expanded neural spines occurs in a range of unrelated tetrapod groups and can no longer be regarded as a viable phyletic character. Molnar (1990) was unable to find any characters that would allow *Metriacanthosaurus* to be classified further than Theropoda *inc. sed.*

Other Late Jurassic carnosaurs include *Allosaurus, Antrodemus, Ceratosaurus* and *Dryptosaurus* from the Morrison Formation of North America (Kimmeridgian–Portlandian?), *Allosaurus, Ceratosaurus* and *Elaphrosaurus* from the Tendaguru beds of Tanzania (Late Kimmeridgian), *Yuangchuanosaurus* and *Szechuanosaurus* from the Late Jurassic of North Szechuan, China, and 'Megalosaurus' from the Kimmeridgian and Portlandian of northern France, Portugal, Wiltshire and Dorset (Encombe Bay–Chapman's Pool site, see below). The only other Oxfordian carnosaur known is *Eustreptospondylus divesensis* Walker (1964) from the Vaches Noires, near Dives, Normandy (a cranium). *Megalosaurus nicaeensis* (Ambayrac, 1913) from the Oxfordian of Monaco turns out to be a pliosaur (Buffetaut, 1982). *Metriacanthosaurus parkeri* fills a gap in the phylogeny of the carnosaurs.

The ichthyosaur and plesiosaur remains from Furzy Cliff have only been recorded briefly (Cope, 1974), and they have not been described. The ichthyosaur consisted of 34 dorsal centra and 5 caudals, with neural spines and ribs. Plesiosaur remains were isolated vertebrae and possibly teeth. The ichthyosaur has tentatively been identified as *Ophthalmosaurus* sp., a genus common in the Late Jurassic.

Conclusions

Furzy Cliff represents Britain's best Oxfordian reptile site. In view of the limited reptile sites of this age elsewhere, it is also one of the best in the world. The dinosaur *Metriacanthosaurus* is represented by good postcranial remains and occupies a unique position in carnosaur evolution. The ichthyosaur *Ophthalmosaurus* is the only British Oxfordian ichthyosaur known, and one of the few known from that stage worldwide. Thus, the small number of finds to date from Furzy Cliff are of great importance, and the site has potential for further discoveries, hence its conservation value.

LATE JURASSIC (KIMMERIDGIAN) OF ENGLAND

Reptiles have been recorded from 60 sites in the Kimmeridge Clay between Dorset and Yorkshire (listed above). The faunas are dominated by marine forms (plesiosaurs, ichthyosaurs, and marine crocodiles; see Figures 7.5 and 7.8), but some localities (e.g. Weymouth) have also yielded significant remains of terrestrial reptiles, including important dinosaurs and turtles. Kimmeridge Bay, the type locality for the Kimmeridgian Stage, has produced the largest fauna, which includes the type specimens of six species. The five GCR localities (Figure 5.1), at Smallmouth Sands, Weymouth (SY 669764–SY 672771), Roswell Pits, Ely (TL 555808–TL 551805), Chawley (SP 475043), Gaulter Gap–Broad Bench, Kimmeridge Bay (SY 9179) and Encombe Bay (SY 937773–SY 955771), provide good coverage of rocks of Early to Late Kimmeridgian age, and cover the best known fossil reptile localities.

SMALLMOUTH SANDS, WEYMOUTH, DORSET (SY 669764–SY 672772)

Highlights

Smallmouth Sands has produced one of the most diverse assemblages of Kimmeridge Clay reptiles anywhere in the world. Its fauna of four species of turtles and three of pterosaurs is unique and, of its total fauna, six species are known only from this site.

Introduction

The Kimmeridge Clay south-west of Weymouth has yielded a large selection of marine and terrestrial reptiles, including many type specimens. Most of the finds appear to have been made on Smallmouth Sands and possibly also in the railway cutting behind. Little in the way of large finds has been collected recently because the enclosure of Portland Harbour has reduced erosion, but the relevant beds could be re-excavated. In addition, specimens are occasionally found offshore from this site and in degraded Kimmeridge Clay beds west of the classic site.

The Kimmeridge Clay outcrop in the Weymouth district forms a tract along the northern shore of the Isle of Portland which is continued across the floor of Portland Harbour. It is best exposed between Sandsfoot Castle and Portland Ferry bridge at Small Mouth, where it forms a series of low sea cliffs. The earliest good account of these cliff exposures was provided by Waagen (1865). The Kimmeridge Clay south of Sandsfoot Castle includes the lowest beds in the *mutabilis* Zone, and shows good sections of the *cymodoce* and *baylei* Zones (Arkell, 1933, p. 454), the three lowest Kimmeridgian zones. The section between Sandsfoot Castle and Smallmouth bridge has been described by several authors (Damon, 1884, p. 77; Blake and Hudleston, 1877, pp. 269–70; Salfeld, 1914, pp. 201–3; Arkell, 1933, pp. 385, 454; 1935, pp. 80–1; 1947c, pp. 56, 88; Birkelund *et al.*, 1978, p. 35; Cox and Gallois, 1981, pp. 4, 9), but most interest has focused on the Corallian.

Description

The section, based on Arkell (1933), Cox and Gallois (1981) and Cope (*in* Cope *et al.*, 1980b, p. 80) is:

Thickness (m)

Lower Kimmeridgian
mutabilis Zone
  Clays with nodules
    (?bed 13 (in part)–17 of
      Damon, 1884)   ?20+
cymodoce Zone
  Black Head Siltstone   0.5
  Shales with *D. delta*
    (bed 13 (in part) of Damon,
      1884; bed 24 (in part) of
      Salfeld, 1914)   2.0+
  Wyke Siltstone
    (bed 12 of Damon, 1884; bed

Thickness (m)

23 of Salfeld, 1914)	1.0
baylei Zone	
pale grey mudstones with thin, tabular clay ironstones at base (beds 8-11 of Damon, 1884; beds 21-22 of Salfeld, 1914)	?6+
dark grey mudstone with *D. delta* (? bed 7 of Damon, 1884; bed 20 of Salfeld, 1914; bed 13 of Arkell, 1933, 1947c)	3.0
Nanogyra nana Bed (bed 7 (in part) of Damon, 1884; bed 19 of Salfeld, 1914; bed 12 of Arkell, 1933, 1947c)	0.25
Rhactorhynchia inconstans Bed (bed 6 of Damon, 1884; bed 18 of Salfeld 1914; bed 11 of Arkell, 1933, 1947c)	0.7
Upper Oxfordian (Corallian)	
Westbury Iron Ore Beds (beds 1-5(?) of Damon, 1884; beds 13-17 of Salfeld, 1914; beds 8-10 of Arkell, 1933, 1947c)	

The beds dip south-west, and the clays with nodules of the *mutabilis* and higher ammonite zones are indicated largely by nodules washed ashore from the floor of Portland Harbour (Cope, *in* Cope *et al.*, 1980b, p. 80). Higher units of the Kimmeridge Clay occur in the harbour and on the north shore of the Isle of Portland.

The reptiles appear to have come from lower units of the Kimmeridge Clay and probably also from the top of the Corallian and from material washed out of Portland Harbour. Damon (1884, p. 77) noted that 'gigantic saurian remains have been found. Among others, *Gigantosaurus megalonyx*, Hulke in his bed 12. This was described as 'gritty clay. . . 3ft', and it is almost certainly equivalent to the Wyke Siltstone, thus *cymodoce* Zone. Damon (1884, p. 77 further noted 'saurian remains' below a 'layer of large flattened septaria', his bed 14, possibly equivalent to the main *Xenostephanus*-rich beds (Cox and Gallois, 1981, p. 5) at the base of the *mutabilis* Zone. Damon (1884, p. 77) also stated that his bed 16 (another horizon higher in the *mutabilis* Zone) also 'contains saurian bones'. BMNH R1798, a partial skull and mandible of *Kimmerosaurus langhami* (no detailed collection data available), most probably came from the cliff exposure between Sandsfoot Castle and the old Portland Ferry Bridge (Damon, 1884; Brown *et al.*, 1986, p. 226).

In his description of *Cetiosaurus humerocristatus*, 'a very large saurian limb-bone', Hulke (1874a, p. 16) noted that 'it was enveloped in large septarian masses, which stuck so closely to it that thin laminae of the surface of the bone were unavoidably detached in stripping the matrix from it'. The nodules probably indicate that the bone came from one of Damon's septarian beds of the *mutabilis* Zone, or from the harbour. Hulke (1874a, p. 16) noted further that 'the bone has been much fissured, and cemented together by spar; and some parts have been distorted by squeezing; but the general figure is well preserved'. The other fossil bones collected here have also been broken and disarticulated.

Other remains, including fragments of juvenile and mature turtles, including the type of *Pelobatochelys blakei*, were noted by the early authors as coming from the junction between the lowest Kimmeridge Clay and the highest Corallian (Oxfordian) horizons. These beds, the Westbury Iron Ore Beds, have formed the subject of many studies, and Blake (1875), for example, concluded that they were 'passage beds'. Other fossils (now lost) seem to have been collected from the uppermost Corallian beds at Sandsfoot itself (Blake and Hudleston, 1877). The ichthyosaur *Brachypterygius extremus* appears to have come from the lowest Kimmeridgian zone or, more probably, from the same 'Passage Beds' (Delair, 1986, p. 133). The fossils from the uppermost Corallian Beds at Sandsfoot show close affinities with those from the immediately overlying Kimmeridgian zones (Arkell, 1935; Brookfield, 1978b; Delair, 1986).

Finds of bones have also been made in Portland Harbour. A jaw fragment of a megalosaurid dinosaur was dredged up in the 1980s (Powell, 1988), and associated ammonites indicated the *autissiodorensis* Zone (top of the Early Kimmeridgian).

The reptile remains from Smallmouth Sands most frequently consist of isolated limb bones, vertebrae or teeth, although partially articulated specimens have been found (e.g. the ichthyosaur paddle described by Boulenger (1904b) and Delair (1987), and several connected series of ichthyosaur vertebrae).

Fauna

Mansel-Pleydell (1888) and Delair (1958, 1959, 1960, 1986) have summarized the reptiles from

'Weymouth'. Some synonymizing can be done as a result of later work, but most of the material has not been studied recently. Most of the Kimmeridge Clay specimens labelled 'Weymouth' may come from the Smallmouth Sands section, on the basis of Damon (1884, pp. 69, 77), outcrop distribution and labels on certain specimens. Repository numbers are given for type specimens, and an estimate of total numbers of each specimen of each species preserved in major collections is appended.

	Numbers
Testudines: Cryptodira: Thalassemyidae	
Acichelys (Eurysternum) sp.	4
Pelobatochelys blakei Seeley, 1875	
Type specimens: BMNH 41235, 44177-8, R2	7
Pelobatochelys sp.	2
Tropidemys langi Rütimeyer, 1873	4
Testudines: Cryptodira: Plesiochelyidae	
Plesiochelys sp.	4
Archosauria: Crocodylia: Thalattosuchia	
Dakosaurus maximus (Plieninger, 1846)	4
Metriorhynchus sp.	3
Steneosaurus sp.	8
Archosauria: Pterosauria	
Rhamphorhynchus manselii (Owen, 1874)	
Type specimen: BMNH 41970	11
Rhamphorhynchus pleydelli (Owen, 1874)	
Type specimen: BMNH 42378	7
Rhamphorhynchus sp.	10
'Ornithocheirus sp.'	1
Pterodactylus suprajurensis Sauvage, 1873	1
Archosauria: Dinosauria: Saurischia: Theropoda: Megalosauridae	
megalosaurid	1
Archosauria: Dinosauria: Saurischia: Sauropoda	
Pelorosaurus humerocristatus (Hulke, 1874)	
Type specimen: BMNH 44635	2
Archosauria: Dinosauria: Ornithischia: Ornithopoda	
'hypsilophodontid'	1
Archosauria: Dinosauria: Ornithischia: Stegosauria: Stegosauridae	
Dacentrurus armatus Owen, 1875	1

	Numbers
Sauropterygia: Plesiosauria: Elasmosauridae	
Colymbosaurus trochanterius (Owen, 1840)	7
Cimoliasaurus brevior Lydekker, 1889	
Type specimen: BMNH 41955	1
Sauropterygia: Plesiosauria: Cryptoclididae	
Kimmerosaurus langhami Brown, 1981	1
Sauropterygia: Plesiosauria: Pliosauridae	
Pliosaurus brachydeirus Owen, 1841	5
Pliosaurus sp.	7
Liopleurodon macromerus (Phillips, 1871)	2
Ichthyopterygia: Ichthyosauria	
Brachypterygius extremus (Boulenger, 1904)	
Type specimen: BMNH R3177	1
Macropterygius thyreospondylus (Owen, 1840)	4
'Ichthyosaurus sp.'	27

Interpretation

The four genera of turtles from Weymouth are all cryptodires according to the classifications of Gaffney (1975b, 1976, 1979a) and Młynarski (1976). The cryptodires, which retract their head in a vertical plane, are the commonest forms today. The cryptodires arose in the Early Jurassic of North America, but only became reasonably abundant in the Late Jurassic of Europe and North America. *Tropidemys*, *Acichelys (Eurysternum)* and *Pelobatochelys* are grouped together in the Thalassemyidae and *Plesiochelys* in the Plesiochelyidae, both of which families arose in the Late Jurassic, and are known elsewhere from Germany and Switzerland, as well as the Portlandian of the Isle of Portland (see below).

Acichelys is represented by several remains of carapace and limbs in the BMNH, but these have not been described. *Pelobatochelys blakei* Seeley (1875) was established on the basis of 'fragments of a chelonian carapace' which could include the remains of one or more animals. A restoration showed a broad low carapace about 0.5 m long. The genus was characterized by the broad vertebral scutes which were strongly fluted underneath and by a pointed midline ridge along the neural plates. It is known only from Dorset and from the Smallmouth section in particular.

The material was reviewed by Lydekker (1889b, pp. 152-5) and Delair (1958, pp. 54-5). *Plesiochelys* is represented by some carapace remains and limb bones. The specimens presently ascribed to this genus in the BMNH were initially named *Tropidemys langi* and 'generically undetermined specimens' (Lydekker, 1889b, pp. 156-8) and require restudy. *Tropidemys langi* is represented by isolated carapace elements which were tentatively identified as such by Lydekker (1889b, pp. 156-7). These turtles are all of significance as some of the earliest cryptodires known, but skulls are lacking, and much current taxonomic work depends on cranial characters.

The crocodile remains from Smallmouth consist largely of teeth, and these have been ascribed (Lydekker, 1889a, pp. 94, 100-1; 1890a, p. 233) to *Dakosaurus, Metriorhynchus, Steneosaurus* and *Teleosaurus*, genera well known from the Kimmeridgian. Some vertebrae, scutes and skull fragments of *Steneosaurus* and *Metriorhynchus* have also been collected. The teleosaurs *Steneosaurus* and *Teleosaurus* were medium-sized, long-snouted, marine, fish-eating crocodiles. They are distinguished on the characters of skull shape and tooth arrangement. The metriorhynchid *Dakosaurus* was much larger (up to 4 m body length) and had a relatively short snout. *Metriorhynchus* had a longer snout and was highly adapted for aquatic life; both fore and hind limbs were shortened and paddle-like.

Several remains of pterosaurs from Weymouth have been described. They consist generally of articular ends of limb bones and skull elements, all of which are readily identifiable as pterosaurian. Owen (1874a, pp. 8-11) described two new species, *Pterodactylus manseli* and *P. pleydelli* on the basis of a humerus (proximal end) and wing phalanx and humerus (distal end) and wing phalanx, respectively. Some carpal bones were called '*Pterodactylus* sp. incert.' Lydekker (1888a, pp. 40-1) enumerated these specimens, and others, in the BMNH collections. He noted that there were three species present, distinguished from each other by size, *P. pleydelli* having 'somewhat inferior dimensions' to *P. manseli*, and 'species c' 'of considerably large size than either of the preceding forms'. Lydekker (1891, pp. 41-2) reinterpreted some of the bones he had earlier identified as metacarpals as pterosaur quadrates, and referred one of them to *Pterodactylus suprajurensis*, a species previously described from France. He also noted that *P. manseli* and *P. pleydelli* probably belonged to

Rhamphorhynchus. Mansel-Pleydell (1888, p. 33) recorded that these specimens came from Kimmeridge, but Delair (1958, p. 70) confirmed Weymouth as the source. Wellnhofer (1978, p. 49) places the two Weymouth species in 'Pterodactylidae *incertae sedis*', but states that both were 'probably remains of rhamphorhynchids'. The lack of diagnostic skull, limb and vertebral remains means that the Weymouth specimens cannot even be confidently assigned to one or other of the two suborders of Pterosauria.

Theropod dinosaurs are represented by a fragment of megalosaurid maxilla (Powell, 1988). The sauropod dinosaur *Pelorosaurus humerocristatus* was based on 'a very large saurian limb-bone adapted for progression upon land', a left humerus originally 1.5 m long (Hulke, 1874a). Lydekker (1888a, p. 152) also referred the dorsal portion of a right pubis from Weymouth to this species. Damon (1884, pp. 69, 77) mentioned specimens of *Gigantosaurus megalonyx* Seeley (1869) from the gritty clay bed (his bed 12) in the Kimmeridge Clay, and Delair (1959, p. 82) suggested that this could include the type humerus of *P. humerocristatus*.

Galton (1975, p. 745) described a dentary tooth from the Kimmeridge Clay of Weymouth (University of California Museum of Paleontology) as that of a hypsilophodontid. Later, he (1980b, p. 85) tentatively suggested that the locality assignment might be wrong because he had seen a strikingly similar tooth from the Late Cretaceous of Wyoming. A stegosaur vertebra (BMNH 15910) has been ascribed to *Dacentrurus armatus* (Galton, 1985b), a stegosaur well known from the Kimmeridge Clay elsewhere in England. This consists of the neural arch of a mid-caudal vertebra, and the short neural spine is a diagnostic character for the species.

The plesiosaur remains from Weymouth are generally isolated vertebrae and limb bones. Most have been ascribed to the cryptoclidid *Colymbosaurus trochanterius* (Owen, 1840), the best-known and most abundant Kimmeridgian plesiosauroid (Brown, 1981), which attained a maximum length of 6.6 m. Lydekker (1889a, p. 243) established the species *Cimoliasaurus brevior* on the basis of six associated centra of immature middle cervical vertebrae from Weymouth; the relative length of the vertebrae was supposed to be diagnostic. Brown (1981, p. 322) noted this species as a *nomen dubium*. A third plesiosaur, represented by a fragmentary skull (BMNH R1798), consisting of an incomplete mandible, the squamosals, and fragments of the quadrates, jugals and postorbitals, was assigned to *Kimmerosaurus langhami* by Brown *et al.* (1986). The species had been named by Brown (1981) on the basis of a partial skull, from the Upper Kimmeridge Clay west of Freshwater Steps. Brown *et al.* (1986, p. 233) discuss the possibility that *Kimmerosaurus* might be synonymous with *Colymbosaurus*, which is the only other cryptoclidid known from the British Kimmeridgian, but the evidence for synonymy is ambivalent and both names are tentatively retained.

The pliosauroids are largely represented by vertebrae and teeth from Weymouth. These have been identified (Lydekker, 1889a, pp. 125, 128, 142, 147) as *Pliosaurus brachydeirus* and *Pliosaurus* sp., and some vertebrae as *Liopleurodon (Stretosaurus) macromerus*.

Ichthyosaurs are represented by abundant remains of vertebrae, limb bones, skull fragments and teeth that were ascribed to various species of '*Ichthyosaurus*' by Lydekker (1889a, pp. 25, 27, 30, 35–40). The most complete specimen (BMNH 44637) consists of 40 vertebrae of a medium-sized individual (Lydekker, 1889a, p. 38). McGowan (1976, p. 670) regards *Macropterygius thyreospondylus* and *M. trigonus* as *taxa dubia* since they were based upon poor material. However, they may be definable (A. Kirton, pers. comm., 1981). Boulenger (1904b) named the new species *Ichthyosaurus extremus* on the basis of a right anterior paddle characterized by its great breadth and by the humerus contacting the wrist bone (the intermedium) directly, and Huene (1922, pp. 91, 97–8) assigned it to the new genus *Brachypterygius*. Boulenger (1904b) did not know its locality, but suggested, on H.B. Woodward's advice, that it came from the Lower Lias of Weston, Bath, while Andrews (1910, p. 54) proposed that it was of Kimmeridgian age. Delair (1960, pp. 68–9) pointed out the surprising fact that the label on the specimen clearly states its provenance as 'Kimmeridge Clay of Smallmouth Sands'. Delair (1986, pp. 131–3) recognized that an isolated left forelimb in Woodspring Museum, Weston-super-Mare (WESTM 78/219) and the type specimen (BMNH R3177) in fact belonged to the same individual.

Comparison with other localities

Sites comparable to the Weymouth section, in having significant terrestrial faunas, include Swindon Brick and Tile Works (Lower Kimmeridge Clay) (SU 142838; *Plesiochelys, Bothriospondylus, 'Megalosaurus', Dacentrurus armatus*, crocodiles, ichthyosaurs, plesiosaurs, pliosaurs); Chawley Brick Pit, Cumnor Hurst, near Oxford, Oxfordshire (SP 475043; *Camptosaurus*, ichthyosaurs, plesiosaurs, pliosaurs); Ely, Cambridgeshire (probably one of several pits at TL 555808; *Thalassemys, Pelorosaurus*, crocodiles, ichthyosaurs, plesiosaurs, pliosaurs); Wootton Bassett, Wiltshire (SU 0638: *Dacentrurus armatus*) and Gillingham, Dorset (ST 809258; *Dacentrurus armatus*).

Of the turtles, *Pelobatochelys* is known also from the Late Kimmeridgian of Encombe Bay (see below), but is restricted to Dorset. *Tropidemys* is known best from the Late Jurassic of Switzerland and Germany and *Eurysternum* from Bavaria and Switzerland (Młynarski, 1976, p. 36). *Thalassemys*, a close relative, has been recorded from Devizes and Ely, as well as from localities in the Late Jurassic of Switzerland and Germany.

Various crocodiles are well represented at Kimmeridge Bay, Wootton Bassett, Swindon, Shotover Hill, Garsington, Cottenham and Ely. *Steneosaurus* and *Teleosaurus* are recorded from all stages of the Jurassic of Europe, *Dakosaurus* from the Late Jurassic to Early Cretaceous of Europe, and *Metriorhynchus* from the Callovian–Kimmeridgian of England and France (Steel, 1973).

Pterosaurs are rare in the British Kimmeridgian. Specimens from Kimmeridge Bay have been identified as '*Rhamphorhynchus* sp.' and *Germanodactylus* sp., and one from Swindon as '*Ornithocheirus* sp.'. These, and other, Late Jurassic pterosaur genera are better known from sites in Germany, France, East Africa and Wyoming, but these overseas sites are younger.

The sauropod *Pelorosaurus* is known from the Kimmeridgian of Kimmeridge Bay (q.v.) and Cottenham, Stretham and Ely, Cambridgeshire. Various species have also been described from the Wealden of Sussex and the Isle of Wight, and from the Kimmeridgian of Boulogne-sur-Mer and Wimille (near Boulogne) (Steel, 1970, pp. 68, 70). *Dacentrurus* is better known from sites in the Kimmeridgian at Gillingham, Wootton Bassett and Swindon, but these are no longer accessible. A specimen from the Late Kimmeridgian of Le Havre, France was destroyed during World War 2 (Buffetaut *et al.*, 1991).

Most of the ichthyosaurs, plesiosaurs and pliosaurs from Weymouth are known from other

British Kimmeridgian sites, so these will not be enumerated. The type of *K. langhami* and the only other referred material (BMNH R10042) is known only from Freshwater Steps (SY 924773).

Conclusions

The Lower Kimmeridge Clay of the Smallmouth Sands section has yielded one of the most varied Kimmeridgian reptile faunas known. It is the best site for turtles (four species) and pterosaurs (three species). The material includes type specimens of one turtle, two pterosaurs, one sauropod, one ichthyosaur and one plesiosaur. *K. langhami* is only the second occurrence of a plesiosauroid from the British Kimmeridgian. One of the best sites in Europe for Kimmeridgian age terrestrial reptiles. The importance of this faunal richness and diversity combined with some potential for future finds give its considerable conservation value.

ROSWELL PITS, ELY, CAMBRIDGESHIRE (TL 555808, TL 552807)

Highlights

Roswell Pits, Ely are famous for the remains of sauropod dinosaurs and of pliosaurs which have been found there. The site includes a mix of terrestrial and marine Kimmeridgian age reptiles, including turtles, crocodilians, dinosaurs, plesiosaurs and ichthyosaurs.

Introduction

The Roswell Pits, or Roslyn Hole, at Ely, excavated in Lower Kimmeridge Clay, have produced a large range of fossil reptiles that includes several type specimens and finds are still being made.

The large pit at Roswell (TL 555808) was excavated in the 19th century, largely for clay to repair the banks of the local fen dykes, and it was mentioned as a source of several reptiles by Seeley (1869a, pp. 100–1), Whitaker *et al.* (1891, p. 15) and Gallois (1988, pp. 40–2). H.B. Woodward (1895, pp. 170–1) noted that the large pit of 'Roslyn Hole' exposed 'about 30 feet of dark shales and clays, in places bituminous and arena-ceous, with thin ochreous layers and bands of septaria'. He believed that 'both lower and upper portions of the Kimmeridge Clay are present'.

Description

Whitaker *et al.* (1891, pp. 15–16) described a section in the Roswell Pits, but this does not provide enough evidence for a link to be made with a modern zonation scheme. *Nanogyra virgula* is recorded in abundance in several beds in the upper 5–6 m of the section, and this suggests a position for these units in the *eudoxus* Zone (Early Kimmeridgian). Whitaker *et al.* (1891) and H.B. Woodward (1895) believed that the fossil evidence pointed to the presence of 2–3 m of Upper Kimmeridge Clay at the top of the section.

Arkell (1933, pp. 468–9) ascribed all the beds to the Lower Kimmeridge Clay: 'the opinion formerly held, that at the Roslyn (or Roswell) Pit, 1 mile north-east of Ely, some 8 ft of paper-shales and clays with *Orbiculoidea latissima* are above the limit of *Exogyra virgula*, is not sustained by recent work. Dr Kitchin and Dr Pringle inform me that they have carefully examined the highest beds there and found that their position is not above the zone of *Aulacostephanus pseudo-mutabilis*.' These beds occupy about 26 ft (8 m) and the main *N. virgula* bed is within 3 ft (1 m) of the top; Arkell (1933) also noted that the 'lowest 4 ft [1.3 m] of clay and shale exposed ... are marked by a band crowded with *Astarte supracorallina*, and from this level Dr Pringle has recorded *Pararasenia desmonota* (Oppel), characteristic of the *mutabilis* zone' (Pringle, 1923, p. 135). *Nanogyra virgula* is abundant throughout the *eudoxus* Zone, and particularly near the top, in the area of The Wash (Gallois and Cox, 1976), in the Warlingham borehole (Callomon and Cope, 1971), in the Oxfordshire–Buckinghamshire area (Cope, *in* Cope *et al.*, 1980b) and in Dorset (Cox and Gallois, 1981, pp. 5, 15). The ammonite zones present, then, appear to be from the *mutabilis* and *eudoxus* Zones. The second Roswell Pit (TL 552807) was excavated in the 1930s (Gallois, 1988). Several recent finds of teeth and vertebrae have been made on the north side of this pit.

The total thickness of Lower Kimmeridge Clay recorded by Arkell (1933) at Roswell is 8 m of *eudoxus* Zone and 1.3 m of *mutabilis* Zone, thus 9.3 m altogether. Gallois (1988, pp. 40–1) gives a section in the modern pit (TL 552807), comprising about 12 m of *eudoxus* Zone and 1.55 m of *mutabilis* Zone. Gallois and Cox (1976) note, from borehole data, that the Kimmeridge Clay thins southwards along its outcrop from an estimated 84 m in The Wash to 42.3 m at Denver Sluice (TF 591011), 20 km north of Ely.

The exact provenance of the reptiles has not been recorded. It can only be assumed that most of them came from sediments of the *eudoxus* Zone which are thicker. Recently found specimens have all come from these upper portions of the section. Specimens are generally fragmentary – isolated teeth, vertebrae or limb bones. However, some associated series of vertebrae and partial skulls are also known. The bones are generally well preserved, but may lack delicate processes. There has clearly been some postmortem transport and disturbance of the skeletons.

Fauna

Newton (*in* Whitaker *et al.*, 1891, pp. 16-18) reviewed a collection of reptiles and fishes from Roswell Pits, and H.B. Woodward (1895, p. 171) specifies Roswell (Roslyn) pit as having yielded a large number of reptile and fish remains. It is also assumed that most of the reptiles labelled 'Ely' came from these pits (Gallois, 1988, p. 40). There were no other extensive pits at Ely, and early authors (e.g. Seeley, 1869a, pp. 92-108) specified nearby sites separately (e.g. Chettisham, Littleport, Stretham, Haddenham, Cottenham).

Type specimens are indicated below, and an estimate of the number of specimens from Roswell Pits is given for each species, based on collections in the BMNH and CAMSM.

	Numbers
Testudines: Cryptodira: Thalassemyidae	
Thalassemys hugii Rütimeyer, 1873	5
Archosauria: Crocodylia: Thalattosuchia	
Dakosaurus lissocephalus Seeley, 1869	
Type specimen: CAMSM J.29419	1
Dakosaurus maximus (Plieninger, 1846)	1
Dakosaurus sp.	3
Metriorhynchus hastifer	
(Deslongchamps, 1868)	1
Metriorhynchus sp.	2
Steneosaurus sp.	1+
Archosauria: Dinosauria: Saurischia:	
Sauropoda	
Pelorosaurus humerocristatus	
(Hulke, 1874)	4

	Numbers
Gigantosaurus megalonyx	
Seeley, 1869	
Paratype specimen: cast of	
claw (CAMSM)	1
Sauropterygia: Plesiosauria: Elasmosauridae	
Colymbosaurus trochanterius	
(Owen, 1840)	6
Colymbosaurus sp.	12
Sauropterygia: Plesiosauria: Pliosauridae	
Pliosaurus brachydeirus Owen, 1842	33
Pliosaurus brachyspondylus	
(Owen, 1840) (Figure 7.2)	
Neotype: CAMSM J.29564	4
Pliosaurus sp.	15
Liopleurodon macromerus	
(Phillips, 1871)	7
Ichthyopterygia: Ichthyosauria	
Macropterygius trigonus (Owen, 1840)	2
Ichthyosaurus sp.	24

Interpretation

The turtle *Thalassemys hugii* is represented by limb bones and carapace elements. This marine turtle, with a shell length of up to 1.2 m, is well known from the Late Jurassic of Germany and Switzerland (Młynarski, 1976, p. 35). The remains in CAMSM include those of *Enaliochelys chelonia* Seeley, 1869 (p. 108, *nomen nudum*)

The crocodilian remains consist of teeth, jaws and vertebrae, which have been ascribed to the genera *Dakosaurus*, *Metriorhynchus* and *Steneosaurus*. Seeley (1869a, pp. 92-3) erected the species *D. lissocephalus* on the basis of a partial skull, and he ascribed several vertebrae, ribs, limb bones and skull bones to it. He did not specify how this differed from the well-known *D. maximus*, and a restudy will probably show that the two forms are identical (Steel, 1973, p. 42). *Metriorhynchus hastifer* is represented by the anterior end of a snout which was identified by Watson (1911b, p. 9). *Dakosaurus* and *Metriorhynchus* are metriorhynchids, highly aquatic forms. *Steneosaurus*, represented by a vertebra and other remains, is a long-snouted marine fish-eating crocodilian. Without skull material, the identification is not certain.

Large sauropod dinosaurs are represented by several postcranial skeletal elements. The proximal end of a right tibia and a late caudal centrum (casts of CAMSM specimens in BMNH: Lydekker,

1890a, p. 241) were ascribed to *Pelorosaurus humerocristatus*. Seeley (1869a, pp. 94–5) erected the species *Gigantosaurus megalonyx* on the basis of various vertebrae and limb bones from Ely and other sites nearby. Lydekker (1890a, p. 241) ascribed some of these to *P. humerocristatus* and some to *P. manseli*. It is not clear whether all of the remains described by Seeley (1869a) pertain to the same species and which of them he regarded as the type specimen. Hulke (1869a, p. 389) mentioned a cast of the 'large tibia' in the BMNH, as well as 'great ungual phalanx from Ely', and introduced Seeley's name *Gigantosaurus megalonyx* in a footnote. Steel (1970, p. 70) suggests that *G. megalonyx* may be included in *P. humerocristatus*, while McIntosh (1990, p. 356) regards it as a *nomen dubium*. There is clearly a taxonomic problem here, exacerbated by the shortage of comparable material.

Among the commonest remains from Roswell Pits are plesiosaur vertebrae, teeth, and limb bones. The common Kimmeridgian species *Colymbosaurus trochanterius*, up to 6 m long, is well represented. All three Kimmeridgian pliosaurs recognized as valid by Tarlo (1960) are present in the Ely fauna. The Ely material formed the basis for much of the revision carried out by Tarlo (1958, 1959b, 1959c, 1960). *Pliosaurus brachydeirus* is represented by teeth, jaw remains, vertebrae and limb bones (Figure 7.2). It is distinguished from *P. brachyspondylus* on the basis of characters of the vertebrae. This latter species was based on some vertebrae from Headington Pits, near Oxford, which have since been lost, and Tarlo (1959b) selected as neotype a closely matching vertebra from Ely. More important was a skeleton collected at Ely in 1889 (CAMSM J.35991) which consisted of 'numerous teeth, the complete mandible, the greater part of the vertebral column, two limb bones, and fragments of limb girdles' (Figure 7.2). Tarlo (1959b) described this specimen in detail and used it in a revision of pliosaur taxonomy. Characters of the jaw, limbs and limb girdles showed that there

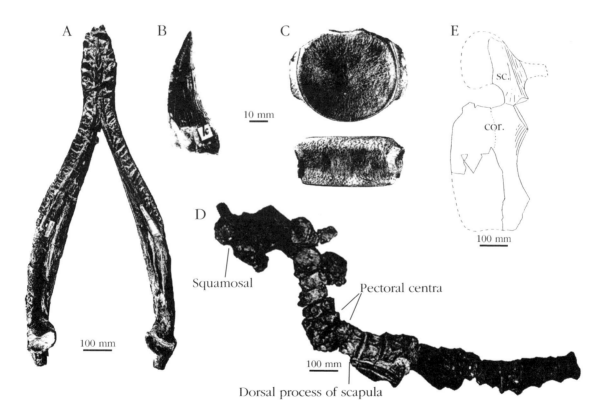

Figure 7.2 The pliosaur *Pliosaurus brachyspondylus* (Owen, 1840), an associated skeleton found in 1889 in the Kimmeridge Clay of Roswell Pits, Ely. (A) The lower jaw in crown view; (B) tooth; (C) cervical vertebra in anterior and ventral views; (D) vertebral column, and associated elements, in dorsal view; (E) reconstructed shoulder girdle. After Tarlo (1959a).

were two genera of Kimmeridgian pliosaur and Tarlo (1959c) erected the genus *Stretosaurus* for *P. macromerus*, but this has since been synonymized with *Liopleurodon*. Some teeth, vertebrae and limb elements of this form are also known from Ely.

Roswell Pits have yielded several ichthyosaur vertebrae, teeth, skull elements and limb bones. Lydekker (1889a, p. 27) mentioned some specimens of *Macropterygius trigonus*, but other bones have not been identified.

Comparison with other localities

Reptiles from the Kimmeridge Clay are known from many other sites in the Wash area, largely old brick-pits, in Northamptonshire, Cambridgeshire and Norfolk, all of which have produced smaller faunas of similar marine and non-marine reptiles (listed at the beginning of this chapter).

Thalassemys hugii is best known from the Late Jurassic of Switzerland and Germany (Młynarski, 1976, p. 35). *Dakosaurus maximus* is also abundant in the Late Jurassic of Switzerland and Germany (Baden-Württemburg), as well as Boulogne-sur-Mer and other sites in France (Steel, 1973, p. 42). It is also known from the Kimmeridgian of Dorset, Oxfordshire and Norfolk, and reworked in the Lower Greensand of Potton, Bedfordshire. *Metriorhynchus hastifer* is known otherwise from the Kimmeridgian of northern France (Steel, 1973, p. 46). The dinosaur *Pelorosaurus* is known from the Wealden of Sussex and the Isle of Wight, and the Kimmeridgian of Weymouth, Kimmeridge Bay, Boulogne-sur-Mer and Wimille (Steel, 1970, pp. 68, 70). *Colymbosaurus trochanterius* is known from several sites between Dorset and Norfolk (Brown, 1981, pp. 315–16; 1984). The pliosaurs are known from a similar range of sites, with good material from Kimmeridge Bay and former sites at Swindon, Shotover and Stretham (Tarlo, 1958, 1959b, 1959c, 1960). Ichthyosaurs have been recorded from many localities with good representation at Kimmeridge Bay, Weymouth, Swindon and Shotover.

Conclusions

The Roswell Pits, Ely are important for the range of Kimmeridgian reptiles that they have yielded. The sauropod dinosaurs and pliosaurs are of particular significance, and there are type specimens of three species from this site. The sauropods are important because of their rarity elsewhere in the European Kimmeridgian, and the pliosaurs from Roswell formed the basis for the taxonomic status of the group in the Late Jurassic. The selection of reptiles present at Ely differs from other good faunas from Dorset and Oxfordshire in the dominance by pliosaurs, and in the apparent absence of pterosaurs. The high conservation value of the site lies in its having yielded one of the richest and most varied Kimmeridgian reptile faunas, being the best site in the northern part of the outcrop of the Kimmeridge Clay and having considerable potential for future finds.

CHAWLEY BRICK PITS, CUMNOR HURST, OXFORDSHIRE (SP 475042)

Highlights

Chawley Brick Pits, Cumnor Hurst are famous for the dinosaur *Camptosaurus*, an important genus that proves a close link with North America during the Late Jurassic. The site has also produced other typical Kimmeridgian-age reptiles, plesiosaurs and ichthyosaurs.

Introduction

Chawley Brick Pits are important for their fauna of marine and terrestrial reptiles, and for the dinosaur *Camptosaurus* (Figure 7.3) in particular. The former exposure of Kimmeridge Clay here, although greatly reduced in area, could be re-excavated to produce further finds.

The brickyard was formerly worked on the north side of Cumnor Hurst and displays the only exposure of Kimmeridge Clay in the area. Parts of the Cretaceous (Lower Greensand–Gault) are also seen. The section formerly spanned the Early and early Late Kimmeridgian, but this is now much reduced. The geology was described by Richardson *et al.* (1946), Arkell (1947a), McKerrow and Baden-Powell (1953) and Cope (1967, 1978; *in* Cope *et al.*, 1980b). A sketch of a photograph of Chawley Brick Pit in the late 19th century is given by Pringle (1926, p. 98). The dinosaur and other reptile finds have been discussed by Phillips (1871), Prestwich (1879, 1880), Hulke (1880a), Lydekker (1888a, 1889c, 1890a) and Galton and Powell (1980).

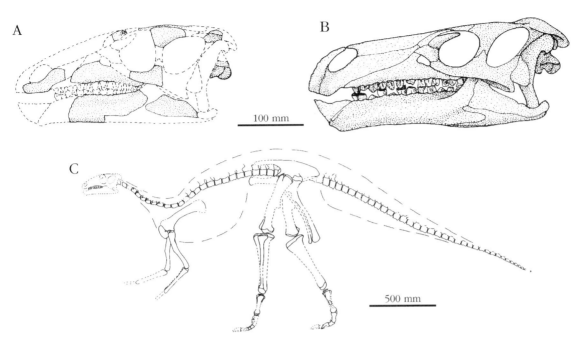

Figure 7.3 The ornithopod dinosaur *Camptosaurus*. (A) Partial restoration of the skull of *C. prestwichii* (Hulke, 1880) showing the known fragments of bone; (B) restored skull of the North American *C. dispar* Marsh, 1879; (C) restoration of the skeleton of *C. prestwichii* (Hulke, 1880): the bones present include parts of the skull, much of the vertebral column, forelimbs and hindlimbs. After Galton and Powell (1980).

Description

The section after Richardson *et al.* (1946, pp. 100–2), Arkell (1947a, pp. 106–7), McKerrow and Baden-Powell (1953, p. 97) and Cope (1967, 1978; *in* Cope *et al.*, 1980b) is:

	Thickness (m)
Northern Drift	3.0
Lower Gault	3.0
Lower Greensand	6.1
Kimmeridge Clay	
?pectinatus Zone	
Shotover Grit Sands: bluish sandy clay (loam) weathering white to a depth of 1 m below junction, with brown weathering grey nodules	*c.* 4.3
(break in sequence)	
wheatleyensis Zone	
Wheatley Nodule Clays; dark shaly clays with big cementstone	

	Thickness (m)
crackers, some crowded with *Pectinatites (Virgatosphinctoides)* cf. *wheatleyensis, P. (V.) tutcheri*, etc. and bivalves	seen to 2.4
(break in sequence)	
eudoxus Zone, ?lower *autissiodorensis* Zone	
dark clays with *Nanogyra virgula, Aulacostephanus eudoxus*, aptychi and reptile bones	seen to 3.0

The fossil reptiles apparently came from the *eudoxus* Zone ('*pseudomutabilis* Zone' Arkell, 1947a, p. 106). Prestwich (1879, 1880) gave a detailed account of the finding of the dinosaur *Camptosaurus*. The skeleton was broken up in collecting, but nearly all elements were found associated. The fossil was found when a tramway was driven into the side of the hill 'in a thin 3 inch [75 mm] sandy seam intercalated in the clay'. This sand occurred about 34 ft (10.4 m) below the

Lower Greensand, according to Prestwich (1880), and his section can be matched with Arkell's (1947a) section to confirm that the dinosaur came from the *eudoxus* Zone. Prestwich (1880) found *Ichthyosaurus* vertebrae and ribs and *Pliosaurus* and *Dakosaurus* teeth in the clay below (also *eudoxus* Zone), and *Plesiosaurus* vertebrae in the clay above (*eudoxus* or *wheatleyensis* Zone).

Fauna

Fossil reptiles from the Chawley Brickpits are preserved in the BMNH and OUM. Numbers of specimens and type specimens are indicated.

Numbers

Archosauria: Dinosauria: Ornithischia:
  Ornithopoda
  Camptosaurus prestwichii
    (Hulke, 1880)
      Type specimen: OUM J.3303   1
      'dinosaur limb'   1
Sauropterygia: Plesiosauria:
  Elasmosauridae
  Colymbosaurus trochanterius
    (Owen, 1840)
      Including 'type specimens'
        of *Plesiosaurus*
        validus Phillips, 1871,
        OUM J.2854-6   *c.* 5
Sauropterygia: Plesiosauria: Pliosauridae
  Pliosaurus brachydeirus Owen, 1841   1
  Pliosaurus sp.   8
  Liopleurodon macromerus
    (Phillips, 1871)   3
Ichthyopterygia: Ichthyosauria
  Macropterygius thyreospondylus
    (Owen, 1840)   2
  Macropterygius trigonus (Owen, 1840)   3
  ichthyosaur indet.   5

Interpretation

The ornithopod *Camptosaurus prestwichii* (see Figure 7.3) was named *Iguanodon prestwichii* by Hulke (1880a). Seeley (1888c) placed it in the new genus *Cumnoria* because of its differences from *Iguanodon*, but Lydekker (1888a, p. 196) questioned the validity of the new genus and returned the species to *Iguanodon*. Then Lydekker (1889c, p. 46; 1890a, p. 258) noted its provisional assignment to *Camptosaurus* because of close resemblance to *C. leedsi* Lydekker, 1889

from the Oxford Clay of Peterborough, which Galton and Powell (1980) confirmed. They redescribed the fragmentary skull and nearly complete skeleton and compared it with other species of *Camptosaurus* from Europe and North America. *C. prestwichii* was about 3.5 m long and it is characterized by features of the skull and teeth. It is more gracile than the better-known *C. dispar* (Marsh, 1879) from the Morrison Formation (Late Kimmeridgian) of Wyoming. Galton (1980c) stressed the palaeogeographical importance of *C. prestwichii*: similarity of the North American and European species indicates a former land connection across the North Atlantic in the Late Jurassic which was probably broken by Kimmeridgian times, hence allowing the two species to diverge slightly.

The plesiosaurs and ichthyosaurs from the Chawley Brick Pits need little comment since they are better represented elsewhere in the British Kimmeridge Clay. Compared with other sites, Chawley lacks crocodiles and turtles. The plesiosaurs, represented by vertebrae and skull remains, probably all belong to *Colymbosaurus trochanterius* (Owen, 1840), the commonest valid Kimmeridgian genus (Brown, 1981). The species *P. validus* Phillips (1871) was based on vertebrae from Shotover, Cumnor and Baldon (Phillips, 1871, pp. 370–2), but Brown (1981, p. 324) regarded the species as a *nomen dubium*. The rarer ichthyosaur remains include several species regarded as *taxa dubia* by McGowan (1976), but a detailed revision is required (A. Kirton, pers. comm., 1981).

Comparison with other localities

The plesiosaurs, pliosaurs and ichthyosaurs from Chawley are typical of other British Kimmeridgian Clay sites. The dinosaur *Camptosaurus* has been recorded elsewhere in the European Kimmeridgian only from Portugal (a femur, Galton, 1980c). Dinosaur bones have been recorded elsewhere in the British Kimmeridge Clay from Kimmeridge Bay (*Pelorosaurus*); Smallmouth Sands, Dorset (*Pelorosaurus, Dacentrurus*); Gillingham, Dorset (*Dacentrurus*); Foxhangers, Wiltshire (?*Megalosaurus*); Rodbourne, Wiltshire (ankylosaur); Wootton-Bassett, Wiltshire (*Dacentrurus*); Swindon, Wiltshire (*Bothriospondylus, Dacentrurus*), Cottenham, Cambridgeshire (*Pelorosaurus*) and Ely, Cambridgeshire (*Pelorosaurus, Dacentrurus*), all but the first two and the last are lost as collecting sites.

Conclusions

Camptosaurus prestwichii is unique in several respects. It is the only ornithopod dinosaur from the British Kimmeridgian, and in fact by far the best preserved dinosaur of any kind from that stage in Britain. It is one of only two ornithopod skeletons described from the Late Jurassic outside North America and East Africa, and it is of palaeogeographic importance in confirming land links between North America and Europe during the Late Jurassic (Galton, 1980c). *C. prestwichii* is one of only two European specimens accepted as belonging to the typically North American genus *Camptosaurus*. The specimen is important also in that a detailed account of its discovery and taphonomy has been published (Prestwich, 1879, 1880).

This historical importance and the potential for future finds with re-excavation give the site its conservation value.

KIMMERIDGE BAY (GAULTER GAP–BROAD BENCH), DORSET (SY 898789–SY 908787)

Highlights

Gaulter Gap–Broad Bench includes the famous fossil reptile sites of Kimmeridge Bay (Figure 7.4). Nearly 20 species of crocodilians, pterosaurs, dinosaurs, plesiosaurs and ichthyosaurs have been found there, including the original specimens of seven species.

Introduction

The Kimmeridge Clay of Kimmeridge Bay, Dorset, is world-famous for its marine reptiles (see Figure 7.8). Several fine ichthyosaur skeletons have been collected, as well as skulls, vertebrae and limb bones of a variety of other fossil reptiles. The cliffs and foreshore reefs of Kimmeridge Bay are subject to continuing erosion, and several finds have been made in recent years. It clearly has considerable potential for future discoveries. The geology of the site has been recorded by many authors, but in most detail by Cope (1967; *in* Torrens, 1969a; *in* Cope *et al.*, 1988b). The fossil reptiles have been described by Owen (1869, 1884a), Hulke (1869a, 1869b, 1870a, 1870b, 1870c, 1871a, 1871b, 1872a, 1874a), Woodward (1885), Lydekker (1888a, 1889a), Huene (1922), Tarlo (1960), McGowan (1976), Brown (1981) and Unwin (1988a).

Description

The Kimmeridge Clay in Kimmeridge Bay is a 138 m sequence of grey to dark grey-blue, ammonite-bearing mudstones and shales with sporadic cementstone (limestone/dolostone) bands. The dominant argillaceous units comprise alternations of homogeneous and sometimes blocky mudstones and finely laminated, fissile, bituminous shales. Mudstone units appear to be quite structureless but, after weathering, a certain degree of mottling may be seen. The base of the mudstone units weathers out more sharply than the upper sections and the upper boundaries appear to be transitional to the bituminous shales. These are rather thinner than the mudstone units, between 0.1 and 0.4 m thick, the mudstones measuring between 0.1 m and about 1 m in thickness. Erosion of the mudstones and shales is rapid, but the cementstone bands resist erosion and stand out.

At Kimmeridge Bay, the beds dip south-east and there are several faults. The general sequence, based on Cope (1967; *in* Torrens, 1969a; *in* Cope *et al.*, 1980b) is:

	Thickness (m)
Late Kimmeridgian (formerly Mid Kimmeridgian)	
wheatleyensis Zone	
Grey Ledge Stone Band	0.7
scitulus Zone	
Upper Cattle Ledge Shales	10.8
Cattle Ledge Stone Band	0.5
Lower Cattle Ledge Shales	15.0
Yellow Ledge Stone Band	0.4
	27.4
elegans Zone	
Hen Cliff Shales	21.5
Double band of cementstone with shale	1.1
	22.6
Lower Kimmeridgian	
autissiodorensis Zone	
Maple Ledge Shales	22.5
Maple Ledge Stone Band	0.3
Gaulter's Gap Shales	32.0
Washing Ledge Stone Band	0.35
Washing Ledge Shales (upper part)	8.0
	63.15

Figure 7.4 The Kimmeridgian of Kimmeridge Bay, showing dipping limestone and shale units, facing south. (Photo: M.J. Benton.)

	Thickness (m)
eudoxus Zone	
Washing Ledge Shales (lower part)	5.0
The Flats Stone Band	0.5
Shales	3.0
Nannocardioceras Bed	0.02
Shales	1.0
Shales	seen to 15.0
	24.52

The sediments of the *Aulacostephanus eudoxus* and *A. autissiodorensis* Zones are exposed between Broad Ledge and Maple Ledge. The named stone bands reach shore level as follows: The Flats at Broad Bench (SY 897787) and at The Flats (SY 905792); Washing Ledge (SY 907791); Maple Ledge (SY 909789). Hen Cliff, between Clavell Tower and Cuddle, exposes the *elegans, scitulus* and *wheatleyensis* Zones (the *Pectinatites* Zones). The stone marker bands include the Yellow Ledge which reaches the shore at Yellow Ledge (SY 912782), and the Cattle Ledge and the Grey Ledge higher in the cliff.

The fauna of the Kimmeridge Clay is restricted to the mudstones and bituminous shale units and in both units is essentially the same, being dominated by infaunal bivalves, including *Lucina* and *Protocardia*. There is also an encrusting epifauna, but this is restricted, consisting only of oysters (*Liostrea* and *Nanogyra*). Minor elements include *Discina, Lingula, 'Gervillia', Entolium* and aporrhaid gastropods. The other biofabrics in the mudstones are quite different from those in the bituminous shales, indicating different post-mortem histories of the fauna and thus different environmental conditions (Aigner, 1980). This limited biota indicates somewhat oxygen-depleted bottom conditions, and that the reptiles must have occupied a mid-water zone.

Most specimens bear the locality and horizon description 'Kimmeridge'. However, a few more specific references indicate that reptiles have been found at all points round Kimmeridge Bay and in Hen Cliff (see Figure 7.6), and at various levels in the *Aulacostephanus* Zones and *Pectinatites* Zones. Arkell (1933, p. 451) considered that most specimens came from the Early

Kimmeridgian and noted that the types of *Steneosaurus manseli* and *Ichthyosaurus enthekiodon* were found embedded in reefs exposed at low water in the bay. Delair (1986, p. 133) notes parts of a disarticulated specimen of the ichthyosaur *Ophthalmosaurus* from 'a level approximately 2.5 feet below the base of the *wheatleyensis* Subzone' at Rope Lake Head. Hulke (1869a, p. 386) reported a large 'saurian humerus' (the type of *Ornithopsis manseli* Hulke, 1869a) from 'amongst the layers of shale immediately above the band of cement-stone which rises from East to West on the west side of Clavell's Tower, between Kimmeridge Bay and Clavell's Head'. This probably refers to the Maple Ledge Shales above the Maple Ledge Stone Band (thus *autissiodorensis* Zone), at a site around SY 909789. Clarke and Etches (1992) record a fine mandible of *Liopleurodon macromerus* from the *autissiodorensis* Zone of 'Kimmeridge Bay'. Hulke (1871b, p. 442) described a crocodile snout (the type of *Teleosaurus megarhinus* Hulke, 1871) that had fallen from the cliff in the bay. The exact locality is uncertain, but it must have come from one of the *Aulacostephanus* Zones. A specimen of *Plesiosaurus* collected in 1971 (CAMMZ T962) is recorded to have come from 5 ft below The Flats Stone Band at Broad Bench (SY 898789), thus in shales of the *eudoxus* Zone. Likewise, Unwin (1988a) reports vertebrae and limb elements of the pterosaur *Germanodactylus* from '5 m below the Flats Stone Band (*eudoxus* Zone . . . at Charnel (NGR 899789))'.

Several finds have been reported from the early portion of the Late Kimmeridgian. Hulke (1870c) described some *Plesiosaurus* vertebrae (the types of *P. manseli* Hulke, 1870) from the cliffs east of Clavell Tower, thus probably *scitulus* Zone. Brown (1981, p. 315) ascribes these to the *pectinatus* Zone (Late Kimmeridgian), but this is unlikely since that zone only appears in the cliffs between Rope Lake Head and Encombe Bay, 2–3 km south east of the Clavell Tower. Cope (1967, p. 10) mentions further remains from the Late Kimmeridgian, the anterior part of a skeleton of *Ophthalmosaurus* 12 ft (4 m) above the Cattle Ledge Stone Band (*scitulus* Zone), and a pliosaur tooth about 5 ft (1.5 m) above the Yellow Ledge Stone band (*scitulus* Zone). A limb bone of *Pliosaurus* (DORCM G187) was found on the foreshore below Clavell Tower.

All of the finds appear to have been made in the shales. Hulke (1869b, p. 390) characterized the matrix of a *Steneosaurus* as 'hard pyritic clay-

stone'. The skeletal elements are frequently disarticulated, but not particularly worn; teeth may be in place and delicate bone processes unbroken. The only nearly complete skeletons found appear to be those of ichthyosaurs (Hulke, 1871a; Cope, 1967, p. 10; Macfadyen, 1970, p. 126). The pterosaur *Germanodactylus* consisted of partially disarticulated bones, some of which were heavily crushed (Unwin, 1988a).

Fauna

Mansel-Pleydell (1888) and Delair (1958, 1959, 1960) gave long lists of reptiles from Kimmeridge Bay, totalling 33 or more species. As a result of recent revisions of the ichthyosaurs (McGowan, 1976; Angela Kirton, pers. comm.), pliosaurs (Tarlo, 1960) and the plesiosaurs (Brown, 1981), many of these species have been synonymized. A revised list of the reptiles is given here, with the repository numbers of type specimens. The approximate numbers of specimens of each species in major British collections are appended in order to give an impression of the relative abundance of each form.

Numbers

Archosauria: Crocodylia: Thalattosuchia
   Dakosaurus maximus
     (Plieninger, 1846)
     Type specimen of *Steneosaurus*
     manseli Hulke,1870; BMNH 40103   1
   Machimosaurus mosae Sauvage
     and Lienard, 1879   1
   Steneosaurus megarhinus
     (Hulke, 1871)
     Type specimen: BMNH 43186   1
Archosauria: Pterosauria
   '*Rhamphorhynchus* sp.'   1
   Germanodactylus sp.   1
Archosauria: Dinosauria: Saurischia:
   Sauropoda
   Pelorosaurus manseli (Hulke
     ms., Lydekker, 1888)
     Type specimen: BMNH 41626   1
Sauropterygia: Plesiosauria: Elasmosauridae
   Colymbosaurus trochanterius
     (Owen, 1840)
     Includes type of *Plesiosaurus*
     manseli Hulke,
     1870: BMNH 40106   6
   '*Plesiosaurus brachistospondylus*'
     Hulke, 1870

	Numbers
Type specimen: BMNH 45869	3
'*Plesiosaurus*' sp.	4
Sauropterygia: Plesiosauria: Pliosauridae	
Pliosaurus brachydeirus Owen, 1841	2
Pliosaurus brachyspondylus (Owen, 1840)	1
Pliosaurus sp.	9
Liopleurodon macromerus (Phillips, 1871)	8
Ichthyopterygia: Ichthyosauria	
Macropterygius dilatatus (Phillips, 1871)	1
Macropterygius ovalis (Phillips, 1871) Type specimen: OUM J.12487	1
Macropterygius trigonus (Owen, 1840)	1
Nannopterygius enthekiodon (Hulke, 1870) Type specimen: BMNH 46497, a	1
Ichthyosaurus sp.	1

Interpretation

The Kimmeridge Clay marks a period of widespread clay sedimentation in north-west Europe in environments that were clearly fully marine. The thick series of clays and bituminous shales are considered to have been deposited in calm bottom waters, and anaerobic conditions may have prevailed in a stratified water column (Aigner, 1980), an environment similar to the present-day Black Sea. The sediments are essentially terrigenous in origin, indicating considerable erosion from a nearby landmass (?the London–Ardennes island and Cornubia), although there are no obvious plant macrofossils.

The remains of crocodiles from Kimmeridge Bay are mainly partial skulls with occasional associated vertebrae. *Dakosaurus*, a large animal, often up to 4 m long, is represented by a partial skull of a relatively short-snouted form (BMNH 40103). Hulke (1870a) described it as the type specimen of *Steneosaurus manseli*. Owen (1884a) referred this species to a new genus, *Plesiosuchus*, but Woodward (1885) included it in *Dakosaurus*.

Machimosaurus and *Steneosaurus* are long-snouted teleosaurs, common in marine deposits from the Early Jurassic to the Early Cretaceous of Europe and other continents. *M. mosae* was based on specimens from the Kimmeridgian of Issoncourt (near Verdun), and Lydekker (1888a)

referred to this species the occipital region of a cranium and an associated mandible from Kimmeridge Bay (BMNH R1089). These specimens were originally figured by Owen (1869) as *Pliosaurus trochanterius*, but Deslongchamps (1869, p. 329) realized their crocodilian nature and ascribed them to *Metriorhynchus*, a determination followed by Woodward (1885). The species *M. mosae* was apparently based on heterogeneous material, some of it probably mosasaurian, and it is probably invalid (Krebs, 1967; Steel, 1973, p. 25). Thus the Dorset material could be ascribed to the type species, *M. hugii* Meyer, 1837. *Steneosaurus megarhinus* (Hulke, 1871b) was based on a slender rostrum, 430 mm long and with greatly expanded premaxillae each containing five alveoli (Figure 7.5A). Hulke originally ascribed this to *Teleosaurus*.

The pterosaur '*Rhamphorhynchus*' is represented by a single specimen (BMNH R1936), and *Germanodactylus* likewise (Etches Collection K96; Unwin, 1988a). Both finds extend the ranges of these taxa from Germany to England, and into even earlier Kimmeridgian strata than those at Solenhofen. Pterosaurs are better known from the Kimmeridge Clay of Weymouth, but the Kimmeridge Bay *Germanodactylus* is the oldest of the family Germanodactylidae in the world, and possibly the oldest pterodactyloid pterosaur (Unwin, 1988a).

The sauropod *Pelorosaurus* is represented by a large humerus (BMNH 41626). This 'stupendous bone' (Hulke, 1869a), when pieced together, had a length of 0.8 m. Hulke (1869a, 1874a) considered that it belonged to a great crocodile. He named it *Ischyrosaurus* Hulke, 1874, and Lydekker (1888a, p.152) later placed it with the large sauropod dinosaurs in *Ornithopsis manseli* Lydekker, 1888. *Ornithopsis*, and many other generic names given to large dinosaur bones from the Late Jurassic and Early Cretaceous of Europe, have been synonymized with *Pelorosaurus* (Steel, 1973, pp. 68, 70; McIntosh, 1990). *Pelorosaurus* is placed in the Brachiosauridae, but familial assignment of such isolated elements is clearly problematic.

The most abundant remains from Kimmeridge Bay are plesiosaurs (Figure 7.5B). *Colymbosaurus trochanterius* (Owen, 1840), of which *Plesiosaurus manseli* Hulke, 1870 is a synonym (Lydekker, 1889a; Brown, 1981, pp. 316–17, 337), is the largest Late Jurassic plesiosauroid, with an estimated total length of 6.15 m for the type specimen of *P. manseli* (BMNH 40106). The

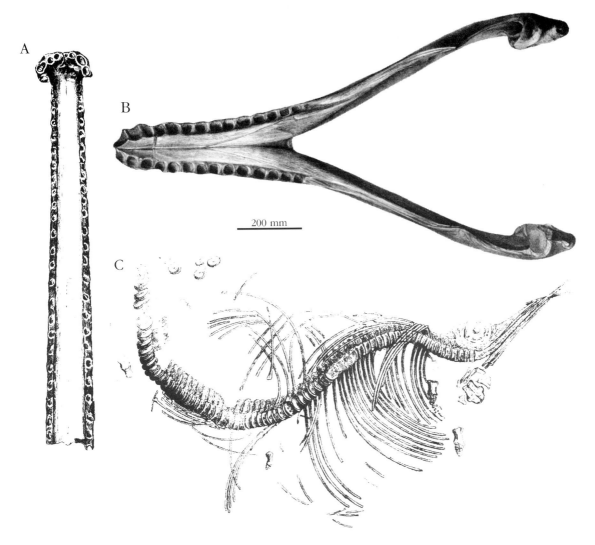

Figure 7.5 Kimmeridgian reptiles from Kimmeridge Bay, Dorset. (A) The elongate slender snout of the marine crocodile *Steneosaurus megarhinus* (Hulke, 1871) in ventral view; (B) lower jaw of *Colymbosaurus trochanterius* (Owen, 1840) in crown view; (C) skeleton of the ichthyosaur *Nannopterygius (Ichthyosaurus) enthekiodon* (Hulke, 1871). (A) after Hulke (1871b); (B) after Owen (1861); (C) after Hulke (1871a).

specimen consists of an extensive series of vertebrae, pieces of two humeri and several paddle bones. Other specimens of *C. trochanterius* from Kimmeridge Bay are a series of vertebrae and isolated limb bones. The type specimen of *P. brachistospondylus* Hulke, 1870 consists of five compressed dorsal vertebrae, some rib fragments and finger elements. Hulke (1870a) regarded the great height and breadth of the vertebrae as unique, but Brown (1981, p. 322) considered the species a *nomen dubium*.

The pliosaurs are represented largely by jaw fragments and teeth, belonging to all three Kimmeridgian species recognized by Tarlo (1960)

as valid. *P. brachydeirus* is distinguished from *P. brachyspondylus* on the basis of details of the vertebrae and from *Liopleurodon macromerus*, with its short mandibular symphysis, as well as characters of the vertebrae and tooth arrangement. The assignments of many of the Dorset specimens were made before Tarlo's work and revisions may be necessary.

Finally, at least five specimens of ichthyosaurs have been collected. Phillips (1871, p. 339) described the new species *Ichthyosaurus ovalis* on the basis of some vertebral centra from Kimmeridge Bay which have never been figured (identified as OUM J.12487 on labels). McGowan

(1976) recorded it under *taxa dubia* because no type material was designated in the original description. Hulke (1870b) described some ichthyosaur teeth and jaw fragments as the new genus *Enthekiodon*, but he did not name a species. Hulke (1871a) then reported a very fine ichthyosaur skeleton from Kimmeridge Bay which he named *I. enthekiodon* (Figure 7.5C). It showed the skull, vertebral column, ribs and limb fragments. The skull is 0.62 m long and the body about 2.8 m long. It had a medium-length snout and was distinguished from other forms by characters of the vertebrae and the very small paddles. Huene (1922, p. 98) established a new genus, *Nannopterygius*, for this species and McGowan (1976, p. 671) considered it to be a valid taxon. Hulke (1872a) described further ichthyosaur teeth and skull fragments from Kimmeridge Bay.

The preserved reptile fauna from Kimmeridge Bay, then, consists mainly of medium to large-sized marine forms: crocodilians, plesiosaurs, pliosaurs and ichthyosaurs. All of these were probably fish-eaters, although they may also have fed on the abundant cephalopods. The crocodilians could doubtless walk and feed on land. Rare bones of large sauropod dinosaurs were washed in. Turtles have not been found and pterosaur remains are rare, although these forms are rather well known from the Kimmeridge Clay around Weymouth.

Comparison with other localities

Marine reptiles have been recorded from all 60 sites in the Kimmeridge Clay of Dorset, Wiltshire, Oxfordshire, Buckinghamshire, Cambridgeshire, Norfolk, Lincolnshire and Yorkshire (listed at the start of the chapter). However, most sites have yielded little more than a few vertebrae or limb elements of one or two species. Sites comparable to Kimmeridge Bay include Smallmouth Sands, Weymouth (SY 6697–SY 672771), and other sites listed under that locality.

The crocodile *Dakosaurus* (Jurassic–Early Cretaceous of Europe) occurs in the Kimmeridge Clay of Ely, Shotover, Norfolk and Boulogne-sur-Mer (Steel, 1973, pp. 42–4). *Machimosaurus* has been recorded from the Kimmeridgian of Shotover, Boulogne-sur-Mer, Verdun and Hanover, as well as the Late Jurassic of Portugal and the Portlandian of Switzerland and Austria (Steel, 1973, p. 25). *Steneosaurus* is known from the Jurassic of many localities in Europe and elsewhere. It has been recorded from the Kimmeridgian of Shotover, Moulin Wibert at Bologne-sur-Mer and Cap de la Hève, Normandy (Steel, 1973, pp. 26–34). The dinosaur *Pelorosaurus* is well known from the Wealden of Sussex and the Isle of Wight, and the Kimmeridgian of Weymouth, Ely, Boulogne-sur-Mer and Wimille, near Boulogne (Steel, 1970, pp. 68, 70).

Colymbosaurus trochanterius is known from several former localities in Wiltshire, Oxfordshire, Cambridgeshire and Norfolk (Brown, 1981, pp. 315-16). The pliosaurs are known from a similar range of sites, with good material from Swindon, Shotover, Ely and Stretham (Tarlo, 1958, 1959b, 1959c, 1960). Ichthyosaurs have been recorded from many localities, with good representation at Weymouth, Swindon, Shotover and Ely. Kimmeridgian plesiosaurs are known from France, the Moscow Basin, Greenland and Sichuan (China), and Kimmeridgian ichthyosaurs from France (Boulogne), Germany and Argentina (McGowan, 1976).

Conclusions

Kimmeridge Bay has yielded more type specimens of reptiles than any other Kimmeridgian site. It is particularly important for the plesiosaurs, ichthyosaurs and crocodilains, many of which are best represented here. Because of the sporadic occurrences of Kimmeridgian plesiosaurs and ichthyosaurs elsewhere in the world, Kimmeridge Bay is particularly important, and it has figured prominently in recent reviews of marine reptiles (Tarlo, 1960; McGowan, 1976; Brown, 1981). The pterosaurs *Rhamphocephalus* and *Germanodactylus*, although more poorly preserved than the original material from Bavaria, are significantly older (Unwin, 1988a). The conservation value lies in the international importance of the site and its considerable potential for future finds.

ENCOMBE BAY, SWYRE HEAD– CHAPMAN'S POOL, DORSET (SY 937773–SY 955771)

Highlights

Swyre Head to Chapman's Pool includes an important array of late Kimmeridgian reptile sites.

These have produced various species of turtle, pterosaur, dinosaur, plesiosaur and ichthyosaur, including the plesiosaur *Kimmerosaurus* and a new theropod dinosaur.

Introduction

The Upper Kimmeridge Clay exposed between Swyre Head and Chapman's Pool also known as Encombe Bay or Egmont Bight (Figure 7.6) has produced a range of fossil reptiles, ichthyosaurs, plesiosaurs, crocodilians and turtles. Many of the specimens have been collected recently. The cliffs are subject to continuing erosion and the section has good potential for future finds. The geology has been described in detail by Cope (1967; *in* Torrens, 1969a; 1978; *in* Cope *et al.*, 1980b), and the occurrence of the reptiles has been reviewed by Taylor and Benton (1986). Reptiles from these sections have been described by Brown (1981), Delair (1986) and Clarke and Etches (1992).

Description

The Kimmeridge Clay in this section covers the upper part of the Late Kimmeridgian. It consists of a sequence of grey and bituminous shales and clays with stone bands towards the base (Figure 7.6A and B). The sequence according to Cope (1967; *in* Torrens, 1969a; 1978; *in* Cope *et al.*, 1980b) is:

	Thickness (m)
Portland Sand, Massive Bed	
Upper Kimmeridgian	
fittoni Zone	
Hounstout Marl	21.00
Hounstout Clay	8.35
Rhynchonella and *Lingula*	
Beds (upper part)	8.00
	37.35
rotunda Zone	
Rhynchonella and *Lingula*	
Beds (lower part)	15.00
rotunda Shales	13.50
rotunda Nodule Bed	1.80
Shales and clays	4.25
Hard bituminous shales	1.25
	35.80

	Thickness (m)
pallasioides Zone	
Clays and shales (9 individual subunits, Cope, 1978)	30.00
pectinatus Zone	
paravirgatus Subzone	
Grey shales	12.10
Hard shale	0.60
Shales	6.10
Freshwater Steps Stone Band	0.40
	19.20
eastlecottensis Subzone	
Shales	8.80
Middle White Stone Band	0.45
Shales and mudstones	8.90
White Stone Band	0.95
	19.10

The beds have an apparent easterly dip, and the ammonite zones occur in sequence from north-west to south-east (Figure 7.6A): *pectinatus* Zone (cliff below Swyre Head–Egmont Bight; SY 937773–SY 947772), *pallasioides* Zone (Freshwater Steps–Chapman's Pool; SY 942772–SY 955771) and the *rotunda* Zone, above this, between Egmont Bight and continuing past Chapman's Pool. The succeeding *fittoni* Zone and the Portlandian occur higher in Hounstout Cliff. The White Stone Band and Middle White Stone Band come to beach level below Swyre Head and 250 m east of that, respectively. The Freshwater Steps Stone Band reaches beach level at Freshwater Steps. The shales and clays of the *pallasioides* and lower *rotunda* Zones (i.e. between the top of the *pectinatus* Zone and the *rotunda* Nodule Bed) are sometimes known as the Crushed Ammonoid Shales.

The reptile specimens have apparently been collected largely at beach level either in the wave-cut platform west of Freshwater Steps, or in Chapman's Pool (Figure 7.6A). Although Hounstout Cliff is accessible, no reptile remains have been reported above the lower parts of the *rotunda* Zone.

Specific localities in the *pectinatus* Zone include the 'ledges below Swyre Head, SY 939773' for a partial *Teleosaurus* specimen (DORCM G.347, label), thus *eastlecottensis* Subzone. Arkell (1947c, p. 78) noted that the White Stone Band occasionally contains 'saurian vertebrae and bones'. A plesiosaur centrum (DORCM G.172) is noted as '*pectinatus* Zone,

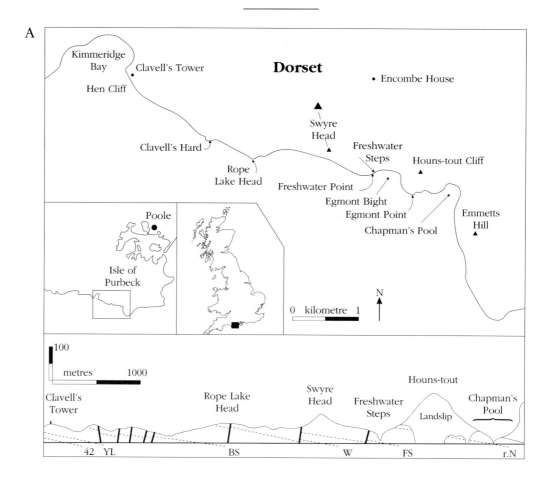

Zones	Description	Code
Upper Kimmeridgian		
fittoni		
rotunda	*rotunda* Nodules (L)	r.N
pallasioides		
pectinatus	White S.B. (B); Freshwater Steps S.B. (M)	W FS
hudlestoni		
wheatleyensis	Blackstone (T)	BS
scitulus	Yellow Ledge S.B. (B); Cattle Ledge S.B. (U)	YL
elegans	Blake's Bed 42 (B)	42
Lower Kimmeridgian		
autissiodorensis		
eudoxus		
mutabilis		
cymodoce		
baylei		

Figure 7.6 (A) Locality map and vertical section of the Swyre Head–Chapman's Pool Kimmeridge Clay site on the Isle of Purbeck, Dorset. The beds dip gently southwards, and the shales and mudstones are punctuated by distinctive limestone beds ('stone bands') which have been named. These may also be matched with the (B) tabulation of the ammonite zones of the Kimmeridgian. Abbreviations: (42) Blake's Bed 42; (BS) Blackstone; (FS) Freshwater Steps Stone Band; (r.N) *rotunda* Nodules; (SB) Stone Band; (W) White Stone Band; (YL) Yellow Ledge Stone Band; (in the zonal chart, B, L, M, U and T refer to basal, lower, middle, upper and topmost parts of the zones). After Taylor and Benton (1986); based on Cope (1967, 1978); Cope *et al.* (1980b); Cox and Gallois (1981).

above highest White Stone Band, west of Freshwater Steps on shore ledge', thus above the Freshwater Steps Stone Band in shales of the *paravirgatus* Subzone.

Recent finds confirm the occurrence of reptiles in the *pectinatus* Zone. A vertebra and ribs of a crocodilian were found by R.A. Langham in the White Stone Band, thus base of the *eastlecottensis* Subzone, to the west of Freshwater Steps. The same collector also found some limb bones of a pterosaur just above the Freshwater Steps Stone Band at Freshwater Steps, thus *paravirgatus* Subzone. Finally, P.A. Langham found some turtle remains (BMNH R8699) from a horizon in the shales just above the Freshwater Steps Stone Band, about 300 m west of Freshwater Steps, thus *pectinatus* Zone also.

Brown (1981, p. 301) reported a skull and isolated teeth of the plesiosaur *Kimmerosaurus langhami* (BMNH R8431) from 'Endcombe Bay' (also known as Egmont Bay) in the Crushed Ammonoid Shales (Figure 7.7). R.A. Langham (pers. comm. to M.J.B., 1982) gave further information on the find stating that it came from a location '*in situ* in shale at the base of the cliff approximately 270 m west of Freshwater Steps', thus perhaps shales of the *eastlecottensis* Subzone at SY 924773. If the find site is generally in the vicinity of Encombe Bay then the specimen could, in fact, come from the upper *pectinatus* Zone, the *pallasioides* Zone, or the *rotunda* Zone; Brown (1981, p. 301) suggested the *rotunda* Zone. However, Brown *et al.* (1986) revise the horizon as 'about 2 m above the Middle White Stone Band' in the upper part of the *eastlecottensis* Subzone of the *pectinatus* Zone (Cope *et al.*, 1980b; Cox and Gallois, 1981). A second partial skull and mandible with some associated postcranial remains (BMNH R10042) belonging to *K. langhami* was reported by Brown *et al.* (1986, pp. 225–34) from the type locality and horizon, *in situ* about 3 m east of the site of R8431. This was collected by P.A. Langham in 1976.

Other records include phalanges of a pliosauroid (DORCM G639) from 'below Encomb(e) House at . . . SY 942772', thus just west of Freshwater Steps, and probably the *pectinatus* Zone. Some plesiosaur vertebrae and a rib (DORCM G5093; BGS(GSM)) came from around SY 940773, also presumably *pectinatus* Zone. A partial ichthyosaur skeleton (BMNH R8693) came from a water-worn platform exposed at low tide, 400 m east of the Yellow Ledge, thus *scitulus* Zone, much lower down. Clarke and Etches

(1992) note a plesiosaur limb bone from a higher horizon, the *rotunda* Zone, at Chapman's Pool. Other plesiosaur and ichthyosaur specimens are not so well localized (Taylor and Benton, 1986).

All of the finds, as at Kimmeridge Bay, appear to have been made in the shales; Brown (1981, p. 304) notes that BMNH R8431 was preserved in a clay matrix. The preservation of this skull was generally good, and surface ornament was visible. Parts of the skull were slightly crushed and the dentary somewhat 'eroded'. Other specimens from this area are generally isolated postcranial elements (vertebrae and limb bones) or slightly disturbed partial skeletons. Fuller details are given by Taylor and Benton (1986).

Fauna

Testudines: Cryptodira: Thalassemyidae
  Pelobatochelys sp.
    BMNH R8699
Archosauria: Crocodylia: Thalattosuchia
  '*Teleosaurus* sp.'
    DORCM G.347
  Dakosaurus/Metriorhynchus
    R.A. Langham collection
Archosauria: Pterosauria
  Unnamed
    R.A. Langham collection
Archosauria: Dinosauria: Theropoda
  Gracile theropod (OUM)
Sauropterygia: Plesiosauria: Cryptoclididae
  Kimmerosaurus langhami Brown, 1981
    Type specimen: BMNH R8431; also BMNH R10042
Sauropterygia: Plesiosauria: Elasmosauridae
  '*Colymbosaurus* sp.'
    DORCM G.172, G.184
Sauropterygia: Plesiosauria: Pliosauridae
  Pliosaurus sp.
    DORCM G.186, G.639; Etches collection
Ichthyopterygia: Ichthyosauria
  Grendelius sp.
    BRSMG
  '*Ophthalmosaurus* sp.'
    DORCM G.8, BMNH R8693

Interpretation

The turtle *Pelobatochelys* is represented by a partial carapace, about 0.4 m long, with remains of limbs (BMNH R8699). The genus is known only from Dorset and was founded on carapace plates from Weymouth. If this undescribed specimen from Encombe Bay belongs to *Pelobatochelys*, the remains include the first record of its limbs.

The partial skeleton of 'Teleosaurus' (vertebrae, ribs, jaws; DORCM G.347) may belong to one of several Kimmeridgian crocodile genera (e.g. *Dakosaurus, Machimosaurus, Steneosaurus, Teleosaurus*). Exact identification depends on snout length and features of the skull roof which is not preserved.

A gracile theropod dinosaur is represented by a partial skeleton of the hip region in the OUM.

The only fossil reptile from Encombe Bay that has been described is *Kimmerosaurus langhami* (Brown, 1981, pp. 300-14; Brown *et al.*, 1986). The type specimen (BMNH R8431; Figure 7.7) consists of the posterior part of a skull roof, an occiput, partial braincase, partial lower jaws and 11 isolated teeth. The referred material from Freshwater Steps (BMNH R10042) consists of a braincase, mandible, atlas–axis complex and five cervical vertebrae. The skull is 0.3 m long. *Kimmerosaurus* differs from all other plesiosaurs by the nature of the teeth, which lack the usual longitudinal ridges, and are greatly recurved and elliptical rather than circular in cross-section. The skull is the most lightly built of all species known from the Late Jurassic and there is no sagittal crest on the parietals, a clear difference from all other plesiosaurs. *Kimmerosaurus* is one of only five genera of Late Jurassic plesiosauroids recognized as valid by Brown (1981), and one of only two species from the Kimmeridgian. The other, *Colymbosaurus trochanterius* Owen, known from five postcranial skeletons and a number of isolated propodials, is the longest and heaviest English plesiosauroid, measuring 6 m from the tip of the snout to the end of the tail. Brown (1981) referred the two genera to different families, tentatively placing *Kimmerosaurus* with

Cryptoclidus in the Cryptoclididae and *Colymbosaurus* in the Elasmosauridae. The possibility that *Kimmerosaurus* might be synonymous with *Cryptoclidus* was discussed by Brown *et al.* (1986). Among the available material, the only elements shared by both forms are the anterior cervical vertebrae (in *Kimmerosaurus*, only in specimen R10042), and these appear closely comparable. Should this be the case, the wider taxonomic status of Elasmosauridae and Cryptoclididae would need to be reviewed.

The elasmosaurids have very long necks, produced by increases both in the number of cervical vertebrae and in the lengths of centra, particularly among the anterior cervicals. The anterior cervicals possess a further distinguishing character, in the development of a lateral keel and an articular face which has either a single shallow concavity or an open V-shape (Brown, 1981). The cryptoclidids, by contrast, have medium-length necks (28–32 cervical vertebrae), and the anterior cervical centra have a deep concavity with a convex rim.

The other plesiosaur remains from Encombe probably belong to *Colymbosaurus* (DORCM G172, G184, G5093) and *Pliosaurus* (DORCM G186, G639) respectively.

The ichthyosaur remains (BMNH R8693; DORCM G.8) could belong to one of several genera that occur elsewhere in the British Kimmeridgian (e.g. *Macropterygius, Grendelius, Nannopterygius, Ophthalmosaurus*). Identification is based on the shape of the skull (e.g. snout length, shape, size and position of openings) or on features of the paddles. A new specimen of *Grendelius* in the BRSMG will be described shortly (McGowan, in prep.). The taxonomy of Late Jurassic ichthyosaurs is controversial (McGowan, 1976; A. Kirton, pers. comm., 1981), and fragmentary remains are hard to identify.

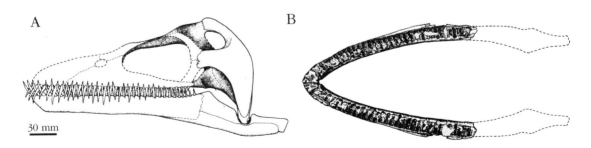

Figure 7.7 The plesiosauroid *Kimmerosaurus langhami* Brown, 1981, from the Upper Kimmeridge Clay of Egmont Bay. (A) Restoration of the skull in lateral view; (B) lower jaws viewed from above. After Brown (1981).

Comparison with other localities

Late Kimmeridgian reptile sites are rare, and none has been highly productive. Some ichthyosaurs, plesiosaurs and crocodilians have come from units equivalent to those described here at Ringstead Bay, Dorset (SY 7581), and rare remains from the Hartwell Clay of Buckinghamshire (*pallasioides* Zone). A referred specimen of *Kimmerosaurus* (BMNH R1798) came from Weymouth, probably from a cliff exposure between Sandsfoot Castle and the old Portland Ferry Bridge and therefore Early Kimmeridgian in age (Damon, 1884; see Smallmouth Sands report). The disused Kimmeridgian pits on Shotover Hill, Oxfordshire (SP 558065, SP 560066, SP 562066, SP 564066, etc.) have yielded some reptiles from the *pectinatus* Zone (Shotover Grit Sands, Shotover Fine Sands), as well as more abundantly from the Early Kimmeridgian.

Conclusions

The whole coast section from Swyre Head to Chapman's Pool (Encombe Bay) represents the best British Late Kimmeridgian (=Early Tithonian) reptile site. It has produced a selection of marine reptiles that have not yet been described in full. One undescribed turtle may be the first specimen of the poorly known genus *Pelobatochelys* with limb remains. The two partial skulls and some postcranial remains of *Kimmerosaurus langhami* Brown, 1981 show that this was a plesiosaur with several unique features that may form part of a lineage separate from the commoner plesiosaurid-elasmosaurid and pliosaur groups. Marine faunas of this age are rare elsewhere in the world, with similarly isolated remains known from France and Germany. At the same time in North America and Africa, the only known faunas are of terrestrial organisms. This potential for future discoveries gives the site its conservation value.

LATE JURASSIC (PORTLANDIAN) OF ENGLAND

Portlandian (=Upper Tithonian) reptile sites are generally rare in Britain, the only productive ones being on the Isle of Portland and near Swanage, both in Dorset, and in Buckinghamshire. Apart from their taxonomic significance, many of the faunas are interesting from the viewpoint of their palaeoecology in that they occur in a variety of facies ranging from lacustrine and lagoonal to shallow marine, which in addition cross the Jurassic/Cretaceous (Portlandian–Berriasian) boundary (Figure 7.8). The sites selected as GCR sites cover the main Isle of Portland localities, Durlston Bay (Swanage) and a quarry near Hartwell (Buckinghamshire) important for its remains of dinosaurs.

ISLE OF PORTLAND REPTILE SITES

Highlights

The Isle of Portland is riddled with old quarries and cliffs that have been the source of a diverse array of fossil marine reptiles (Figures 7.9 and 7.10). The fossil turtles have proved to be particularly important, including some of the oldest known well preserved marine turtles.

Introduction

The latest Jurassic Portlandian limestone of the Isle of Portland has long been famous for its marine reptiles. Many skeletons and individual bones of these reptiles (Figure 7.11), plesiosaurs and ichthyosaurs were collected in the 19th century when the equally famous stone quarries were more active than today, but there have been some important finds in recent years. The island has yielded type specimens of five species of reptile, and it is the best source of latest Jurassic marine reptiles in the world. Recent finds from several quarries show its continuing potential.

The Portland Beds of the Isle of Portland has been described by Damon (1884, pp. 79-97), H.B. Woodward (1895, pp. 196-202), Strahan (1898, pp. 60-71), Arkell (1933, pp. 492-7; 1947c, pp. 118-22), Cope (*in* Torrens, 1969a, pp. A53-A57), Cope and Wimbledon (1973), and Wimbledon (*in* Cope *et al.*, 1980b, pp. 88-9). Portland reptiles have been described by Owen (1842b, 1869, 1884b), Lydekker (1889a, 1889b, 1890a), Gaffney (1975a, 1976), McGowan (1976) and Brown (1981), and summarized by Delair (1958, 1959, 1960, 1966, 1992).

Figure 7.8 Some typical marine reptiles of Late Jurassic times in southern England. (A) The pliosaur *Liopleurodon*, one of the largest marine predators of all time at 12 m long. (B) The ichthyosaur *Ophthalmosaurus*, which was 2–4 m long. (C) The plesiosauroid *Cryptoclidus*, which was 4 m long. These animals occur typically in the Oxford Clay and Kimmeridge Clay faunas. Drawn by John G. Martin, copyright City of Bristol Museums and Art Gallery.

Figure 7.9 Nicodemus Nob on the eastern side of the Isle of Portland, showing the partly overgrown quarried clifline. Upper parts of the Portland sequence are exposed. (Photo: M.J. Benton.)

Figure 7.10 Broadcroft Quarries on the Isle of Portland, showing large blocks quarried for building stone. Fossil reptiles have been found in most of the inland and cliffline quarries. (Photo: M.J. Benton.)

A   B

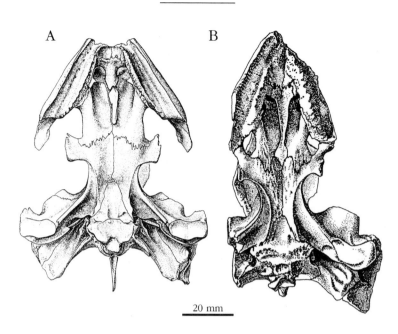

20 mm

Figure 7.11 Turtles from the Portlandian of the Isle of Portland. (A) *Plesiochelys planiceps* (Owen, 1842), skull in partially restored ventral view; (B) *Portlandemys mcdowelli* Gaffney, 1975, partial skull in ventral view. In both cases, the toothless jaws are directed to the top, and the palate and braincase extend to the bottom. The top of both skulls is missing. After Gaffney (1975a).

Description

The sequence, based on Arkell (1947c) and Wimbledon (*in* Cope *et al.*, 1980b) is:

Thickness (m)

Purbeck Limestone Formation
  'Lulworth Beds': algal limestone
    (caps) with dirt beds and plant
    remains, passing up into
    argillaceous and laminated
    limestones   *c.* 30
Portland Stone Formation
  Portland Freestone Member:
    Roach: cream-coloured oolite
      with moulds of bivalves,
      gastropods (northern half
      of island); Whit Bed Freestone:
      buff oolite; Curf and Chert:
      soft chalky micrite and
      micrite full of chert (northern
      half of island); Little Roach:
      shelly oolitic limestone;
      Base Bed Freestone: soft, white
      oolitic limestone   9
  Cherty Beds: limestones, predominantly
    micrites, with nodules and beds of
    chert, with giant ammonites   15

Thickness (m)

Basal Shell Bed: hard micrite with
  a rich fauna (bivalves, gastropods,
  ammonites, echinoids, bryozoa,
  fish and reptiles)   2.7
Portland Clay   4–6
Portland Sand
  West Weare Sandstones: brown
    and grey dolomites and sandy
    cementstones   15
  Cast Beds   1.3
  Exogyra Bed: 'stiff' marl/limestone
    packed with *Nanogyra nana*   2.5
  Upper Black Nore Beds: black
    silts/silty clays with lines of
    light grey limestone nodules   12
  Black Nore Sandstone: hard, black
    argillaceous sandstone with
    intensely hard concretions   2

Most specimen labels, and published descriptions, have little more locality and horizon data than 'Portland'. However, several authors have pointed to the Cherty Beds and the Basal Shell Bed as source horizons. Damon (1884, p. 86) noted 'Saurian remains' from the Cherty Beds of the Verne district (north-eastern part of the Isle) and a partial skeleton of a plesiosaur in the

Museum of the Royal Engineer's Office at The Verne Works. Cox (1925) recorded the plesiosaur *Cimoliasaurus portlandicus* in the Basal Shell Bed. Savage (1958) noted a specimen of *C. portlandicus* 'probably from the Whit or Base Bed Freestone'. Centra, probably of ichthyosaurs, are not uncommon in the West Weare Sandstones at the top of the Portland Sand (W.A. Wimbledon, pers. comm. to M.J.B., 1992)

Several recently collected specimens have locality data. A centrum of *C. portlandicus* (DORCM G.177) is labelled 'Little Beach, NE of the Verne', thus a specimen that has probably fallen from The Verne. A further partial plesiosaur skeleton (DORCM G.181) is labelled 'below coastguard station, E. side, H. tide level – SY 706722 – coll. 1966-8'. Some bone fragments (BGS(GSM) Zm 7700-40) have an associated sheet with a geological section and the words 'bones and shells' inscribed against the (?) Basal Shell Bed and Cherty 'Series'. Delair (1966, p. 61) mentions these specimens as coming from the lower part of the Portland Stone, '2 or 3 feet above the Portland Sand'. Dr J.N. Carreck (pers. comm. to M.J.B., 1982) lists isolated discoveries of bones from the following horizons: a dinosaur vertebra from the Whit Bed at Parkfield Quarry (?=Perryfield Quarries, SY 695712, or disused quarries, unnamed on 6-inch OS map at SY 692714, near Park Road; a specimen in Portland Museum, noted by Delair (1992), parts of a plesiosaur femur from the Whit Bed at Bottom Coombe Quarry (SY 694715), a goniopholid crocodilian tooth from the Skull Cap (Lower Purbeck) or top of the Portland Stone, a large part of a plesiosaur skeleton and a turtle cranium from the base Bed Freestone, and a large ?femur from the Lower Purbeck (Great Dirt Bed/Cap) of Wakeham Quarries (SY 698713). The present location of all but one of these specimens is unknown. Delair (1992) also notes 'characteristic megalosaurid metatarsals' on show in Portland Museum, from the Whit Bed of the Bath and Portland Stone Co.'s Quarry.

On the basis of these scraps of information, it may be concluded that most of the fossil reptiles from Portland came from the Portland Stone, and from a variety of localities. A selection of exposures is chosen for the GCR site since no particular quarry has proved to be the single main source.

Fauna

Fossil reptiles from Portland are preserved in several major collections. A list of species recorded is given, with repository numbers of type specimens. An indication is also noted of the number of specimens of each species preserved in major collections (especially BMNH, CAMSM, DORCM and OUM).

Numbers

Testudines: Cryptodira	
Plesiochelys planiceps (Owen, 1842)	
Type specimen: OUM J.1582	3
Pleurosternon portlandicum	
Lydekker, 1889	
Type specimen: BMNH 44807	1
Portlandemys mcdowelli	
Gaffney, 1975	
Type specimen: BMNH R2914	4
Archosauria: Dinosauria: Saurischia: Theropoda	
'Megalosaurid'	2
Archosauria: Dinosauria: Saurischia: Sauropoda	
Pelorosaurus sp.	1
Sauropterygia: Plesiosauria: Elasmosauridae	
Colymbosaurus portlandicus (Owen, 1869)	
Type specimen: BMNH 40640	33
Sauropterygia: Plesiosauria: Pliosauridae	
?Pliosaurus brachydeirus Owen, 1841	2
Ichthyopterygia: Ichthyosauria	
Macropterygius thyreospondylus (Owen, 1840)	2

Interpretation

The turtle remains from Portland have proved to be of some interest recently. *Plesiochelys planiceps* (Owen, 1842) was based on a single cranium and partial carapace (Owen, 1842b, pp. 168–70). It was a relatively large animal (skull 90 mm long) with the temporal fossa completely roofed by postfrontal and parietal bones and a deep notch immediately behind the maxilla (Figure 7.11A). Owen (1884b, vol. 2, pl. 8, figs 1–2) figured the skull without further description. Lydekker (1889b, pp. 232–3) erected the new genus *Stegochelys* for this form. The type specimen was mentioned as lost by Lydekker (1889b) and Parsons and Williams (1961), but Delair (1958, p. 55) located it, and Gaffney (1975a, 1976) redescribed it as *Plesiochelys*. Parsons and Williams (1961) tentatively referred some further turtle skulls from the Isle of Portland to *P. planiceps*. Gaffney (1975a, 1976) further described these specimens, and named them *Portlandemys mcdowelli* (Figure 7.11B). Gaffney (1975a, 1975b, 1976, 1979a, p. 281) ascribed the genera *Plesiochelys* and *Portlandemys* to the Plesiochelyidae, a family that he places in the Chelonioidea (living and extinct marine turtles).

The third Portland turtle is *Pleurosternon portlandicus* Lydekker, 1889b (pp. 215-16), a species based on a carapace which is only 250 mm long. Młynarski (1976, p. 120) mentioned the species briefly and referred the family Pleurosternidae to Testudines *inc. sed.*

Dinosaurs from the Isle of Portland are represented by some metatarsals and a damaged vertebra of a 'megalosaurid' (Delair, 1992), and by a tooth named *Ornithopsis* sp. by Delair (1959, p. 83). *Ornithopsis* is currently ascribed to *Pelorosaurus* (Steel, 1970, p. 68).

Pliosaurus portlandicus Owen, 1869 (pp. 8-12) was described on the basis of a right hind paddle with a 370 mm long femur. Lydekker (1889a, pp. 227-30; 1890a, pp. 274-5) ascribed *P. portlandicus* to *Cimoliasaurus* and listed many specimens of vertebrae and limb bones. Brown (1981, pp. 314-17, 324) synonymized *C. portlandicus* with many other Late Jurassic plesiosaurs as *Colymbosaurus trochanterius* (Owen, 1840). Savage (1958) reported a recent find of a femur of *C. portlandicus*. Lydekker (1890a, pp. 271-2) and Delair (1959, p. 70) note the head of an ischium and the distal portion of a propodial (BMNH R1679, R.1680) from Portland as *Pliosaurus brachydeirus*, but Tarlo (1960), in a review of British Late Jurassic pliosaurs, does not comment on these.

The ichthyosaur *Macropterygius thyreospondylus* is represented by a caudal vertebra from the 'Portland Oolite' (BMNH R1684; Delair, 1960, pp. 66-7). McGowan (1976) considered this species a *taxon dubium* because of the 'inadequate' type material, but it may be valid (A. Kirton, pers. comm. to M.J.B., 1982).

Comparison with other localities

Reptiles are relatively rare in the Portland Beds of Britain (listed near the beginning of the chapter). Of the turtles from the Isle of Portland, the genus *Plesiochelys* is known also from the Kimmeridge Clay of England, Switzerland, Bavaria and Hanover, the Portlandian (?) of eastern France, the Purbeck of Durlston Bay, the 'Upper Jurassic' of China, and the Wealden of the Isle of Wight and Sussex (Lydekker, 1889b; Gaffney, 1975a; Młynarski, 1976, pp. 55-7). *Portlandemys* is unique to the Isle of Portland. *Pleurosternon* is also known from the Purbeck Beds of Swanage (Lydekker, 1889b), and elsewhere in the Late Jurassic and Early Cretaceous of western Europe and Asia (?) (Młynarski, 1976, p. 120). *Colymbosaurus portlandicus* is unique to the Portlandian of England but, if synonymized as *C. trochanterius*, it is well known from many localities in the Kimmeridge Clay also (Brown, 1981). Delair (1966, pp. 66-7) regarded the Portland *M. thyreospondylus* as the only English Portlandian ichthyosaur, although there are some ichthyosaur vertebrae from Swindon (OUM J.1585-6; BGS(GSM) old number). *M. thyreospondylus* occurs in the Portlandian of the region of Boulogne (Huene, 1922, p. 91). The Portlandian fauna from the Isle of Portland is similar in many ways to the preceding Kimmeridgian faunas, but overall species diversity seems reduced, and turtles are perhaps relatively a little more diverse.

Conclusions

The Isle of Portland has yielded the best faunas of marine Portlandian reptiles in the world. Other marine faunas of this age are known from southern England, but the range of material is less. Better known faunas from elsewhere in the world (e.g. Morrison Formation, USA; Tendaguru, Tanzania) are dominated by terrestrial forms such as dinosaurs. The turtles from Portland include good skull material, and have formed the basis of recent reviews of early turtle anatomy and taxonomy. The plesiosaur material is good, and appears to have closest affinities with Kimmeridgian species. The sites include the best sources for marine reptiles of latest Jurassic age anywhere in the world and, with their continuing potential for new finds, they therefore have high conservation value.

BUGLE PIT, HARTWELL, BUCKINGHAMSHIRE (SP 793121)

Highlights

Bugle Pit, near Aylesbury is the source of a small number of important reptile specimens. These include teeth of a variety of dinosaurs, which are particularly important since most other Late Jurassic sites in England are marine.

Introduction

The Bugle Pit, Hartwell has produced teeth of megalosaur and sauropod dinosaurs and some other reptile species, and it has been the best source of dinosaur remains in the British Portlandian. The Bugle Pit is now filled in, but a new section was exposed a short distance to the south by the NCC in 1984 (Radley, 1991), which only shows the top of the Portland Stone and the Purbeck Beds, but it has potential to produce more finds after further excavation.

The Bugle Pit, named after the Bugle Horn Inn nearby, was first mentioned by Morris (1856). Geological sections have been published by at least 20 authors, including H.B. Woodward (1895, pp. 223–4), Arkell (1947a, p. 126), Barker (1966), Wimbledon (*in* Cope *et al.*, 1980b) and Radley (1991). The reptile remains were described by Hudlestone (1887), Lydekker (1893a) and A.S. Woodward (1895).

Description

The section after Barker (1966) is as follows:

Thickness (m)

Purbeck Limestone Formation
   ?anguiformis Zone
     BP 19–21. Fine-grained limestone
       and grey marl   2.08
     BP 10–18. Marls, grey and greenish
       and bands of pale earthy
       limestone; dark clay at base;
       fishes and ostracods in lower
       units   2.93
     BP 9. Laminated, blue-hearted
       cementstone with plant, insect
       and fish remains along partings   0.23–0.25
Portland Stone
   kerberus Zone
   Creamy Limestones
     BP 8. Tough, highly bituminous,
       shaly marl with large oysters and
       other bivalve casts   0.20
     BP 7. Hard, fine-grained limestone with
       a band of trigoniid casts at the base   0.30

Thickness (m)

BP 6. Marly shales and black shale
   with a layer of bivalves near the
   base   0.15–0.20
BP 5. Blue-hearted, marly limestone
   with large bivalves   0.76
BP 4. Brown clay with serpulids   0.08
BP 3. Blue-hearted, rather soft,
   marly limestone, trigoniids etc.   0.91
BP 2. Hard, blue-hearted limestone
   with oysters, bottom 0.07 m fossil
   casts   0.60
Crendon Sand
BP 1. Yellow-brown sand   seen to 0.23

Woodward (1895) and Arkell (1947a) continued the section down to the Upper Lydite Bed (now probably *glaucolithus* Zone), but these lower units have not been seen in Bugle Pit itself.

There are no records of the units from which the reptiles came, but the Purbeck facies seem to be the most likely source (Radley, 1991, p. 242): fishes occur in beds BP 9, 11a, 11b, 11c (H.B. Woodward, 1895; Barker, 1966). The ostracod record (Barker, 1966) indicates a predominance of freshwater and euryhaline conditions of deposition in beds BP 9–12, 14–21 (Purbeck Limestone Formation).

There is an old quarry at Stone (SP 783121), close to Hartwell, and there may have been small pits in the grounds of Hartwell House. However, there is no evidence of other Purbeck quarries in the immediate district and it seems likely that any Purbeck or Portland fossils labelled 'Hartwell' came from the once extensive Bugle Pit.

Fauna

The dinosaur teeth have been described by Lydekker (1893a) and Woodward (1895). These specimens were donated by their collector, J. Alstone, to the BMNH. Several reptile specimens in BUCCM probably came from the Bugle Pit. Delair (pers. comm., 1982) has a list of many reptiles from Bugle Pit.

Testudines: Cryptodira
   'Turtle'
     BUCCM Lee Coll. 3948
Archosauria: Dinosauria: Saurischia: Theropoda
   Megalosaurus sp.
     BMNH R2566, R2567, R2821 (teeth)

Archosauria: Dinosauria: Saurischia: Sauropoda
　Pelorosaurus sp.
　　BMNH R2004, R2005, R2565 (teeth)
Archosauria: Dinosauria: Ornithischia:
　Ornithopoda
　Iguanodon sp.
　　BUCCM 467.22
Archosauria: Dinosauria: Ornithischia: Stegosauria
　?Stegosaur
　　BUCCM 9.43

Interpretation

The environment of deposition of the Portland Stone is essentially marine: the Creamy Limestones are marginal marine shelly limestones and mudstones that shallow upwards. The overlying Purbeck facies are restricted marine and non-marine marls, clays and fine-grained limestones, with evidence for two erosive episodes that resulted from emergence (Radley, 1991).

The turtle is represented by a poor scapula and is unidentifiable. It is labelled 'Purbeck Hartwell' and probably came from the Bugle Pit.

Hudleston (1887) recorded dinosaur bones from the Bugle Pit. Lydekker (1893a) described two sauropod teeth, a large crown with part of the root (55 mm long), and a smaller crown of similar appearance (25 mm long). The teeth are broad (35 and 15 mm respectively) and spatulate in shape. The slightly concave inner surface has rugose enamel and a mid-line ridge, and there is an offset ridge on the outer surface. Lydekker (1893a) compared these teeth with some very similar specimens from the Portlandian of Boulogne-sur-Mer. Lydekker (1890a, p. 241) had ascribed these to *Pelorosaurus humerocristatus* (Hulke, 1874), a species which had been based on a humerus from the Kimmeridge Clay of Smallmouth Sands, Weymouth. Earlier Lydekker (1888a, pp. 151–2; 1890a, p. 241) had ascribed a partial pubis, a fibula, a tibia, a phalanx and a caudal vertebra to this species. Lydekker's assignment of the Bugle Pit teeth to this species may have been the presence of similar limb bones and teeth in the Wealden (*Pelorosaurus conybeari* Mantell, 1850). A.S. Woodward (1895) reported one more sauropod tooth 'of the same animal'.

Two teeth from the Kimmeridgian of Portugal had been assigned to *P. humerocristatus*, but the assignment is doubtful (Steel, 1970, p. 70). Other similar teeth, now ascribed to species of *Pelorosaurus* (Steel, 1970, p. 70) include:

Oplosaurus armatus Gervais, 1852 (=*Hoplosaurus armatus*, Lydekker, 1890a, p. 243) (BMNH R964; Wealden, Isle of Wight), *Ornithopsis hulkei* Seeley, 1870 (BMNH R751, R964; Lydekker, 1888a, pp. 146–8; Wealden, Isle of Wight), *Pelorosaurus conybeari* Mantell, 1850 (Lydekker, 1890a, pp. 240–1; Wealden, Kent); *Pelorosaurus (Iguanodon) precursor* (Sauvage, 1876) (Kimmeridgian of Wimille, near Boulogne-sur-Mer; 'Lusitanian' (Oxfordian–Kimmeridgian) of Ourem, Portugal).

The megalosaur teeth were described by A.S. Woodward (1895). Two of the specimens differ merely in size; both are high-crowned, compressed only on the posterior margin, which is clearly serrated, and without serrations on the anterior part, which is distinctively worn. The third specimen is shorter, broader and more laterally compressed. The tooth is less worn and serrations occur on both posterior and anterior borders, although only on the upper third of the latter. The teeth are 36, 30 and 31 mm long, respectively and 14, 12 and 16 mm broad. Woodward (1895) did not attempt to identify the species represented, and characterized them simply as 'megalosaurian'.

Ornithischian dinosaurs are represented by an *Iguanodon* toe bone (BUCCM 467.22) from 'Portland Stone of Mr Lee's pit, Hartwell', and a possible stegosaur (BUCCM 9.43) from 'Kimmeridge Clay of Bugle Pit, Hartwell'. The latter specimen (13 fragments of limb bones and other elements) cannot have the exact provenance stated. The Kimmeridgian is not represented in the Bugle Pit. The nearest site in the Hartwell Clay is Lockes' Pit, Hartwell (SP 805125), over 1 km to the north-east.

Comparison with other localities

Several localities in the Hartwell Clay (Upper Kimmeridgian, *pallasioides* Zone), have yielded tetrapod faunas that are comparable to those from Bugle Pit (e.g. *Megalosaurus*, Lydekker, 1893a; *Hoplosaurus*, Woodward, 1895a); these include Lockes Pit, Hartwell (1 km north-east of the Bugle Pit, filled in; basal Glauconitic Beds); Ward and Cannons, ?Beirton (extant) (basal Glauconite Beds, Portland facies well represented); and Websters and Cannon's (Hill's) Pit on Bierton Road, Aylesbury (overgrown).

In the course of discussion of the sauropod teeth, comparable specimens have been noted from the Kimmeridgian of Ourem, Portugal and of

Wimille and Boulogne-sur-Mer, France. The Wimille (Mont-Rouge) quarries, with Late Portland (=Portland Stone) equivalents, termed 'Wealden' (decalcified Portland sands, or equivalent to Upper–Middle Purbeck?) yielded bones of sharks, bony fishes, the turtles *Plesiochelys* and *Tropidemys*, an elasmosaurid tooth, teeth of the crocodilians *Steneosaurus, Goniopholis, Theriosuchus* and *Bernissartia*, a pterosaur and isolated dinosaur teeth (theropod, sauropod, ornithopod and nodosaurids) in a recent dig (Cuny *et al.*, 1991). Others have come from the Wealden of the Isle of Wight and the Weald. The Bugle Pit specimens appear to be the only Portlandian examples of *Pelorosaurus* known, and one of only a few European sauropods of this age. Other sauropods of Kimmeridgian to Portlandian age come from the Morrison Formation of the western United States (Kimmeridgian–Portlandian); *Haplocanthosaurus, Brachiosaurus, Camarasaurus, Apatosaurus, Diplodocus, Barosaurus* (Steel, 1970).

'*Megalosaurus*' is known from the Purbeck of Durlston Bay (*M. (Nuthetes) destructor*, Owen, 1854) and Swindon, and the Kimmeridgian of several sites near Boulogne-sur-Mer and Cap de la Hève, in France, from Pembal, in Portugal, and Foxhangers, Wiltshire as well as the Portlandian of Boulogne-sur-Mer (*M. insignis* Deslongchamps, 1870) (Steel, 1970). Other Portlandian carnosaurs include *Elaphrosaurus* from the Morrison Formation.

In the English Portlandian, dinosaur remains are known sparsely from the Isle of Portland, Dorset (see above), source of an '?*Ornithopsis* sp.' tooth and possible megalosaur remains, and from Garsington, Oxfordshire, source of caudal vertebrae named *Cetiosaurus longus* by Owen (1841), and referred to by Phillips (1871, p. 390; OUM, specimens not found). Recent digs at Chicksgrove Quarry, Wiltshire have also produced a variety of dinosaurian remains and teeth of 12 crocodilian, dinosaurian and pterosaurian taxa (W.A. Wimbledon, pers. comm. to M.J.B., 1992).

Conclusions

Bugle Pit, Hartwell has yielded a sparse, but important, dinosaur fauna. Portlandian dinosaur remains are rare in Britain and many are controversial or poorly identified. The sauropod teeth are some of the few recorded Portlandian sauropods from Europe, and they will be useful in comparisons with North American sauropods of the same age. The megalosaur and ornithischian teeth are also some of the few known from rocks of this age. Thus this small but significant reptile fauna and the potential for further discoveries with re-excavation gives the site its conservation value.

DURLSTON BAY, DORSET (SZ 035772 – SZ 039786)

Highlights

Durlston Bay is one of Britain's richest fossil reptile sites, the source of over 40 species of reptiles living in the earliest Cretaceous (Figure 7.12). The small reptiles are especially important, and Durlston Bay is unique worldwide for its diverse early lizards, its turtles, its small crocodilians and its pterosaurs.

Introduction

The mile stretch of coastal sea cliffs between Peveril Point and Durlston Head displays the finest sections of the Purbeck Limestone Formation in Britain which comprise the type locality for the formation (Figures 7.13 and 7.14). The Purbeck Limestone Formation here is famous for its exceptionally diverse fauna, which includes mammal remains unique to the Early Cretaceous of Britain (to be dealt with in a subsequent volume in the GCR series). The Durlston Bay section is also well known for its reptiles (Figures 7.15 and 7.16) and is arguably Britain's most important fossil reptile site. The reptile fauna is large and diverse, containing abundant lizards, turtles, crocodilians, pterosaurs and dinosaurs (36 species; 29 type species). The small size of many of the animals, and their fine preservation, give this locality a unique position in comparison with other reptile faunas of the same age worldwide.

Most of the reptiles from Durlston have been obtained from the natural cliff exposures, but some remains, especially the turtles, came largely from underground stone workings, the products of now extinct quarrying operations. Although the latter source for reptiles is no longer available, the extensive cliff exposures (Figure 7.13) have continued to yield new specimens. Many finds were made by Beckles in the course of his

Figure 7.12 The latest Jurassic and earliest Cretaceous sequences in Durlston Bay, showing marine limestones in the northern part of the section. (Photo: J.L. Wright.)

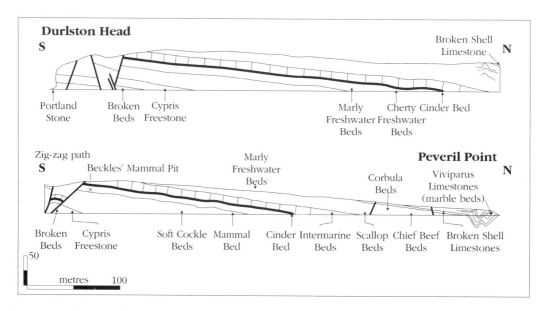

Figure 7.13 Cliff profiles of Durlston Bay showing the type section of the Durlston Beds (after Strahan, 1898).

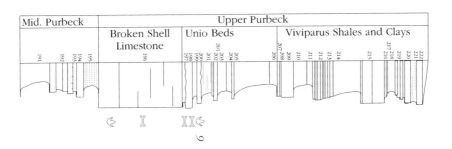

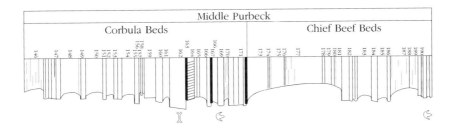

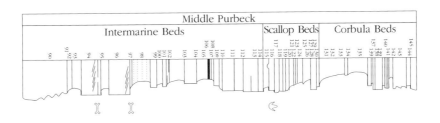

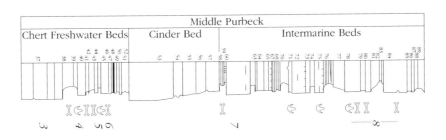

Durlston Bay North - Main Section

1 metre

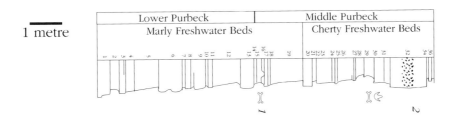

Figure 7.14 Sedimentary log of the reptile-bearing units at Durlston Bay. Bone and footprint symbols indicate fossiliferous horizons. Supplied by W.A. Wimbledon.

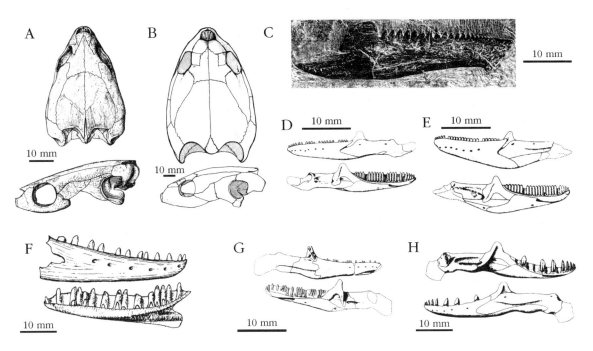

Figure 7.15 A small selection of the Purbeck menagerie from Durlston Bay: turtles, sphenodontid, and lizards. (A) The skull of the cryptodire turtle *Mesochelys durlstonensis* Evans and Kemp, 1975, in dorsal and lateral views; (B) the skull of the cryptodire turtle *Dorsetochelys delairi* Evans and Kemp, 1976, in dorsal and lateral views; (C) the sphenodontid *?Homoeosaurus*, partial left lower jaw; (D) the lizard *Paramacellodus oweni* Hoffstetter, 1967, left lower jaw in lateral and medial views; (E) the lizard *Becklesius hoffstetteri* (Seiffert, 1973), left lower jaw in lateral and medial views; (F) the lizard *Saurillus obtusus* Owen, 1854, anterior end of right lower jaw in lateral and medial views; (G) the lizard *Pseudosaurillus becklesi* Hoffstetter, 1967, right lower jaw in lateral and medial views; (H) the lizard *Dorsetisaurus purbeckensis* Hoffstetter, 1967, left lower jaw in lateral and medial views. (A) After Evans and Kemp (1975); (B) after Evans and Kemp (1976); (C) after Boulenger (1891); (D)–(H) after Estes (1983), based on various sources.

excavation of the cliff in an area, just north of the Zigzag Path, that is well south of the outcrop of the Mammal Bed at beach level. Reports of the Durlston Bay vertebrates include Owen (1842b, 1853, 1854, 1855a, 1861b, 1871, 1874b, 1878a, 1878b, 1879a, 1879b), Mantell (1844), Owen and Bell (1849), Seeley (1869a, 1875a, 1893a, 1893b), Lydekker (1888a, 1889b), Boulenger (1891), Watson (1911b), Andrews (1913), Huene (1926), Nopcsa (1928), Hoffstetter (1967), Joffe (1967), Delair (1969b), Seiffert (1973), Evans and Kemp (1975, 1976), Gaffney (1976, 1979b), Galton (1978, 1981a), Estes (1983) and Howse (1986). Dinosaur footprints have been reported recently from Durlston Bay (Ensom, 1983, 1984b, 1985b, 1987c, 1987d; Nunn, 1990), as well as a new sphenodontid jaw bone (Evans, 1992c).

The Purbeck section at Durlston Bay has been described with varying degrees of accuracy by many authors (Austen, 1852, pp. 9-16; Bristow and Fisher, 1857, pp. 245-54; Strahan, 1898, pp. 91-6; Arkell, 1933, pp. 521-9; Clements, *in* Torrens, 1969a, figs A35-7; 1993). Cope and Clements (*in* Torrens, 1969a, pp. A57-A64) and Macfadyen (1970, pp. 134-52) give details of the history of research on the stratigraphy and palaeontology of the Durlston section. Parts of the coastline were formerly mined and quarried opencast for building stone, paving flagstone and gypsum.

Description

The section of the Purbeck Limestone Formation at Durlston Bay (Figures 7.13 and 7.14) is based on Wimbledon and Hunt (1983, p. 270, fig. 2) and new data from W.A. Wimbledon (pers. comm. to M.J.B., 1992), and using the numbering scheme of Clements (*in* Torrens, 1969a), where bed numbers are prefixed DB. A detailed map of the northern part of Durlston Bay, showing the occur-

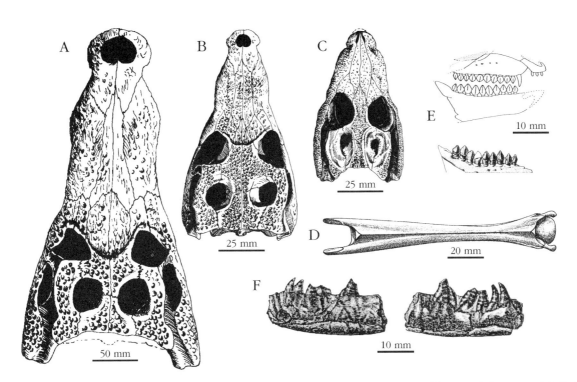

Figure 7.16 A small selection of the Purbeck menagerie from Durlston Bay: crocodilians and dinosaurs. (A) Skull of the crocodilian *Goniopholis simus* Owen, 1878, in dorsal view; (B) skull of the crocodilian *Nannosuchus gracilidens* Owen, 1879, in dorsal view; (C) skull of the crocodilian *Theriosuchus pusillus* Owen, 1879, in dorsal view; (D) elongate cervical vertebra of the pterosaur *Doratorhynchus validus* Owen, 1870, dorsal view; (E) the ornithopod dinosaur *Echinodon becklesi* Owen, 1861, partially restored snout region, and detail of lower jaw; (F) jaw fragment of the theropod dinosaur *Nuthetes pusillus* Owen, 1854, in lateral and medial views. (A)–(C) After Joffe (1967); (D) after Howse (1986); (E) after Galton (1978); (F) after Owen (1854).

rence of numbered Purbeck beds, has been published by Nunn (1992).

The Purbeck Limestone Formation is generally taken to span the Jurassic-Cretaceous (Portlandian–Berriasian) boundary, with the base of the Cretaceous taken in the *Cypris* Freestone, quite low in the formation (Allen and Wimbledon, 1992). Ammonites have not been found in these beds, and the stratigraphy is based on palynomorphs, ostracods and gastropods; the position for the boundary between the Cretaceous and Jurassic has been in dispute, but an integrated approach to correlation has made possible the positioning of the base of the Berriasian in the section.

Fossil reptiles have been recorded from several levels in the section (Figure 7.14) between the Mammal Bed (14 and 16) at the base of the Middle Purbeck beds and the *Unio* Beds and above (197+) in the Upper Purbeck beds:

197–224 *Unio* Beds, Upper *Cypris* Clays and Shales: Delair (1966, p. 60) noted specimens of the crocodile *Goniopholis* (BGS(GSM) Zm.7702) and of turtles (BGS(GSM) Zm. 7703–4) 'in and just above the *Unio* Bed at Peveril Point'.

200 (Bed 6 of Austen; Bed 81 of Bristow; bed 221 of Clements) Crocodile Bed: 'teeth of crocodile'; plants, coprolites, fish, turtles, crocodiles (*Goniopholis*).

196 (Bed 9 of Austen; Bed 78 of Bristow; bed 220 of Clements) Broken Shell Limestone Member (Soft Burr): fishes and turtles. Ensom (1983) noted dinosaur footprints in this bed.

191 in the Chief Beef Beds: Ensom (1983) noted dinosaur footprints from this horizon.

131–174 (Beds 24–44 (in part) of Austen; Beds 59–70 of Bristow; beds 154–174 of Clements); *Corbula* Beds: insects, fishes, turtles and footprints (West and El-Shahat, 1985).

58–114 Intermarine Beds (Beds 45–70, Turtle Beds of Austen; Beds 45–57, Intermarine Beds of Bristow; beds 112–144 of Clements) Upper Building Stones: DB133 (Bed 52 of Austen; Bed 54 of Bristow, Red Rag) yields fishes, turtles and coprolites. Beds ?78-80 (Bed 61 of Austen; Bed 50d of Bristow; bed 124 of Clements) in the Roach (Freestone Quarry) includes the pink bed with reptile footprints. Bed 61 (Bed 69 of Austen; Bed 45d of Bristow; bed 113 of Clements) contains remains of fishes, fresh-water tortoises (*Pleurosternon*), pterosaurs and crocodiles. Evans and Kemp (1975) ascribe the type specimen of *Mesochelys durlstonensis* to the '?Upper Building Stones'. Most of the larger crocodiles described by Owen (1878b) and the turtles (Owen, 1853) apparently come from the massive limestones of the Upper Building Stones. Ensom (1985b, 1987c) and Nunn (1990) note dinosaur footprints from beds 71, 74, 75, 78, and 96 (but see Figure 7.14 herein).

Beds 20-52 (Beds 72-88 of Austen, Beds 25-42 of Bristow). Cherty Freshwater Beds: Austen (1852) notes 'bones' in his beds 72 (=51) and 81 (=37) and 'turtle' in his beds 83 and 84 (=30-32). The Feather Bed (Bed 74 of Austen, Bed 40 of Bristow, bed 108 of Clements =45-48) yielded postcranial remains of the dinosaur Nuthetes (Owen, 1854, 1878a; Delair, 1959, p. 80), granicones (Owen, 1878a, 1879b), and the 'dwarf' crocodilians *Nannosuchus* and *Theriosuchus* (Owen, 1879a, 1879b; Joffe, 1967). Owen (1879b) also referred the small crocodilians *Goniopholis tenuidens, Oweniasuchus major, O. minor* and *Nannosuchus gracilidens* to the Feather Bed. However, Owen (1861b) stated that the type *Nuthetes* jaw came from Bed 93 of Austen (1852), namely the Mammal Bed (bed 14-16). The Under Feather (Bed 76 of Austen; Bed 38 of Bristow; 106 of Clements =43) yielded *Iguanodon hoggi* Owen, 1874b (p. 3). Ensom (1984b) reported dinosaur footprints from DB94, exposed south of the point where the Cinder Bed is present at shore level (SZ 03607835). A related pod of limestone (DORCM G7261) was discovered in the carbonaceous shale below (DB93). Ensom (1984b) noted isolated impressions on the base of DB103 and DB102 (Clements 1969).

The Mammal Bed ('Dirt Bed') of Beckles' excavations (Bed 93 of Austen; Bed 22 of Bristow; bed 83 of Clements): always equated with bed 14-16 (Figure 7.14) of the shore section, has yielded plant remains, ostracods, gastropods, bivalves, lizards (Macellodus, *Saurillus* etc; Owen, 1854, 1855a, 1861b; Hoffstetter, 1967), dinosaurs (*Echinodon*; Owen, 1861b; Galton, 1978; *Nuthetes*, type jaw; Owen, 1861b), and mammals (18 species). Certain authors (e.g. H.B. Woodward, 1895, p. 251; Macfadyen, 1970, p.

137) referred the dwarf crocodilians to the Mammal Bed, and such remains are abundant from about same level in the southern half of the bay (*fide* W. A. Wimbledon, 1993).

Fauna

Many thousands of identifiable reptile specimens have been collected from Durlston Bay, and there is no point in attempting to list them all. About 41 species have so far been recognized. Collections in the following institutions were examined: BMNH, BGS(GSM), CAMMZ, CAMSM, DORCM and OUM. Many specimens are labelled merely 'Swanage', and they could have come from some of the inland quarries. Type specimens are indicated, as well as an estimate on the numbers of specimens of each species in major British collections. The numbers are probably rather too high because of the lack of recent reviews of most groups and, in the case of the lizards, because of the existence of such recent reviews!

Numbers

Testudines: Cryptodira: Pleurosternidae	
Mesochelys durlstonensis Evans and Kemp, 1975	
Holotype: CAMMZ T.1041	1
Pleurosternon bullocki (Owen, 1842)	
Type specimen: BMNH R911	70
Pleurosternon sp.	47
Tretosternon punctatum Owen, 1842	
Type specimen: lost	16
Testudines: Cryptodira: Plesiochelyidae	
Plesiochelys belli (Mantell, 1844)	2
Plesiochelys emarginata Owen, 1853	
Type specimen: DORCM G.16/ BMNH 46317(?)	7
Plesiochelys latiscutata (Owen, 1853)	
Type specimen: DORCM G.20	3
Plesiochelys sollasi Nopsca, 1928	
Type specimen: OUM J.13796	1
Plesiochelys sp.	13
Testudines: Cryptodira: *inc. sed.*	
Dorsetochelys delairi Evans and Kemp, 1976	
Holotype: DORCM G.23	1
'Chelonian indet.'	17
Testudines: Pleurodira	
Platychelys ?anglica Lydekker, 1889	
Type specimen: BMNH 48357	1

Durlston Bay

Interpretation

Turtles are the commonest remains from Durlston. A 'petrified tortoise' was recorded in 1809 (Anon., 1809). Owen (1842b), and Owen and Bell (1849, pp. 62–6) described *Platemys bullocki* on the basis of a turtle supposedly from the London Clay of Sheppey. It was later shown to have come from Durlston (Lydekker, 1889b, p. 209). Meanwhile, Owen (1853, pp. 1–9) erected the genus *Pleurosternon*, with the new species *P. concinnum*, *P. emarginatum*, *P. ovatum* and *P. latiscutatum*, all from Durlston. The first three belong to *P. bullocki* (although *P. emarginatum* only in part; Lydekker, 1889b, pp. 206–15), as do four invalid species named by Seeley (1869a, pp. 86–8). *P. bullocki* was a medium-sized (400 mm long), probably freshwater, turtle. The carapace was oval and relatively low. The relationships of the Pleurosternidae are uncertain because of the general absence of skull material (Gaffney, 1975b; Młynarski, 1976). Gaffney and Meylan (1988) regard the family as the basal cryptodires in their cladogram.

Owen (1842b, pp. 165-7) also described *Tretosternon punctatum* on the basis of carapaces from the Purbeck Limestone Formation of Durlston and from the Wealden. The type specimens were lost, but a few others were identified (Lydekker, 1889b, pp. 141-3). *Tretosternon* was a moderate-sized form with a thick sculptured armour and a very flat, broad carapace about 500 mm long. It has been placed in the Pleurosternidae (Romer, 1956) and the Dermatemydidae (Cryptodira), but a restudy is required (Młynarski, 1976).

Several species of *Plesiochelys* have been described from Durlston: *P. latiscutata* (Owen, 1853), *P. emarginata* (Owen, 1853, in part), *P. belli* (Mantell, 1844) (see Lydekker, 1889b, p. 194) and *P. sollasi* Nopsca, 1928. *Plesiochelys* is a thick-shelled form, with a low carapace which is round to oval in outline. The limbs are adapted for aquatic and terrestrial life. Most of the species are defined on characters of the carapace, but their general relationships are not certain. All are moderate in size: *H. latiscutata* was 400 mm long, *H. emarginata* some 500 mm and *H. sollasi* had a carapace about 450 mm long and about 455 mm wide. Romer (1956) classed *Plesiochelys* in the Plesiochelyidae, with *Pleurosternon* in the Amphichelydia. Gaffney (1975a, 1975b, 1976) argued for the abolition of the Amphichelydia and placed the Plesiochelyidae in the Chelonioidea, a group of cryptodires (the large suborder of turtles that withdraw their heads in a vertical plane), but without reference to the Purbeck species. On the other hand, Młynarski (1976) associated the Plesiochelyidae with the Dermatemydidae (a group of primitive cryptodires). Gaffney and Meylan (1988) place the Plesiochelyidae in their Eucryptodira, between the Baenidae and the more derived cryptodires.

Platychelys (?) *anglica* Lydekker, 1889b (pp. 217-18) was a small turtle; the species was erected on a single carapace from Durlston. *Platychelys* is classed in the second turtle suborder, the Pleurodira (side-necked turtles) (Gaffney, 1975b; Młynarski, 1976; Gaffney and Meylan, 1988), but Lydekker's assignment of such a poor specimen may be incorrect.

Two recently described turtles are important. *Mesochelys durlstonensis* Evans and Kemp, 1975 is based on an excellent skull (Figure 7.15A) and partial skeleton. Evans and Kemp (1975) suggested that *Mesochelys* was related to the North American Late Jurassic *Glyptops*, a primitive cryptodire. The second specimen, *Dorsetochelys delairi* Evans and Kemp, 1976, also a cryptodire, is based on a good skull (Figure 7.15B). Evans and Kemp (1976) considered that it represented a group related to both Glyptopsidae and Baenidae. Gaffney (1979b) stressed the primitive nature of both genera and their importance in the classification of cryptodire turtles. *Mesochelys* is placed by Gaffney and Meylan (1988) in the Pleurosternidae, the basal cryptodire family.

A sphenodontid rhynchocephalian is represented only by three jaw fragments (Boulenger, 1891; Delair, 1960, pp. 77-8, Evans, 1992c) (BMNH R1765, R4808, DORCM G10831). These show the characteristic triangular acrodont teeth and squared-off symphysis found in the extant *Sphenodon*, and the new specimen has been ascribed to *Opisthias*, a sphenodontid known from the Late Jurassic of North America (Evans, 1992c). The earlier specimens (Figure 7.15C) are assigned to *Homoeosaurus* (Boulenger, 1891), a genus better known from the Late Jurassic of France and Germany, and a comparison of all the material is required.

The lizards from Durlston Bay are of particular significance in representing some of the earliest known types. Owen (1854) described several jaw fragments as *Macellodus brodiei*. He also referred some associated dermal scutes to *Macellodus*, but these probably pertained to a dwarf crocodile. Owen (1855a, 1861b) then described further jaws as *Saurillus obtusus*, but Lydekker (1888a, p. 289) synonymized this taxon with *Macellodus*.

The NHM, London later acquired the Beckles collection of 170 lizard specimens from Durlston Bay. Hoffstetter (1967) restudied these and erected five new genera and seven new species. He could not locate the type specimens of *Macellodus* and *Saurillus*, but considered that they were quite distinct. The new species were nearly all based on dentary or maxilla fragments, and the diagnostic characters were based on jaw shape and tooth morphology. The seven genera of Purbeck lizards recognized by Hoffstetter (1967) are: *Macellodus* and *Paramacellodus* (jaws 25 mm long; teeth tubular, peg-like, with rounded ends), *Saurillus* and *Pseudosaurillus* (jaws 12-25 mm long, teeth peg-like and pointed),

Becklesisaurus (jaw 40 mm long, teeth peg-like, with rounded ends), *Durotrigia* (poorly known, teeth with multiple points) and *Dorsetisaurus* (jaw 40 mm long; teeth flattened, leaf-shaped and pointed). Most of the genera are represented by assorted skull bones, vertebrae and limb bones in addition to jaws, but there is not sufficient to reconstruct a complete skull or skeleton in any specimen. The lizards are referred to the extant groups Scincomorpha and Anguimorpha.

Seiffert (1973) reviewed Hoffstetter's (1967) work when he described new lizards from the Oxfordian of Guimarota, Portugal. He noted that Hoffstetter's (1967) interpretation of *Macellodus* differs from Owen's (1854, 1861b): Owen (1854) clearly showed the teeth as compressed, spade-shaped, with striations, and 8–10 mm wide, whereas Hoffstetter's (1967) neotype has peg-like rounded teeth without striations, and is 2 mm wide. Seiffert (1973) referred the '*Macellodus*' material of Hoffstetter (1967) to a new form, *Becklesisaurus hoffstetteri* Seiffert, 1973, and noted that '*Macellodus*' differs from *Becklesisaurus* only in the size of the jaws. Seiffert (1973) also expressed doubt about the validity of some other Hoffstetter taxa. Estes (1983) supported these views and erected the family Paramacellodidae for some taxa, and renamed others in Hoffstetter's (1967) Dorsetisauridae. Estes' (1983) list of taxa is as follows:

Family Paramacellodidae Estes, 1983
  Paramacellodus oweni Hoffstetter, 1967
    (Figure 7.15D)
    (=*Saurillus robustideus, Becklesisaurus scincoides*).
  Becklesius hoffstetteri (Seiffert, 1973) (Figure 7.15E)
    (=*Macellodus brodiei* of Hoffstetter 1967)
  Saurillus obtusus Owen, 1854 (Figure 7.15F)
  Pseudosaurillus becklesi Hoffstetter, 1967
    (Figure 7.15G)
  Pseudosaurillus sp.
    (=*Saurillus obtusus* of Hoffstetter, 1967)
Family Dorsetisauridae Hoffstetter, 1967
  Dorsetisaurus purbeckensis Hoffstetter, 1967
    (Figure 7.15H)
  D. hebetidens Hoffstetter, 1967
Sauria *incertae sedis*
  Durotrigia triconodeas Hoffstetter, 1967
Not lizard
  Macellodus brodiei Owen, 1854 is
    crocodilian

The Purbeck beds of Durlston Bay have yielded nine crocodilian species. The best-represented forms belong to the Family Goniopholididae, which includes terrestrial–aquatic forms typical of the Purbeck and Wealden: broad-faced, with stout and rounded skulls, and with moderately long snouts reminiscent of modern crocodiles. *Goniopholis crassidens* was one of the first crocodilians recorded from Durlston, and was referred to as the 'Swanage crocodile' by Mantell (1837); the type skull and skeleton was described from one of the Swanage quarries (inland or coastal?) by Owen (1842b). *G. crassidens* was of large size, estimated at about 6 m long, with a 0.6 m long skull. The teeth are characteristic of the species, being remarkably stunted and thimble-shaped in outline. Owen (1878b) redescribed the type specimen from Swanage and reported abundant material from the Wealden. The species *G. simus* Owen, 1878 (Figure 7.16A) was smaller, about 2.5 m long, with more slender teeth and less tapering head. The type specimen also came from a 'Swanage quarry', and the remains include a fine 0.4 m skull. *Petrosuchus laevidens* Owen, 1878 was erected on the basis of a 0.25 m long partial skull and a mandible also from 'the Middle Purbecks, now quarried at Swanage'. The animal is estimated to have been of moderately large size. The skull is characterized by possessing slender teeth and a distinct angle between the slender rostrum and the temporal region that is almost as abrupt as that of a gavial. Watson (1911b) noted that the lower jaw and the skull described as associated by Owen (1878b) in fact belonged to different animals. He retained the lower jaw as the type of *Petrosuchus laevidens* Owen, 1878, and ascribed the poorly preserved skull to the new species *Pholidosaurus decipiens* Watson (1911), a member of the Family Pholidosauridae, an advanced long-snouted aquatic group. Andrews (1913) further described and figured *P. decipiens*.

Owen (1879b) described a partial fragmentary mandible from the Middle Purbeck as *Goniopholis tenuidens*, and further jaw remains as *Brachydectes major* Owen, 1879 and *B. minor* Owen, 1879. These were all distinguished on characters of the teeth and tooth arrangement. Woodward (1885) and Lydekker (1888a, pp. 79–83) accepted the validity of Owen's three species of *Goniopholis*. Woodward (1885, p. 506) noted that *Brachydectes* was pre-occupied and renamed it *Oweniasuchus*, and Lydekker (1888a, pp. 85–6) accepted the two species as valid. These five species are all small short-snouted forms (skulls 0.25–0.40 m long with stout teeth) and they are placed in the Goniopholididae. Steel (1973, pp. 15–19) accepted the validity of all five species, and of *Petrosuchus laevidens*.

The best-known Purbeck crocodilians are the so called 'dwarf', or small, crocodilians from the Feather Bed (?45–48). Owen (1879a, 1879b) described two forms, *Nannosuchus gracilidens* (Figure 7.16B) and *Theriosuchus pusillus* (Figure 7.16C), on the basis of good skulls and some post-cranial remains, noting (Owen, 1879b) the similarity of *Nannosuchus* to *Goniopholis*, and that *Theriosuchus* differed from the latter form in several respects. The skulls are 40–170 mm long, broad and short-snouted. *Nannosuchus* was like a miniature *Goniopholis*, but with long, slender, curved teeth adapted for catching fish. *Theriosuchus pusillus*, based on a nearly complete skeleton about 450 mm long, was discovered by Beckles in the Mammal Bed. Owen stated that its scattered teeth, scutes, vertebrae and limb bones are very numerous, and that a few skulls (about 90 mm long), mandibles and considerable portions of naturally articulated skeletons have also been found. The teeth of *T. pusillus* vary in shape and are consequently more specialized than those of any other Purbeck crocodilian in approaching a heterodont condition. Owen (1879a) argued at length that these small crocodilians captured the shrew-sized Purbeck mammals, and drowned them, just as crocodiles do today with larger mammals. Joffe (1967) re-examined the 'dwarf' crocodiles, and concluded that *Nannosuchus* was a juvenile *Goniopholis simus* and that *Theriosuchus* was a juvenile atoposaurid (a group of small, short-snouted crocodiles restricted to the Late Jurassic of the northern hemisphere).

A few pterosaur remains are recorded from Durlston Bay. Owen (1870, pl. 19, fig. 7) figured a phalanx from Swanage (Acton Quarries, Langton Matravers: SZ 990783) under the name *Ornithocheirus validus*. Although he appended no description, this is regarded as a valid characterization of the species. Seeley (1875a) erected the genus *Doratorhynchus* for a lower jaw and cervical vertebra (Figure 7.16D) from Durlston Bay. The jaw is long (300 mm+) with small close-set teeth. He associated these remains with Owen's specimen as *D. validus* (Owen, 1870). Lydekker (1888a, p. 26) returned the species to *Ornithocheirus*, and Wellnhofer (1978, p. 58) listed it among 'Ornithocheiridae *incertae sedis*'. Howse (1986) suggested that

the *Doratorhynchus* vertebra (CAMSM J5341) is an elongated cervical, the oldest evidence of an azhdarchid pterosaur, a group of giant forms known otherwise only from the Late Cretaceous.

The Purbeck dinosaurs from Durlston are represented by limited, but important, material. A large theropod tooth (BMNH 44806) was ascribed by Lydekker (1888a, p. 163, 1890b) to *Megalosaurus dunkeri*, which Huene (1926) referred to *Altispinax*, a genus known otherwise from the Wealden of Germany, England and Belgium. This genus, like so many, is based on such an agglomeration of odds and ends from different sites that its true affinities cannot be determined (Molnar, 1990). It now seems that *Nuthetes* is also a megalosaur. The type species, *N. destructor*, was described by Owen (1854) on the basis of a small, partial, left mandibular ramus (Figure 7.16F) with pointed, recurved, double-rooted teeth with serrated edges from the Mammal Bed. Owen (1854) classified the specimen as a lizard, and ascribed to it some small scutes and limb bones from the Feather Bed. Owen (1861b, 1878a, 1879b) supplemented this description and argued that certain small, conical, granulated objects (granicones) found in the Feather Bed also were dermal ossifications of *Nuthetes* since they were found mixed with *Nuthetes* fragments. These conical dermal ossicles, of up to 14 mm height and 8 mm across the base, are now considered as belonging to an unknown ornithischian dinosaur. Lydekker (1888a, p. 247) and Seeley (1893a, 1893b) noted the dinosaurian character of *Nuthetes* and Swinton (1934, p. 214) identified it as a megalosaur. This assignment has been accepted by Romer (1956, p. 599), Delair (1959, p. 79), Steel (1970, p. 34) and Galton (1981a, p. 253), although Molnar *et al.* (1990) term it a carnosaur *taxon dubium*.

Several ornithischian dinosaurs have been described from Durlston Bay. Owen (1861b) noted some small fragmentary jaws with leaf-shaped teeth (Figure 7.16E) as *Echinodon becclesii* (ever since quoted as *E. becklesii*, but Owen (1861b) consistently misspelt the collector's name as S.H. 'Beccles'). Owen (1861b) interpreted the jaws as 'lacertilian', but noted similarities with dinosaurs. Lydekker (1888a, p. 247) noted the similarity of the teeth to those of *Scelidosaurus*, and *Echinodon* has generally been associated with it in the Stegosauria (Delair, 1959, p. 88; Steel, 1970, pp. 48–9). Galton (1978), on the other hand, argued that *Echinodon* is a fabrosaurid, one of a group of small bipedal ornithopods, and he later (Galton, 1981a) suggested that the granicones probably belong to *Echinodon* since an American fabrosaurid is known with small dermal ossicles possessing a similar structure. However, Coombs *et al.* (1990, p. 434) argue that *Echinodon* may well be a basal thyreophoran, related to *Scelidosaurus*, as had long been suspected.

Owen (1874b, pp. 3-4) described a small single imperfect mandible with teeth as *Iguanodon hoggii* on the basis of its tooth striations. Although clearly an *Iguanodon*, the differences from the better-known Wealden species have been regarded as slight (Delair, 1959, p. 85; Steel, 1970, p. 17). Nevertheless, it is probably a valid species (Norman and Weishampel, 1990, p. 530), and the oldest *Iguanodon* known. It is estimated to have been about 2.5 m long, small for an *Iguanodon*.

Marine reptiles have been described from Durlston, but their remains are poor. Lydekker (1889a, p. 227) noted a small, imperfect limb bone of a plesiosaur (BMNH 21974), and Delair (1969b) reported a series of postcranial remains of an ichthyosaur from the Purbeck of Swanage (OUM J.13795). The horizons of these marine reptiles are not known.

Finds of dinosaur trackways have been noted from Durlston Bay, and the earliest record found by P. Ensom (pers. comm., 1993) is in the minutes of the Purbeck Society for 1861, while W.T. Ord noted the discovery of tridactyl impressions 'near the Coastguard Station', observed during an excursion by the Bournemouth Natural History Society in 1912. A single print was found at nearby Peveril Point by G. Tyler in 1967 (Sarjeant, 1974, p. 357), and Delair and Sarjeant (1985, p. 148) suggested that this may have been an isolated fallen block related to the 1912 discovery. Ensom (1983, 1984b, 1985b, 1987c) and Nunn (1990) reported the discovery of poorly preserved tridactyl footprints preserved on the sole of DB103 in the Cherty Freshwater Beds at Durlston Bay.

Comparison with other localities

Durlston Bay is by far the richest Purbeck reptile site. The other recorded Purbeck locations producing significant finds of reptiles are listed near the beginning of the chapter.

The turtles *Dorsetochelys* and *Mesochelys* are unique to Durlston, although the latter genus is similar to *Glyptops* from the Late Jurassic (Kimmeridgian-Portlandian) of Wyoming, Utah and Colorado (Gaffney, 1979b). *Pleurosternon* is best known from Durlston, with some specimens also from the Isle of Portland (Portland Stone) and Asia (?) (Młynarski, 1976, p. 120). Other pleurosternids come from the Early Cretaceous of Kelheim, Bavaria (*Helochelys*) and the Late Jurassic to Early Cretaceous of China (*Changyuchelys*) (Młynarski, 1976, pp. 119-20). *Tretosternon* is well known from the English Wealden and from other parts of western and central Europe (Młynarski, 1976, p. 60). *Plesiochelys* is well known from the Late Jurassic of Solothurn, Switzerland and Szechuan, China and the Wealden of the Isle of Wight (Młynarski, 1976, pp. 55-6).

The sphenodontid *Homoeosaurus* is best known from the Kimmeridgian and Portlandian of Germany and France, and *Opisthias* from the Late Jurassic Morrison Formation of North America (Fraser and Benton, 1989). Comparable lizards to the Durlston genera are known from the Oxfordian of the Guimarota coal-mine, near Leiria, Portugal (Seiffert, 1973) and the Morrison Formation of Wyoming (Prothero and Estes, 1980). Kimmeridgian and Portlandian lizards belonging to other groups are also known from France (Bugey), Germany (Franconia, Bavaria), Spain (Catalonia) and Manchuria (Hoffstetter, 1967; Estes, 1983).

Goniopholid crocodiles were widely distributed in the Late Jurassic of Europe and North America, and worldwide in the Cretaceous. *Goniopholis* is well known from the Wealden of southern England and the Late Jurassic Morrison Formation of North America, as well as from many scrappy remains elsewhere (Steel, 1973, pp. 15-18; Buffetaut, 1982). *Oweniasuchus, Petrosuchus* and *Theriosuchus* are restricted to Swanage, except for a partial jaw and tooth of *Oweniasuchus* from Portugal (Steel, 1973, pp. 18-19). *Pholidosaurus* is better known from the Wealden of northern Germany (Buffetaut, 1982). Teeth of *Bernissartia* have been found at the Sunnydown Purbeck site (Ensom *et al.*, 1991), but not in Durlston Bay.

The pterosaur *Doratorhynchus* is unique to Durlston and, if correctly determined by Howse (1986), is the world's oldest azhdarchid, a group known otherwise only from the Late Cretaceous.

Megalosaurus is widely distributed throughout the Jurassic and Early Cretaceous of Europe (Lower Lias-Wealden of England, Portugal,

France, Monaco, Germany, Transylvania) and Morocco (Steel, 1970, pp. 33–6), but this distribution is inflated by the identification as megalosaurids of a range of theropod taxa, most of which are indeterminate (Molnar, 1990). Early thyreophorans such as *Echinodon* are known from the Early Jurassic of Lyme Regis (*Scelidosaurus*), China (*Tatisaurus*) and Arizona (*Scutellosaurus*) (Coombs *et al.*, 1990). *Iguanodon* is known from the Early Cretaceous of southern England, Belgium, Portugal, possibly Bohemia, Mongolia and the Lakota Formation of North America (Norman and Weishampel, 1990).

Dinosaur footprints are preserved as moulds and casts in the limestones of the Middle and Upper Purbeck in several quarries on the Isle of Purbeck (Delair, 1960, 1963, 1966, 1982b; Charig and Newman, 1962; Walkden and Oppé, 1969; Delair and Lander, 1973; Delair and Brown, 1975; Ensom, 1982a, 1982b, 1984a, 1984b, 1985a, 1985b, 1986a, 1986b, 1987b, 1987c, 1988; West and El-Shahat, 1985; Newman, 1990). The prints have been attributed mainly to *Iguanodon* or some similar ornithopod, but also to *Megalosaurus* after the discovery of three-toed footprints in a quarry at Herston, near Swanage (Charig and Newman, 1962; Newman, 1990), and to sauropods (Ensom, 1987b).

Conclusions

The Purbeck beds of Durlston Bay have yielded one of the most important Mesozoic terrestrial faunas in the world. The 10 species of turtles are nearly all unique to Durlston, and they represent the earliest members of several important lineages. Durlston is the best early lizard site in the world, having produced so far a more diverse and better preserved fauna than other comparable Late Jurassic and earliest Cretaceous sites. The crocodilians include several genera unique to Durlston, and the small juvenile ('dwarf') specimens are unique. The pterosaur *Doratorhynchus*, if correctly determined, is the oldest azhdarchid, otherwise a Late Cretaceous group. The few dinosaur remains include the smallest known megalosaur, *Nuthetes*, an unusual armoured ornithischian, *Echinodon*, and perhaps the oldest known specimen of *Iguanodon*. All of these taxa have been restudied recently, or are in need of revision, and new finds continue to be made. The Durlston fauna occurs in marine and non-marine rocks which occupy a unique position at the Jurassic–Cretaceous boundary, it has yielded many unique genera (29 type species), and the range of small- to medium-sized reptiles gives it a position of international significance in vertebrate palaeontology and its high conservation value.

Chapter 8

British Cretaceous fossil reptile sites

Cretaceous stratigraphy and sedimentary setting

INTRODUCTION: CRETACEOUS STRATIGRAPHY AND SEDIMENTARY SETTING

The Cretaceous System in Britain (Figures 8.1, 8.2) is represented by two broad phases of deposition which relate to palaeogeography. Earth movements during the Late Jurassic uplifted most of north-west Europe to form land. In the British region, there were initially two main basins of deposition, in the East Anglia–North Sea area, and in the Wessex–Weald region and northern France. Facies of the Early Cretaceous were deposited subaerially or in relatively shallow-water marine and freshwater environments, represented by lagoonal, fluvial and lacustrine sediments of the Purbeck and Wealden, and by shallow-marine shelf facies of the Lower Greensand, Gault and

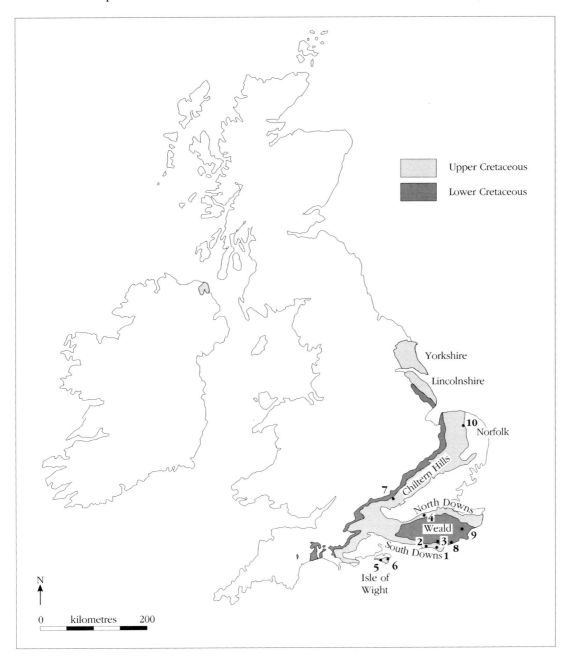

Figure 8.1 Map showing the distribution of Cretaceous (Lower and Upper) rocks in Great Britain. GCR Cretaceous reptile sites: (1) Hastings; (2) Black Horse Quarry, Telham; (3) Hare Farm, Brede; (4) Smokejacks Pit, Ockley; (5) Brook–Atherfield Point, Isle of Wight; (6) Yaverland; (7) Wicklesham Pit, Faringdon; (8) East Wear Bay, Folkestone; (9) Culand Pits, Burham; (10) St James's Pit, Norwich.

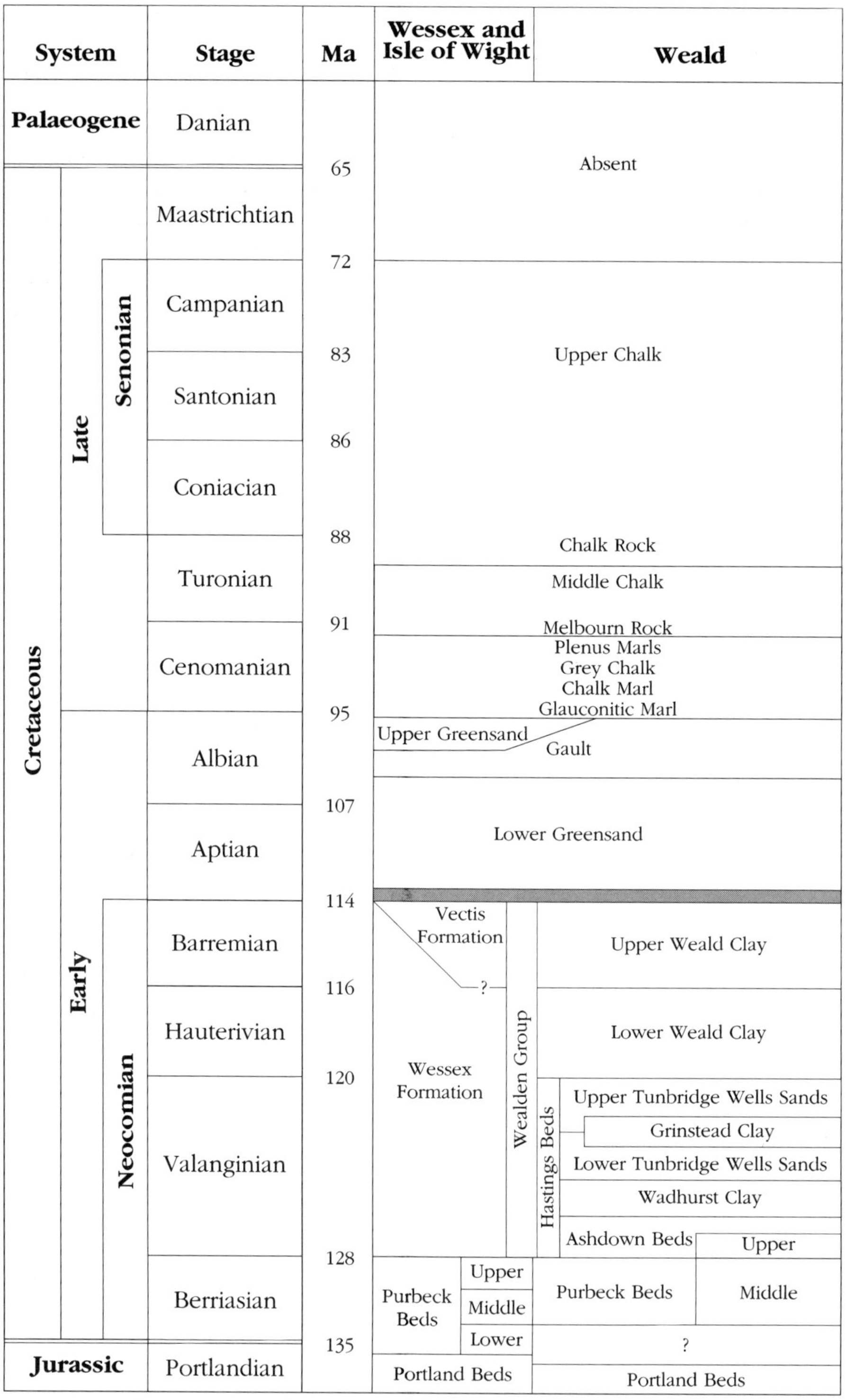

System	Stage	Ma	Wessex and Isle of Wight	Weald		
Palaeogene	Danian					
Cretaceous — Late — Senonian	Maastrichtian	65	Absent			
	Campanian	72				
	Santonian	83	Upper Chalk			
	Coniacian	86				
	Turonian	88	Chalk Rock			
			Middle Chalk			
	Cenomanian	91	Melbourn Rock / Plenus Marls / Grey Chalk / Chalk Marl / Glauconitic Marl			
Early — Neocomian	Albian	95	Upper Greensand / Gault			
	Aptian	107	Lower Greensand			
	Barremian	114	Vectis Formation	Wealden Group — Upper Weald Clay		
		116	— ? —	Lower Weald Clay		
	Hauterivian	120	Wessex Formation	Hastings Beds — Upper Tunbridge Wells Sands		
				Grinstead Clay		
	Valanginian			Lower Tunbridge Wells Sands		
				Wadhurst Clay		
				Ashdown Beds	Upper	
	Berriasian	128	Purbeck Beds — Upper / Middle / Lower	Purbeck Beds	Middle	
		135		?		
Jurassic	Portlandian		Portland Beds	Portland Beds		

Figure 8.2 Summary of Cretaceous stratigraphy, showing global stage nomenclature and some major southern British formations. Based on Harland *et al.* (1990).

Cretaceous stratigraphy and sedimentary setting

Upper Greensand. Following a major transgression in mid Cretaceous times, seas flooded most of the British area, leaving small patches of land only in the mountainous areas of North Wales, eastern Ireland, southern Scotland and the Scottish Highlands. Late Cretaceous history in Britain is dominated by the predominantly coccolith limestone facies of the Chalk.

The Cretaceous has been zoned on the basis of ammonites and belemnites, but the relative, or complete, absence of these fossils from much of the sequence gives a poorer overall macrofossil stratigraphic resolution than for the Jurassic. Where ammonites are absent, for example in the Purbeck–Wealden facies in Britain, alternative biostratigraphic indicators (e.g. pollen, spores, ostracods) have been used. Even in the marine Late Cretaceous Chalk facies, selective preservation, probably because of sea-floor dissolution, has limited the ammonites to discrete horizons (Kennedy, 1969), and schemes of correlation have involved the use of inoceramids, belemnites, brachiopods and echinoderms. Micropalaeontological dating, using Foraminifera in particular, are used in the absence of macrofossils.

Late Jurassic to Early Cretaceous earth movements led to the development of regressive facies over much of northern Europe and England, and in Britain the base of the Cretaceous System falls in the non-marine Purbeck Beds and within the Norfolk–Lincolnshire marine sequence (Allen and Wimbledon, 1991). The succeeding Wealden Group consists of lagoonal, fluvial and lacustrine deposits which outcrop over an extensive area of Sussex, Surrey and Kent (the Weald area) and on the Isle of Wight and in Dorset. The Wealden of the Weald sub-basin (Berriasian–Barremian) falls into two divisions: the lower sand-dominated Hastings Beds and the upper Weald Clay.

The Hastings Beds consist of predominantly sandy, but often argillaceous, deposits which reach a maximum thickness of *c.* 400 m in the centre of the Weald; within it two major cycles of sedimentation can be identified (=Ashdown Beds + Wadhurst Clay, Lower Tunbridge Wells Sand + Grinstead Clay, and, less well developed, Upper Tunbridge Wells Sand + Lower Weald Clay). The base of each cycle commences with clays and siltstones which gradually coarsen upwards into cross-bedded sandstones. The uppermost beds may include pockets and lenses of bone-rich gravel. These pass upwards into cross-laminated siltstones with the horsetail *Equisetites* and then return to argillaceous rocks forming the base of the following cycle. These sediments have, in the past, been interpreted as deltaic in origin, but the more recent work of Allen (1976, 1981) indicates that they were deposited in lagoonal to lacustrine mudplain environments in which salinity was controlled by the rates of run off of surface freshwater and evaporation. The occurrence of soil horizons, dinosaur footprints and the remains of *in situ* horsetail roots and stems are testimony to the maintenance of shallow-water conditions of deposition throughout.

The Weald Clay, above the Hastings Beds, with a maximum thickness of 450 m, was deposited almost exclusively in mudplain environments, with occasional localized influxes of coarser sediment (Allen 1976, 1981). The incoming of brackish water fossils toward the top of the Weald Clay documents the initial phases of the main mid-Cretaceous transgression and subsequent deposition of the Lower Greensand across southern and eastern England during the Aptian and Albian stages.

The Wealden Group (Berriasian–Aptian; Kerth and Hailwood, 1988) in the Wessex sub-basin comprises the Wessex Formation (formerly Wealden Marls) and the overlying Vectis Formation (formerly Wealden Shales). The Wessex Formation is a red-bed sequence, consisting of an alternation of varicoloured, but mainly red, mudstones with subordinate sandstones. The unit thins from about 530 m below the Isle of Wight to 70 m in Dorset. Sedimentological and palaeoecological evidence indicates that the Wessex Formation was deposited on an alluvial plain crossed by a perennial meandering river system (Stewart, 1981a, 1981b, 1983; Daley and Stewart, 1979). The Vectis Formation comprises mainly grey mudstones and siltstones, usually organized in thin fining-upwards cycles. This unit is about 60 m thick on the Isle of Wight, but thins westwards into Dorset, and it is absent in some sections. Sedimentological and palaeoecological data suggest that the Vectis Formation was deposited in a shallow coastal lagoon which was subject to increasing salinity and storm frequency towards the top (Stewart *et al.*, 1991; Wach and Ruffell, 1991).

The Lower Greensand Group consists of a complex series of mudstone and sandstone facies with a rich marine fauna (bivalves, gastropods, brachiopods, echinoids, ammonites, crustaceans, corals), and is assumed to have been laid down in marine and nearshore marine environments, with frequent estuarine intercalations in the Isle of

Wight (Wach and Ruffell, 1991). Lower Greensand deposition over much of southern and south-east England was terminated by a further transgression which, during the early Albian, led to widespread development of basinal marine mudstone facies (the Gault Clay Formation). These argillaceous deposits are often highly condensed, and phosphatic nodule horizons may be present. Westwards the facies passes laterally into the Upper Greensand Formation, a variable, often bioturbated deposit of glauconitic sands. This unit contains marine fossils, such as bivalves, ammonites and serpulid worms. In Cambridgeshire Albian fossils are reworked into the Cenomanian Cambridge Greensand. Further north (from Norfolk into the North Sea) the Gault passes laterally into the condensed carbonate sequences of the Carstone and Red Chalk, or Hunstanton Red Rock.

Transgression, initiated in the Aptian, continued until near the close of the Cretaceous and brought changes in sedimentation which led to massive developments of coccolith ooze that now forms the Chalk. Subsequent sedimentation was occasionally interrupted when regressive phases led to deposition of 'nodular chalk' and associated hardgrounds.

At the end of the Cretaceous (late Maastrichtian) there was a substantial marine regression in Britain, and much of Europe. This coincided with a major phase of extinction that affected many groups of invertebrates and vertebrates; among marine invertebrates, the ammonites, belemnites, inoceramids and rudists became extinct.

REPTILE EVOLUTION DURING THE CRETACEOUS

The Cretaceous Period is known for its highly diverse dinosaur faunas. In Britain the best represented forms are the ornithischians which occur abundantly in the Wealden Group of southern England. These include the well-known ornithopods *Iguanodon* and *Hypsilophodon*, and the armoured ankylosaurs (e.g. *Polacanthus*). The sauropods and theropods were also important elements in Cretaceous terrestrial ecosystems, and theropods include the unusual scavenging or piscivorous form *Baryonyx* from the Weald Clay. The Wealden Group gives Britain an enviable record of Early Cretaceous dinosaurs, arguably the best in the world. Comparable faunas are known

from North America (especially the Cloverly Formation of Montana and Wyoming), Europe (the Wealden of France, Belgium and north Germany, and equivalent units in Spain and Portugal), Mongolia (mainly Mid-Cretaceous in age), and sparse faunas from South America, Africa and Australia.

British records of Mid- and Late Cretaceous dinosaurs are less satisfactory because of the shift to marine sedimentation. Worldwide, however, dinosaurs showed major advances in the Late Cretaceous. New groups of ornithopods, particularly the duck-billed hadrosaurs, came to dominate terrestrial faunas and their relatives, the horned ceratopsians, also became diverse elsewhere. The sauropods were only patchily represented during Late Cretaceous times, and the stegosaurs had declined dramatically. Ankylosaurs witnessed a modest radiation, and carnivorous theropods, large (e.g. *Tyrannosaurus rex*) and medium-sized (e.g. *Struthiomimus, Stenonychosaurus*), are known from several parts of the world. Late Cretaceous dinosaur faunas are best known from North America (the midwest states of Montana, the Dakotas, Colorado, Wyoming, Texas and the province of Alberta, as well as some eastern states) and Mongolia. Some significant Late Cretaceous dinosaur faunas are also becoming better known from South America, India, China and Romania.

Among other terrestrial reptile groups, such as turtles, crocodilians, lizards and snakes, major evolutionary steps took place. The turtles diversified on land and in the sea, and many modern families appeared. Lizards also diversified on land, giving rise to many modern groups, as well as some extinct ones, most notable of which were the large marine mosasaurs and their relatives. In addition, snakes arose from 'lizards' during the Early Cretaceous, and some early constricting (non-poisonous) groups became established. Crocodilians diversified mainly on land and in fresh waters, while the marine metriorhynchids of the Jurassic declined. Many new crocodilian groups appeared, including the mammal-like terrestrial notosuchians, the giant sebecosuchians, both of these mainly in southern continents, and the modern eusuchians. Species of true crocodile and alligator are known from the Late Cretaceous.

In the air pterosaurs had become greatly advanced, and by the end of the Cretaceous occupied a variety of adaptive zones as highly efficient fish-eating soarers, as well as insectivorous forms using flapping flight. Cretaceous pterosaurs were all pterodactyloids, the advanced clade, and their size was, on the whole, much larger than the sparrow- to seagull-sized Jurassic pterosaurs. British Cretaceous pterosaur records are patchy, and not comparable in quality with the finer Early Cretaceous forms from Brazil (Santana Formation) and Mongolia, or the forms from the marine Late Cretaceous of the mid-American seaway area (Kansas, Texas). These animals were accompanied by birds which had arisen from advanced theropod dinosaurs during the Late Jurassic. Birds have a very weak Cretaceous record, with good representation only of the Late Cretaceous coastal forms in Kansas and Texas, and little in Britain.

Marine reptiles show very considerable changes in the Cretaceous. Ichthyosaurs never again achieved the importance they had in the Jurassic, and remains are patchily distributed in many parts of the world through the period, with the last ones seemingly being Cenomanian in age. Plesiosaurs also dwindled in significance, although several groups, especially giant pliosaurs and long-necked elasmosaurs, lasted right to the end of the Cretaceous, and are represented especially in southern continents and in Texas. Cretaceous ichthyosaur and plesiosaur fossils are rare in Britain. The main Cretaceous marine group was the mosasaurs, giant marine lizards, which became top carnivores, possibly as a result of the decline of the pliosaurs. Mosasaurs are patchily represented in the British Chalk, although they are better known in the type Maastrichtian of the Netherlands and in Belgium, in the United States and in parts of north Africa. At the end of the period the mosasaurs, with the other large marine reptiles of the Jurassic and Cretaceous (e.g. ichthyosaurs and plesiosaurs), which had started to decline earlier, also disappeared. The end-Cretaceous mass extinction event is best known, however, for the demise of the dinosaurs, although by very latest Cretaceous times the group seems to have been somewhat depleted both in numbers and diversity.

BRITISH CRETACEOUS REPTILE SITES

British Cretaceous localities have provided good material of many typical reptile groups, particularly of ornithischian dinosaurs, which are known from several localities in the Early Cretaceous rocks of the Weald of Sussex, Surrey and Kent, and the Isle of Wight. Saurischian dinosaurs, the theropods and sauropods, are rare. Important finds of pterosaurs are known from the Gault of Folkestone, the Cambridge Greensand and also from the Middle Chalk where they are associated with well preserved remains of lizards, snakes and turtles. Terrestrial turtles and crocodilians are also known from the Wealden, and the Cambridge Greensand. The marine plesiosaurs and ichthyosaurs are also represented in most of the sequence, and mosasaurs are known from a few localities in the Chalk.

The strength of the British Cretaceous record lies in the relatively well-dated and rich Early Cretaceous terrestrial faunas of the Wealden; this provides the richest and best view of Early Cretaceous vertebrates anywhere in the world. Some Mid- and Late Cretaceous faunas are good, but they represent mainly marine components of the reptilian faunas, and there are better faunas elsewhere. The British record is of no value in depicting Late Cretaceous terrestrial reptilian evolution.

EARLY CRETACEOUS: WEALDEN (BERRIASIAN–BARREMIAN)

The lagoonal, lacustrine and fluvial deposits of the Wealden Group of the Weald and the Isle of Wight are famed for their dinosaur faunas which are the most varied in Europe (Figures 8.3, 8.5 and 8.8). The Brook–Atherfield section on the south-west coast of the Isle of Wight exceeds the contemporaneous dinosaur-rich sediments of Mongolia and the United States of America both in the abundance and variety of the material.

The Wealden of the Weald (Berriasian–Barremian) is well known for its fossil reptiles, and specimens have come from many localities, most of which are inland extractive sites and no longer accessible. Dinosaurs, crocodilians and pterosaurs are known from all Wealden formations, but they occur most frequently in the Hastings Beds. The succeeding Weald Clay has yielded fewer remains, but has recently produced the unusual theropod dinosaur *Baryonyx*. Well-recorded reptile sites include the following:

WEST SUSSEX: Loxwood (TQ 0331; *Iguanodon*, Murchison, 1829, pp. 103–5); Rudgwick Brickworks (TQ 083344; dinosaur; Horsham Museum); Longbrook Brickworks (TQ 117188; pterosaur, crocodilian, fishes; Wells Collection); Itchingfield (TQ 123287; *Iguanodon*); Southwater (TQ 1526; pterosaur); Horsham (TQ 1730; *Goniopholis, Iguanodon*, hypsilophodontid); Henfield Brickpit (TQ 218143; *Iguanodon*;

Young and Lake, 1988, p. 23); Bolney (TQ 2622; *Hylaeosaurus*; Pereda-Suberbiola, 1993); Wivelsfield (TQ 3420; *?Iguanodon*; various localities with fossil fishes; Young and Lake, 1988, p. 23); Balcombe Quarry (TQ 3030; *Iguanodon*); Philpots Quarry (TQ 355322; *Iguanodon*; Allen 1976, 1977); Tilgate Forest (TQ 2735, exact localities uncertain; *Plesiochelys, Tretosternon, Goniopholis, Suchosaurus, Heterosuchus, Pelorosaurus, Iguanodon, Hylaeosaurus*, pterosaurs); Cuckfield (TQ 300256, original *Iguanodon* quarry; *Archaeochelys, Plesiochelys, Tretosternon, Cimoliasaurus, Goniopholis, Suchosaurus, Heterosuchus, Ornithocheirus, Iguanodon, Valdosaurus, Hylaeosaurus, 'Megalosaurus', Pelorosaurus, Pleurocoelus*, including type specimens of eight species, but now largely filled in; Mantell, 1825, 1827, 1833, 1850a, 1850b; Murchison 1829; Galton, 1981b, p. 32; locality on Whiteman's Green determined by Swinton, 1970, pp. 29–30; also Topley, 1875, pp. 91–5; White, 1924, pp. 8–10); Keymer Tile Works (TQ 325189; microvertebrates; crocodilian and dinosaurian remains; Young and Lake, 1988, p. 24; A. Ross, pers. comm., 1993).

EAST SUSSEX: Hamsey Brick Works (TQ 398159, pterosaur; Martin Collection); Berwick Brick Pit (TQ 523070; *Plesiochelys, Leptocleidus*, Andrews, 1922; White, 1928, pp. 29–30); Pevensey (TQ 6404; *Iguanodon*); Burwash (TQ 6724; *Plesiochelys, Goniopholis*); Brightling (TQ 6821; *Goniopholis*); Bexhill (TQ 7407; *Iguanodon*, pterosaur, *'Iguanodon'* footprints at TQ 74460705, TQ 741071, TQ 738070, TQ 70950640; Beckles, 1854; Tylor, 1862; Delair and Sarjeant, 1985; Lake and Shephard-Thorn, 1987, p. 20; Delair, 1989; Woodhams and Hines, 1989); Little Galley Hill, Bulverhythe (TQ 767079; *'Iguanodon'* footprints; Beckles, 1854; Tylor, 1862; White, 1928, pp. 25, 28, 53; Ballerstedt, 1914; Sarjeant, 1974, p. 531; Delair and Sarjeant, 1985, pp. 142–3); Crowhurst Pit, Rackwell Wood (TQ 764124; *Goniopholis, Iguanodon*, Sweeting, 1925; White, 1928, pp. 65–6); Brede (TQ 8218; several sites, including Hare Farm Lane, TQ 832184, q.v.; *Goniopholis, Saurosuchus, Iguanodon*; Topley, 1875, pp. 62–3; Allen, 1949); Knellstone, Udimore (TQ 8819; *Iguanodon*, etc.; Allen, 1949, p. 279); Peasmarsh, Waterfall Wood (TQ 8621; *?Heterosuchus*, etc., Allen, 1949); Tighe Farm (TQ 936266; bone bed; Lillegraven *et al.*, 1979, p. 27; K.A. Kermack, pers. comm.).

Sites around Hastings are detailed in the Hastings report.

KENT: Brenchley (TQ 6741; *'Plesiosaurus'*); Tunbridge Wells (TQ 5839; *Megalosaurus*); Southborough (TQ 5842; *Thecospondylus*; Seeley, 1882b); New Barn (TQ 6168; 'turtle').

SURREY: Harting Combe, near Haslemere (*'Iguanodon'* footprints; Delair and Sarjeant, 1985, p. 146); Clockhouse Brickworks (TQ 175386; *Iguanodon*, microvertebrates; Jarzembowski, 1991a).

In the Isle of Wight the Wealden beds are represented by predominantly argillaceous facies of the Wealden Marls and the Wealden Shales (Wessex and Vectis formations), which are seen best in coast sections between Brook and Atherfield Point on the south-west coast and at Yaverland on the south-east coast.

Outside the Wealden of the Weald and the Isle of Wight Early Cretaceous reptiles are known from the Spilsby Beds (Portlandian–Berriasian) of Spilsby, Lincolnshire (TF 4066), from Speeton, Yorkshire (TA 1180; various marine reptiles from Valanginian and Hauterivian; Drake and Sheppard, 1909; R. Rawson, pers. comm., 1981; J.W. Neale, pers. comm., 1982), and from Ridgway Hill, Dorset (SY 6788; *Iguanodon, Plesiosaurus* and *Pliosaurus*; Reid, 1899), Swanage (*fide* Buckland), and Upwey. In Europe, comparable Wealden faunas are known from Belgium (Bernissart coal mines; Casier, 1978; Norman, 1980), France (Buffetaut *et al.*, 1991), Germany (Hannover; Norman *et al.*, 1987), and North America (Cloverly Formation, Wyoming; Ostrom, 1970; upper parts of the Morrison Formation, Wyoming; Dodson *et al.*, 1980).

Six early Cretaceous GCR sites are selected (Figure 8.1), including three sites in the Hastings Beds of East Sussex, one in the Weald Clay of Surrey, and two in the Wealden of the Isle of Wight:

1. Hastings, East Sussex (TQ 831095–TQ 853105). Early Cretaceous (Berriasian–Valanginian), Hastings Beds (Ashdown Beds, Tunbridge Wells Sand).
2. Black Horse Quarry, Telham, East Sussex (TQ 769142). Early Cretaceous (Valanginian), Hastings Beds.
3. Hare Farm, Brede, East Sussex (TQ 832184). Early Cretaceous (Valanginian), basal Wadhurst Clay.

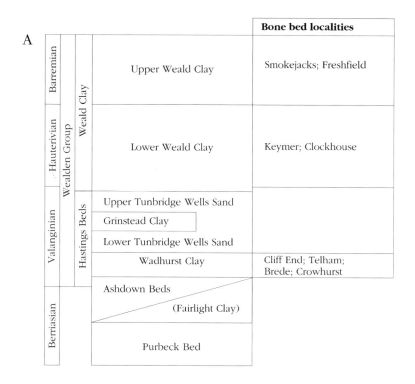

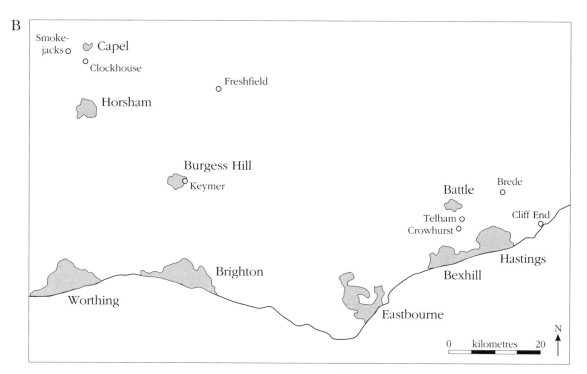

Figure 8.3 The Wealden of the Weald. (A) Summary stratigraphic succession, showing the relative temporal position of the bone beds; (B) map of some key Wealden reptile sites. Courtesy of E. Cook.

4. Smokejacks Pit, Ockley, Surrey (TQ 113373). Early Cretaceous (Barremian), Weald Clay.
5. Brook-Atherfield Point, Isle of Wight (SZ 375842–SZ 452788). Early Cretaceous (Barremian–Early Aptian), Wessex and Vectis formations (Sudmore Point Sandstone–Chale Clay).
6. Yaverland, Sandown, Isle of Wight (SZ 613850–SZ 622835). Early Cretaceous (Barremian–Early Aptian), Wealden Marls (Perna Beds Member).

HASTINGS, EAST SUSSEX (TQ 831095–TQ 853105)

Highlights

The Early Cretaceous sandstones and shales that outcrop along the coast and foreshore east of Hastings have been famous for 150 years for specimens of dinosaurs, crocodilians, turtles, and footprints. More recent discoveries include rare mammal teeth, and other small bones, from the Cliff End Bone Bed.

Introduction

The Hastings Beds (Early Cretaceous: Berriaisian–Valanginian, Figure 8.2) of the cliff sections east of Hastings have been renowned for 150 years for finds of fossil reptile bones and footprints (Figure 8.4), and this century for microvertebrate remains in the Cliff End Bone Bed. Bones were first found at Hastings about 1830, and the brick pit and stone quarries around the town produced many specimens, but these quarries and pits are no longer accessible. The coast is subject to continuous erosion, and dinosaur footprints and a variety of reptile bones have been reported from the 1850s to the present day. Parts of the section are obscured by landslips, and there is limited access to the beach.

The stratigraphy of the Wealden of Hastings has been described by several authors (e.g. Beckles, 1856; Topley, 1875; White, 1928; Allen 1976;

Figure 8.4 The cliff at Cliff End, east of Hastings, with the Cliff End Bone Bed near the top of the section. Fossil footprints and reptile bones have been found at, and in the vicinity of this locality. (Photo: E. Cook.)

Lake and Shephard-Thorn, 1987). Accounts of reptiles and footprints have been given by Hulke (1885), Seeley (1887e), Lydekker (1892, 1893b), Ballerstedt (1914), Delair and Sarjeant (1985), Delair (1989) and Woodhams and Hines (1989), and the Cliff End Bone Bed has been described by Allen (1949) and Clemens and Lees (1971).

Description

The succession, in outline is (Lake and Shephard-Thorn, 1987):

	Thickness (m)
Hastings Beds	
Tunbridge Wells Sand	
Fine-grained, yellowish sandstones and silts with impersistent seams of mottled silty clay	up to 50
Wadhurst Clay	50–57
Grey mudstones interlaminated with thin siltstones.	
Also: calcareous sandstone beds (Tilgate stone), sandstone channel fills, soils and near the base:	
Cliff End Bone Bed	
Cliff End Sandstone	
Top Ashdown Pebble Bed	10
Ashdown Beds	180–200
The upper 30–50 m are chiefly sandstones, while the strata below are dominantly massive mottled spherosideritic clays with subordinate sandstone beds.	
Near the base:	
Lee Ness Sandstone	1–2

The geology of the Ecclesbourne–Fairlight section has been described by Allen (1962), Stewart (1981b) and Lake and Shephard-Thorn (1987). The sections immediately east of Hastings Old Town show the top Ashdown Beds and the cliff is topped by the Cliff End Sandstone, the lowest unit of the Wadhurst Clay. The lowest exposed unit, the Lee Ness Sandstone, is seen on the foreshore at low tide. These units appear throughout the section, repeated by faults and with varying dips as a result of the WNW–ESE-trending Battle–Fairlight anticline. Individual horizons show lateral facies variation – the Cliff End Bone Bed does not occur throughout.

Most reptile remains do not have accurate local-ity or horizon data. Bones and footprints seem to have been found at all levels in the section, and from several sites, which may be identified on the basis of the literature and museum labels.

1. Ecclesbourne Glen (TQ 837099). Tylor (1862) recorded fossil footprints 'upon detached blocks of sand-rock which had fallen in large masses from the upper part of the cliff a little west of Ecclesbourne Glen'. Various crocodilian and dinosaur bones (e.g. BMNH R605–9, R1637) are labelled 'Ecclesbourne Glen'. Several specimens in Hastings Museum (GG94–101) are recorded from the Wadhurst Clay and Ashdown Beds of Ecclesbourne.
2. Lee Ness Ledge (TQ 867108). Beckles (1854, p. 457), Allen (1976, p. 393) and Lake and Shephard-Thorn (1987) note 'casts of the footprints of *Iguanodon*' on the undersurface of blocks of the Lee Ness Sandstone. These may be the same specimens described *in situ* by Woodhams and Hines (1989), who report iguanodontid and theropod footprints at three levels 'near Lee Ness', two of these horizons being near the base of the Lee Ness Sandstone itself, and one 5–6 m lower.
3. Fairlight Cove–Cliff End (TQ 876116). White (1928, p. 30) noted 'a few footprints of *Iguanodon*, on a slab of light-grey sandstone . . . on the shore close to the base of the cliff at Cliff End. Allen (1976, p. 393) notes large dinosaur footprints at the top of the Ashdown beds at Cliff End. Footprints have been seen along this section in fallen blocks (P. Allen, pers. comm. to M.J.B., 1982). Tylor (1862) also recorded footprints from the following sites: TQ 835097, TQ 854104, TQ 860107, and Lake and Shephard-Thorn (1987, p. 21) noted some at Goldbury Point (TQ 877114). Several turtles, crocodilians and dinosaurs are labelled 'Fairlight West' or 'Cliff End' (BMNH R1954, R4416, R4434, R4439, HASTM GG80–6, 88, 92, 105–7, 313). Source horizons are the 'Ashdown Sands' and the 'Fairlight Clays' (upper and lower portions of the Ashdown Beds respectively). In general, bones and footprints may be found anywhere along the section where there is fresh exposure.

Iguanodontid footprints figured by Ballerstedt (1914, figs 2, 4) 'aus dem Wealden von Hastings' may come from this area too: one of them is a photograph by C. Dawson, presumably the archaeologist associated with the Piltdown find, and with Tielhard de Chardin, who was involved with the first collections from the Cliff End Bone Bed.

Reptile remains are also known from the Cliff End Bone Bed (less than 5% of all bones: Patterson, 1966; two teeth, K.A. Kermack, pers. comm. 1982). The Cliff End Bone Bed fauna consists largely of sharks' teeth, and those of the actinopterygian fish *Lepidotes*, together with rare mammal teeth. The Bone Bed, exposed in the cliffs at TQ 887129 (Figure 8.4), 2.5 m above the Cliff End Sandstone (Lake and Shephard-Thorn, 1987, pp. 67–71), is a poorly sorted cross-bedded coarse sandstone or fine-grained conglomerate of quartz and chert pebbles and abraded fragments of sideritic mudstone, with abundant fragments of fishes and reptiles as well as mollusc debris.

Fauna

As already indicated, some specimens are labelled as coming from Ecclesbourne, Fairlight or Cliff End. The majority, however, are labelled 'Hastings' and although many probably came from the cliffs east of the town, some must have been found in the old quarries and brickpits. In the following list, only those specimens definitely recorded from the cliffs are listed, and numbers of all 'Hastings' material are given.

	Numbers
Testudines: Cryptodira	
Tretosternon bakewelli	
(Mantell, 1827)	
HASTM GG92, 96	3
Plesiochelys brodiei Lydekker, 1889	3
Archosauria: Crocodylia:	
Neosuchia: Goniopholididae	
Goniopholis crassidens Owen, 1842	
BMNH R605, R607	10
Goniopholis sp.	
BMNH R608; HASTM GG80–2,	
84–6, 88, 105–7, 313	17
Archosauria: Crocodylia: Neosuchia:	
Pholidosauridae	
Suchosaurus sp.	
BMNH R4416, R4439	2
Archosauria: Crocodylia: Neosuchia:	
Bernissartiidae	
Bernissartia sp.	1
Archosauria: Crocodylia: Neosuchia:	
inc. sed.	
Heterosuchus valdensis Seeley, 1887	
Type specimen: BMNH 36555	1

	Numbers
Archosauria: Pterosauria:	
Pterodactyloidea	
Ornithocheirus sp.	
HASTM GG100, 101	2
Archosauria: Dinosauria:	
Saurischia: Theropoda	
Megalosaurus dunkeri Dames, 1884	
BMNH R1954	4
Megalosaurus sp.	
HASTM GG98	2
Archosauria: Saurischia: Sauropoda	
Cetiosaurus brevis Owen, 1842	4
Ornithopsis hulkei Seeley, 1870	5
Archosauria: Dinosauria: Ornithischia	
Iguanodon sp.	
BMNH R1637	20+
'?stegosaur'	
BMNH R4434	1
Polacanthus sp.	1
Sauropterygia: Plesiosauria	
Cimoliasaurus valdensis	
Lydekker, 1889	
BMNH R609	3
'plesiosaur'	
HASTM GG94, 95	2

Interpretation

The Tunbridge Wells Sand has been interpreted as a fluvio-deltaic deposit, the Wadhurst Clay and Grinstead Clay as pro-deltaic or lagoonal in origin, and the Ashdown Beds as fluvial (Lake and Shephard-Thorn, 1987). The environments of deposition in the Wealden of the Weald had formerly been interpreted as largely deltaic (e.g. Allen, 1959, 1962; Taylor, 1963), but Allen (1976, 1981) revised his former theory in favour of a model (Figure 8.8) in which the normal Wealden environment was a variable-salinity mudplain, periodically transformed into a sandy braidplain by powerful overloaded streams, the salinity changes being controlled by the rate of freshwater runoff. Allen (1981) argued that many of the rivers were braided in their proximal portions, whereas Stewart (1981a, 1981b, 1983) emphasized evidence for meandering streams. The climate was warm, with marked wet and dry seasons and 'herds of dinosaurs travelled freely across the basin and maintained themselves in it'.

The Cliff End Bone Bed was interpreted as a high-energy deposit by Allen (1949) and correlated by him with the Telham Bone Bed, exposed near Battle, and with other occurrences of the Cliff End Bone Bed inland (Figure 8.3). Allen (1949) regarded this bone bed as a correlatable event horizon, restricted to the most eastern part of East Sussex, and neighbouring parts of Kent, and lying on top of the 'Tilgate Stone' horizon (Lake and Shephard-Thorn, 1987, p. 28).

The turtles *Tretosternon* and *Plesiochelys* are represented by fragmentary remains of the carapace, plastron and limbs. Such remains are relatively common in the Wealden of the Weald, but they are inadequate for a proper understanding of their anatomy and relationships.

The crocodilians are more abundant. *Goniopholis*, represented by numerous vertebrae, limb bones, teeth, jaws and scutes, was a moderate- to large-snouted aquatic crocodilian. The genus is known from the Late Jurassic and Early Cretaceous of Europe and North America (Steel, 1973). *Suchosaurus*, an aquatic medium-sized pholidosaur, is represented by some teeth. *Bernissartia* (partial skeleton) was a small (1 m long) animal with characteristic heterodont teeth, conical and pointed in the anterior part of the jaws, and rounded and blunt further back. The genus is known also from the Wealden of Bernissart, Belgium and the Isle of Wight and the Early Cretaceous of Galve, Spain (Buffetaut, 1975; Norell and Clark, 1990).

The type of the crocodilian *Heterosuchus valdensis* (Figure 8.5A) probably came from the cliff section at Hastings. The specimen consists of a water-worn slab containing about 12 vertebrae of a small crocodilian (Seeley, 1887e). Further material from the Wealden of Sussex and the Isle of Wight was ascribed to this form by Lydekker (1888a), and Woodward and Sherborn (1890, p. 231) identified as *Heterosuchus* sp. specimens from the Isle of Wight and from the Middle Purbeck of Durlston Bay. The genus has been synonymized with *Hylaeochampsa* Owen, 1874, erected for Isle of Wight material (Steel, 1973, p. 53), but this cannot be demonstrated since the two taxa are based on non-overlapping material (Buffetaut, 1983; Clark and Norell, 1992). Indeed, Clark and Norell (1992) argue that the taxon is a *nomen dubium*, since it lacks diagnostic material. They regard it as a neosuchian, possibly a eusuchian, on the basis of its procoelous vertebrae and the well-developed condyles on the trunk vertebrae. If it is a eusuchian, as *Hylaeochampsa* is, then it is one of the oldest in the world.

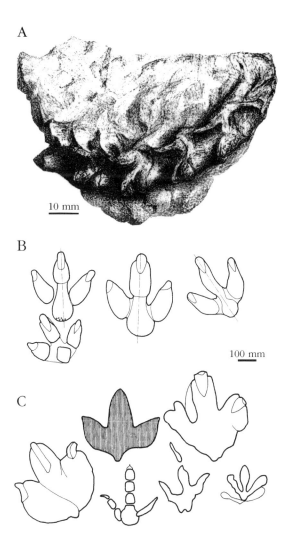

Figure 8.5 Fossil reptile remains from the Early Cretaceous Hastings Beds of Hastings. (A) Sequence of dorsal vertebrae of the crocodilian *Heterosuchus valdensis* Seeley, 1887; (B) iguanodontid footprints; (C) theropod footprints from the foreshore. (A) After Seeley (1887c); (B) and (C) after Woodhams and Hines (1989).

A large carnivorous dinosaur, generally ascribed to *Megalosaurus* is represented by teeth and limb bones from Hastings. The generic assignment is unlikely, since *Megalosaurus* is typical of the Mid Jurassic. There is a problem over the definition of the two Wealden 'species', *M. dunkeri* and *M. oweni*, and Huene (1923) ascribed these to the new genus *Altispinax*, but the specimens are too incomplete for certain assignment. Molnar (1990) regards *M. dunkeri* as a 'problematic carnosaur' and *M. oweni* as a *nomen dubium*.

Some large bones, ribs and vertebrae from Hastings have been named *Cetiosaurus* and *Ornithopsis* (Lydekker, 1892, 1893b). While these assignments may or may not be correct, there seem to have been at least two large sauropods in the Wealden (Ostrom, 1970).

The commonest dinosaur remains from Hastings are teeth, jaws, vertebrae, ribs and limb bones of the large bipedal ornithopod *Iguanodon* (Hulke, 1885). The specific assignment is difficult and awaits revision (Norman, 1980, 1986). Several species of *Iguanodon* were named from the Hastings area, including *I. hollingtonensis* Lydekker (1889), based on a partial skeleton from Hollington Quarry, St Leonards, near Hastings (TQ 795115), and *I. dawsoni* Lydekker (1888), both based on limb bones and vertebrae from Shornden Quarry, Hastings (TQ 802106, TQ 803104) (Lydekker, 1888a, 1889e). The status of these taxa is currently unclear. Norman and Weishampel (1990, p. 530) synonymize *I. holling-tonensis* with *I. fittoni*, and accept *I. dawsoni* as valid, but do not elucidate the differences of these rather poorly known taxa from the typical *I. atherfieldensis* Hooley, 1924 and *I. bernissarten-sis* Boulenger, 1881, the sympatric small and large forms respectively found in most of the Wealden of Europe.

Armoured dinosaurs are represented by a '?stegosaur' tooth and a ?*Polacanthus* spine. The English ankylosaurs *Hylaeosaurus* and *Pola-canthus* are known from the Wealden of the Weald, the Isle of Wight, and the Upper Greensand of Charmouth, Dorset, but dermal elements such as the spine are hard to identify.

Pterosaurs are relatively uncommon, with only a few wing bones of '*Ornithocheirus*' known. Plesiosaurs, typically marine animals, are also uncommon; some vertebrae and limb bones of *Cimoliasaurus* suggest that they may have wandered into coastal fresh waters at times.

The iguanodontid and theropod footprints from Hastings (Tagart, 1846; Beckles, 1854; 1856; Tylor, 1862; White, 1928; Delair and Sarjeant, 1985; Lake and Shephard-Thorn, 1987, pp. 19–21; Woodhams and Hines, 1989) are large (0.3–0.6 m long), tridactyl (three-toed) imprints (Figure 8.5B, C). The 'toes' are broad, short and curved to a point and there is a broad heel impression. They are generally seen as casts on the underside of sandstone beds, or as wave-eroded hollows in silts on the present foreshore. Theropod prints have narrower toes than the iguanodontid prints, and they should have evidence of sharp claws if preservation is good enough. They are much rarer than the iguanodontid prints.

Comparison with other localities

In the Hastings area several quarries in the Wadhurst Clay and Ashdown beds have yielded similar fossil reptiles. These include St Leonards (0.02–0.1 m thick bed in Hall and Co.'s Quarry behind the church just off the West Marina; TQ 797088; *Tretosternon, Goniopholis, Pleurocoelus, Iguanodon,* 'stegosaur', *Ornithocheirus, Cimo-liasaurus,* Parish, 1833; Topley, 1875, p. 61; White, 1928, p. 47; Allen, 1949, p. 276); Hollington Quarry ('quarry at Rose Cottage', Topley, 1875, p. 61; TQ 795115; *Tretosternon, Goniopholis, Megalosaurus, Iguanodon, Hylaeo-saurus,* 'stegosaur', *Cimoliasaurus,* Lydekker, 1889f, p. 355; 1890b, pp. 40–3; White, 1928, pp. 66, 71); Little Ridge Farm Quarry (TQ 809127; *Iguanodon*); Shornden Quarry (TQ 802106; *Iguanodon*); Silver Hill/Tivoli Brickworks (TQ 799115; *Iguanodon*); Bucks Hole Quarry (TQ 806110, TQ 806112; *Iguanodon, Cimo-liasaurus*); Ore (?TQ 826108; *Goniopholis, Megalosaurus, Iguanodon, Cimoliasaurus*). Unfortunately, none of these sites is still extant.

The Cliff End Bone Bed is currently exposed (Lake and Shephard-Thorn, 1987, pp. 37, 39) near the steps from the Undercliff to Watchbell Street, Rye (TQ 91952018), and formerly in a brickpit near Baldslow (TQ 810133). Bone beds which may be equivalent to the Cliff End Bone Bed are seen at Reyson's Farm, near Brede (TQ 832192) and West Ascent, St Leonards (TQ 79820885).

Conclusions

The most varied faunas of Early Cretaceous dinosaurs are known from the Wealden of Europe. One of the best of these faunas is that from the Hastings Beds in their type area, and the fossils include skeletons and footprints. Moreover, this is the only extensive, eroding coastal setting in these non-marine strata, which therefore has considerable potential for future finds. Previous finds include a selection of terrestrial and aquatic reptiles – two genera of turtles, four genera of crocodilians, one genus of theropod, two of sauropods, three of ornithischians, one genus of pterosaur and one plesiosaur. Also, further collecting from bone-rich horizons – such as the Cliff End Bone Bed – may yield new genera of smaller reptiles: lizards, snakes, turtles.

The conservation value lies in the combination of this potential for future discoveries and the importance of the fossil faunas recovered from the site over the past 150 years.

BLACK HORSE QUARRY, TELHAM, EAST SUSSEX (TQ 769142)

Highlights

Black Horse Quarry, Telham is the main site of the Telham Bone Bed, a sediment which produced specimens of turtles, crocodilians, pterosaurs, dinosaurs, and a plesiosaur. This bone bed has produced relatively small bones, which supplement the larger elements found in coastal sites.

Introduction

Black Horse Quarry, Telham Hill, near Battle, east Sussex was formerly a well-known source of Early Cretaceous reptile remains. The dissociated bones were found in a thin bone bed, the Telham Bone Bed (Figure 8.2, 8.3), for which this is the type locality. Although not currently exposed, it could be re-excavated; the quarry is presently 6–7 m deep in parts. The site has been described by Binfield and Binfield (1854), Topley (1875), Woodward and Sherborn (1890), White (1928) and Lake and Shephard-Thorn (1987).

Description

The section at Black Horse Quarry (Boyd Dawkins, *in* Topley, 1875, pp. 63–4; thicknesses approximate) was:

	Thickness	
	ft	in
Surface soil	1	0
Rust-coloured grey and white shales with indurated layers	3	0
Rust and slate-coloured shales with ironstone	3	0
Rust and slate-coloured shales without ironstone (*Cyrena*)	6	0
Slate-coloured shales with a layer of a lighter colour (*Cyrena* and plants)	3	0
Shale and clay (*Cypridea*, *Cyrena* and vegetable matter)	3	6

	Thickness	
	ft	in
Grey clay with nodules; the 'bone bed' [0–4 inches] in its lower part	2	0
Calcareous grit [Tilgate Stone], fine grained and hard, dug for roads	2	6
Calcareous grit [Tilgate Stone], blue on the unweathered surface	2	0

The beds are within the Wadhurst Clay of the Hastings Beds (Valanginian) and the Telham Bone Bed has been regarded as equivalent to the Cliff End Bone Bed as exposed east of Hastings (q.v.; Allen, 1949; Lake and Shephard-Thorn, 1987). It lies above the main 'Tilgate Stone' horizon, and 6–10 m above the Top Ashdown Pebble Bed. According to the section given here, the bone bed is 0–0.1 m thick and occurs about 5.2 m (17 ft) below the soil surface. Binfield and Binfield (1854) noted insect remains 10–13 ft (*c.* 3–4 m) above the 'Calcareous grit' (Jarzembowski, 1976).

Boyd Dawkins (*in* Topley, 1875, p. 64) described the bone bed as 'composed of a mass of coprolites, bones, teeth, scutes and ganoid scales . . . It is conglomeratic in character and contains pebbles of white quartz, which vary in size from a pigeon's egg to a pea, and are all much worn and highly polished. Very few organic remains are perfect, but the great bulk of them have been reduced to the conditions of pebbles. The only perfect bones that have been found consist of the hard and solid phalanges of the larger reptilia . . . In the interior of one long dinosaurian bone there were fragments of jet . . . The condition of all these remains is precisely identical with those from the Crag 'coprolite' beds, and the bone-beds of the Rhaetic and Carboniferous rocks'.

Fauna

Many Wealden reptiles are labelled 'Battle' or 'Telham' and Black Horse Quarry must have been the source of most of these. Some specimens (BMNH R2845–6) bear the label 'Lambert's Quarry, Black Horse', probably in reference to the former owner. Boyd Dawkins (*in* Topley, 1875, p. 64) listed reptile remains that he had identified from Black Horse Quarry. In the following list, numbers of specimens in the collections of BGS(GSM), BMNH and HASTM are given as an approximate guide to relative abundance:

Numbers

Testudines: Cryptodira
Plesiochelys sp.   5
Tretosternon bakewelli
(Mantell, 1827)   3
Archosauria: Crocodylia: Neosuchia
Goniopholis crassidens (Owen, 1842)   8
Suchosaurus cultridens (Owen, 1842)   3
Archosauria: Pterosauria:
Pterodactyloidea
Ornithocheirus? clifti (Mantell, 1844)   ?
Archosauria: Dinosauria: Saurischia:
Theropoda
'Megalosaurus' dunkeri Dames, 1884   1
'Megalosaurus' sp.   2
Archosauria: Dinosauria: Saurischia:
Sauropoda
?Cetiosaurus sp.   1
Pleurocoelus valdensis Lydekker, 1890   1
Archosauria: Dinosauria: Ornithischia
Iguanodon sp.   10
Hylaeosaurus armatus Mantell, 1833   1
Sauropterygia: Plesiosauria:
'Plesiosaurus' sp.   4

Interpretation

Allen (1949, pp. 279-82) interpreted the Telham Bone Bed as a river deposit, pebbles and bones of which were rolled along for some distance before deposition. Allen (1976, pp. 393, 406) equated the Telham Bone Bed tentatively with either the Broad Oak Top Pebble Bed or the Cliff End Pebble (Bone) Bed (both low in the Wadhurst Clay Formation). The Bone Bed facies all appear to occur in 'the muddier parts of the shoreface, beneath a metre or so of water'.

The turtles *Plesiochelys* and *Tretosternon* are represented by broken carapace and plastron pieces not adequate for proper identification. These turtles were moderate to large in size (0.3-1 m plastron length). They are both classed as chelydroids (Młynarski, 1976, pp. 55, 60), or *Plesiochelys* may be a chelonioid (Gaffney 1975b).

The crocodilians *Goniopholis* and *Suchosaurus* are based on fairly common teeth and vertebrae. These were both long-snouted aquatic forms, although the latter genus is essentially known only from teeth.

Both Topley (1875, p. 64) and Woodward and Sherborn (1890, p. 255) noted a pterosaur in the Black Horse Quarry fauna, but the specimen(s)

have not been located. *Ornithocheirus clifti* was initially interpreted as a bird because of its hollow limb bones and its exact relationships are uncertain (Wellnhofer, 1978, p. 58).

The carnivorous dinosaur *Megalosaurus* is represented by vertebrae and teeth. Most of these have been named *M. dunkeri*, a species known also from the Wealden of Hannover (Germany), as well as other places in the south of England. *Megalosaurus* is typical of the Mid Jurassic, and Huene (1926) renamed this species *Altispinax* on the basis of the high neural spines on the vertebrae.

Herbivorous dinosaurs include the large sauropods *?Cetiosaurus* and *Pleurocoelus*, the former represented by vertebrae, the latter by teeth from Black Horse Quarry. These generic assignments are probably incorrect – the Wealden sauropods urgently require restudy (Ostrom, 1970). *Iguanodon*, the commonest Wealden dinosaur, is recorded from Black Horse Quarry on the basis of teeth, vertebrae, limb bones and phalanges. *Hylaeosaurus*, an armoured ankylosaur has been identified tentatively on the basis of a vertebra.

Comparison with other sites

The nearest exposures of the Telham Bone Bed are at Rackwell Wood, Crowhurst (TQ 764124; Sweeting, 1925; White, 1928, pp. 65-6; Lake and Shephard-Thorn, 1987, p. 38), in Crowhurst Park (TQ 781138; R.D. Lake, pers. comm. to M.J.B., 1982), and at Maplehurst Wood (TQ 81001307; Lake and Shephard-Thorn, 1987). Allen (1949, pp. 279-82) noted further exposures at Baldslow (TQ 8013), Brede (Post Office; Kicker Wood; Reyson's Farm; Cat's Nest; Broadlands, TQ 826183; ?TQ 8320; TQ 832192; ?TQ 837192); Peamarsh (Waterfall Wood; TQ 8621); Udimore (Knellstone; TQ 8819); Stone (Stone Hole Quarry; Tighe Farm; ?TQ 9428, TQ 936266). The Telham Bone Bed is apparently equivalent to the Cliff End Bone Bed (Allen, 1976) which is seen in blocks on the foreshore at Cliff End, Pett (TQ 887130).

Most of the turtles, crocodilians, pterosaurs and dinosaurs in the Black Horse Quarry fauna are common in the Wealden of southern England, and indeed many of them in Early Cretaceous sediments elsewhere in the world. New excavations are required, and more extensive series of specimens are needed, for more precise identifications, and for fuller comparisons, of the taxa present.

Conclusions

Black Horse Quarry, Telham has provided good collections of Wealden reptiles. Although the bones are disarticulated, and generally water-worn, the remains are relatively abundant in the thin bone bed. This is the type locality of the Telham Bone Bed, and it is the best site for fossil reptiles in that unit, an attribute that in combination with its potential for re-excavation gives the site its conservation value.

HARE FARM LANE, BREDE, EAST SUSSEX (TQ 83141844)

Highlights

Hare Farm Lane, Brede is the best site for the Brede Bone Bed. It has produced specimens of dinosaurs and crocodilians, and there is potential for future significant discoveries.

Introduction

The Brede Bone Bed, another of the bone beds within the Wadhurst Clay (Figure 8.2, 8.3), is most readily accessible in Hare Farm Lane, Brede, and this site has yielded the most complete fauna (Allen, 1949). Literature recording the reptile fauna is limited to Lydekker (1890) and Allen (1949, 1976), but the diversity could be greatly enhanced by re-excavation of this lane-side cutting.

Description

The Brede Bone Bed has been described in some detail at the Hare Farm Lane locality (Allen, 1949, pp. 276–9): 'The bone bed comprises thin lenticles of buff sand up to 2 feet long, 1 foot wide and 2 inches thick . . . The lenticles cut across the current-bedding of the surrounding siltstones and shales, and on top are bevelled off to a common level . . . Rootlets from the overlying soil-bed pass through the bone-bed . . . The buff sand constituting the bone-bed . . . is non-pebbly, rather argillaceous and poorly sorted, and always contains bivalve casts (including *Neomiodon medius*). The detritus includes quartz and glau-conite, mixed with large quantities of comminuted scales, teeth and bone varying in size from finest powders to fragments over two inches long . . .' It lies in the basal Wadhurst Clay (Valanginian) between the Top Ashdown Pebble Bed and the Brede *Equisetum lyelli* Soil Bed. Lake and Shephard-Thorn (1987, p. 29) note that the Brede Bone Bed may reflect localized concentrations of material, and is probably not laterally correlatable over long distances like the Cliff End/Telham Bone Bed.

Fauna

Allen (1949, p. 278) listed the fauna of the Bone Bed as molluscs (*Neomiodon, Viviparus*), fishes (*Lepidotus, Hybodus*) and reptiles ('chelonian fragments, crocodilian teeth, bone'). A few specimens labelled 'Brede' are preserved in the BMNH and HASTM. These may or may not have come from Hare Farm Lane.

Archosauria: Crocodylia: Neosuchia
   Goniopholis crassidens Owen, 1842
     BMNHR3373
   Suchosaurus sp.
     BMNHR4415
Archosauria: Dinosauria: Ornithischia:
   Iguanodontidae
   Iguanodon fittoni Lydekker, 1888
     BMNHR1627
   Iguanodon sp.
     HASTMEJB3

Interpretation

The turtle remains noted by Allen (1949) have not been further described. *Goniopholis* is represented by a partial mandible and *Suchosaurus* by a tooth: these were aquatic genera. *Goniopholis* had a long-snouted skull up to 0.7 m long.

Iguanodon, one of the commonest Wealden reptiles and the commonest dinosaur, is represented by a phalanx (*Iguanodon* sp.) and a partial skeleton (ascribed to *I. dawsoni* by Lydekker, 1890b). This specimen, consisting of a partial pelvis, several dorsal and caudal vertebrae, a partial hindlimb and other elements, apparently differed from other species of *Iguanodon* on the basis of characters of the pelvis in particular. D.B. Norman (pers. comm., 1983) considers that BMNH R1627 belongs to *I. fittoni* and differs from *I. dawsoni*, two species accepted provisionally by

Norman and Weishampel (1990) as valid (see Hastings report).

Comparison with other localities

The Brede Bone Bed was noted by Allen (1949, p. 276) at St Leonards-on-Sea (TQ 79820885; ?Cliff End Bone Bed, Lake and Shephard-Thorn, 1987, p. 39), Stubb Lane, Brede (TQ 82171853; Lake and Shephard-Thorn, 1987, p. 37), Ludley Hill, Beckley (TQ 8521), and possibly also Oxenbridge Hill, Iden (?TQ 9225). Allen (1949, p. 280) noted a bone bed supposedly equivalent to the Telham Bone Bed at Reyson's Farm (TQ 832192)

Conclusions

Hare Farm Lane is the type locality of the Brede Bone Bed. The fauna reported to date is small. Earlier dinosaur finds from Brede, including a good specimen of *Iguanodon fittoni*, point to the potential of this site and that it is a key Wealden bone bed site, hence its conservation value.

SMOKEJACKS PIT, OCKLEY, SURREY (TQ 113373)

Highlights

Smokejacks Pit, Ockley is famous as the site which yielded *Baryonyx*, the meat-eating dinosaur with a giant hook claw. This recent dinosaur find supplements earlier discoveries of *Iguanodon* and crocodilians at Smokejacks, and bones and teeth have now been found at several levels.

Introduction

Smokejacks Pit at Wallis Wood, Ockley, near Dorking, Surrey (Figure 8.1) has been operated as a private brickworks for some time and is presently under the ownership of the London Brick Company (bought from the Ockley Brick Company). The quarry exposes sections in the Weald Clay, in the Barremian (Early Cretaceous). Remains of the dinosaur *Iguanodon atherfieldensis* were collected in 1945-6, and further material of this species, and of *I. bernissartensis*, has been collected since. The discovery, in

1983 of the theropod *Baryonyx walkeri*, with its 300 mm long claw (Figure 8.9), has brought the locality to attention. Since 1983 fossil reptile material has been recovered from several horizons within the quarry. The pit is still operational and, because work is relatively slow, fresh bones are occasionally seen and collected.

The first finds of dinosaurs from Smokejacks Pit were reported by Rivett (1953, 1956) who collected 'upwards of 100 bones' belonging to several individuals of *Iguanodon atherfieldensis*. Further specimens were excavated by the BMNH, but not reported. In January 1983 William Walker discovered an ironstone nodule in the pit which contained a giant claw. A rescue excavation by staff from the BMNH led to the discovery in the following month of a partial skeleton of a theropod dinosaur, which Charig and Milner (1986, 1990) named *Baryonyx walkeri*. The specimen has taken nearly 10 years of laborious preparation because of the impenetrability of the ironstone matrix.

Subsequent collecting by a variety of people has produced evidence of reptiles at several horizons and in several parts of the pit. John Cooper of Worthing found scattered large *Iguanodon* bones in 1987 and these were excavated with the assistance of the BMNH. Further material of the same individual was excavated by David Cooper, M.J.B. and colleagues from Bristol University in 1992 (Figures 8.6 and 8.7). Allen (1976, 1981) and Jarzembowski (1991a) have given preliminary descriptions of the site.

Description

Smokejacks Pit shows a section in the clays and subordinate sandstones of the Weald Clay, which lies above the Hastings Beds in the Early Cretaceous (Barremian) (Rawson *et al.*, 1978). Topley (1875, p. 106) mentioned the Weald Clay of Smokejacks Farm, and included it in his 'No 5. Sand and Sandstone with Calcareous Grit'. The sediments currently exposed in the pit represent about 27 m of Weald Clay below BGS bed no. 5c (Alfold Sand Member; Allen, 1976), which suggests that it is in the Upper Weald Clay, and is dated as early Barremian (Jarzembowski, 1991a).

Rivett (1956) recorded several general sections in the brickpit, the one for the central part being:

Figure 8.6 Excavation of a partial *Iguanodon* skeleton at Smokejacks Brickpit in summer, 1992. (Photo: M.J. Benton.)

	Depth (in)
Topsoil turning to clay	0–12
Red Clay	12–36
Sandy Clay	36–60
Reddish Clay	60–72
Thinly bedded sandstone	72–108
Sandy clay and sandstone	108–174

The original discoveries of *Iguanodon* bones were in the 'harder rock' 'from 6–12 feet from the surface'. Rivett (1956) later gave the depth beneath the surface as '8 to 15 feet'. Nevertheless, it seems that they came from the two sandstone beds listed above. Rivett (1953, 1956) noted further that iron pyrites was found associated with, and impregnating, some of the bones, and that the bones were 'often resting on, or embedded in, the sand beds.' Some of the bones were broken and water-worn and they had evidently been transported, but most were relatively well preserved. Rivett (1953, 1956) also noted the presence of rounded pebbles, which he interpreted as gastroliths.

Other fossils known from Smokejacks Pit include plants (ferns, conifers), with some specimens interpreted as angiosperm-like vegetative parts (Hill *et al.*, 1992). If this interpretation is correct, these could be the oldest angiosperm macrofossils known. The invertebrate fauna includes ostracods, conchostracans, egg cases of cartilaginous fishes, molluscs and insects. The insect fauna consists of cockroaches, beetles, true flies, bugs, termites, crickets, grasshoppers, lacewings, scorpionflies, wasps and dragonflies (Jarzembowski, 1991a, 1991b).

The excavation of *Baryonyx* provides information on the taphonomy of large vertebrate remains from the Weald Clay in Smokejacks Pit. The bones located *in situ*, although largely disarticulated and scattered, all came from within the confines of an area measuring 5 m by 2 m, and the general arrangement of the bones demonstrated that none was far from its natural position; most of the pieces of skull, pectoral girdle and forelimbs were located at one end and most of the pelvic girdle and hindlimbs at the other. In general, the bones of *Baryonyx* were not found to be

Figure 8.7 Dr Glenn Storrs consolidates an *Iguanodon* vertebra in Smokejacks Brickpit, part of the partial skeleton excavated in summer, 1992. (Photo: M.J. Benton.)

distorted or crushed to any significant extent. This may relate to the mode of fossilization, for most of the remains became encased in ironstone, presumably not long after burial. However, the few bones preserved in clay also appear to be unaffected by compaction. It may be significant that many of the bones appear to have become disarticulated prior to fossilization.

There are other modes of fossil reptile occurrence in Smokejacks Pit. Teeth and scutes of crocodilians and dinosaurs have been found seemingly isolated, and fish remains occur in lenses of siltstone. Others, such as the *Iguanodon* skeleton excavated in 1987–92, are remains of a single large dinosaur skeleton, but disarticulated and scattered over a wider area, in this case some 200 bones recovered from an area measuring approximately 7 m by 4 m. The sedimentary situation in this case seems to represent overbank deposits produced during a flood (E. Cook, pers. comm., 1993).

Fauna

The Rivett collection is housed in the BMNH. Rivett (1953, 1956) ascribed his finds to

Iguanodon and some he identified as belonging to a large sauropod. Other material from Smokejacks is also in the BMNH.

Archosauria: Dinosauria: Crocodilia:
    Goniopholididae
    Isolated teeth of ?goniopholidids.
Archosauria: Dinosauria: Saurischia: Theropoda:
    Baryonychidae
    Baryonyx walkeri Charig and Milner, 1986
        Type specimen: BMNH R9951
Archosauria: Dinosauria: Saurischia: Sauropoda:
    Titanosauridae
    Titanosaurus-like sauropod (Rivett, 1953)
Archosauria: Dinosauria: Ornithischia:
    Ornithopoda: Iguanodontidae
    Iguanodon bernissartensis Boulenger, 1881
        BMNH and David Cooper collection
    Iguanodon atherfieldensis Hooley, 1925
        BMNH R6432–640

Interpretation

Allen (1976, p. 414; 1990) interpreted the Weald Clay, including that at Smokejacks Brickworks, as having been deposited in an alluvial and lagoonal mudplain with short-lived sand channels (Figure 8.8). Salinities varied from freshwater to nearly marine. All facies were liable to exposure, as shown by large footprints in sandstone (at Capel, TQ 18294048), suncracks and mudflake conglomerates, as well as soil beds and the presence of horsetails. The fauna is terrestrial and aquatic (fresh–brackish), containing numerous insect remains in addition to the reptiles, but with freshwater aquatic insects and fishes.

Crocodilian teeth and scutes are found, scattered about the site, and possibly coming from various horizons. They have been collected by many visitors to the site, but not curated or studied yet.

Iguanodon atherfieldensis, a large, herbivorous, bipedal or facultatively quadrupedal ornithopod dinosaur, is represented by vertebrae and limb bones, mostly the remains of small or immature animals. *I. atherfieldensis* is far more gracile than the other well-known forms of *Iguanodon* from the Hastings Beds (*I. dawsoni, I. anglicus, I. fittoni*) and it is notable for its distinc-

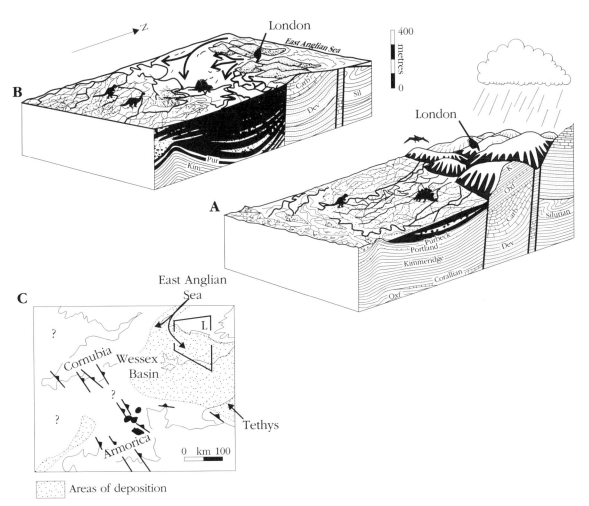

Figure 8.8 Sedimentological process models for the formation of the Wealden of the Weald. (A) Arenaceous formations; (B) argillaceous formations; (C) regional setting. Uplift of the London horsts, to the north of the basin of deposition, produced an area of high relief and an extensive source of sediment (A). Braided alluvial sand plains expanded southwards from the uplands, and the lowlands supported diverse floras and faunas, including dinosaurs (A). Downfaulting and denudation of the London horsts reduced relief and the rate of sediment supply (B), and the Weald area became a brackish–freshwater lagoonal–alluvial mudplain. Again, abundant vegetation grew around the lakes, and a diverse fauna of fishes, insects and reptiles inhabited the area.

tive postcranial morphology (Norman, 1986). The large contemporary form *I. bernissartensis*, found rarely in the Wealden Marls of the Isle of Wight (Barremian–Early Aptian), but better known from the Early Cretaceous on the continent (Norman, 1980, 1987; Norman *et al.*, 1987), is represented by the 1987–92 specimen found by David Cooper.

To date more than half of the skeleton of *Baryonyx walkeri* has been recovered (Figure 8.9). This includes parts of the skull (conjoined premaxilla, anterior left maxilla, conjoined nasals, lacrimal, frontals, anterior braincase and occiput), lower jaw (left dentary with some associated post-dentary elements), axial skeleton (axis, one cervical vertebra, some dorsal vertebrae, a caudal vertebra, cervical ribs, dorsal ribs, gastralia, chevrons) and limb skeleton (both scapulae, both coracoids, ?clavicle, fragments of ilia, pubes and ischium, both humeri, phalanges of the manus including unguals [?large claw], portions of left and right femur, left fibula, right calcaneum, and elements of the pes). Teeth are present in both upper and lower jaws and also in isolation. Associated with the remains of *Baryonyx* were

fish teeth and scales, an isolated humerus from a small individual of *Iguanodon*, and a small claw. Polished lithic fragments also found associated have been interpreted as probable gastroliths.

Baryonyx possesses some unique characters: an extremely narrow snout with a spatulate expansion at the tip and a slight downturn of the premaxilla seen in lateral view; a long low external naris situated far back from the front of the snout; and a probable mobile articulation involving a loose 'hinge' between the premaxilla and maxilla. *Baryonyx* possesses an unusually high number of marginal teeth (32 alveoli in the lower jaw, compared with the usual theropod count of 16), and in the postcranial skeleton, the upward bend of the neck seen in all other theropods is not developed. The femur indicates a bipedal stance, but the massive humeri demonstrate that there must have been a degree of quadrupedality, and *Baryonyx* is regarded as a facultative quadruped, again a feature unknown in other theropods.

The mode of life of *Baryonyx* is difficult to determine. Charig and Milner (1986, 1990) argued

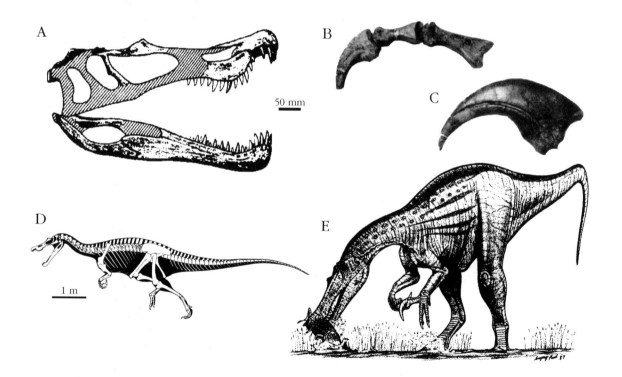

Figure 8.9 The most famous recent British dinosaur discovery, the enigmatic theropod *Baryonyx walkeri* Charig and Milner, 1986 ('Superclaw'), from the Early Cretaceous Weald Clay of Smokejacks Pit, Ockley, Surrey. (A) Skull bones as preserved; (B) normal digit, presumably from the hand; (C) the claw; (D) restoration of the skeleton; (E), imagined life appearance. (A) and (C) after Charig and Milner (1990); (B), (D) and (E) after Milner (1987).

Figure 8.10 A reconstructed scene in the Wealden of southern England, combining elements of the faunas from the Weald and from the Isle of Wight. The small ornithopod *Hypsilophodon* (bottom left) looks up at its larger relative *Iguanodon*, just behind. A mammal and a turtle stand in the bottom right, while behind them the theropod *Baryonyx* prepares to eat a fish. Behind it, a small theropod runs towards a small herd of the sauropod *Pelorosaurus*. Based on a painting by Graham Rosewarne in Benton (1989). Reproduced with permission of Quarto Publishing plc.

for ichthyophagy on the basis of the enlarged claw, the numerous finely serrated teeth, the superficially crocodilian-like appearance of the skull, and the fish scales in its gut region. They envisioned *Baryonyx* as a quadrupedal predator crouching on river banks and using the large claw (presumably on the hand) like a gaff (Figure 8.10), in a way comparable to the method used by grizzly bears today. Kitchener (1987) took a contrasting approach, suggesting that the combination of the flexible snout tip, large sharp talon, the powerfully developed forelimbs and the narrow snout could be adaptations toward a scavenging lifestyle. However, Reid (1987) was not convinced by a carrion-feeding habit for the animal.

The unusual characters of *Baryonyx* have presented problems in classification. Charig and Milner (1986) considered that its specializations merited erection of a new theropod family which they named Baryonychidae. The only other material directly comparable with *Baryonyx* consisted of two fragmentary snouts from the Aptian (late Early Cretaceous) of Niger previously ascribed to the mandibular symphysis of a spinosaurid dinosaur. Buffetaut (1989, 1992) noted that, although there were some differences between *Spinosaurus* and *Baryonyx*, they share several characters, particularly the structure of the teeth and jaws. These characteristics suggested that they were closely related to each other and might indicate the inclusion of *Baryonyx* in the family Spinosauridae. Charig and Milner (1990) accepted the similarity of *Baryonyx* to the fragmentary skull specimens from Niger and southern Morocco, but argued that the latter were not spinosaurids. Molnar (1990) referred *Baryonyx* provisionally to 'problematic carnosaurs'.

Comparison with other localities

Reptiles are rare in the Weald Clay. The only other important site is at Berwick, East Sussex – the Cuckmere Brick Co. pit (TQ 523070) which has yielded the turtle *Plesiochelys* sp., and much of the skull and skeleton of a plesiosaur, the type specimen of *Leptocleidus superstes* Andrews, 1922. Other Weald Clay sites include Clockhouse, Rudgwick and Keymer (see above). The so-called spinosaurids from the Gadoufauna of Nigar (Aptian) and from southern Morocco (Early Cretaceous) are the nearest relatives of *Baryonyx*.

Conclusions

The general rarity of fossil reptiles in the Weald Clay makes Smokejacks Pit important. Furthermore, the abundance of the remains of *Iguanodon* collected in the 1940s and 1990s suggests that there are pockets containing concentrations of bones. This seems to be corroborated by the taphonomic data obtained from the *Baryonyx* excavation. *Baryonyx walkeri* is known only from Smokejacks Pit and is unique among all other theropod dinosaurs. The unusual features of *Baryonyx* have not only resulted in the establishment of a new genus and species for the animal, but also a new family of theropods, the Baryonychidae. *Baryonyx* is the most dramatic new dinosaur discovery from Europe for a long time. The site has tremendous potential for further finds and this contributes significantly to its conservation value.

BROOK–ATHERFIELD, ISLE OF WIGHT (SZ 375842-SZ 452788)

Highlights

Brook-Atherfield, Isle of Wight is one of the most important dinosaur sections in Europe. Over the past 200 years, dozens of nearly complete skeletons of dinosaurs have been excavated, representing about 20 species, some of them unique to the site. In addition, many species of turtles, crocodilians and pterosaurs have also been found. The dinosaur fauna is of importance because most of the specimens are well localized, and the fauna is the richest in the world for the Early Cretaceous.

Introduction

The Wealden Group of the south-west coast of the Isle of Wight (Figure 8.11) is world-famous for their rich reptile faunas (Figures 8.14 and 8.15). They have yielded abundant material in the past and good finds are made frequently because of continuing erosion. This section is currently the best source of dinosaur material in Britain and it is just as rich as the well-known deposits in North America and Mongolia.

The section between Compton Bay and Atherfield Point has been described by White

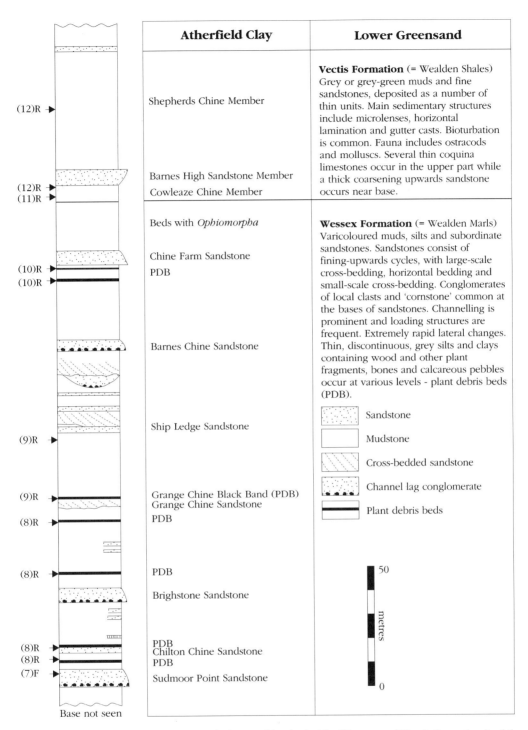

Figure 8.11 Summary sedimentary log through the Wealden beds (the Wessex and Vectis formations) of the south-western coast of the Isle of Wight between Sudmoor Point and Atherfield Point. Known reptile bone-bearing horizons are noted (R), as are footprint beds (F), and the numbers 7–12 match those used in the text in the locality descriptions. After Stewart (1981b).

(1921, pp. 5–15), Daley and Stewart (1979), Stewart (1981b), Simpson (1985), Stewart *et al.* (1991) and Wach and Ruffell (1991). The exposed portions are dated as mostly Barremian, but may range up to Early Aptian (Kerth and Hailwood,

1988; Hughes and McDougall, 1990; Allen and Wimbledon, 1991). The section is best known for its dinosaurs, having yielded remains of about 100 individuals belonging to 15 or so species, although Swinton (1936a) recognized 22 valid

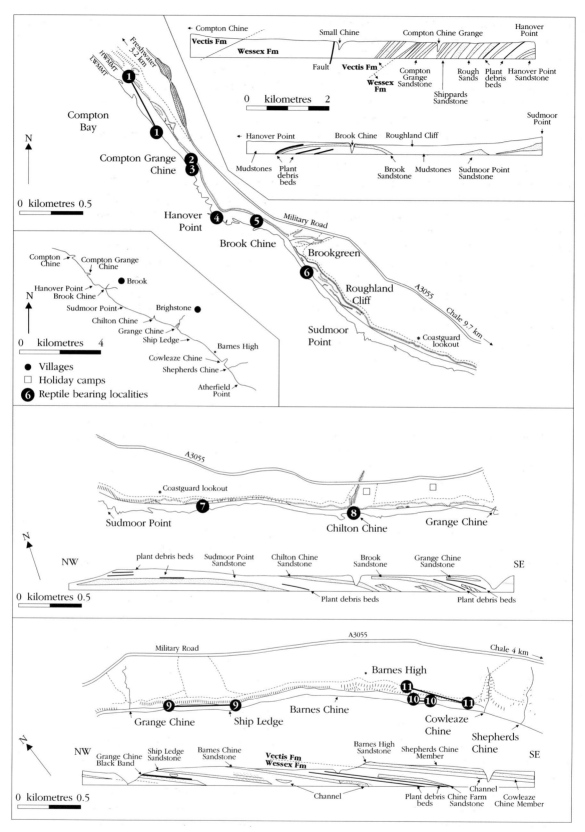

Figure 8.12 Maps and diagrammatic cliff views of the coastal section from Compton Chine to Atherfield Point, on the south-western coast of the Isle of Wight. Fossil reptile localities are indicated as 1–11, corresponding to the sites described in the text. After Stewart (1981b).

species. The reptiles from the Isle of Wight section have been described by numerous authors, including Buckland (1835), Owen (1842b, 1855b, 1858, 1859b, 1864, 1874c, 1876), Mantell (1849), Wright (1852), Beckles (1862), Fox (1866, 1869), Hulke (1870d, 1871c, 1873, 1874b, 1874c, 1874d, 1874e, 1876, 1878, 1879a, 1879b, 1880b, 1882a, 1882c, 1882d), Huxley (1870b), Seeley (1870a, 1875c, 1882a, 1883, 1887b, 1887c, 1887d, 1888d, 1892, 1901), Lydekker (1887a, 1888a, 1888b, 1889a, 1889b, 1889d, 1890a, 1890c, 1890d, 1891), Andrews (1897), Hooley (1900, 1907, 1912, 1913, 1925), Nopcsa (1905a, 1905b, 1928), Huene (1923, 1926, 1929b), Swinton (1936a, 1936b), Galton (1969, 1971a, 1971b, 1973, 1974, 1975, 1976a, 1977, 1981a), Ostrom (1970), Blows (1978, 1982, 1987), Buffetaut and Ford (1979), Buffetaut and Hutt (1980), Charig (1980), Norman (1980, 1986, 1990b), Delair (1982c), Hutt *et al.* (1989), Howse and Milner (1993), Pereda-Suberiola (1993), Radley and Hutt (1993) and Radley (1993), and Insole and Hutt (1994).

Description

The Wealden Group along the Brook–Atherfield section (Figures 8.11 and 8.12) is exposed in the core of the Brighstone Anticline, the hinge of which is difficult to locate, but lies within Brook Bay. The Wealden Group and Atherfield Clay Formation (part) are to be seen at both ends of the section, and the oldest in the Brook Chine area. The section, on the southern limb of the anticline, is summarized from White (1921), with refinements from Simpson (1985), and formation and member names from Stewart (1978), Daley and Stewart (1979), Simpson (1985) and Wach and Ruffell (1991).

Thickness (m)

LOWER GREENSAND
Atherfield Clay Formation (=Atherfield Group)
    Chale Clay Member (=Atherfield Clay) (beds 3–6 of Simpson (1985, p. 27, fig. 4): pale bluish-grey silty clay with numerous small round or irregular clay–ironstone nodules, some forming discrete bands: highly fossiliferous, containing small teeth (presumably derived) of *Hybodus* and *Lonchidon*, pyritized wood and bivalves.   19

Thickness (m)

Perna Beds Member: Upper Sandstone (bed 2 of Simpson, 1985, p. 27, fig. 4): Hard, coarse-grained, greenish calcareous sandstone in which marine fossils (bivalves, brachiopods, corals, rare ammonites and burrows) occur   0.54
Lower Clay and Atherfield Bone Bed (bed 1 of Simpson, 1985, p. 27, fig. 4): grey-brown, passing into dark blue, sandy clay with many bivalves (including *Panopea*, *Aetostreon* and *Mulletia*), echinoids, brachiopods, but no indigenous ammonites; at the base is a thin layer (10–100 mm) of coarse quartz grit, bone fragments, fish teeth, phosphate nodules, rolled Jurassic ammonites and reptile remains (Atherfield Bone Bed)   0.85
.... Disconformity ....

WEALDEN BEDS
Vectis Formation (=Wealden Shales)
    Shepherd's Chine Member: grey or grey-green muds and fine sandstones, deposited as a number of thin cyclic units; impersistent ironstone lenses; several thin coquina limestones, and other beds with ostracods, plants and fishes   45
    Barnes High Sandstone Member (=Sandstone of Cowleaze Chine and Barnes High of White (1921): massive, cross-bedded, yellow sandstone, with bands of *Filosina*, overlying thin-bedded sandstone with shale   7
    Cowleaze Chine Member: blue shales containing bivalves, overlying white sand and clay   8
Wessex Formation (= Wealden Marls)
    Beds with *Ophiomorpha*: at the very top, red sand with bones (*Hypsilophodon* Bed, 1 m); then reddish-brown mudstones, laminated in places, with mudcracks, calcareous nodules, burrows and rootlets, inter-bedded with medium-grained, cross-laminated sandstones; includes, about the middle, a new fossiliferous bed (Radley and Hutt, 1993)   14
    Chine Farm Sandstone: white and yellow sand, with fragments and large trunks

Thickness (m)

of carbonized wood ('lignite')	3
Clays/marls: pale-blue and purple clays, with two plant debris beds near the top (9 m), overlying 'hard green bed, containing lignite and bones' (0.7 m), followed by deep-red marls (2 m) and purple and mottled marls (10 m)	22
Barnes Chine Sandstone: sandstone with clayey beds	4
Deep-red marls, purple below	9
Pebbly sandstone: channel fill	1
Clays/marls/sands: green and white clays with purple and red marl and white, sandy interbeds	20
Ship Ledge Sandstone: fine, white sandstone.	1
Mottled marls	8+
Grange Chine Black Band (Black Band of Brixton Chine; White, 1921, p. 14): plant debris bed with bivalves and bones.	0.8
White, sandy marl (1 m) overlying 'mottled red marls of Brixton (=Grange) Chine, with a plant debris bed near the middle' (29 m); the Grange Chine Sandstone occurs to the west of Grange Chine near the top	30
Marls/sandstones: green sandy bed with bones (0.7 m), overlying red and white sandstones interbedded with marl and a (0.1 m) bed of fragmented bone and pebble bed at the base (5 m), overlying mottled marls (15 m)	30
(?) Brighstone Sandstone: pebbly band with carbonized wood and pebbles of sandstone (top of east bank of Chilton Chine)	0.7
Chilton Chine Sandstone: cross-bedded sandstone (near the bottom of Chilton Chine);	4
Marls/sandstones: mottled marls, purple marls with white calcareous concretions, and red marls passing down into cross-bedded, white sandstone and marl; plant debris beds near base.	13
Sudmoor Point Sandstone: massive sandstone with irregular bands of bone; 0.2–0.6 m of gravel at base, with bones; *Iguanodon* footprints near the top	6
Deep red and purple marls	seen to 6

Unlike most other British fossil reptile locali-
ties, there is a large amount of information about provenances of finds made in the Compton Bay–Atherfield section. The information given below is extracted particularly from White (1921), other sources (cited below) and from museum labels. Unusually, there has always been a tradition among collectors of recording the locations of fossil reptile finds with a degree of precision encountered nowhere else in Britain. Nearly all the specimens have a label designation such as 'Brook Bay' or 'Cowleaze Chine', which restricts the provenance to a particular part of the stratigraphic column, and further collector information such as 'at beach level' or 'in a 6ft thick sandstone' is sometimes sufficient to iden-tify the exact horizon. The records given below are arranged geographically from north-west to south-east along the section (Figure 8.11), thus descending stratigraphically from Compton Bay to Sudmoor Point, and then ascending to Atherfield Point.

Compton Bay
(There is a fault at about SZ 371849).

1. 'White sandy clay, with bones' (2 m thick), above 4 m of 'deep red marls' immediately north-west of the fault (White, 1921, p. 9). This bone-bearing horizon, at about SZ 370850 (?) in the cliff, lies 82 m below the Perna Bed (28 m Wealden Shales, 54 m Wealden Marls). A recent find of *Polacanthus* (1979) by William Blows was probably from this bed (BMNH R9293; Blows, 1982, 1987). Blows (1987) records that the remains lay scattered within a confined pocket exposed near a ship-wreck, and only visible at low tide on the beach at SZ 347854. The site occurs in the lowest bed of the Vectis Formation and repre-sents the first recorded find of *Polacanthus* from this stratigraphic unit (A. Insole, pers. comm. to W. Blows, 1987). The sediment con-taining the remains was a pale grey, non-fissile, massive clay otherwise generally devoid of fos-sils. A femur of ?*Dryosaurus* (BMNH R8670) from Compton Bay (? this bed) is mentioned by Galton (1975, p. 750).

2. Sandstone containing '*Iguanodon*' footprints (Beckles, 1862) at about SZ 376842 on the beach ('600 yards west of Hanover Point': White, 1921, p. 14). This sandstone is in the Wessex Formation (repeated by the fault) just above the lignite band north-west of

Shippard's (Compton Grange) Chine (?Compton Grange Sandstone of Stewart's unpublished section).

3. Plant debris bed at SZ 377840, about 200 m west of Hannover Point (=locality IV.2 of Daley and Insole, 1984, p. 6; bed CH12 of Stewart, 1978). Buffetaut and Ford (1979) reported the discovery of crocodilian teeth (*Bernissartia*, Figure 8.14C) and other vertebrate remains beneath a fossil tree trunk in the cliff face. They stated that the tree trunk occurred 'at beach level in the second of the three 'lignitic bands' depicted by Osborne White (1921, fig. 1, p. 12).' White (1921) illustrates three lignitic bands, none of which is anywhere near the site mentioned by Buffetaut and Ford (1979). The map reference is probably correct since these latter authors state that the site was 'midway between Compton Grange Chine and Hanover Point', and thus in the Wessex Formation, and probably in the region of White's (1921, p. 9) 16 ft (5 m) 'White Sandstone (east of Compton Grange Chine)' or the 'variegated marl' (30 ft, 9 m) below.

4. Hanover Point sandstones: '*Iguanodon*' footprints are to be seen on reddish and grey sandstones on the foreshore reef at Hanover Point and to the north-west in bed CH8 of Stewart (1978), and abundantly in the overlying red mudstones (Daley and Insole, 1984, p. 10). Beckles (1862, p. 443) described such prints from 'the shore at low water, between Brook Point [i.e. Hanover Point] and the Chine to the west of it.' A specimen of *Iguanodon* was excavated on the foreshore reef at Hanover Point in 1984 (S. Hutt, pers. comm. to M.J.B.). Various other dinosaur remains have been recorded from Hanover Point (in IWCMS), but most seem to have come from localities in Brook Bay just to the south-east (see below).

Brook Bay

5. Hanover Point to Brook Chine: A specimen of *Iguanodon* was collected in 1872 between the cliff and the 'pine raft' (Seeley, 1875c; Blows, 1978, pp. 26–34), and there are several further dinosaur remains in the IWCMS from 'Hanover Point'. Buckland (1835, p. 428) recorded *Iguanodon* vertebrae 'along a quarter mile of this shore [near Brook], but most abundantly at a spot called Bull-face Ledge near Brook Point, where the iron-stone is abundantly loaded with prostrate trunks of fossil trees.' Mantell (1846, p. 94) further noted that many hundreds of bones had been collected along this stretch of shore where they had been eroded from beds of sandy clay with *Unio* immediately above the 'pine raft'. These sandstones are probably equivalent to those seen at Hanover Point and immediately to the west of it, since the same beds are seen at both sides of Hanover Point because of the sharp angle in the coastline here. Hulke (1882a, p. 135) described some *Iguanodon* remains from 'a bed of hard nodules intercalated between the red and purple clays below and the iron-stained flint-gravel which caps the cliff west of Brook Chine . . . A few yards east of where this nodule-bed touches the cliff-foot, the cliff is cut through by a small gully worn by a little rill. In the east bank of this gully were the fossils.' Hulke describes the nodule bed as apparently dipping west and passing beneath the sand seawards towards the 'pine raft'. The source bed, then, is probably close to those described by Buckland (1835) and Mantell (1846) on the coastal strip between Hanover Point and Brook Chine (SZ 379837–SZ 385835). Seeley (1882a, p. 367) further described a dinosaur coracoid 'from the cliff midway between the pine raft and Brook Chine, at about 10 feet above high-water mark'. Andrews (1897) reported an *Iguanodon* cranium found 'on the shore near Brook Point'. Most other specimens labelled as 'Brook' or 'Brook Bay' probably came from this section, and this includes material described by Seeley (1883, 1887b, 1888d) and Lydekker (1887a, 1890c, 1890d). Delair (1989) notes Victorian finds of '*Iguanodon* ichnites', in sandstones on the shore west of Brook Point.

6. Brook Chine to Sudmoor (Sedmore) Point: parts of the cliff have collapsed along this section, and exposure is poor, except at Sudmoor Point. Some of the specimens labelled as 'Brook' may have come from this section, but there are no specific records.

7. Sudmoor Point to Chilton Chine: Sudmoor Point Sandstone: tridactyl '*Iguanodon*' and '*Megalosaurus*' footprints have been recorded in the sandstone between Sudmoor Point and

Chilton Chine by several authors (Beckles, 1862, p. 444 ('Southmore'); White, 1921, p. 7; Blows, 1978, pp. 44–58; Insole, *in* Daley and Stewart, 1979; Delair, 1989). The recent finds, from the Sudmoor Point Sandstone were made from a foreshore ledge at low tide level just west of Chilton Chine. These consisted of over 30 imprints of different shapes and sizes and constituting portions of 10 separate tracks (Blows, 1978). Insole (1982) regards all the tracks as being iguanodontid. Limb bones of *Valdosaurus* were found recently west of Chilton Chine (Radley, 1993).

8. Sudmoor Point to Chilton Chine: several bone-bearing horizons occur in the marls, sandstones and plant debris beds above the Sudmoor Point Sandstone (White, 1921, p. 14). Hulke (1870d) described a large vertebra whose locality was considered to be 'a bed which occurs near the top of the high cliff between Brooke and Chilton', and this could lie either to the west or east of Sudmoor Point. Buffetaut and Ford (1979) noted the occurrence of *Bernissartia* teeth 'from the *Unio* bed on the cliff at Sudmore Point' (?exact horizon). Galton (1975, p. 750) noted an ornithopod femur (BMNH R8670) from a 'bone bed between high and low water, Clinton Chine' (?Chilton Chine), and thus probably a bed just below the Chilton Chine Sandstone. Hulke (1879a) described a centrum from the cliff near Chilton, which could refer to a location to the east or west of the chine. There is further localized material from these beds in the IWCMS. The locality is a small conglomeratic lens, rich in *Margaritifera* ('*Unio*'), between the Chilton Chine Sandstone and Sudmoor Point Sandstone (Bed SS3 of Stewart, 1978; A. Insole, pers. comm., 1993).

Brighstone (or Brixton) Bay

9. Brighstone Bay (Grange [Brighstone] Chine to Barnes Chine): the upper portion of the Wealden Marls sequence is exposed between Grange Chine and Barnes Chine and there are several plant debris beds with bones – in particular the Grange Chine (Brixton Chine) Black Band at the top of the east side of Grange Chine. The iguanodontid dinosaur *Vectisaurus valdensis* was collected in a clay at the cliff-foot, '300 yards east of the flagstaff near Brixton Chine' (Hulke, 1879b). The flagstaff

was at the small headland east of Grange Chine (SZ 427813) (S. Hutt, pers. comm.), so that the skeleton was found on Ship Ledge at about SZ 429812, probably in the marls below the Ship Ledge Sandstone. A theropod was collected in 1978 by William Blows from the mottled red and blue marls above the Grange Chine Black Band at SZ 423815 (W. Blows, pers. comm.), and several IWCMS specimens have also been found here. Several specimens bear the labels 'Jolliffe's Road, Brixton' or 'Jolliffe's Road, Barnes Chine' (e.g. BMNH R5226-7, R5338, IWCMS 3306), but this name cannot be found on 6-inch OS maps. A trackway of trifid impressions was noted from a low intertidal locality between Brook and Brighstone by Beckles (1862). Other finds from Brighstone (Brixton) Bay are not localized further (Wright, 1852, p. 89; Hulke, 1874b, 1874c).

10. Barnes Chine-Cowleaze Chine (upper portion of Wealden Marls): White (1921, p. 13) mentions a 'lignite bed' with bones 12 m above the Barnes Chine Sandstone which is 'seen in the top of Barnes Chine' and reaches beach level to the east of Barnes High. A second plant debris bed, a few metres higher has also yielded bones. Several specimens have been recorded from these beds. Hulke (1882b) noted a good skeleton of *Polacanthus* found 'in a bed of blue shaley clay, a short distance east of Barnes Chine. The bed is easily recognized by the large quantities of lignite which it contains.' A theropod femur (BMNH R5194) is labelled 'Wealden from bone bed under Barnes High, Brighstone Bay, found on beach.' Galton (1973) suggested that this was the *Hypsilophodon* Bed (base of Wealden Shales; top of cliff at Barnes High), but it is more likely to have been one of the plant debris beds which outcrop at beach level. Blows (1978, pp. 34–42) described the excavation of an *Iguanodon* pelvis from one of these beach-level lignite beds between Barnes High and Cowleaze Chine. Delair (1982c) reported a spine of ?*Polacanthus* 'from the uppermost of the two lignite bone beds in the Wessex Formation (Wealden Marls), exposed in the low foreshore cliff below the south-east face of Barnes High, Isle of Wight (SZ 439805)'. Further bones have been found in these plant debris beds (IWCMS 5122, 5129, 5136-9). Two recent finds have been made in the top bed of the Wealden Marls, a 14 m thick bed of

red and mottled mudstones underlain by massive white and yellow sandstones. Buffetaut and Hutt (1980) reported a crocodilian, *Vectisuchus*, from the base of the bed at Barnes High, and a partly articulated *Iguanodon* (IWCMS 5126) was found about 10 m below the top and 400 m west of Cowleaze Chine (SZ 441804) (Insole, 1980). Several further specimens have been collected from these beds recently, including the new sauropod, from SZ 437807 (Radley, 1993; Radley and Hutt, 1993).

11. Barnes Chine–Cowleaze Chine (*Hypsilophodon* Bed): the *Hypsilophodon* Bed is one of the best known units of the sequence (Figure 8.13). It can be traced from the top of the cliff just west of Barnes Chine (SZ 434808) to beach level just west of Cowleaze Chine (SZ 443801). Owen (1855b, p. 2) noted a skeleton from 'about a hundred yards west of Cowleaze Chine.' Huxley (1870b) described specimens from the bed 'which forms the floor of Cowleaze Chine and rises to the top of the sea cliff at Barne's High'. Hulke (1873) reported *Hypsilophodon* remains 'from the same Cowleaze bed' and further specimens (Hulke, 1874d) from the same unit 'in a block of sandy clay-stone.' Owen (1874b, p. 13) quoted from a letter by Fox: 'this slab was found in the fallen cliff, about 150 yards east of 'Barnes High', directly fronting the den of my *Polacanthus* . . . The skull and broken jaw were found about 60 yards further eastward' (SZ 437806, SZ 438806). Hulke (1882c, p. 1036) described the bed in some detail: 'The rock varies much often within the space of a few yards. Generally the upper 3ft of it consist of a cap of grey sandstone resting on sandy clay; this is succeeded by about the same depth of mottled-red and blue clay lying on the bands of sandstone. The *Hypsilophodon* remains are almost restricted to the lower half of the bed.' He mentioned the only other bones from the bed: rare remains of *Goniopholis* (?) and turtles. White (1921, p. 13) gave the relevant section as:

	Thickness	
	ft	in
White sand and clay	2	6
White rock	2	6
Red sand, with bones (*Hypsilophodon* Bed)	3	0

He noted that near Cowleaze Chine the 'white rock' was a pale, calcareous, silty stone containing *Unio* and bones, and that remains of *Hypsilophodon* had also been found in the marls a little below the *Hypsilophodon* Bed in Brixton Bay (White, 1921, p. 15). Galton (1974, pp. 15-18) gave more details of the *Hypsilophodon* Bed and of its lateral variation. He noted finds of bones both in the bed itself and in the white rock above, and emphasized that the locality designation usually given, 'Cowleaze Chine', is rather inappropriate since specimens came from sites 100–900 m west of the chine. Several recent finds have been made in the *Hypsilophodon* Bed (IWCMS 5123-4) and the 'White Rock' (IWCMS 5143, 5165, 5180). Insole (1980) noted remains of *Hypsilophodon* in red-mottled grey marls 'immediately beneath the *Hypsilophodon* Bed about 200 metres west of the Chine' (i.e. Cowleaze Chine, thus about SZ 442802).

12. Barnes Chine–Atherfield Point (Vectis Formation): Hooley (1912) reported a partial *Iguanodon* skeleton from 8 ft (2.5 m) above the *Hypsilophodon* Bed at the base of the blue shales 150 yards west of Cowleaze Chine. White (1921, p. 15) noted bones of *Iguanodon*, *Goniopholis* and *Ornithodesmus* from the shales above the *Hypsilophodon* Bed and in the Barnes High Sandstone. Buffetaut and Ford (1979) recorded teeth of *Bernissartia* 'in the Wealden Shales overlying the *Hypsilophodon* Bed at Cowleaze Chine.' A partial *Iguanodon* skeleton (BMNH R5331) is labelled 'from the shales between the grey sandstone and purple-coloured marls overlying the *Hypsilophodon* Bed, 300 yards west of Cowleaze Chine'. The exact horizon of another partial *Iguanodon* skeleton ('lignite band, 100 yards west of Cowleaze Chine'; probably from a plant debris bed within the White Rock; A. Insole, pers. comm., 1993) is uncertain. Hooley (1900) reported a fossil tortoise from 'about 10 feet above low water-mark opposite Shepherd's Chine' (SZ 446798). Further bones are labelled 'Wealden Shales, Sheperd's Chine' (IWCMS 4128, 4199-200). Hooley (1913) noted two specimens of *Ornithodesmus* from a rock fall at Atherfield, and the label (BMNH R3877-80) indicates a locality 20 yards west of Shepherd's Chine (SZ 447789). Many of the other fossil reptiles collected by Hooley are labelled 'Tie Pits, Atherfield' (BMNH speci-

Figure 8.13 The *Hypsilophodon* Bed at Cowleaze Chine, high in the Wealden sequence. Stephen Hutt points to the horizon from which several complete skeletons of *Hypsilophodon* have been excavated. (Photo: M.J. Benton.)

mens), which probably refers to the broad area of collapsed and pitted cliffs between the coastguard station and Atherfield Point. These include a partial skeleton of *Goniopholis* found about 80-90 ft (25-28 m) below the top of the Vectis Formation (Hooley, 1907), thus just below the middle of the Shepherd's Chine Member. There was a small brickpit immediately west of Atherfield Point, and in the upper part of the Vectis Formation, which probably yielded these older specimens, as well as some new finds of *Iguanodon* (A. Insole, pers. comm., 1993). The *Iguanodon* (IWCMS 5196) came from the 'Diplocraterion Band' of the Shepherd's Chine Member and is encrusted in oysters, pyritized and marked with some predatory scratches (J. Radley, pers. comm., 1993). Stewart *et al.* (1991, p. 125) note plesiosaur remains from black mudstones near the top of the Shepherd's Chine Member. Tridactyl footprints have been found recently loose on the shore between Cowleaze Chine and Atherfield Point (SZ 444801–453792; Radley, 1993).

The preservation of the reptile remains from the Compton Bay–Atherfield section is variable. Bones found *in situ* are in various degrees of articulation or are isolated elements, and they may be crushed or virtually unaffected by compaction. The well-recorded (Blows, 1987) new specimen of *Polacanthus* (Figure 8.15G) is atypical of the preservation at this locality, being semi-articulated and in good condition, with the delicate processes of most elements intact. There appear to be two modes of preservation: well mineralized (pyrites, baryte, etc.) black bones in organic facies, such as the plant debris beds and Vectis Formation shales; and, poorly mineralized pale-coloured bones, found in overbank muds and channels (J. Radley, pers. comm., 1993).

Fauna

Large numbers of reptiles from various sites in the Compton–Atherfield section are preserved in British museums, especially BMNH and IWCMS. Type specimens are noted, and an estimate is given of the numbers of specimens of each species in major collections. Clearly there is much more material in other collections, but the figures will give an impression of relative abundance. Reptiles from all horizons are treated together since most occur throughout the succession (except *Hypsilophodon*).

Numbers

Testudines: Cryptodira: Pleurosternidae	
Helochelydra Nopsca, 1928 (no species name)	
Type specimen: BMNH R171	1
Testudines: Cryptodira: Plesiochelyidae	
Plesiochelys brodiei Lydekker, 1889	
Type specimen: BMNH R1444 (cast)	2
Plesiochelys valdensis Lydekker, 1889	
Type specimen: BMNH 28967	1
Plesiochelys vectensis Hooley, 1900	
Type specimen: BMNH R6683	1
Plesiochelys sp.	2
'chelonian'	1
Archosauria: Crocodylia: Neosuchia: Goniopholididae	
Goniopholis crassidens Owen, 1841	12
Goniopholis minor Koken, 1887	1
Goniopholis sp.	c. 60
Oweniasuchus sp.(?)	1
Vectisuchus leptognathus Buffetaut and Hutt, 1980	
Type specimen: Staatl. Mus. Naturk. Stuttgart 50984	1
Archosauria: Crocodylia: Neosuchia: Pholidosauridae	
Pholidosaurus meyeri (Dunker, 1844)	3
Suchosaurus cultridens Owen, 1841	1
Suchosaurus sp.	1
Archosauria: Crocodylia: Neosuchia: Atoposauridae	
Theriosuchus sp.	1
Archosauria: Crocodylia: Neosuchia: Bernissartiidae	
Bernissartia sp.	(40 teeth)
Archosauria: Crocodylia: Neosuchia: Eusuchia	
Hylaeochampsa valdensis (Seeley, 1887)	1
Hylaeochampsa vectiana Owen, 1874	1
Type specimen: BMNH R177.1	
Hylaeochampsa sp.	2
'crocodilian'	1
Archosauria: Pterosauria: Pterodactyloidea: Ornithodesmidae	
Ornithodesmus latidens Seeley, 1901	

Numbers

Type specimen: BMNH R176	3
Ornithodesmus sp.	2
'pterosaur'	1
Archosauria: Dinosauria: Saurischia: Theropoda	
Aristosuchus pusillus (Owen, 1876)	
Type specimen: BMNH R178	?5
Calamospondylus foxi Lydekker, 1889	
Type specimen: BMNH R901	1
Ornithodesmus cluniculus Seeley, 1887	
Type specimen: BMNH R187	1
Thecocoelurus daviesi (Seeley, 1888)	
Type specimen: BMNH R181	1
'coelurosaur'	2
Megalosaurus dunkeri Koken, 1887	1
Megalosaurus sp.	21
?Allosaurid	1
Archosauria: Dinosauria: Saurischia: Sauropoda	
Astrodon valdensis (Lydekker, 1889)	
Type specimen: BMNH R1730	3
Cetiosaurus sp.	4
'diplodocid'	1
Pelorosaurus hulkei (Seeley, 1870)	22
(?)*Titanosaurus valdensis* Huene, 1929	
Type specimen: BMNH R151	2
'sauropod'	5
brachiosaurid	1
Archosauria: Dinosauria: Ornithischia: Ornithopoda: Hypsilophodontidae	
Hypsilophodon foxi Huxley, 1870	
Type specimen: BMNH R197	26
'hypsilophodontid'	1
Valdosaurus canaliculatus Galton, 1975	
Type specimen: BMNH R185, R186	4
Archosauria: Dinosauria: Ornithischia: Ornithopoda: Iguanodontidae	
Iguanodon atherfieldensis Hooley, 1925	
Type specimen: BMNH R5764	1
Iguanodon bernissartensis Boulenger, 1881	23
Iguanodon gracilis (Lydekker, 1888)	
Type specimen: BMNH R142	7
Iguanodon sp.	105
Vectisaurus valdensis Hulke, 1879	
Type specimen: BMNH R2494	4
Archosauria: Dinosauria: Ornithischia: Ankylosauria: Nodosauridae	
Polacanthus foxi Hulke, 1882	
Type specimen: BMNH R175	5
Polacanthus sp.	c. 30
'nodosaur'	4

Numbers

Sauropterygia: Plesiosauria
  '*Plesiosaurus* sp.'   10

Interpretation

Stewart *et al.* (1991) interpret the Wealden Group on the west coast of the Isle of Wight (Figures 8.11 and 8.12) as a sequence that records a shift from terrestrial deposition to fully marine. The lower unit, the Wessex Formation, is a fluviatile/coastal plain unit; the Vectis Formation above was deposited in a lagoon that was shallow and temporarily emergent, and the overlying Atherfield Clay Formation consists of marine units. Climatic conditions were seasonal, with wet and dry seasons in warm temperate to subtropical latitudes (Stewart, 1981b). The Wessex Formation contains numerous coarse sandstones deposited in channels, as well as overbank mudstones (marls), and a number of thin plant debris beds (carbonized wood with dinosaur and crocodilian bones, fish remains, plant cones and, occasionally, bivalve shells) represent reworked terrestrial fossils from flood events (Daley and Stewart, 1979).

The Vectis Formation is divided by Stewart *et al.* (1991) into four facies: fine sandstones, heterolithic sand/silt and mudstones, parallel-laminated mudstones and black mudstones, which occur cyclically through the sequence. The cyclicity may relate to advance and retreat of deltaic sand bodies into the lagoon, of which the Barnes High Sandstone Member may be a major example. Mollusc and ostracod associations give measures of salinity. These authors note that salinity and the frequency of storms increase towards the top of the Vectis Formation, and the sequence is terminated by the Atherfield Clay Formation, representing the major Aptian marine transgression.

Turtles are relatively uncommon in the Wealden of the Isle of Wight. Fewer than ten specimens are known, compared with many hundreds of crocodilians and dinosaurs. The genus *Helochelydra* Nopsca, 1928 belongs to *Tretosternon* Owen, 1842 (Młynarski, 1976, pp. 60-1). All other forms have been referred to the genus *Plesiochelys*, a well-known Late Jurassic and Cretaceous form of disputed affinities (Gaffney, 1976; Młynarski, 1976). The species *P. brodiei* and *P. valdensis* were erected by Lydekker (1889d, pp. 236-9) on the basis of well-preserved carapaces (also Lydekker, 1889b, pp. 199-201). Hooley (1900) erected the third species, *P. vectensis*, again on the basis of a carapace. The species are distinguished by minor differences in the shapes of various plates in the carapace. An examination of the illustrations suggests, for example, that *P. valdensis* and *P. vectensis* may be identical.

A variety of small and large crocodilians is known from the Isle of Wight, and with a variety of terrestrial and aquatic adaptations. *Goniopholis*, which is well known in the Late Jurassic and Early Cretaceous of Europe and North America, is represented on the Isle of Wight by many specimens. Lydekker (1890a, pp. 229-30) mentioned some material of *G. crassidens* from the Isle of Wight (Figure 8.14A), and Hooley (1907) described a relatively complete skeleton from Atherfield. The skull was 540 mm long and it was capable of a gape of over 1 m. *Oweniasuchus* and *Vectisuchus* are also goniopholids. *V. leptognathus* has been described on the basis of a partial skeleton and skull (Figure 8.14B), which is characterized by a long slender snout (Buffetaut and Hutt, 1980). *Pholidosaurus* and *Suchosaurus* are pholidosaurids, a largely aquatic group. The goniopholids were ecological counterparts of today's crocodilians and alligators, and the pholidosaurids of gavials (Buffetaut, 1982, pp. 29-38). Buffetaut (1983) has also noted the occurrence of *Theriosuchus*, based on odd teeth and a skull fragment. *Theriosuchus* is an atoposaurid (Benton and Clark, 1988, p. 321), previously known only from the Purbeck (q.v.).

More advanced crocodilians from the Compton–Atherfield section include *Bernissartia* and *Hylaeochampsa*. *Bernissartia*, a small crocodilian with button-like teeth (Figure 8.14C) for crushing molluscs, has recently been identified from several locations (Buffetaut and Ford, 1979). *Hylaeochampsa* was a 2 m long crocodilian known from the Purbeck and Wealden of England; Owen (1874c) described *H. vectiana* on the basis of a partial skull with large orbits (Figure 8.14D), and Lydekker (1888a, p. 75) referred some Isle of Wight material to *H. valdensis* (Seeley, 1887). Both *Bernissartia* and *Hylaeochampsa* are of some importance, the latter being the oldest known eusuchian (Benton and Clark, 1988, p. 323; Clark and Norell, 1992), the former being close to the origin of the Eusuchia (Norell and Clark, 1990), and each is placed in its own family.

Remains of pterosaurs are rare, but significant.

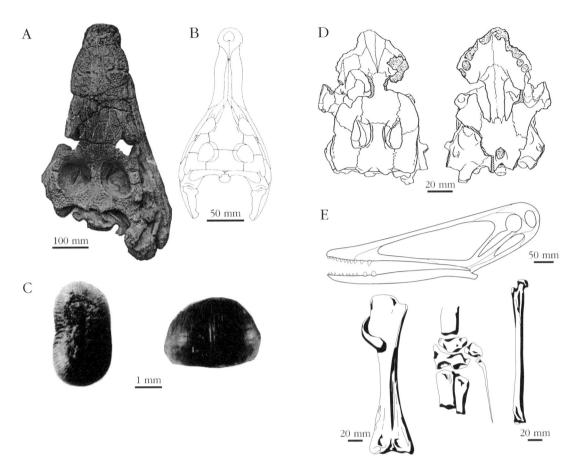

Figure 8.14 Typical non-dinosaurian reptiles from the Early Cretaceous Wealden of the south-western coast of the Isle of Wight. (A) The crocodilian *Goniopholis crassidens* Owen, 1841, skull in dorsal view; (B) the crocodilian *Vectisuchus leptognathus* Buffetaut and Hutt, 1980, restored skull and lower jaws in dorsal view; (C) teeth of the crocodilian *Bernissartia* sp., in crown and side views; (D) the crocodilian *Hylaeochampsa vectiana* Owen, 1874, skull in dorsal and ventral views; (E) the pterosaur *Ornithodesmus latidens* Seeley, 1901, restoration of skull, humerus, wrist, and femur. (A) After Hooley (1907); (B) after Buffetaut and Hutt (1980); (C) after Buffetaut and Ford (1979); (D) after Clark and Norell (1992); (E) after Wellnhofer (1978), based on several sources.

Seeley (1887b) described a sacrum from Brook as *Ornithodesmus cluniculus* (BMNH R187) and interpreted it as that of a bird. Lydekker (1888a, p. 42) suggested that it was, in fact, a pterosaur, but Howse and Milner (1993) have reinterpreted it as a theropod dinosaur (see below). Seeley (1901, p. 173) later named a partial pterosaur skeleton and skull from Atherfield (BMNH R176) as *O. latidens* (Figure 8.14E) and Hooley (1913) described it in detail. Wellnhofer (1978, pp. 54–5) suggested that both species may be the same. *Ornithodesmus* was a large animal (skull 560 mm long (?), estimated wing-span 5 m) and it is placed in its own family.

Four species of carnivorous theropod, three 'coelurosaurs', and one carnosaur have been described. *Calamospondylus oweni* was described by Fox (1866) on the basis of some pelvic remains (Figure 8.15A), and is probably the same as *Aristosuchus pusillus,* which was described by Owen (1876) on the basis of some sacral and lumbar vertebrae and a claw. Owen (1876) regarded the remains as those of a crocodilian and ascribed his new species to *Poikilopleuron*, a genus known from the Mid Jurassic of France. Seeley (1887c) noted that *Poikilopleuron* was very like *Megalosaurus*, and that the Isle of Wight animal was a 'coelurosaur' for which he erected the new genus *Aristosuchus*. Lydekker (1888a, pp. 157–9) agreed with this, and Huene (1926) amplified the original description. Galton (1973) ascribed a partial femur from Barnes High to *A. pusillus*. *Calamospondylus foxi* Lydekker (1889a) was probably rather similar, but it was based on only two cervical vertebrae. Lydekker (1891) figured more material which he ascribed to *C. foxi*. The third Isle of Wight 'coelurosaur', *Thecocoelurus daviesi* Seeley, 1888 was described on the basis of the anterior third of a cervical vertebra. Seeley

(1888d) referred this to *Thecospondylus*, a genus erected on the internal mould of a sacrum from Kent (Figure 8.15B). Lydekker (1888a) referred the specimen to the genus *Coelurus*, and Huene (1923, p. 455; 1926) erected the new genus *Thecocoelurus* for it. In conclusion, three genera

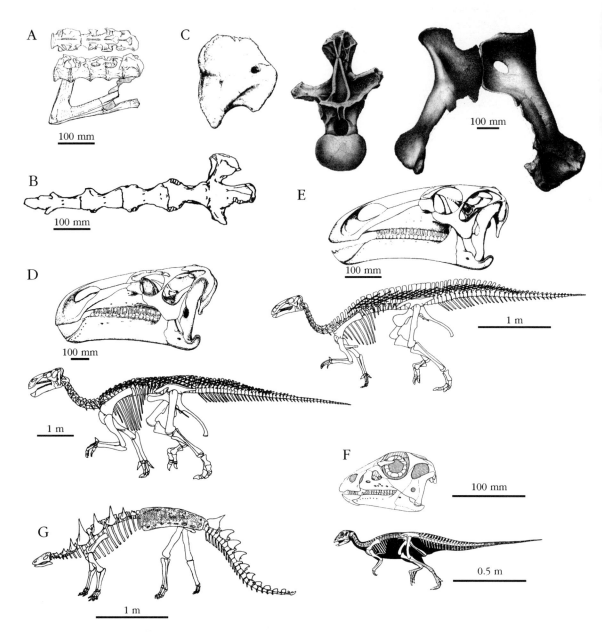

Figure 8.15 Typical dinosaurs from the Early Cretaceous Wealden of the south-western coast of the Isle of Wight. (A) The theropod dinosaur *Calamospondylus oweni* Fox, 1866, sacrum and pubis in dorsal and lateral views; (B) the theropod dinosaur *Thecospondylus horneri* Seeley, 1882, natural cast of the sacral cavity; (C) the sauropod dinosaur *Pelorosaurus hulkei* (Seeley, 1870), a dorsal vertebra in anterior view, a coracoid, and the pubis and ischium; (D) the large ornithopod *Iguanodon atherfieldensis* Hooley, 1925, skull and skeleton; (E) the large ornithopod *Iguanodon bernissartensis* Boulenger, 1881, skull and skeleton; (F) the small ornithopod *Hypsilophodon foxii* Huxley, 1869, skull and restored skeleton; (G) the ankylosaur *Polacanthus foxi* Hulke, 1881, skeleton. (A) After Seeley (1887c); (B) after Seeley (1882a); (C) after Hulke (1880b, 1882d), Seeley (1882); (D) and (E) after Norman (1980, 1986); (F) after Galton (1974); (G) after Blows (1987).

of 'coelurosaur' have been named from the Isle of Wight section, but each is based on miserable material, and there may only be one or two forms present (Ostrom, 1970, pp. 130-1, 140). Norman (1990a, p. 282) wisely termed all of these as *nomina dubia*. *Ornithodesmus cluniculus* has been reinterpreted (Howse and Milner, 1993) as a fourth small theropod, specifically a maniraptoran and possibly a troodontid, the earliest record of that family, if confirmed.

The carnosaur '*Megalosaurus*' is represented by some teeth, claws and vertebrae (Lydekker, 1889a, pp. 44-5, 166; 1891, pp. 244-5), a fragmented skeleton and two partial skeletons (Hutt *et al.*, 1989). The first skeleton discussed by Hutt *et al.* (1989) (BMNH R10001/IWCMS 6348) appears to share certain characters with *Megalosaurus nethercombensis* (Waldman, 1974) from the Inferior Oolite of Dorset. The second skeleton (IWCMS 6352) consists of cervical and dorsal vertebrae, ilia, sacral vertebrae, complete paired pubes and other elements. The pubic symphysis in this form is extraordinarily enlarged and the animal may belong to a new carnosaur species (Hutt *et al.*, 1989, p. 140). A comparison with North American carnosaurs has more recently suggested allosaur affinities for the new specimen, and it is being described by Stephen Hutt.

Sauropods are also rare on the Isle of Wight, being represented by incomplete material, but this did not deter early workers from erecting numerous genera and species, which gives the false impression of a diverse fauna. Lydekker (1890c) described *Pleurocoelus valdensis* on the basis of some teeth and a vertebra from Sussex and a vertebra from Brook Bay. The species has been referred to the genus *Astrodon* (Steel, 1970, p. 67; Galton, 1981a, p. 252), but McIntosh (1990, p. 348) is uncertain of the validity of the latter genus. The meagre remains indicate a relatively small sauropod (vertebrae 100-130 mm long compared with 500 mm in *Diplodocus*). Several vertebrae from the Isle of Wight were referred by Lydekker (1888a, pp. 139-41) to *Cetiosaurus brevis* Owen, 1842, but this species is invalid since the type specimen belongs to *Iguanodon* (Steel, 1970, p. 64; Ostrom, 1970, p. 129). Lydekker (1887a, 1888a, pp. 135-6) described two partial caudal vertebrae from the Isle of Wight as *Titanosaurus* sp., and Huene (1929b) erected the new species *T. valdensis* for these. Ostrom (1970, p. 130) confirmed the titanosaurid nature of these, and McIntosh (1990, p. 351) ascribed them to *Macrurosaurus semnus* Seeley, 1869, known

also from the Cambridge Greensand. An unusual caudal chevron has been identified as 'diplodocid' (Charig, 1980). The '1992 sauropod' (Radley, 1993; Radley and Hutt, 1993), consisting of vertebrae and limb bones, appears to be a brachiosaurid that would have been about 15 m long.

The commonest sauropod in the Wealden of the Isle of Wight, and of the Weald, is *Pelorosaurus*, and numerous isolated vertebrae, teeth and limb bones have been described from the Compton-Atherfield section (Figure 8.15C), and ascribed to the genera *Chondrosteosaurus, Eucamerotus, Ornithopsis* and *Pelorosaurus* (Wright, 1852; Seeley, 1870a; Hulke, 1870d, 1879a, 1880b, 1882d; Owen, 1876; Lydekker, 1888a, pp. 146-51). Steel (1970, pp. 68, 70) synonymized these and numerous other Late Jurassic and Early Cretaceous genera with *Pelorosaurus*, and he ascribed all the Isle of Wight material to *P. hulkei* (Seeley, 1870). This animal had 85 mm long peg-like teeth and 350 mm long vertebrae. McIntosh (1990, pp. 348-9) accepted the validity of *Pelorosaurus conybeari* (Melville, 1849) and *Chondrosteosaurus gigas* Owen, 1876 from the Isle of Wight. Ostrom (1970, pp. 129-30, 140) considered that there may be a minimum of two Wealden sauropods.

The commonest dinosaurs on the Isle of Wight are the ornithopods *Iguanodon* and *Hypsilophodon*. *Iguanodon* was recorded from Brook Bay and Yaverland by Buckland (1835) and Mantell (1846). Further material from the Isle of Wight was described by Owen (1842b, 1855b, 1858, 1859b, 1864), Hulke (1871c, 1874b, 1874e, 1876, 1878, 1882a), Seeley (1875c, 1882a, 1883, 1887d), Lydekker (1888a, pp. 201-40, 1888b), Andrews (1897) and Hooley (1912, 1925). The species currently recognized from the Isle of Wight (Figure 8.15D and E) are *I. bernissartensis* (including *I. gracilis*) and *I. atherfieldensis* (Norman and Weishampel, 1990, p. 530), although Steel (1970, pp. 17-19) and Ostrom (1970, pp. 131-4) had accepted others as valid. The various species attained lengths of 5-8 m, and they may have fed on vegetation from trees.

Hypsilophodon, a small bipedal herbivore 1.5-2.5 m long (Figure 8.15F), was originally considered to be a juvenile *Iguanodon* (Mantell, 1849; Owen, 1855b; Fox, 1869). A good skull was described as *H. foxi* by Huxley (1870b). Numerous further finds were made (Hulke, 1873, 1874d, 1882c; Lydekker, 1888a, pp. 193-5; Nopcsa, 1905a). Since then, several studies on the anatomy, lifestyle and relationships of *Hypsilophodon* have been published (e.g. Swinton, 1936b; Galton, 1969, 1971a, 1971b, 1974, 1975). It has been variously interpreted as a tree-percher and as an active cursorial biped, the latter being the current view.

The other ornithopods from the Wealden of the Isle of Wight are less well known. *Vectisaurus valdensis* was described (Hulke, 1879b) on the basis of six vertebrae and an ilium. Galton (1976a) referred a further three specimens (vertebrae, pelvis and dentary) to the species and concluded that it was an iguanodontid. However, Norman (1990b) argued that *Vectisaurus* is a juvenile *Iguanodon atherfieldensis*. Finally, Galton (1975) erected the species *Dryosaurus? canaliculatus* for two small femora (previously referred to *Hypsilophodon foxi* by Lydekker, 1888a) and later made this the holotype of the genus *Valdosaurus* (Galton, 1977; Galton and Taquet, 1982).

Most of the Isle of Wight ankylosaurs have been referred to *Polacanthus foxi* (Figure 8.15G), but a few were classified as *Hylaeosaurus armatus*, a form originally described from the Wealden of Cuckfield. Fox (1866) reported a skeleton of an armoured reptile, lacking the skull, from Brighstone Bay and mentioned Owen's new name *Polacanthus*. However, Owen never described the specimen, and Hulke (1882b) supplied a detailed account, with the name *P. foxi*. Further descriptions of '*Hylaeosaurus*' and of *Polacanthus* from the Isle of Wight are those of Hulke (1874c), Lydekker (1888a, 1890d), Seeley (1892), Nopcsa (1905b), Blows (1982, 1987), Delair (1982c) and Pereda-Suberbiola (1991). Nopcsa (1928) erected the genus and species *Polacanthoides ponderosus* for a partial skeleton from Atherfield. Most authors noted the close similarity of *Polacanthus* and *Hylaeosaurus* (Hulke, 1882b; Lydekker, 1888a; Seeley, 1892; Ostrom, 1970, pp. 134–5; Coombs, 1978; Coombs and Maryanska, 1990), although Steel (1970), Blows (1987), and Pereda-Suberbiola (1993) argued for the validity both genera. Ostrom (1970, pp. 135, 141) suggested that *Polacanthoides* may be distinct from the other two genera, but others (Coombs, 1978; Coombs and Maryanska, 1990) have synonymized *Polacanthus* and *Polacanthoides* with *Hylaeosaurus*.

The plesiosaur remains (teeth, vertebrae and limb bones) from Tie Pits, Atherfield (BMNH R5180–5, 7–8), Brook (IWCMS 1586) and Compton Bay (BMNH R5186) do not appear to have been described, although they are mentioned by Stewart *et al.* (1991, p. 125), and it is consequently hard to assess their significance in the fauna.

Footprints variously ascribed to *Iguanodon* and '*Megalosaurus*' have been reported from several locations along the section (e.g. Beckles, 1862; Blows, 1978; Delair, 1989; S.H. Hutt, pers. comm.). They are found as trackways, or isolated prints weathered out in sandstone units on the foreshore. They are generally large three-toed prints, and resemble specimens from the Purbeck beds of Swanage and the Wealden of the Sussex coast (see above). Newer finds include four-toed casts from Brook, which may have been produced by a sauropod or an ankylosaur (J. Radley, pers. comm., 1993).

Comparison with other localities

The nearest comparable Wealden locality to the Compton–Atherfield section is the stretch of coast at Yaverland (see below) which exposes similar rocks and has yielded *Suchosaurus, Pelorosaurus, Iguanodon, Yaverlandia* and *Polacanthus*. The exposed Isle of Wight Wealden is largely, or wholly, Barremian in age (mid-Early Cretaceous), whereas reptile localities in the Wealden of the Weald are generally Valanginian (earliest Early Cretaceous). The exception in the Weald is Smokejacks Pit, Ockley (TQ 113372) which is in the Weald Clay (Hauterivian/Barremian in age) (*Iguanodon, Baryonyx*, ?crocodilians).

The turtles *Tretosternon* and *Plesiochelys* are well known from the latest Jurassic (Purbeck) and the Cretaceous of Europe (Młynarski, 1976, pp. 55, 60).

The crocodilian *Bernissartia* is known from the Wealden of Belgium, Sussex and eastern Spain, as well as possibly the Early Cretaceous of Texas (Buffetaut and Ford, 1979; Norell and Clark, 1990) and the ?latest Jurassic of Wimille, northern France (Cuny *et al.*, 1991). *Hylaeochampsa* may also be known from the Wealden of Sussex, but the synonymy is uncertain (Clark and Norell, 1992). *Goniopholis* occurs widely in the Late Jurassic and Cretaceous of Europe, North and South America, while *Oweniasuchus* is known from the Purbeck of Swanage, and ?Early Cretaceous of Portugal (Steel, 1973). *Vectisuchus* is restricted to the Isle of Wight. *Pholidosaurus* occurs in the Purbeck of Swanage and the Wealden of Germany, as well as the ?Late

Cretaceous of Brazil; *Suchosaurus* has been reported from the Early Cretaceous of the Weald and of Portugal (Steel, 1973). *Theriosuchus* is known best from the Purbeck of Swanage.

The Isle of Wight 'coelurosaurs' are hard to compare with relatives elsewhere because of the inadequate material. The theropod *Megalosaurus* has been reported from all parts of the world and from earliest Jurassic to latest Cretaceous. *M. dunkeri* is reputed to come from the Purbeck, Wealden and Lower Greensand of southern England (Steel, 1970, pp. 43–5). The sauropod *Pleurocoelus* is known from the Wealden of Sussex, and the Early Cretaceous of Maryland and Texas, USA (McIntosh, 1990). Diplodocids are known from the Late Jurassic of North America, Tanzania and China (Charig, 1980), *Pelorosaurus* is known from the Early Cretaceous of England, as is *Macrurosaurus* (McIntosh, 1990).

Of the ornithischians, *Vectisaurus*, is restricted to the Isle of Wight. The genus *Hypsilophodon*, however, is known from the Early Cretaceous of Spain (Las Zabacheras Beds, Teruel) and reputedly also from the Early Cretaceous Lakota Formation of North America (*H. wielandi*: Galton and Jensen, 1979), but Sues and Norman (1990, p. 500) note this last taxon as *nomen dubium*. Other hypsilophodontids are known from the Kimmeridge Clay of Weymouth (?), the Late Jurassic Morrison Formation of North America and Tendaguru Beds of Tanzania, the early to mid-Cretaceous of Montana, USA (Cloverly Formation), Antarctica, Victoria (Otway Group) and New South Wales (Griman Creek Formation), Australia, and the Late Cretaceous of Montana, Wyoming, South Dakota and Colorado, USA and Alberta and Saskatchewan, Canada (Sues and Norman, 1990). *Iguanodon* is best known from the Wealden of southern England and Belgium, but it has also been reported from the Purbeck beds of Swanage, the Wealden of Germany, the Lower Greensand of southern England, and the Early Cretaceous of Spain, Mongolia and North America (Norman and Weishampel, 1990, p. 530). *Valdosaurus canaliculatus* is known from the Wealden of Tilgate Forest, Sussex, and from Cornet, Bihor, Romania and the species *V. nigeriensis* from the El Rhaz Formation (Aptian), Gadoufaoa, Niger, West Africa (Sues and Norman, 1990, p. 500). *Valdosaurus* is closely similar to the hypsilophodontid *Dryosaurus* (e.g. Galton, 1977), a form known from the Late Jurassic of western North America and Tanzania.

The ankylosaur *Polacanthus* ranges from the Wessex Formation to the Lower Greensand (Ferruginous Sands) (Barremian to Lower Aptian) mostly from the Isle of Wight, with one specimen known from the mainland. This block came from the Upper Greensand (Albian) at Charmouth, Dorset, and contained parts of four disarticulated, but associated, dorsal vertebrae, a rib section and portions of flat dermal armour (sacral shield). If *Hylaeosaurus* is a synonym then the range extends to the Wealden of the Weald area. Further, if *Hoplitosaurus* is synonymous with *Polacanthus*, as Pereda-Suberbiola (1991) suggests, the range expands to include the Early Cretaceous Lakota Formation of South Dakota (source also of the North American *Iguanodon* and *Hypsilophodon*).

Plesiosaurs are rare in the Wealden. Isolated bones have also been registered from Ridgeway Hill, Dorset (SY 6785), Cuckfield and Hastings, Sussex (Lydekker, 1889a, pp. 188–90, 224–7), Berwick, Sussex (TQ 5205: Andrews, 1922), Telham, Sussex (TQ 769142) and Brenchley, Kent (TQ 6741).

Wealden dinosaur footprints are also known on the Isle of Wight from Yaverland, and from several sites along the Sussex coast from Bexhill to Cliff End (Beckles, 1854; Tylor, 1862; White, 1928; Delair and Sarjeant, 1985; Delair, 1989; Radley, 1993), as well as from the former West Germany (Bückeburg and Bad Rehburg, Niedersächsen) and Belgium (Bernissart) (Haubold, 1971, pp. 79, 86–9).

Conclusions

The Wealden section between Compton Bay and Atherfield Point is one of the most famous sources of dinosaurs in the world, and Britain's best. A large reptile fauna is known, including turtles and plesiosaurs, but the archosaurs are best represented. The seven genera of crocodilians include a good selection of aquatic goniopholids and pholidosaurids, as well as some forms close to the origin of the modern crocodilians, the eusuchians (*Hylaeochampsa*, *Bernissartia*). Remains of pterosaurs (*Ornithodesmus*) may represent a unique group. Of the dinosaurs, fragmentary 'coelurosaur' and sauropod remains are known, but the best represented dinosaurs are the ornithopods *Hypsilophodon*, *Iguanodon* (two species), *Valdosaurus*, and the armoured ankylosaur *Hylaeosaurus*. *Valdosaurus* and *Iguanodon* are of biostratigraphic importance, providing evidence of a land connection between northern Europe and Africa across Tethys during the Early Cretaceous, and *Hypsilophodon* and *Hylaeosaurus* relate the Isle of Wight dinosaur fauna with the Early Cretaceous faunas of Dakota in North America.

The international importance of finds from this site and the continuing potential for significant future discoveries give it a very high conservation value.

YAVERLAND, SANDOWN, ISLE OF WIGHT (SZ 613850–SZ 622853)

Highlights

Yaverland is an important Early Cretaceous dinosaur site, especially as the location which yielded *Yaverlandia*, the oldest pachycephalosaur (bone-headed dinosaur) known. Both dinosaur bones and footprints are still found at Yaverland.

Introduction

The cliff section and beach at Yaverland are a well-known source of Wealden dinosaurs. Remains were reported as long ago as 1835 and finds are still being made. Although rather overshadowed by the Compton–Atherfield section on the western side of the Isle of Wight, Yaverland is an important supplementary source of reptiles and these include the unique pachycephalosaur *Yaverlandia* (Figure 8.16).

The section in the Wealden at Yaverland (Sandown Bay) has been described by Reid and Strahan (1889, p. 17) and White (1921, pp. 15-19). The beds lie on the northern limb of the Sandown Anticline, the hinge zone of which occurs at Sandown Fort. The reptiles have been described by Buckland (1829c, 1835b), Mantell (1846, 1854), Reid and Strahan (1889), Gibson (1858), White (1921) and Galton (1971c).

Description

A summary of White's (1921) section is given, with recent stratigraphic nomenclature from Daley and Stewart (1979) and Simpson (1985).

Thickness (m)

Lower Greensand
  Perna Beds
    (bed 2) Calcareous sandstone — 0.1
    (bed 1) Thick, blue, sandy clay;
      Atherfield Bone Bed at base
      with a derived vertebrate fauna
      including *Hybodus* and *Lonchidon*
      (Patterson, 1966) — 1.15
      .... disconformity ....
Vectis Formation (=Wealden Shales)
  Shales, grey or blue, with ostracods
    and bivalves interbedded with thin
    (0.12 m) beds of shelly limestone
    and rare ironstones — *c.* 37
  Sandstone, yellow and white
    (equivalent of the Barnes High
    Sandstone) — *c.* 2.4
  Shales, grey with molluscs and
    ostracods with a 0.1 m band of clay
    ironstone in the middle — *c.* 10.5
Wessex Formation (=Wealden Marls)
  Marl, grey with coloured mottlings
    and irregular bands of large
    calcareous nodules in upper part — *c.* 3.0
  Silt, pale greenish-grey; hard and
    vesicular in places, with one or
    more bands of rolled concretions;
    much pyritized carbonized wood,
    Margaritifera (some phosphatized),
    reptilian bones, scales of *Lepidotus*,
    etc. — 0.45
  Clays, green, red and variegated
    (behind the old sea-wall) — *c.* 6.0
  Clays and marls, variegated, with
    bands of cross-bedded sand — seen to 15.0

Reptile remains have been described from the shore and cliff. Buckland (1829c; 1835, pp. 425-8) noted isolated dinosaur bones 'in the iron sand which forms the shore, a little east of Sandown Fort, between high and low water'. The Sandown Fort noted here is not the current one, which was built in the 1870s, but an earlier structure some 500 m to the south-west (approximately SZ 605846): there have been no exposures of bedrock at this site in living memory (A. Insole, pers. comm., 1993). This would place the site on the southern limb of the Sandown Anticline and within the Wessex Formation. Mantell (1846, p. 95) noted an *Iguanodon* tibia from the same site (also Mantell, 1847, pp. 137-8). Reid and Strahan (1889, p. 17) implied that Buckland's and

Mantell's bones came from the conspicuous yellow and white sandstone in the Wealden Shales, and Buckland's (1835) reference to 'iron sand' could be interpreted in such a way. However, no finds have been made since in that unit, nor in any of the iron-stones in the Vectis Formation (White, 1921, p. 18). It must be assumed that these early discoveries were not *in situ*, and that they were washed up on the beach (at about SZ 620853). Gibson (1858) refers to this eastern end of the section as Buckland's site and notes that 'large vertebrae and other portions of bone are frequently found, but always much rolled and broken.'

Gibson (1858) reported an *Iguanodon* femur in a low cliff of 'Weald Clay' exposed by a storm 'a little to the west of Sandown Fort . . . lies immediately above the ferruginous sandstone in which Dr Buckland discovered the metacarpal bone. The clay-bed in which the bone was found is near the centre of the arch which . . . is formed by the Wealden in Sandown Bay, dipping slightly westward . . . about half-a-mile' from Buckland's beach site. Therefore, Gibson's find was on the western limb of the anticline at about SZ 612849, thus in the low cliff currently exposed, or possibly a portion now covered by the concrete sea-wall. Mantell (1854, pp. 98-9, 226) noted *Iguanodon* remains from the foot of a small cliff that forms the sea boundary of Yaverland Farm, but this farm no longer exists and the site is hard to identify.

White (1921, pp. 15-19) described the occurrence of reptiles in two silty plant debris beds within the Wessex Formation which are to be seen in the cliff 10–50 m west of the old sea-wall (now collapsed), thus at about SZ 616851. He mentions bones and teeth of *Iguanodon* and carnivorous reptiles in the lower bed and from the top of a multiple plant debris bed unit, which is occasionally exposed on the beach east of the old wall, thus at about SZ 617852; it includes a basal bone-rich lag deposit (J. Radley, pers. comm.). The higher plant debris bed 'is rich in the remains of reptiles (chiefly *Iguanodon*, including *I. mantelli* Meyer, together with less common *Goniopholis crassidens* Owen and turtle (*Plesiochelys*?), and is the principal source of the wave-washed bones usually to be seen on the shore between Yaverland sea-wall and Redcliff . . . This bed comes down to the beach about 70 yards east of the sea-wall', thus at about SZ 619852. The type specimen of the pachycephalosaur *Yaverlandia* came 'from the Upper Silty Bed north of the sea wall below Yaverland Battery' (Galton, 1971c, p. 41). Numerous

remains in IWCMS are labelled 'silty beds', 'upper silty bed' or 'lower silty bed', and finds are regularly made in these units in the cliff and on the beach.

Dinosaur footprints were found at Yaverland in early April 1979, a large series of iguanodont trackways exposed on the shore (Delair, 1989). Subsequently, several horizons have been found to contain footprints (Radley, 1993).

The bones from Yaverland are generally isolated and in reasonable condition if collected *in situ*, but often much abraded if picked from the beach. Few articulated elements have been collected, although J. Radley (pers. comm., 1993) notes occasional articulated vertebrae. The finds range in size from 10 mm crocodilian teeth to 1.5 m long dinosaur limb bones (Gibson, 1858).

Fauna

The main repositories for Yaverland material are the BMNH and IWCMS. Very few of the specimens have been described, and the names are taken from the museum labels. An estimate of the numbers of specimens of each form is given:

Numbers

Testudines: Cryptodira
  Plesiochelys sp. — 2
  Tretosternon bakewelli
   (Mantell, 1833) — 1
Archosauria: Crocodylia: Neosuchia
  Goniopholis crassidens Owen, 1841 — 11
  Suchosaurus cultridens Owen, 1841 — 1
  'crocodilian' — 2
Archosauria: Dinosauria: Saurischia:
  Theropoda
  Megalosaurus dunkeri Koken, 1887 — 2
  Megalosaurus sp. — 6
Archosauria: Dinosauria: Saurischia:
  Sauropoda
  Cetiosaurus brevis Owen, 1842 — 1
  Pelorosaurus hulkei (Seeley, 1870) — 1
Archosauria: Dinosauria: Ornithischia:
  Ornithopoda: Iguanodontidae
  Iguanodon bernissartensis Boulenger,
   1881 — 2
  Iguanodon mantelli Meyer, 1832 — 1
  Iguanodon sp. — c. 40
Archosauria: Dinosauria: Ornithischia:
  Pachycephalosauria: Pachycephalosauridae
  Yaverlandia bitholos Galton, 1971
   Type specimen: IWCMS 1530 — 1

British Cretaceous fossil reptile sites

Numbers

Archosauria: Dinosauria: Ornithischia:
  Ankylosauria: Nodosauridae
  Polacanthus foxi Hulke, 1882   2
  Polacanthus sp.   3
Sauropterygia: Plesiosauria
  '*Plesiosaurus* sp.'   2

Interpretation

The range and relative numbers of reptiles recorded from Yaverland are similar to those from the Compton–Atherfield section. Turtles are not common, and consist of partial carapaces ascribed to the typical Wealden genera *Plesiochelys* and *Tretosternon*. Crocodilians are more abundant, with several finds of teeth, scutes and vertebrae of the aquatic metamesosuchians *Goniopholis* and *Suchosaurus*.

Among the dinosaurs, several limb bones and vertebrae of the carnivore '*Megalosaurus*' have been collected, and the material is more extensive than that from the Compton–Atherfield section. However, the smaller 'coelurosaurs' do not appear to be represented. A couple of large limb bones in the BMNH have been ascribed to the sauropod genera *Cetiosaurus* and *Pelorosaurus*.

As in the Compton–Atherfield section, the most abundant remains are those of ornithopod dinosaurs. The commonest genus is *Iguanodon*, with numerous finds of limb bones, vertebrae, teeth and a partial jaw (IWCMS 3866). Several of these have been ascribed to species of *Iguanodon*, but the taxonomy of that genus is in some confusion (Norman and Weishampel, 1990). The most important specimen from Yaverland (Figure 8.16) is the type specimen of the pachycephalosaur *Yaverlandia bitholos* Galton, 1971. This has a thickened skull cap (14.2 mm thick, 45 mm long: original skull length *c.* 70 mm) which is a feature characteristic of the group. This may

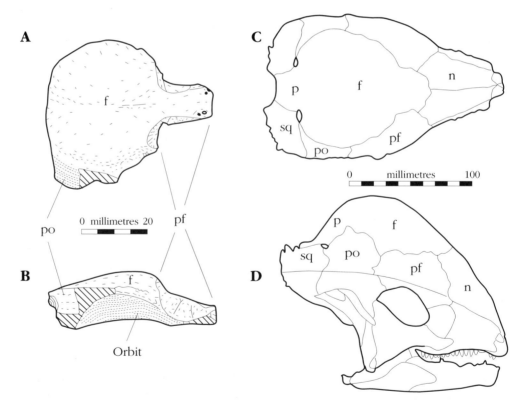

Figure 8.16 The skull cap of the oldest pachycephalosaurid, *Yaverlandia bitholos* Galton, 1971, from the Early Cretaceous Wessex Formation of Yaverland, Isle of Wight, in (A) dorsal and (B) lateral views. Skull of the Late Cretaceous pachycephalosaurid *Stegoceras* in (C) dorsal and (D) lateral views, for comparison. Abbreviations: f, frontal; n, nasal; p, parietal; pf, postfrontal; po, postorbital; sq, squamosal. After Galton (1971c).

256

have been used as a battering ram in intraspecific competition (Galton, 1971c). A partial brain cast is preserved. *Yaverlandia* is the oldest member of the group and shows characters intermediate between hypsilophodontids and typical Late Cretaceous pachycephalosaurs (Galton, 1971c; Wall and Galton, 1979).

The ankylosaur *Polacanthus* is represented by several dermal spines and scutes, and there are two plesiosaur vertebrae (IWCMS 5059, 5108).

Comparison with other localities

The most productive comparable section is the coast between Compton Bay and Atherfield which exposes similar beds: it has yielded a similar fauna, although '*Megalosaurus*' is apparently more abundant at Yaverland. The west coast fauna includes all the Yaverland genera except the pachycephalosaur *Yaverlandia*, yielding material attributed to 11 genera. Pachycephalosauria are known elsewhere from the Late Cretaceous of North America (*Pachycephalosaurus, Stegoceras, Gravitholus, Ornatotholus, Stygimoloch*), Mongolia (*Tylocephale, Goyocephale, Prenocephale, Homocephale*), Madagascar (*Majungatholus*) and China (*Wannanosaurus*) (Maryanska, 1990). *Stenopelix* from the Early Cretaceous of Germany, formerly regarded as a pachycephalosaur, is probably something else (Wall and Galton, 1979), which makes *Yaverlandia* the earliest member of the group.

Conclusions

Yaverland is important as a supplementary site yielding the same fauna of dinosaurs, and other fossil reptiles, as the west coast Compton-Atherfield section. It is unique as the site of *Yaverlandia*, the oldest pachycephalosaur dinosaur known. Yaverland still frequently yields good dinosaur specimens and footprints, and the combination of these attributes gives its conservation value.

EARLY CRETACEOUS (APTIAN–ALBIAN)

The Aptian and Albian stages in Britain are important for reptile faunas that include a variety of marine and terrestrial forms, and significant finds have come from the Lower Greensand (Aptian-Early Albian), Gault Clay (Albian), Upper Greensand (Late Albian), and from the areally restricted Cambridge Greensand (remanié latest Albian material in a basal Cenomanian matrix). Another reworked deposit, the Lower Greensand of Potton, Bedfordshire, contains reworked fossils from the Late Jurassic or Wealden. Reptile remains in the Lower and Upper Greensand are usually fragmentary and sparse. The Gault Clay has yielded abundant and well-preserved remains, particularly from the cliff sections of Folkestone.

Lower Greensand reptile sites include the following:

DORSET: Punfield Cove, Swanage (SZ 032798; *Iguanodon*, '*Megalosaurus*', sauropod; Buckland, 1835; Strahan, 1898, pp. 122-32; Delair, 1966, p. 58; Rawson *et al.*, 1978, pp. 41-2).

BUCKINGHAMSHIRE: Brick Hill (SP 9131, ?exact locality; *Dakosaurus*, plesiosaur, ichthyosaur).

BEDFORDSHIRE: Potton (TL 2249; various localities on Old Potton-Sandy railway line; *Dakosaurus, Cimoliasaurus, Pliosaurus, 'Ichthyosaurus', Iguanodon, 'Megalosaurus', Craterosaurus*, etc.; Seeley, 1869a, pp. 74-80, 1869b, 1874a; Nopcsa, 1912; Casey, 1961; Edmonds and Dinham, 1965; Galton, 1981b).

CAMBRIDGESHIRE: Upware, Commissioner's Pit (TL 539708; *Goniopholis, 'Plesiosaurus', Pliosaurus, 'Ichthyosaurus', Iguanodon*; Walker, 1867; Keeping, 1883; Whitaker *et al.*, 1891, pp. 22-32; Casey, 1961; Rawson *et al.*, 1978).

ISLE OF WIGHT: Atherfield (SZ 4579, ?exact locality; plesiosaur, turtle; Atherfield Clay); Blackgang Chine (SZ 484768; *Iguanodon*; Fitton, 1847; Mantell, 1854, pp. 170-3); Sandown (?SZ 625855; ichthyosaur, plesiosaur); Shanklin (?sauropod; Sandrock Series).

SURREY: Godalming (SU 9643; pliosaur; Swinton, 1930).

KENT: Chipstead (TQ 501560; '*Plesiosaurus*'); Maidstone *Iguanodon* Quarry (TQ 746558; *Iguanodon*, turtle, plesiosaur, pliosaur; Bakewell, 1835; Owen, 1841c, p. 452; Bensted, 1860; Topley, 1875, pp. 117-18; Worssam, 1963, pp. 26, 37, 48, 107, 136; Delair, *in* Swinton, 1970, p. 301); Hythe (TR 163352; *Dinodocus,*

Polytychodon; Mackeson, 1840; Owen, 1841c, pp. 449-52; Topley, 1875; Woodward, 1908c; Smart *et al.*, 1966, pp. 77-8); Folkestone (TR 2235; plesiosaur, ichthyosaur; Topley, 1875, p. 422; Smart *et al.*, 1966, pp. 93-6; Padgham, 1972).

The Cambridge Greensand is a remanié deposit of early Cenomanian age, containing reptile bones reworked from the uppermost Albian (*dispar* Zone) (Cookson and Hughes, 1964; Casey, *in* Edmonds and Dinham, 1965; Rawson *et al.*, 1978, pp. 38, 50). The vertebrate remains are associated with abundant phosphate material derived from the Gault, and were collected from former phosphate workings located along a SW–NE line from Whaddon (TL 3447) to Swaffham Fen (TL 5667). Typical source localities may have resembled the sequence at Barnwell (TL 5667), where at least four levels of phosphates are developed, including the Barnwell Hard Band in which abundant vertebrate remains have been found (A.C. Morter, pers. comm.). Seeley (1869a) noted 30 or more Cambridge Greensand sites, and he and others (Owen, 1859c, 1861c; Huxley, 1867b; Seeley, 1869a, 1870b, 1873, 1874c, 1875b, 1876b, 1876c, 1876d, 1879; Lydekker, 1888a, 1889a, 1889b) described 80 or more species of turtles, crocodilians, dinosaurs, ichthyosaurs, plesiosaurs and especially pterosaurs. Seeley's 40 or so pterosaur 'species' have been synonymized to four or five by Unwin (1991). There are few extant exposures of the Cambridge Greensand: it may be seen at Barrington (TL 3949) and Arlesey (TL 185350; M.B. Hart, W.J. Kennedy, pers. comm., 1993).

Upper Greensand (Late Albian) reptiles have been found at these localities:

DORSET: Melbury Down, near Shaftesbury (SP 9020; *Trachydermachelys*; Jukes-Browne and Hill, 1900, pp. 158-61; Andrews, 1920; White, 1923, p. 63; Charmouth, ?exact locality; ichthyosaur; Jukes-Browne and Hill, 1900, pp. 183-9).

SOMERSET: Kilmerton (?exact locality; ichthyosaur).

WILTSHIRE: Shute Farm, Warminster (ST 844411; ichthyosaur, *Polysphenodon*; Jukes-Browne, 1896; Jukes-Browne and Hill, 1900, pp. 237-41); Savernake (SU 2166; plesiosaur; Jukes-Browne and Hill, 1900, pp. 262-5).

ISLE OF WIGHT: St Lawrence Cliff (SZ 5376; *Hylaeochelys*; Owen, 1881; Parkinson, 1881; Jukes-Browne and Hill, 1900, pp. 132-6).

BEDFORDSHIRE: Ampthill (LT 0338; ichthyosaur).

MIDDLESEX: Croydon, London (TQ 3164; ichthyosaur).

KENT: Folkestone (TR 2235; ichthyosaur; Topley 1875, p. 152).

Two Greensand exposures are selected as GCR sites:

1. Wicklesham Pit, Faringdon, Oxfordshire (SU 292943). Early Cretaceous (Late Aptian), Faringdon Sponge Gravels (Lower Greensand).
2. East Wear Bay, Folkestone, Kent (TR 243366). Early Cretaceous (Albian), Lower–Upper Gault.

WICKLESHAM PIT, FARINGDON, OXFORDSHIRE (SU 292943)

Highlights

Wicklesham Pit, Faringdon is a productive Greensand site where the abundant isolated bones and teeth of reptiles are found reworked from older levels. Remains of turtles, crocodilians, ichthyosaurs, plesiosaurs and dinosaurs have been reported.

Introduction

The Lower Greensand (Late Aptian) of Faringdon, Oxfordshire has been known as a source of fossil reptile bones for many years. Wicklesham Pit is the best current source of such bones. The bones occur in the Faringdon Sponge Gravels, beds famous for their invertebrate fossils (Arkell, 1947a; Casey, 1961; Krantz, 1972). The Lower Greensand occurs as several outliers south and east of Faringdon which are surrounded by Late Jurassic (Kimmeridge Clay, Corallian, etc.). Arkell (1947a, pp. 155-60) reviewed the occurrence and geology of the Faringdon Lower Greensand. Fossil reptile specimens from Faringdon have been noted by several authors, but the fauna has never been described.

Description

The sequence of the Faringdon Sponge Gravels has been given by Krantz (1972):

Wicklesham Pit, Faringdon

	Thickness (m)
Iron-rich sands	0–15
Laminated clays	0.9–16.8
Clay-rich sands, with two Fuller's earth bands	15–33
Red Gravels: quartz-rich sandstones, conglomerates with bioclasts, and cross-bedded coarse sands with sponges	2–8.7
Yellow gravels: fine- to medium-grained quartz-rich sandstones, with well-preserved sponges, brachiopods, bryozoa and echinoids	0–10

The Red Gravels and Yellow Gravels together are generally referred to as the Sponge Gravels, but the term is occasionally reserved for the Yellow Gravels alone. The Sponge Gravels are a basal conglomerate of the Lower Greensand which rest unconformably on Kimmeridge Clay or Corallian, and the unit is laterally extensive, extending 10 km or more south-east from Faringdon, according to borehole evidence (Krantz, 1972).

Arkell noted that Wicklesham Pit (photograph: 1947a, pl. 5) exposed an 8 m section of the Red Gravel. At 2–3 m from the bottom, he noted a pebble bed, about 0.3 m thick, 'full of bored mudstone nodules, and black phospathic fragments of ammonites (*Prionodoceras*) derived from the basal Kimmeridge [sic] Clay and Upper Calcareous Grit'. The pit has recently been reworked, and it displays an 8–10 m section of red-brown and brown unconsolidated sands with limestone beds and lenses. The exact localities of the older Faringdon fossils are uncertain – there were several pits operational at one time, including Little Coxwell Pit (SU 285943) and Faringdon Pit (SU 288943), both of which still exhibit sections (Krantz, 1972).

Fossils occur in consolidated and unconsolidated coquinas. These are all stained brown by iron oxide and phosphate, and include bryozoans, sponges, echinoderm spines, brachiopods, bones, phosphatic pebbles, ammonites, belemnites and fish teeth. The invertebrates are generally in good condition – brachiopod valves may still be articulated (though not in life position) – but the colonial organisms, such as sponges, are clearly not in growth position.

Fauna

On a brief visit to Wicklesham pit in 1983, M.J.B. collected fragments of vertebrae and ribs, probably from large marine reptiles. Similar specimens, as well as teeth, have been collected by field parties from Oxford University and Oxford Brookes University, among others (H.P. Powell, A. Kearsley, pers. comm., 1983). Reptile specimens from Faringdon are housed in the BGS(GSM), BMNH, CAMSM and OXFPM.

Testudines:
 'turtle'
  CAMSM B58645 (scute)
Archosauria: Crocodylia
 Dakosaurus
  CAMSM B58636, B58707–9 (teeth)
Ichthyosauria
 '*Ichthyosaurus*'
  CAMSM B58640–2 (teeth); CAMSM B58643–4, B58696 (vertebrae)
Sauropterygia: Plesiosauria
 Colymbosaurus
  CAMSM Zr 2240–5, 2250–3, 52368–9; CAMSM B58703–6 (teeth); BMNH 11901, 46382; CAMSM B5871 (vertebrae; limb bones)
 Pliosaurus:
  CAMSM B58638–9, (teeth); CAMSM B58695 (vertebra)
Archosauria: Dinosauria
 ?sauropod, stegosaur, or ankylosaur: OXFPM (3 unnumb. teeth)

Interpretation

Arkell (1947a) suggested that the Faringdon Sponge Gravels collected as a sand-and-gravel bank on the sea bed some distance offshore in pre-existing hollows in the Kimmeridge Clay. Krantz (1972) noted that, although the unconsolidated sands and gravels accumulated within a channel during a transgressive episode, the contained invertebrate fossils are frequently very well preserved. She resolved this apparent paradox by suggesting that the fossils accumulated mainly on the protected western side of the channel where only the upper layers of the Sponge Gravels were reworked, and that the waters were rich in $CaCO_3$, which prevented the dissolution of calcareous fossils. Krantz (1972) argues that the sponges and bryozoans were torn up from their life positions on neighbouring hardgrounds formed on exposed areas of Corallian rock and were then mixed with previously abraded and sorted shell material. The overall palaeogeographic setting seems most comparable with a forereef deposit (Krantz, 1972). The vertebrate fossils appear to derive mainly from the Kimmeridge Clay, whereas the invertebrates are coeval with the time of deposition.

The reptile remains consist of teeth and vertebrae, which are usually identifiable at least as 'ichthyosaur' or 'plesiosaur'. Fragments of ribs and limb bones are hard to identify. The fauna, as far as can be assessed, is typical of the Kimmeridge Clay, from which most of the fossils apparently derive; marine ichthyosaurs, plesiosaurs and pliosaurs predominate. Some giant marine crocodilians (*Dakosaurus*) are also represented. The 'turtle' scute is more of a rarity, although several Kimmeridge Clay turtles are known. The dinosaur teeth are rather like the sculptured peg-like teeth of stegosaurs or ankylosaurs, but they lack the 'frilled' cutting edge, possibly as a result of abrasion; they seem very small for sauropod teeth. Woodward and Sherborn (1890, p. 219) note the plesiosaur genus *Cimoliasaurus latispinus* (Owen, 1854) from Faringdon. This is based on a vertebra (BMNH 11901) referred to a species originally described from the Lower Greensand of Maidstone (Lydekker, 1889a, pp. 222–3), but the species is likely to be a *nomen dubium*, being founded on scrappy material. *Cimoliasaurus* may be the same as *Colymbosaurus* (Brown, 1981), a long-ranging genus known especially from the Late Jurassic and Early Cretaceous of England, but with some earlier and later records.

Conclusions

The best available Greensand-type mid-Cretaceous site in Britain. A moderately diverse fauna of turtles, crocodilians, ichthyosaurs, plesiosaurs and dinosaurs is present. The site provides crucial evidence on the reworking of material from the Kimmeridge Clay locally and this important remanié fauna and the continuing availability of the site for collection substantiate its conservation value.

EAST WEAR BAY, FOLKESTONE, KENT (TR 243366)

Highlights

East Wear Bay, Folkestone is the most productive British Gault Clay reptile site (Figure 8.17). Abundant specimens of turtles, pterosaurs, ichthyosaurs and plesiosaurs have been reported,

Figure 8.17 The Gault clays at East Wear Bay, Folkestone. (Photo: D.J. Ward.)

Introduction

The Gault on the coast east of Folkestone has been known as a good source of fossil vertebrates for 150 years. The fauna of turtles, ichthyosaurs, plesiosaurs, pliosaurs and pterosaurs (Figure 8.18) includes many good specimens, and the types of two species. The section is currently well exposed, with new portions revealed by marine erosion and land-slipping, and the site continues to yield reptile bones.

The Gault section has been described by many authors, such as De Rance (1868), Price (1875), Topley (1875, pp. 145–7), Jukes-Browne and Hill (1900, pp. 69–83), Smart *et al.* (1966, pp. 56–8, 99–101, 112–13) and Owen (1971, pp. 11–15; 1976). Reptile finds have been discussed by Owen (1874a), Seeley (1877), Woodward and Sherborn (1890) and Persson (1963).

Description

The Gault section is best seen just east of Copt Point. To the west, towards Folkestone Harbour, the underlying Folkestone Beds and Sandgate Beds of the Lower Greensand cropout, and to the north, on the shore of East Wear Bay, the Gault is broken up by landslips. The section at Copt Point (from Price, 1875 and Jukes-Browne and Hill, 1900, p. 71) is as follows:

	Thickness (m)
Upper Gault	
XIII. Pale grey and buff-coloured marl	7.3
XII. Dark, glauconitic sand	1.0
XI. Pale bluish-grey, marly clay	10.8
X. Grey, marly clay	5.1
IX. Hard, marly clay	2.8
Lower Gault	
VIII. Junction bed	0.2
VII. Dark-grey clay	1.9
VI. Mottled, grey clay	0.3
V. Mottled clay	0.5
IV. Light-grey clay	0.1
III. Light buff-coloured clay	1.4
II. Very dark clay	1.3
I. Dark clay and glauconitic sand with nodules at base	3.1
Ia. Yellowish sand with phosphatic nodules	1.9

These lithological divisions of the 30–35 m thick section are readily determined in the field, and

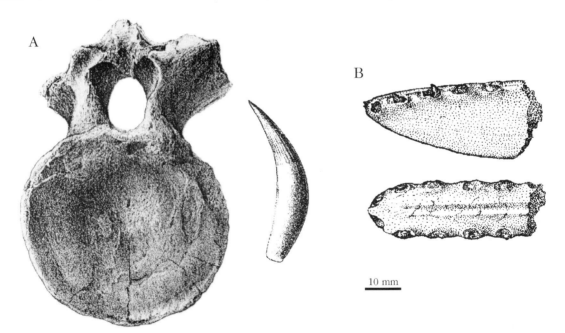

Figure 8.18 Reptiles from the mid-Cretaceous Gault of Folkestone. (A) The elasmosaur *Mauisaurus gardneri* Seeley, 1877, tooth and dorsal vertebra; (B) the pterosaur *Ornithocheirus daviesi* (Owen, 1874), top of snout in medial and ventral views. (A) After Seeley (1877); (B) after Owen (1874a).

Owen (1971, 1976) gives more detailed logs. An ammonite biostratigraphy has been worked out (Jukes-Browne and Hill, 1900; Spath, 1923–43; Smart *et al.*, 1966; Owen, 1971) and the section is dated as Mid–Late Albian (*dentatus* to *dispar* Zones). There is a clear break between the Lower and Upper Gault here, between beds VIII and IX. Fossils include common molluscs, fishes and crustaceans.

Reptiles have been found throughout the whole section, and the horizons are compiled below from Price (1875), Topley (1875, p. 436), Jukes-Browne and Hill (1900), Smart *et al.* (1966, pp. 112-13) and museum records. Most of the museum specimens and described fossils were not localized to an horizon.

XII. Turtle bones and jaw, ichthyosaur and plesiosaur vertebrae (Topley, 1875, p. 436).
XI. Ichthyosaur remains (BGS(GSM)).
X. 'Bones of Chelonians and fish, and the eggs of a species of Crocodilian' (Jukes-Browne and Hill, 1900, p. 79).
IX. Turtle remains (BGS(GSM)), *Polyptychodon* (Price, 1874).
VIII. Turtle remains (BGS(GSM)), ichthyosaur and plesiosaur vertebrae (Topley, 1875, p. 436).
VII. *Polyptychodon* (Price, 1875), turtle (CAMSM).
IV. Pterosaur remains (BGS(GSM)).
II. Large plesiosaur (Smart *et al.*, 1966, p. 112).
I. Ichthyosaur and plesiosaur vertebrae (Topley, 1875, p. 436).
Ia. *Polyptychodon* (Gault/Folkestone Beds junction: BMNH); turtle (BMNH).

The bones are relatively well preserved, with delicate processes intact, and the skeletons are often partly articulated. These facts suggest that the carcasses were often buried where the animals died, without post-mortem transport.

Fauna

Fossil reptiles from the Gault of Folkestone are preserved in several museums: BMNH, BGS(GSM), CAMSM and OUM. In the following list, all species are listed, and type specimens noted where relevant. An estimate for the total number of specimens of each is also given as an approximate guide to the relative abundance of each species.

	Numbers
Testudines: Cryptodira: Chelonioidea: Protostegidae	
Rhinochelys elegans Lydekker, 1889	1
Rhinochelys sp.	24
Cimochelys benstedi (Mantell, 1847)	1
Archosauria: Pterosauria: Pterodactyloidea: Ornithocheiridae	
Ornithocheirus daviesi (Owen, 1874) Type specimen: BMNH 43074	4
Ornithocheirus sp.	6
Ichthyopterygia: Ichthyosauria: Ophthalmosauridae	
Ophthalmosaurus campylodon (Carter, 1846)	6
Ophthalmosaurus sp.	20
Sauropterygia: Plesiosauria: Plesiosauroidea	
Cimoliasaurus cantabrigiensis Lydekker, 1889	1
Mauisaurus gardneri Seeley, 1877 Type specimen: BMNH 47295	1
Sauropterygia: Plesiosauria: Pliosauroidea	
Polyptychodon interruptus Owen, 1841	5
'plesiosaur'	4

Interpretation

The Gault is a low-energy basinal mud unit. The environment of deposition is interpreted as 'a fairly shallow muddy-bottomed sea' (Smart *et al.*, 1966, p. 102). It forms part of the major mid-Cretaceous marine transgression over much of north-west Europe, which began with deposition of the coarse sands of the Lower Greensand (Aptian), followed by deepening of the basin in the early Albian. The Lower Greensand progressively overstepped older Mesozoic deposits, and the Gault Clay Formation was the first unit completely to cover the Palaeozoic London Platform (Owen, 1971).

The turtles from Folkestone, originally ascribed to *Chelone*, *Protostega* and *Rhinochelys*, all probably belong to the last genus. The material consists mainly of carapace and plastron elements, as well as limb bones, and a skull of *R. elegans* (BMNH R27). *Rhinochelys* had a 30–60 mm long skull which is characterized by its short snout and other features. It is classed in the family Protostegidae, a group of turtles mainly from the Late Cretaceous and Early Palaeogene of North America (Collins, 1970). A specimen of a partial plastron (BMNH 47210) from Folkestone is assigned to *Cimochelys benstedi* by Collins (1970, p. 375) and is, she suggests, possibly the postcranial material of *Rhinochelys*.

A few slender bones of pterosaurs have been found at Folkestone. *Ornithocheirus daviesi* is represented by a mandible (BMNH 43074, the type specimen; Figure 8.18B) and some limb bones, while *Ornithocheirus* sp. is also based on limb bones. The type mandible is 47 mm long, has five alveoli on each side and the jaw end is rounded. Referred limb bones include a 220 mm tibia from Folkestone, as well as specimens from the Cambridge Greensand (Owen, 1874a; Lydekker, 1888a, pp. 23–4). Wellnhofer (1978, pp. 56–7) accepts *O. daviesi* as a valid species, but the distinguishing characters are not made clear.

Ichthyosaurs are represented by teeth, vertebrae, and limb elements of *Ophthalmosaurus*. *O. campylodon* is known from the Cambridge Greensand and the Chalk of Kent and Cambridgeshire. Similar material has been found in the Gault of France, Germany and Russia (Lydekker, 1889a, pp. 15–20). *Ophthalmosaurus* was a large genus, with centra 100 mm in diameter and a skull 2.5–3.0 m in length. McGowan (1972) ascribed all Cretaceous ichthyosaurs, including *Ophthalmosaurus*, to *Platypterygius*.

Several species of plesiosaurs have been recorded. *Cimoliasaurus cantabrigiensis* Lydekker, 1889, *C. constrictus* (Owen, 1850) and *C. smithi* (Owen, 1884) are noted by Woodward and Sherborn (1890) in the BMNH, but only a limb bone of '*C. smithi*' has been found at the site. *C. cantabrigiensis* was based on some vertebrae from the Cambridge Greensand and was distinguished on some minor vertebral characters. This species was regarded as dubious by Persson (1963, p. 18), and he tentatively ascribed it to the Rhomaleosauridae. The type specimen of *Mauisaurus gardneri* is represented by a partial skeleton, a tooth, the vertebrae of the neck and back (Figure 8.18A), most limb bones and parts of the pectoral girdle (Seeley, 1877). The animal was large with vertebrae up to 100 mm across, but it had a very long neck, typical of an elasmosaurid. Persson (1963, p. 19) retained *M. gardneri* as a valid species of elasmosaurid.

The larger pliosaur *Polyptychodon* is represented by several teeth and limb bones. Similar material is also known from the Upper Greensand and Cambridge Greensand. The species *P. interruptus* was regarded as valid by Persson (1963).

The reptile fauna is essentially coastal marine, containing turtles, fish-eating ichthyosaurs and plesiosaurs, as well as the top carnivore *Polyptychodon*. Pterosaurs may have been washed in from the land, or they may have died while feeding on fish.

Comparison with other localities

Reptiles are known from several sites in the Gault, but none is as rich as East Wear Bay, Folkestone. Other sites in Kent and West Sussex include Wrotham (TQ 6159; *Iguanodon*); Horish Wood, Maidstone (TQ 786575; turtle, ichthyosaur, *Polyptychodon*, pterosaur; Casey, 1959; Worssam, 1963, pp. 6, 58, 62); Henfield (TQ 2116; *Rhinochelys*; White, 1924, p. 28); Upper Beedon Pit (TQ 205123; ichthyosaur vertebra; White, 1924, pp. 27–8); in Oxfordshire: Towersey, near Thame (SP 7305; ichthyosaur); in Buckinghamshire: Ford (SP 7709; ichthyosaur; Jukes-Browne and Hill, 1900, pp. 277–8), Bishopstone, near Aylesbury (SP 8010; *Ophthalmosaurus*); in Hertfordshire: Puttenham (SP 8814; *Ophthalmosaurus*; Jukes-Browne and Hill, 1900, pp. 280–2); in Cambridgeshire: Barnwell (TL 4658; ichthyosaur vertebrae; Jukes-Browne and Hill, 1900, p. 292).

An age-equivalent horizon to the Gault is the Red Chalk at Hunstanton, Norfolk (TF 673414 to TF 674419; a 1.3 m bed ascribed to the Mid–Late Albian (Rawson *et al.*, 1978) which has yielded teeth, jaws, vertebrae and limb bones of *Ophthalmosaurus*, as well as teeth and vertebrae of *Cimoliasaurus* (Jukes-Browne and Hill, 1900, pp. 302–4). The Red Chalk at West Dereham, Norfolk (TF 6500) has also yielded an ichthyosaur skull.

Similar plesiosaurs are known from the Mid Cretaceous of northern France (Louppy), central Germany (Langelsheim, etc.), Russia (near Moscow), the United States (Kansas), and Australia (Queensland and New South Wales) (Persson, 1963). Other mid-Cretaceous reptile localities include the French Alps (turtles), and the Meuse and Normandy (ichthyosaurs, plesiosaurs) (Buffetaut *et al.*, 1981). During these times, ichthyosaurs and plesiosaurs had waned from their Late Jurassic diversities, and faunas are rather sparse in all parts of the world where they occur. On land, dinosaurs were flourishing, with new groups such as pachycephalosaurs and ceratopsians appearing, and the ornithopods and theropods further diversifying. Pterosaurs and crocodilians were also abundant in terrestrial deposits, and new groups were coming on the scene. The Gault of Folkestone contains few terrestrial elements, other than the rare pterosaur bones, but it shows good examples of the rare mid-Cretaceous ichthyosaurs, plesiosauroids and pliosaurs.

Conclusions

The section at East Wear Bay, Folkestone is Britain's best Gault reptile site. It has yielded one of the best mid-Cretaceous reptile faunas in the world. This international importance and the continuing yield of specimens establishes the site's high conservation value.

LATE CRETACEOUS (THE CHALK)

The Late Cretaceous Chalk facies of Britain (Cenomanian–Maastrichtian) have produced rather sparse remains of mainly marine reptiles, and these are usually represented by isolated elements. However, examples have been found of most reptiles representative of the time, particularly of mosasaurs which had evolved during the Late Cretaceous as top carnivores in the Chalk sea. Ichthyosaurs, by Late Cretaceous times, had dwindled in significance, and the last specimens date from the Cenomanian. Plesiosaurs, in the form of elasmosaurids and pliosaurids, survived through the Late Cretaceous to the end of the period, but in reduced diversity and mainly in the southern hemisphere. Other marine tetrapods of the Late Cretaceous include turtles, notably the giant protostegids of the Niobrara Sea in Kansas and the marine diving hesperornithid birds of the same region. Pterosaurs are found occasionally in marine sediments, and Late Cretaceous forms were essentially the large to very large pteranodontids and azhdarchids. Terrestrial reptiles are rare in the Chalk, which is a marine deposit, so there is little evidence in Britain of the dramatic changes which occurred elsewhere during the Late Cretaceous. Dinosaurs burgeoned, with hadrosaurs, ceratopsians and ankylosaurs becoming especially diverse. Lizards, snakes and crocodilians also radiated, and rare examples have been found in the Chalk.

Reptiles have been found at 50 Chalk localities, based on literature references and museum specimens. These are listed below by county from the south-west to north-east, with zones indicated, where known:

DORSET: Weymouth (Lower Chalk; ?exact locality; *Rhinochelys*; Delair, 1958, p. 54).

SOMERSET: Frome (ST 7747?; *Polyptychodon*).

WILTSHIRE: Norton Ferris (Lower Chalk, *varians* Zone; ST 7936; unidentified bones); Porton Railway Cutting (Upper Chalk, *coranguinum* Zone; SU 1936; *Leiodon*; Jukes-Browne and Hill, 1904, pp. 83–4); Highfield (Upper Chalk, *marsupites* Zone, ?SU 0038; plesiosaur; Jukes-Browne and Hill, 1904, pp. 83–4); Harnham (Upper Chalk, *quadratus* Zone; ?SU 1428; *Leiodon*, plesiosaur; Jukes-Browne and Hill, 1904, pp. 83–4).

HAMPSHIRE: Horsebridge (Upper Chalk, *quadratus* Zone; SU 3430; mosasaur); Shawford waterworks, Southampton (SU 4725; *Rhinochelys, Mosasaurus*); Portsdown (Upper Chalk, *mucronata* Zone; SU 6406; *Leiodon*, Jukes-Browne and Hill, 1904, pp. 59–60).

ISLE OF WIGHT: Shanklin (SZ 5881; *Polyptychodon*).

WEST SUSSEX: Charlton (Upper Chalk; SU 8812; *Chelone*); Arundel, (TQ 0107; *Chelone*); Houghton (Lower Chalk; TQ 0111; '*Cimoliosaurus*', *Polyptychodon*); Washington, near Worthing (TQ 1212; *Coniasurus*; White, 1928, pp. 36, 40); Steyning (Lower Chalk; TQ 1711; '*Cimoliosaurus*', *Polyptychodon*; White, 1928, p. 36).

EAST SUSSEX: Saddlescombe (Middle–Upper Chalk, TQ 27001162; '*Cimoliosaurus*'; White, 1928, pp. 38, 40, 44; Young and Lake, 1988, p. 68); Brighton (?exact locality; *Chelone*); Kemp Town, Brighton (TQ 3303; *Leiodon*; White, 1928, pp. 32, 42, 50, 53, 56–7, 59); Clayton Pit, Falmer (Upper Chalk; TQ 3508; *Coniasaurus, Polyptychodon*; White, 1928, pp. 35–6, 38); Balcombe Pit, Glynde (Lower Chalk; TQ 46050850; *Chelone, Polyptychodon*; White, 1928, pp. 48, 51; Lake and Shephard-Thorn, 1987, pp. 69–70); Southerham Grey Pit, Lewes (Lower–Middle Chalk; TQ 42800900; *Chelone, Protostega, Rhinochelys, Dolichosaurus, Mosasaurus, Polyptychodon*, '*Cimoliosaurus*', *Ornithocheirus*; Jukes-Browne and Hill, 1904, pp. 46–58; Lake and Shephard-Thorn, 1987, p. 68).

SURREY: Dorking (L. Chalk; TQ 160503/TQ 200510?; *Protostega, Mosasaurus, Polyptychodon*;

Owen, 1860); Betchworth (Lower Chalk, *subglobosus* Zone, TQ 205515; *Ornithocheirus*).

KENT: Folkestone (Lower Chalk etc., *?naviculare* Zone; TR 2438 and east; *Rhinochelys, Polyptychodon, Ophthalmosaurus, Acanthopholis*; Huxley, 1867b; Etheridge, 1867; Jukes-Browne and Hill, 1904, pp. 135, 137; Smart *et al.*, 1966, pp. 118-19, 128-9); Lidden Spout, near Folkestone (TR 281387; *Dolichosaurus*); Dover ('Chalk Marl', 'Grey Chalk', Lower Chalk, TR 3141, ?exact locality; some from Round Down Tunnel at TR 297395; *Chelone, Rhinochelys, Polyptychodon, Ophthalmosaurus*; Jukes-Browne and Hill, 1904, pp. 135-43); Ramsgate (Upper Chalk; TR 3865; *Mosasaurus*); Northfleet (TQ 6274; *Chelone*); Gravesend (Upper Chalk, *coranguinum* Zone; ?TQ 6474; *Polyptychodon*; Jukes-Browne and Hill, 1904, p. 166); Offham (TQ 6557; *Polyptychodon*); Snodland (?TQ 697625; *Ornithocheirus*); Halling (Lower–Middle Chalk; TQ 7064, various quarries; *Chelone, Lytoloma, Polyptychodon, Leiodon, Coniasaurus, Ornithocheirus*); Cuxton (Middle Chalk; ?TQ 7066; *Polyptychodon, Mosasaurus*; Jukes-Browne and Hill, 1904, pp. 159-60; Woodward, 1906); Wouldham (TQ 7164; *Chelone,* mosasaur, *Polyptychodon*); Borstal (Lower/Middle Chalk, 'Terebratulina Zone'; ?TQ 7366; *Polyptychodon*; Jukes-Browne and Hill, 1904, p. 160); Rochester (?TQ 7268; *Chelone, Trionyx, Polyptychodon*); Maidstone (Lower–Upper Chalk; ?TQ 7655; *Chelone, Polyptychodon, Ornithocheirus*); Charing (?TQ 942506; *Polyptychodon, Coniasaurus*; Worssam, 1963, p. 79, etc.).

HERTFORDSHIRE: Hitchin (Lower Chalk, *subglobosus* Zone; TL 1829, ?exact locality; '*Iguanodon*', *Ornithocheirus*).

CAMBRIDGESHIRE: Barrington (TL 3949; *Ophthalmosaurus*); Haslingfield (TL 4052; *Polyptychodon*); Hauxton (TL 4352; *Polyptychodon*); Trumpington (TL 4454; *Ophthalmosaurus*); Cambridge (TL 4658, various localities; *Ophthalmosaurus*); Cherry Hinton (Lower Chalk; TL 483557, TL 485558; *Cimochelys, Dolichosaurus, Polyptychodon*); Coldham's Common, Cambridge (TL 4858; turtle); Swaffham Fen (TL 5464; *Ophthalmosaurus*); Isleham (TL 6434; *Ophthalmosaurus*).

NORFOLK: Hunstanton (?Upper Chalk; TF 6740; *Ophthalmosaurus, Polytychodon*); Marham (Lower Chalk, *subglobosus* Zone; TF 712092; *Ophthalmosaurus*); Norwich (Upper Chalk; St James' Pit, Lollard's Pit, Catton Grove Chalk Pit, Whitlingham; TG 242094, TG 241089, TG 228108, TG 272087; *Mosasaurus, Leiodon*; see below).

HUMBERSIDE: Sewerby Cliff, Bridlington (Upper Chalk, *quadratus* Zone, TA 1766; '*Tylosaurus*').

Of these sites, the only significant ones, which have yielded more than a few bones, are Glynde, Southerham, Dorking, Folkestone, Dover, Halling, Burham, Rochester, Charing, Cherry Hinton, Hunstanton and Norwich.

Two sites are selected as GCR sites on the basis of their important Chalk reptile faunas. The first, at Culand Pits, Burham, is well known for its content of exceptionally well-preserved terrestrial reptiles (e.g. pterosaurs and lizards) which are associated with more typical marine mosasaurs and plesiosaurs. The second, at St James's Pit, Norwich, is Britain's best mosasaur locality.

1. Culand Pits, Burham, Kent (TQ 738617). Late Cretaceous (Cenomanian–Turonian), Lower Chalk–Upper Chalk.
2. St James's Pit, Norwich, Norfolk (TG 242094). Late Cretaceous (Campanian), Upper Chalk ('Norwich Chalk').

CULAND PITS, BURHAM, KENT (TQ 738617)

Highlights

The Culand Pits, Burham are Britain's richest Chalk (Late Cretaceous) reptile site (Figure 8.19). In their heyday, they were then source of beautiful specimens of turtles, marine lizards, pterosaurs and plesiosaurs. The specimens include original material of five new species, and the pterosaurs and marine lizards have attracted particular attention.

Introduction

The two Culand Pits on Blue Bell Hill near Burham, the Lower Pit (TQ 737613) and the Upper Pit (TQ 739619), have yielded some of the most important fossil reptiles from the British

Figure 8.19 The rather overgrown Upper Culand Pit, Burham, showing the Middle and Upper Chalk, source of several specimens of fossil turtles, marine lizards, pterosaurs and plesiosaurs. (Photo: M.J. Benton.)

Chalk (Figure 8.20), which form the basis for several descriptive papers. The chalk quarries are still accessible, although no longer working and further finds could only be made with excavation.

Lower Culand Pit is in the Lower Chalk (Cenomanian) and Upper Culand Pit is in the Middle and Upper Chalk (Turonian). Jukes-Browne and Hill (1903, p. 35) give a sketch section which makes this clear. The lowest units of the Chalk, the Chloritic Marl and the Chalk Marl (Cenomanian, *mantelli* Zone), were recorded in the tramway between Burham Brick Pit (TQ 723610) and the Chalk Quarries (Jukes-Browne and Hill, 1903, pp. 46–7). Further details of the quarries are given in a series of papers by Dibley (1900, 1904, 1907, 1918, Dibley and Spath, 1926) and by Dines *et al.* (1954). Kennedy (1969, pp. 482–6) gave a section, with details of the ammonites, for the Lower Chalk. The reptiles have been described by Owen (1842b, 1842e, 1851b, 1852a), Owen (*in* Dixon, 1850), Bowerbank (1846, 1848, 1852), Mantell (1842), Lydekker (1889b), Woodward (1888), Woodward and Sherborn (1890), Seeley (1870b) and Wellnhofer (1978).

Description

A combined section of the Chalk in the two quarries, summarized from Jukes-Browne and Hill (1903, pp. 49, 382; 1904, pp. 158–9) and Dines *et al.* (1954, pp. 32, 37, 42), is:

	Thickness (m)
UPPER CULAND PIT	
Soil	0.3
Upper Chalk (*planus* Zone)	
Very rough, rubbly, hard, crystalline chalk	6.1
Rough, lumpy chalk	4.9
Layer of flints	0.2
Massively bedded chalk	1.5
Layer of flints	0.2
Rough, hard, lumpy chalk	1.1
Layer of flints	0.2
Rather rough and lumpy chalk	0.9
Rather rough, hard chalk with scattered flints	0.9
	16.3

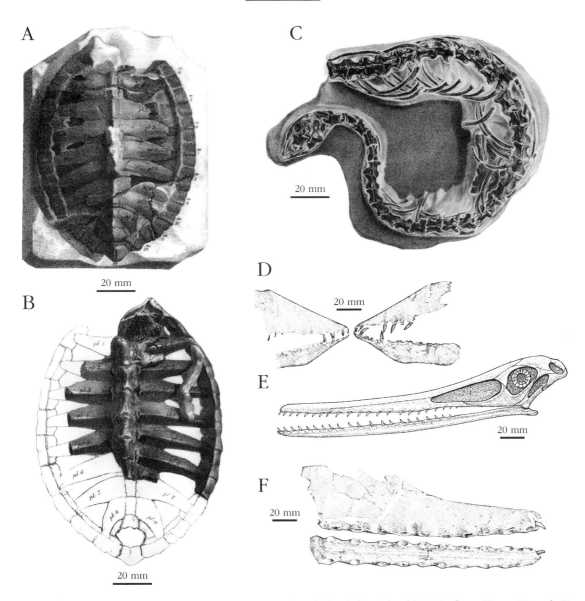

Figure 8.20 Typical reptiles of the Late Cretaceous Middle Chalk of the Culand Pits, Burham, Kent. (A) and (B) The turtle *Chelone (Cimolochelys) benstedi* (Mantell, 1841), carapace in dorsal and ventral views; (C) the elongate marine lizard *Dolichosaurus longicollis* Owen, 1850, crushed skull and anterior part of skeleton; (D) the pterosaur *Ornithocheirus compressirostris* (Owen, 1851), skull in lateral view; (E) *O. cuvieri* (Bowerbank, 1851), anterior part of snout in lateral and crown views; (F) *O. giganteus* (Bowerbank, 1846), anterior part of snout, right and left sides. (A)-(C) After Owen (1851b); (D)-(F) after Wellnhofer (1978), from various sources.

	Thickness (m)		Thickness (m)
Middle Chalk (*lata* and *labiatus* Zones)		Lower Chalk	
Firm, soft, lumpy chalk	1.2	*Plenus* Marls (Belemnite Marls)	
Firm, white, smooth chalk	7.6-9.1	(*gracile* Zone)	
Massive, homogeneous, white chalk	35.7	Yellowish-grey laminated marl	0.3-0.4
	———	Pale yellowish-grey marly chalk	1.8
	c. 46.0	Grey Chalk and *mantelliana* Band	
		(*naviculare* and *rhotomagense*	
		Zones).	
LOWER CULAND PIT			
Middle Chalk and Melbourn Rock	9.1	Beds 6, 7 and 8. Firm white chalk	

Thickness (m)

passing gradually down into	
grey chalk	about 25
Bed 5. Grey marly chalk	about 5

Some museum specimens are labelled 'Lower Chalk' (BMNH 28706) and others as 'Middle Chalk' (BMNH 49008-10) or '*H. subglobosus* Zone, Middle Chalk' (R3735-6), but others lack horizon information. Hence, it is impossible to gain an impression of the vertical distribution of reptile finds through the Chalk in the Culand Pits.

The fossils are generally well preserved and fine detail may be seen (e.g. in the pterosaurs and lizards). Skeletons may be largely articulated (e.g. *Dolichosaurus*), or broken up. Often only isolated teeth and vertebrae of larger forms are found.

Fauna

The abundant reptile remains from the Culand Pits are preserved in the BMNH, CAMSM and MAIDM. These are generally labelled 'Burham' or 'Blue Bell Hill'.

Testudines: Cryptodira
  '*Chelone* sp.'
    BMNH 41642, R1345, R1934
  '*Chelone/Lytoloma* sp.'
    BMNH 49008-10
  '*Chelone (Cimochelys) benstedi*' (Mantell, 1841)
    Type specimen: BMNH 28706
  Puppigerus camperi (Gray, 1831)
    CAMSM B20600-5
  '*Protostega* sp.'
    BMNH R3736
  'chelonian'
    BMNH R3735
Lepidosauria: Squamata: Sauria: Dolichosauridae
  Dolichosaurus longicollis Owen, 1850
    Type specimen: BMNH 49002. Also BMNH 32268
Lepidosauria: Squamata: Sauria: Mosasauridae
  Mosasaurus sp.
    MAIDM unnumb.
Archosauria: Pterosauria: Pterodactyloidea: Ornithocheiridae
  Ornithocheirus compressirostris (Owen, 1851)
    Type specimen: BMNH 39410; others: BMNH 39411, 39416, 49003-4, MAIDM unnumb.
  Ornithocheirus cuvieri (Bowerbank, 1851)
    Type specimen: BMNH 39409

Ornithocheirus giganteus (Bowerbank, 1846)
    Type specimen: BMNH 39412. Others, BMNH 39413-5, 39417
Ornithocheirus sp.
    BMNH 41637, 49005-6,R1357-8, R1935-6, R2644
Sauropterygia: Plesiosauria: Plesiosauroidea
  '*Cimoliasaurus smithi*' (Owen, 1884)
    BMNH 49007
Sauropterygia: Plesiosauria: Pliosauroidea
  Polyptychodon interruptus Owen, 1841
    BMNH 41641, 41644, 46959, 49007, R1217, R1938

Interpretation

The turtle remains from the Culand Pits consist of complete and partial carapaces - hence the difficulties in identification, since turtles are classified mainly on the basis of their skulls. Owen (1842b) described four marginal plates and remains of ribs of a small turtle from Burham (Figures 8.20A, B). He named this fragment *Chelone benstedi* in the explanation to the figure (Owen, 1842b, p. 176; 1842e, p. 412, pl. 39, fig. 5). In a footnote (1842e, p. 412) dated 'April, 1842', Owen ascribed these plates to a skeleton more recently acquired from Burham. Mantell (1842) described this fairly complete carapace and plastron, and noted that it was found in the Lower Chalk (i.e. Lower Culand Pit?). The carapace was oval, 150 mm long and 100 mm wide in the middle. Further turtle remains noted by Mantell (1842, p. 158) included an abdominal plate and a femur. Owen (1842b, p. 176; 1851b, pp. 4-8) further described these specimens. The small size of these turtles led some authors (e.g. Lydekker, 1889b) to consider them to be juveniles, whereas others regarded them as adults (e.g. Woodward, 1888, pp. 275-6).

The other turtles listed above, on the basis of old museum labels, include several forms that may be wrongly identified - *Puppigerus camperi* is a common Eocene species, for example (Moody, 1974). Owen (1851b, pp. 9-11) called these *Chelone camperi* and Lydekker (1889b, p. 31) noted the probable mistake.

The lizard *Dolichosaurus* was described by Owen (*in* Dixon, 1850, pp. 388-95; 1851b, pp. 22-9); Mackie (1863); Lydekker (1888a, p. 275) and Woodward (1888, pp. 281-2). The type specimen (BMNH 49002) is a crushed skull and a series of vertebrae and ribs, with scattered, short limb bones (Figure 8.20C). Owen (1842e, p. 412) had earlier ascribed the posterior portion of the skeleton to *Rhaphiosaurus*, a genus also known from Cambridge. The two parts of the skeleton were later associated and renamed. This lizard had a small head with conical teeth and a long, thin body; the presacral vertebral column consists of 57 vertebrae and is about 450 mm long. The dolichosaurs are elongate marine lizards which swam in a snake-like fashion.

A jaw in MAIDM is referred to '*Mosasaurus gracilis*' Owen, 1850. Mosasaurs were large marine lizards with specialized predatory dentition. They are surprisingly rare at Burham, although more common elsewhere in the English Chalk.

Plesiosaurs are represented by some paddle bones referred to '*Cimoliasaurus smithi* (Owen, 1884)' by Woodward and Sherborn (1890). Lydekker (1889a, p. 215) noted that this species was 'doubtful', since it was based on small proportional characteristics of a dorsal vertebra. The heavier pliosaur *Polyptychodon interruptus*, a common species in the English Chalk, is represented by several teeth, vertebrae and paddle bones from the Culand Pits. Neither of these marine reptiles is known from an articulated skeleton, but comparison with better-preserved fossils elsewhere shows that *Cimoliasaurus* was a long-necked fish-eater, and *Polyptychodon* a shorter-necked, large-headed fish- and reptile-eater.

The most important fossil reptiles from the Culand Pits are the pterosaurs. Owen (1842e) described some hollow limb bones from Kent as those of birds. Bowerbank (1846) discovered a fragment of the jaws and teeth of a definite pterosaur, with portions of the hollow limb bones, in the Lower Chalk of 'Halling', and named them *Pterodactylus giganteus*. Woodward (1888, p. 238) noted that these finds actually came from Burham, as indicated by Bowerbank (1852). Owen (1846, pp. 545-8) reaffirmed the 'bird nature' of his bones and named them *Cimoliornis*. Bowerbank (1848, 1852) described further pterosaur jaw material from Burham as *P. cuvieri*, and argued again that pterosaurs were reptiles. Owen (*in* Dixon, 1850, pp. 401-4; 1851b, pp. 88-104; 1852) finally acknowledged that the hollow bones belonged to pterosaurs and not birds, and he described (1851b) a third species from Burham, *P. compressirostris*, also on the basis of a snout. These three species were subsequently shown (Seeley, 1870b, pp. 28-94, 112-18) to belong to the Cretaceous genus *Ornithocheirus*. More detail of these debates are given by Woodward (1888, pp. 283-5).

The three species of *Ornithocheirus* from Burham are based on partial skulls and skeletons (Figure 8.20D-F). The type of *O. giganteus* is a partial skull, pectoral girdle and other fragments, and other limb bones come from the same site. *O. compressirostris* is based on a partial skull and fragments of limb bones, all from Burham. *O. cuvieri* was also based on a snout and a wing bone. The three species are distinguished on proportional differences of the snout shape and tooth arrangement. The estimated lengths vary from 250 mm to 450 mm. The available material is listed by Lydekker (1888a, pp. 11-13). Wellnhofer (1978, pp. 56-8) reviewed the Burham pterosaurs and regarded all three species as valid.

The fauna at Burham is a mixture of large marine carnivores (*Mosasaurus, Cimoliasaurus, Polyptychodon*), turtles (*Chelone*), a marine lizard (*Dolichosaurus* and ?*Mosasaurus*) and pterosaurs (*Ornithocheirus*), the latter forms probably washed in from land. The vertebrates of the Chalk were reviewed by Woodward (1888).

Comparison with other localities

Most of the genera recorded from the Culand Pits have also been found in other Chalk quarries in southern England (see listing above), but none of the other sites has such a diverse fauna. Similar Late Cretaceous marine faunas are known from the Chalk of Belgium, France, Sweden, and from North America (Texas, Mississippi, Alabama, New Jersey, Kansas, etc.). However, these overseas Chalk localities are dominated by mosasaurs (Russell, 1967), a group that is barely, if at all, represented in the Culand Pits.

Conclusions

The Culand Pits at Burham have yielded the most complete fauna of Chalk reptiles in Britain. The mosasaurs, so typical of certain localities, are rare, but several well-preserved fossils of turtles, lizards, plesiosaurs and pterosaurs have been collected. These include type specimens of five species. This is the best British Chalk reptile site with potential for new finds and a key Late Cretaceous site of international importance, hence its considerable conservation value.

ST JAMES'S PIT, NORWICH, NORFOLK (TG 242094)

Highlights

St James's Pit, Norwich is Britain's best mosasaur locality. Mosasaurs were giant marine lizards, which are well known from North America, the Low Countries and parts of Africa. The remains from St James's Pit are rather fragmentary, but the best in Britain.

Introduction

The Chalk pits of Norwich have long been known as a source of remains of mosasaurs. St James's Pit is the best available site for future finds in Britain. The geology of the site has been described by Jukes-Browne and Hill (1904) and Peake and Hancock (1978). The mosasaur fossils have seemingly never been described.

Description

Several quarries were formerly worked in a strip of Upper Chalk around the north-east side of Norwich. These include Lollard's Pit, Gas Hill (TG 241098), St James's Pit (TG 242094), Kett's Cave (TG 237093) and Catton Grove Pit (TG 228108). The Chalk belongs to the zone of *Belemnitella mucronata* (Late Campanian; Rawson *et al.*, 1978, p. 52). The thick sequence of the 'Norwich Chalk' has yielded abundant fossils, and the chalk contains occasional flints and iron-stained bands (Jukes-Browne and Hill, 1904, p. 259; Peake and Hancock, 1978). There is no information on the exact horizon of the bones.

St James's Pit, currently used as a recreation area, has a sloping north face approximately 50 m high. This is largely covered by sand and gravel from above, but the Chalk may be exposed with only a little digging. The Chalk appears to be well bedded with some pyrite nodules, but no flints.

Fauna

Several teeth, vertebrae and other bones of mosasaurs have been found at Norwich. Specimens are preserved in the BGS(GSM), BMNH, CAMSM and NORCM.

Lepidosauria: Squamata: Sauria: Mosasauridae
  Leiodon anceps Owen, 1841

CAMSM B20608-9, B20611-2 (teeth);
  NORCM 13-65 (pelvis)
Leiodon sp.
  BMNH R2767 (vertebra); BMNH R6376, unnumb. (teeth)
?'*Mosasaurus oweni*' (Hector, 1874)
  CAMSM B20627
Mosasaurus sp.
  BMNH 37000a. 48940d (teeth, vertebrae: Lollard's Pit); BGS(GSM) 5560, 99085; NORCH 36-64 (vertebrae)
'mosasaur'
  BMNH 37000 (vertebrae); BGS(GSM) 114240-3 (various: St James's Pit)

Interpretation

Mosasaurs were large marine reptiles of the Late Cretaceous. Remains are known from various parts of the United States (Gulf Coast, New Jersey, Kansas), Belgium, the Netherlands, France and Sweden (Russell, 1967). Although related to varanid lizards, the mosasaurs achieved large size (up to 17 m long) and were clearly formidable predators. They lived generally in subtropical epicontinental seas of less than 180 m depth, and many species had a wide geographic range. The fragmentary remains from Britain probably belong to species that occur elsewhere in Europe and North America. The specimen ascribed to *Mosasaurus oweni* is probably wrongly assigned; this form was originally described from New Zealand.

Comparison with other localities

Mosasaurs are known from several sites in the English Upper Chalk. In the pits around Norwich, remains are known from Lollard's Pit, Gas Hill (TG 241089; Bayfield, 1864); Catton Grove Chalk Pit (TG 228108; P. Lawrence, pers. comm., 1982); and Whitlingham (TG 272078). Mosasaurs are known from several other English localities, such as Dorking (Surrey) and Halling (Kent), as noted in the Chalk locality lists (q.v.).

Conclusions

Mosasaurs are present at several British Chalk localities, but rarely represented by more than an odd tooth or vertebra. The old chalk quarries at Norwich preserve only mosasaurs, and St James's Pit has the greatest potential for future finds, which gives its conservation value.

Chapter 9

British Caenozoic fossil reptile sites

INTRODUCTION: BRITISH CAENOZOIC STRATIGRAPHY AND SEDIMENTARY SETTING

The British Caenozoic is marked by extensive vulcanism in northern areas (especially Mull and the other Hebridean islands, and Northern Ireland), and by deposition of sediments in a number of basins in the North Sea, the English Channel and the southern Irish Sea. These offshore basins exhibit sections that correlate with onshore outcrops (Figure 9.1) on the Isle of Wight, in the Hampshire Basin and in the London Basin (northern Kent, London, Berkshire, Essex and extending up the coasts of Suffolk and Norfolk). The sediments are all Palaeogene (Palaeocene, Eocene, Oligocene, Figure 9.2) in age, except for a limited Neogene (Pliocene) sequence in Suffolk, and restricted Mio-Pliocene occurrences in Kent, Norfolk and Cornwall (Curry *et al.*, 1978). Younger Caenozoic sediments (Pleistocene, Holocene) are more abundant, being distributed over much of the British Isles and yielding fossil vertebrates especially in East Anglia and southern counties of England.

At the end of the Cretaceous much of Britain lay beneath the Chalk sea, but uplift was taking place. A large part of Britain became land during the Tertiary, with strongest uplift in the north and west, and renewed subsidence in the south-east.

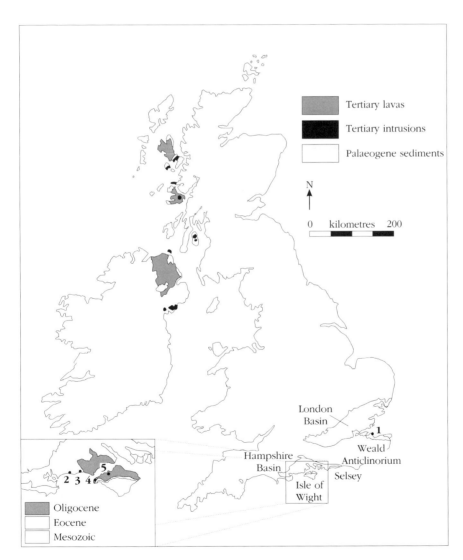

Figure 9.1 Map showing the distribution of Tertiary rocks in Great Britain. Only major divisions are indicated, and an enlargement of the Hampshire and Isle of Wight areas is given. GCR Tertiary reptile sites: (1) Warden Point; (2) Barton Cliff; (3) Hordle Cliff; (4) Headon Hill and Totland Bay; (5) Bouldnor and Hamstead Cliffs.

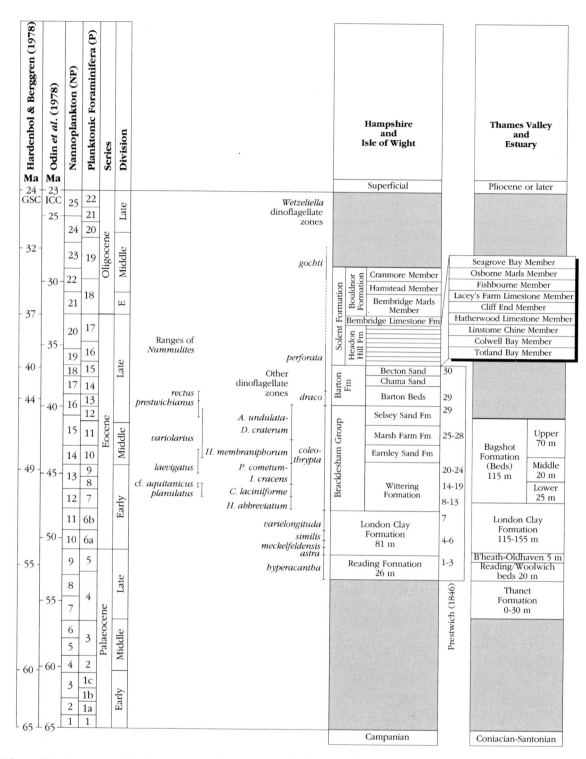

Figure 9.2 Summary of Tertiary stratigraphy, showing global standards and some major British formations. Based on Curry *et al.* (1978).

The sediments of the London and Hampshire Basins record the interplay of marine sediments from the subsiding basins and tongues of terrestrial sediments feeding off the lands to the north. This produces cyclic patterns recording repeated transgressions and regressions.

The earliest transgressive beds, of Late Palaeocene age, occur in eastern Kent (Thanet Formation), and these marine units extended ever further westwards through Berkshire and into

Wiltshire during the Late Palaeocene (Woolwich and Reading Beds) and Eocene (London Clay Formation, Bagshot beds). The Reading Formation (Late Palaeocene) is represented by a sequence of non-marine, red, mottled, kaolinitic clays and sands of fluvial origin that have yielded plant and insect fossils, as well as a marine horizon toward the base (Bone, 1986). The London Clay Formation (Early Eocene) is characterized by monotonous dark-grey to bluish mudstones, some of which are intensely bioturbated, together with more sandy beds near the base. The abundant fossils include diverse molluscs (bivalves, gastropods, nautiloids), crustaceans, fishes and tetrapods, and over 500 species of flowering plants (angiosperms) and gymnosperms, both groups being represented by pollen, logs, fruits and leaves. The Bagshot beds (Early–early Mid Eocene) are composed of decalcified, sparsely fossiliferous, marine sands and continental clays of the London Clay Formation and Bracklesham Group.

The Tertiary sediments of the Hampshire Basin (including the Isle of Wight; Figure 9.2) begin slightly later than in the London Basin, with the Late Palaeocene Reading Formation, the Early Eocene London Clay Formation, and the Early–early Mid Eocene Bracklesham Group (Insole and Daley 1985; Edwards and Freshney 1987). These represent several Late Palaeocene/Early Eocene marine transgressions from the North Sea over East Anglia and south-eastern England as far west as Dorset. The same episode drowned areas of Belgium, the Netherlands, northern France and north-western Germany. The sedimentary sequence in the Hampshire Basin spans from the latest Palaeocene to the Early Oligocene, and it consists of lateral equivalents of the Belgian and Paris basins where the sequences consist of limestones terrigenous clastics including sands, clays and thick deposits of lignite.

The Bracklesham Group includes a variety of marine and continental facies that appear to have been deposited over a long interval of time spanning the Early to early Mid Eocene. The marine units of the Group are restricted to the east of the outcrop, while in the Isle of Wight it is in part continental, and the commonest fossils are leaves and palynomorphs. The succeeding 'Barton Sands' and lower part of the Headon Hill Formation (Totland Bay Member; Late Eocene) are marine in the lower section and broadly continental toward the top. The upper parts of the Headon Hill Formation (Colwell Bay Member to Seagrove Bay Member; latest Eocene), and the Bembridge Limestone and Bouldnor formations (Early–Mid Oligocene), which is confined to the north of the Isle of Wight, consist of mainly continental facies, with rare marine-influenced units in the Cranmore Member of the Bouldnor Formation (Insole and Daley, 1985). This whole sequence, barring the 'Barton Sands' at the base, is placed in the Solent Group (Insole and Daley, 1985).

In Northern Ireland and the Inner Hebrides, rare Tertiary sediments include the eroded remnants of a vast Palaeogene region of plateau lavas (Tertiary Igneous Province), and some associated subsidiary continental sedimentary deposits. In south-west England the Oligocene basins of Bovey and Petrockstow contain fluvially deposited and lacustrine beds. These unusual deposits were the result of penecontemporaneous faulting and local subsidence in the Palaeozoic basement along the line of the Sticklepath–Lustleigh fault zone.

During Neogene time the North Sea Basin continued to subside, and sediments accumulated in the Miocene and Pliocene. The Miocene and Pliocene are largely absent from onshore sites. The notable exceptions are the Coralline Crag and the Red Crag, a combined sequence of about 70 m of the Pliocene age. These are stratified cross-bedded sands containing marine invertebrate fossils which seem to have been deposited in shallow seas by tidal currents, and indicate a cooling of the climate. During Pleistocene times, as is well known, the British Isles experienced a number of cooling and warming episodes. There were as many as six cold phases during the past 2 Ma, with associated glaciation extending, at its maximum, southwards to a line roughly from London to Bristol. Pleistocene vertebrates have been found in cave deposits and in water-laid and glacial deposits. Reptiles are known, of course, exclusively from fluvial and estuarine deposits and from caves, of the warm interglacial periods.

REPTILE EVOLUTION DURING THE CAENOZOIC

After the extinction of the dinosaurs, the pterosaurs and the marine plesiosaurs and mosasaurs at the end of the Cretaceous, and the ichthyosaurs rather earlier in the Late Cretaceous, reptile evolution seemed rather less dramatic than it had been during the Mesozoic Era. The surviving reptiles included lizards, snakes, turtles and crocodilians, and terrestrial faunas were dominated increasingly by mammals. Palaeogene faunas include snakes, lizards and amphisbaenids of essentially modern type. The turtles, crocodilians and choristoderes survived the Cretaceous with little change, but the choristoderes became extinct by Oligocene times. Crocodilians continued to diversify, and several groups of terrestrial and aquatic forms radiated, but diversity plummeted towards the present day. Turtles likewise showed a number of radiations during the Tertiary in various parts of the world, but settled to a diversity pattern similar to that of today by Miocene times.

BRITISH CAENOZOIC REPTILE SITES

Fossil reptiles occur rather sporadically through most of the British Palaeogene succession (Figure 9.2). In the Late Palaeocene there are few recorded occurrences, but remains of trionychid turtles are known from the Blackheath Beds and from the Woolwich Beds. The earliest Palaeogene reptiles have recently been obtained from rare, lignite-rich clay lenses in the Reading Formation. The remains, consisting of crocodilians and turtles, occur with fish debris and are associated with a diverse flora (seeds, other plant remains and rare amber). Early Eocene reptiles are extremely well represented in the London Clay, particularly at Sheppey, where the fauna is dominated by marine forms (turtles and the aquatic snake *Palaeophis*). In the succeeding Bracklesham Group, marine reptiles again dominate. The best Late Eocene reptiles have been obtained from the Totland Bay Member of the Headon Hill Formation (Insole and Daley, 1985) of the western Hampshire coast, and from the whole of the Headon Hill Formation of the Isle of Wight. The Late Eocene faunas are dominated by terrestrial forms (lizards, trionychid turtles, crocodilians, snakes). The Oligocene Hamstead Member of the Bouldnor Formation of the Isle of Wight has produced a fauna dominated by freshwater turtles and crocodilians. The GCR scheduled sites include a selection from these Palaeogene units.

Neogene and Pleistocene reptile localities are sparse and it was hard to determine particular locations that had assessable potential for future finds; hence, none was scheduled. The Pleistocene sites are reviewed at the end of this chapter.

LATE PALAEOCENE AND EOCENE

The latest Palaeocene and Eocene series of Britain have produced significant faunas of terrestrial and marine reptiles. The remains obtained from several of the formations described above are listed by county from the Hampshire and London Basins respectively. The host formations are indicated.

ISLE OF WIGHT: Colwell Bay (Colwell Bay Member; Late Eocene; SZ 3388; 'reptiles', *Crocodilus* sp.; Insole, pers. comm. to M.J.B.); Headon Hill and Totland Bay (Totland Bay Member–Lacey's Farm Limestone Member; Late Eocene, Priabonian; SZ 3085–SZ 3287; lizards, snakes, type of *Vectophis wardi* Ford and Rage, 1980); Cliff End (middle to upper Headon Hill Formation; Late Eocene; SZ 332890; *Trionyx* sp., ?*Ocadia* sp., turtles indet., *Diplocynodon hantoniensis*, ?crocodilian, Insole, pers. comm. to M.J.B.; Gamble, 1981); Fishbourne (Fishbourne Member; Late Eocene; SZ 537941–SZ 556934; trionychid indet., *Ophisaurus*, *Paleryx rhombifer*, cf. *Calamagras*, erycine indet., cf. *Dunnophis*; Rage and Ford, 1980); Binstead (Seagrove Bay Member; Late Eocene; SZ 5792; 'tortoise carapaces'; Mantell, 1854, pp. 79–82, *Crocodilus* sp.); King's Quay, Wootton (Fishbourne Member; SZ 5492; lizards, turtle fragments); Ryde (upper Headon Hill Formation; Late Eocene; SZ 5992; *Trionyx incrassatus* Owen, 1849); St Helens (SZ 6289; tortoise carapaces; Mantell, 1854, pp. 79–82).

HAMPSHIRE: Warblington (Reading Formation; Late Palaeocene, Thanetian; SU 731058; crocodilian teeth; Bone, 1989, p. 151); Bishop's Waltham (Late Palaeocene; SU 5517; *Diplocynodon hantoniensis*); Southampton Docks (Earnley Member, Bracklesham Group; Mid Eocene, Lutetian, zone NP14, top zone P8 [=*coleothrypta* zone, King, 1981]; SU 4112; *Argillochelys* sp., *Trionyx bowerbanki*, *Palaeophis porcatus*, *P. typhaeus*, *Palaeophis* sp.); Barton Cliff (Barton Clay and Becton Sand formations; Late Eocene; SZ 305834–SZ 262922; *Trionyx planus*, *Argillochelys* sp., *Trionyx* sp., '*Podocnemis*', '*Palaeophis*', type of *Argillochelys athersuchi* Moody, 1980); Knight Bros., Higher Brickyard, near Bransgore (Becton Sand Formation, Horizon G; locality?; *Chelone* sp.; Burton, 1933, p. 148); Hordle Cliff (Totland Bay Member; Late Eocene; SZ 2891; lizards, snakes, amphisbaenian, types of turtles *Aulacochelys circumsulcata* Owen, 1849, *Ocadia crassa* Owen,

1849, *O. oweni* Lydekker, 1889, *Geomydia headonensis* Hooley, 1905, *Trionyx barbarae* Owen, 1849, *T. incrassatus* Owen, 1849, *T. planus* Owen, 1849, *T. rivosus* Owen, 1849, *Trachyaspis hantoniensis* Lydekker, 1889, crocodilian *Diplocynodon hantoniensis* Wood, 1844).

WEST SUSSEX: Felpham, near Bognor Regis (Woolwich and Reading beds; Late Palaeocene, Thanetian; SZ 9599; crocodilian skull, turtles; Bone, 1986); West Wittering, Bracklesham Bay (Wittering Formation, Bracklesham Group; Early Eocene; SZ 777973-SZ 793966; pelomedusid indet., *Puppigerus camperi* (Gray, 1831); *Argillochelys* sp., *Erycephalochelys fowleri* Moody and Walker, 1970 [in W9-W15 of Curry *et al.*, 1978]; 'chelonian'; Curry *et al.*, 1978, pp. 243-54; Moody and Walker, 1970; Walker and Moody, 1985); Bracklesham Bay (Earnley Formation, Bracklesham Group, beds E1-E8 of Curry *et al.*, 1978; Early-Mid Eocene), SZ 807961-SZ 823951; *Puppigerus camperi* (Gray, 1831), *Trionyx bowerbanki* Lydekker, 1889, trionychid indet.; Hooker and Ward, 1980, p. 5); Bracklesham Bay (Selsey Formation, Bracklesham Group, S1-S11 of Curry *et al.*, 1978, p. 249; Mid Eocene; SZ 825947-SZ 843932; *Argillochelys* sp., *?Psephophorus* sp., trionychid indet.; Hooker and Ward, 1980); Bracklesham Bay (Bracklesham Group; Early-Mid Eocene; SZ 825947-SZ 843932; *Trionyx* sp., *Palaeophis toliapicus* Owen, 1841, type of *P. typhaeus* Owen, 1850, *Argillochelys* sp., type of *Gavialis dixoni* Owen, 1850, *Lytoloma ?trigoniceps* Owen, 1849, *Psephophorus?* sp., type of *Thalassochelys eocaenica* Lydekker, 1889, type of *Trionyx bowerbanki* Lydekker, 1889, *Crocodilus* sp., *Chelone* sp., *Pseudotrionyx* sp., chelonian indet., *Palaeophis porcatus*); Barton-on-Sea (Barton I-K; Becton Sand Formation; Mid/Late Eocene; SZ 251925-SZ 263923; Hooker and Ward, 1980, p. 6, Burton, 1929; turtle indet., *Puppigerus* sp.); Lymington (Barton Beds; Mid/Late Eocene; SZ 3493; *Crocodilus* sp., *?Diplocynodon*, chelonian); Brockenhurst (SU 3002; *Crocodilus* sp., *Ocadia crassa*, *Trionyx* sp.).

DORSET: Highcliffe (Barton Clay Formation; Mid/Late Eocene; SZ 2192; *Argillochelys* sp., *Puppigerus* sp., trionychid indet.; Burton, 1929, 1933); Creechbarrow (Creechbarrow Limestone Formation, Mid Eocene; SY 921824; crocodilian indet., chelonian indet., cf. *Cadurceryx* sp. indet., ?lizard indet.; Hooker, 1986).

THAMES VALLEY AND ESTUARY: Abbey Wood, Kent (Blackheath Beds; Early Eocene; early Ypresian, Zone MP8-9; TQ 480786; *Trionyx silvestris*, *Trionyx* sp.; Moody and Walker, 1970; Walker and Moody, 1974; Hooker 1991); Dulwich (Woolwich Shell Beds; Late Palaeocene; Thanetian; ?TQ 3374; *Trionyx*; White, 1931, p. 9; Hooker and Ward, 1980); Herne Bay, Kent (Oldhaven Member, London Clay Formation, Early Eocene; TR 198688, TR 203688; *Chelone*, *Trionyx*; Ward, 1979; Hooker and Ward, 1980); Bellfields, Guildford (London Clay; Early Eocene; ?SU 9951; *Crocodilus*); Highgate (London Clay; Early Eocene; TQ 2887; turtle); Isle of Sheppey, Warden Point (London Clay; Early Eocene; TQ 955738-TR 024717; types of *?Palaeaspis bowerbanki* Owen, 1842, *Argillochelys antiqua* Koenig, 1825, *Eosphargis gigas* Owen, 1861, *Chrysemys bicarinata* Bell, 1849, *C. testudiniformis* Owen, 1849, *Dacochelys delabechei* Bell, 1849, *Palaeophis toliapicus* Owen, 1841, *Crocodilus spenceri* Buckland, 1837); Old Haven, Forstall, Kent (London Clay; Early Eocene; TR 0661; *Palaeophis*); Harwich (London Clay; Early Eocene; TM 263317; *Lytoloma, Erquelinnesia, Neurochelys*; Moody, 1980b); Walton-on-the-Naze, Essex (London Clay; Early Eocene; TM 267243; *Eosphargis*, *?Lytoloma*).

Four GCR sites have been selected from this list as those with the best potential for future finds of Palaeogene reptiles. The first, at Warden Point (Ypresian), is of historical and scientific importance for its wealth of predominantly marine reptiles. The other sites are Bartonian and Priabonian in age, and include two localities in the Barton-Hordle coast section and a third at Headon Hill on the Isle of Wight, all important for their large faunas of terrestrial and marine forms.

1. Warden Point, Kent (TQ 955738-TR 024717). Early Eocene (Ypresian), London Clay Formation.
2. Barton Cliff, Hampshire (SZ 305854-SZ 262922). Late Mid Eocene (Bartonian), Barton Clay and Becton Sand formations.
3. Hordle Cliff, Hampshire (SZ 2891). Late Eocene (Priabonian), Totland Bay Member, Headon Hill Formation.
4. Headon Hill and Totland Bay, Isle of Wight (SZ 3085-SZ 3287), Late Eocene (Priabonian), Totland Bay Member-Lacey's Farm Limestone Member, Headon Hill Formation.

WARDEN POINT, KENT
(TQ 955738–TR 024717)

Highlights

Warden Point represents the richest collecting site for London Clay reptiles. Over the years, hundreds of superb specimens have been found, representing 13 species of turtles, one snake and one crocodilian. Most of these 15 reptiles were first found in the London Clay of Sheppey, and the fauna is internationally important as a key shallow marine Early Eocene fauna (Figure 9.3).

Introduction

The London Clay Formation (Figure 9.2) exposed on the northern and north-eastern shores of the Isle of Sheppey has yielded an important fauna of Eocene fossil reptiles. These include crocodilians, snakes and turtles in particular, with type specimens of eight or more species. The fossil turtles from Sheppey have been known for a long time: Parkinson (1811) noted 'two or three fossil tortoises' from Sheppey and figured a plastron and a skull, while Cuvier (1824, pp. 165, 234–5)

described further remains of turtles as well as some crocodilian bones. Specimens are still being found, and the coast of Sheppey has excellent potential for future finds.

The marine London Clay Formation is up to 153 m thick (Davis, 1936), but only the top 52 m are exposed on the Isle of Sheppey. The London Clay Formation (London Clay and Claygate Beds) in the London Basin has been divided into five zones (termed A–E) on the basis of marine molluscs, and a correlation scheme based on lithology, micro- and macrofaunas, has been developed by King (1970, 1981, 1984). The Claygate Beds consist of sparsely fossiliferous alternations of marine sands and clays and are probable lateral equivalents of the highest London Clay sequences at Highgate and Sheppey. Zones A–B of the London Clay are known only from borehole records. Divisions C (12.3 m), D (16.2 m) and E (24.8 m) comprise silty clays with silt and sand partings at some levels, and beds of sandy silt. The geology of the Warden Point section (Figure 9.3) has been described by Davis (1936, 1937) and King (1970, 1981, 1984), and the reptiles by Parkinson (1811), Cuvier (1824), Owen (1841d, 1841e, 1842b, 1850), Owen and Bell (1849), Seeley (1871), Lydekker (1889b, 1889d, 1889g), Mook (1955), Moody (1968, 1974) and Zangerl (1971).

Figure 9.3 The London Clay, exposed at Warden Point, Isle of Sheppey, showing collapsed cliffs and fossil-bearing material on the foreshore. (Photo: D.J. Ward.)

Warden Point

Description

The main fossiliferous horizon is identified as lying in division D: 'an interval 9.5 m–16 m below the base of division E. It can be seen on the foreshore and in the base of the cliff between Eastchurch Gap and Paddy's Point (TQ 997730 to TQ 971735), and rises eastwards to a height of about 15 m O.D. at Warden Point' (King, 1981, p. 53). This bed, probably equivalent to bed C of Davis (1936, 1937) yields fishes, molluscs, brachiopods, bryozoans, crustaceans (including decapods, barnacles and ostracods), annelids, echinoderms, corals, foraminiferans and plants – a mixture of shallow-marine and drifted terrestrial forms.

Most of the published descriptions of fossil reptiles and museum specimens give no more locality information other than 'London Clay, Sheppey'. Warden Point is indicated for a turtle (BMNH R8353) and a snake (BMNH R5886), and Eastchurch for a crocodilian (BMNH R5879).

Davis (1936, p. 334) noted vertebrae of the snake *Palaeophis* 'rarely in the clay at Warden Point', scutes of *Crocodilus spenceri* at Eastchurch (TQ 997730) and vertebrae at Warden Point (TR 021725), as well as 'indeterminate remains' of turtles 'at all points of the section'. Davis (1937) further noted a specimen of the turtle *Lytoloma* from Warden Point. King (1981, p. 53) noted recent finds of vertebrae 'from nodule layers 5–10 m above the base of the exposed section', thus probably below Davis' fossiliferous horizon C mentioned above, and King (1984, p. 145) confirmed that most of the larger specimens probably came from the large phosphatic nodules in layer D. Hooker and Ward (1980, p. 5) note that fossil vertebrates on Sheppey occur at various points in the section from TQ 955738 to TR 024717. Particular fossil localities include Minster (TQ 955736), Royal Oak (TQ 967757), Bugsby's Hole (TQ 974725), Eastchurch Gap (TQ 997730), Barrow Brook (TR 013718) and Warden Point (TR 021725).

Fauna

Fossil reptile specimens from Sheppey are to be found in many British and European Museums. The best collections are in the BMNH and CAMSM. The reptile species represented are (turtles after Moody, 1980a):

Numbers

Testudines: Pleurodira: Pelomedusidae	
?Palaeaspis bowerbanki (Owen, 1842)	
Type specimen: BMNH 37209	1
Testudines: Cryptodira: Cheloniidae	
Argillochelys cuneiceps (Owen, 1849)	
Type specimen: BMNH 41636	10
Argillochelys antiqua (Koenig, 1825)	
Type specimen: BMNH 49465	5
Eochelone brabantica Dollo, 1903	1
Puppigerus camperi (Gray, 1831)	4
Puppigerus crassicostatus (Owen, 1849)	5
Testudines: Cryptodira: Dermochelyidae	
Eosphargis gigas (Owen, 1861)	
Type specimen: BMNH R31	12
Testudines: Cryptodira: Trionychidae	
Trionyx sp.	4
Testudines: Cryptodira: Carettochelyidae	
Allaeochelys sp.	1
Testudines: Cryptodira: Emydidae	
Chrysemys bicarinata (Bell, 1849)	
Type specimen: BMNH 39450	2
Chrysemys testudiniformis (Owen, 1849)	
Type specimen: BMNH 39767	1
Testudines: Cryptodira: *incertae sedis*	
Dacochelys delabechei (Bell, 1849)	
Type specimen: BMNH 39257	2
Pseudotrionyx delheidi Dollo, 1886	2
Lepidosauria: Squamata: Serpentes: Palaeophiidae	
Palaeophis toliapicus Owen, 1841	
Type specimen: BMNH 39447	14
Archosauria: Crocodylia: Neosuchia: Eusuchia: Crocodylidae	
Kentisuchus spenceri Buckland, 1837	
Type specimen: BMNH 19633	21

Interpretation

The London Clay Formation on Sheppey is interpreted by King (1984, p. 121) as a marine deposit laid down in a 'well-oxygenated low-energy shelf environment, varying in depth from *c.* 20 to *c.* 100 metres. Alternation of fine and coarser beds is ascribed to minor sea-level fluctuations. The upper part of the London Clay Formation was deposited in a progressively shallowing environment.' The bulk of the fauna, foraminifera, coelenterates, scolecodonts, serpulids, brachiopods, bryozoans, benthic molluscs, pteropods, ostracods, crustaceans, echinoderms and fishes presumably lived in the water, or in or on the sediment. Of the tetrapods, most of the turtles were indigenous marine forms, but the remainder (as with wood, leaves, pollen and spores, and insects) may have been washed in.

Following early find of turtles (Parkinson, 1811; Cuvier, 1824), further descriptions of Sheppey finds, including the erection of many new species, were given by Owen (1841d, 1842b), Owen and Bell (1849), Seeley (1871), Lydekker (1889b, 1889d, 1889g), Moody (1968, 1974) and Zangerl (1971). Owen, in his several accounts, erected 12 or more species, but many of these have been synonymized: Moody (1980a) gives an updated list of valid species. The turtles belong to several groups: Pelomedusidae, Cheloniidae, Dermochelyidae, Trionychidae, Carettochelyidae, Emydidae and *incertae sedis* (Moody, 1980a). These are mainly medium-sized marine forms, and they all have living relatives.

Palaeaspis is rather poorly known from carapace remains (Młynarski, 1976, pp. 114-15). It is a pleurodire (folds its neck sideways), belonging to the Family Pelomedusidae, which is known from the Cretaceous to the present. Modern forms occur in Africa and South America.

The other Warden Point turtles are cryptodires, forms that fold their necks vertically. *Eochelone* and *Puppigerus* (Figure 9.4A-D), both cheloniid sea turtles, are well known from a fair number of specimens from Sheppey and Belgium (body length 0.6-0.8 m) (Moody, 1974). *Argillochelys*, another cheloniid, is represented by skull and carapace remains which suggest a body length of 200 mm. The Cheloniidae have a record extending back to the Cretaceous, and they live worldwide today. *Eosphargis* (body length 1-1.5 m) is represented by skull and limb remains; it is the oldest undisputed dermochelyid turtle in the world (Benton, 1993). *Eosphargis* is known from Eocene deposits in Denmark, Belgium, and possibly offshore South Africa. The dermochelyids are now cosmopolitan. Carapace fragments of *Trionyx* sp. represent the Trionychidae, soft-shelled turtles, a widespread group of freshwater turtles known from the Cretaceous to the present. *Allaeochelys*, a carettochelyid, is represented by carapace elements. The Carettochelyidae is another marine turtle group that is present today in seas off Asia and North America, and known since the Cretaceous. The two species of *Chrysemys* (Figure 9.4E) are based on carapace remains;

they are the oldest representatives in the world of Emydidae (Benton, 1993).

Pseudotrionyx and *Dacochelys* are noted as *incertae sedis* by Moody (1980a). Młynarksi (1976, pp. 73-4) had suggested that the former might belong to the carettochelyid genus *Allaeochelys*.

The snake *Palaeophis toliapicus* was described by Owen (1841e) on the basis of a partial backbone consisting of 28 vertebrae, as well as some other vertebrae and ribs (Figure 9.4I). The total length of *Palaeophis* is unknown, although Owen noted that its vertebrae were 'as large as those of a Boa Constrictor ten feet in length'. There has been some confusion over whether *Palaeophis* was a snake or a lizard, since complete skeletons are not known. Holman (1979) argues strongly that it was a snake. Further material of this genus is known from Belgium, France and Denmark (Rage, 1984), and the genus has been reported from Late Cretaceous to Late Eocene rocks of Europe, Africa and North America.

Kentisuchus spenceri was based on an incomplete skull of an animal approximately 1.5 m long (Figure 9.4H,I). Earlier, a Sheppey crocodilian (that of Cuvier, 1824) had been named *Crocodilus delucii* Gray, 1831, but the description was inadequate. Owen (1842b) ascribed other material to *K. spenceri*, and later (Owen, 1850) erected the new species *C. toliapicus* for a skull from the London Clay, and *C. champsoides* for an incomplete skull and other material from Sheppey. Lydekker (1887b, 1888a) described more Sheppey material, and reduced *C. champsoides* to synonymy with *C. spenceri*, and Mook (1955) placed them in the new genus *Kentisuchus* (Steel, 1973, pp. 69-70). *K. spenceri* is represented by skull remains, limb bones, vertebrae, ribs and scutes from Sheppey. Some remains from Bognor, Sussex and from Morocco have been ascribed to this species.

Comparison with other localities

The nearest comparable units with the London Clay Formation of Sheppey outside Britain are the Sables de Erquelinnes (Hainaut, Belgium; Late Palaeocene), the Argile d'Ypres (France, Belgium; Early Eocene), and the Sables de Bruxelles (Belgium; Mid Eocene), as well as equivalent-age units in France, Morocco, Nigeria, Mali and the eastern United States. These have yielded abundant specimens of turtles, including many that are conspecific with those from Sheppey, as well as lizards, snakes and crocodilians.

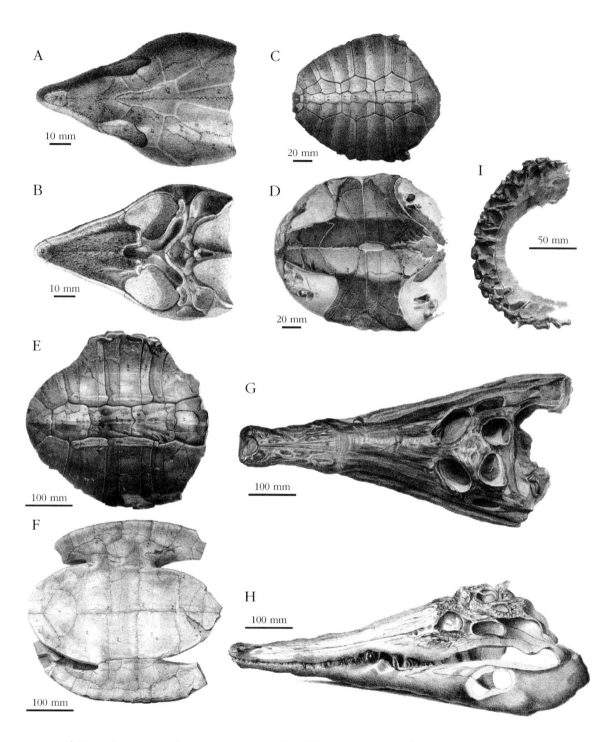

Figure 9.4 Typical reptiles of the Eocene London Clay of Sheppey. (A)-(D) The turtle *Puppigerus camperi* (Gray, 1831), skull in (A) dorsal and (B) ventral views, (C) carapace in dorsal view, (D) plastron in ventral view; (E) *Chrysemys bicarinata* (Bell, 1849), partial carapace in dorsal view; (F) *Platemys bullocki* Owen, 1841, plastron in ventral view; (G) and (H) the crocodile *Crocodilus spenceri* Buckland, 1837, skull in (G) dorsal and (H) lateral views; (I) the snake *Palaeophis toliapicus* Owen, 1841, 30 dorsal vertebrae in side view. (A)-(F) After Owen and Bell (1849); (G) and (H) after Owen (1850b); (I) after Owen (1850c).

Conclusions

The London Clay Formation at Sheppey has yielded Britain's best fauna of Tertiary fossil turtles. The fauna is important for both its relative abundance and diversity, and the good quality of preservation. The locality has been well known for over 150 years and has provided the basis for many important works on the evolution of turtles. The turtle fauna includes numerous type species, as well as the oldest undisputed dermochelyids and emydids in the world.

The international importance of the site and its continuing supply of new specimens define its high conservation value.

BARTON CLIFF, HAMPSHIRE (SZ 218930–252925)

Highlights

Barton Cliff has yielded the most productive Mid Eocene reptile fauna in Britain. The material includes specimens of 10 species of turtles, a lizard and a snake, and these are associated with rich fossils of marine shellfish, plants, birds and mammals.

Introduction

The stretch of sea cliffs in Christchurch Bay between Chewton Bunny (on the Hampshire–Dorset border) eastwards to Becton Bunny, known as Barton Cliff, has produced a good fauna of reptiles of Mid–Late Eocene age. Fossils from the Barton Beds have been collected for more than two centuries and the site has excellent potential for further finds.

The stratigraphy of the marine Barton Beds (Figure 9.2) at Barton Cliff has been discussed in detail by Gardner *et al.* (1888), Burton (1929, 1933), Hooker (1986) and Edwards and Freshney (1987). The succession was divided by Gardner *et al.* (1888) into the Lower, Middle and Upper Barton Beds on the basis of faunal changes through the sequence. Burton (1929) provided vertebrate and invertebrate faunal lists and lettered the Barton Beds A1–L based on different lithologies and faunal content. Hooker (1986) formally designated the Barton Clay Formation and erected the new unit, the Becton Sand Formation,

for the Barton Sand of earlier stratigraphic schemes. Reptiles from Barton have been described by Burton (1929, 1933), Hooker (1972, 1986) and Moody (1980a, 1980b).

Description

The Barton Clay Formation (*sensu* Hooker, 1986, pp. 203–5) is exposed in the cliff section between Friar's Cliff, Mudeford in the west to just east of Barton-on-Sea in the east (SZ 194927–SZ 242927). The beds (*c.* 40–60 m thick) consist of grey to brown silty, usually shelly, sometimes moderately to very sandy, clay, occasionally with some subordinate clayey, sandy silts. There are several layers of calcareous phosphatic and sideritic nodules. The faunal list is large, including a fauna of shark teeth and teleosts, malacostracan crustaceans, ostracods, foraminifera, brachiopods, molluscs (bivalves and gastropods), asteroids and ophiuroids, marine mammals, turtles and land-derived mammals, birds and reptiles (Burton, 1929; Hooker, 1986). An associated flora of fruits, seeds, cones and wood indicates the close proximity of land, and the marine aspect of the fossils and the sediments suggest a predominantly low-energy near-shore marine environment for the formation.

The Becton Sand Formation (*c.* 25 m) (Hooker, 1986, p. 205) occurs in the cliff section to the west of Sea Road Gap, and may be traced eastwards to Long Mead End (Taddiford Gap) at the eastern end of Beacon Cliff (SZ 229931–SZ 262922). The lithology is fine sand which is clayey and silty at the base of the formation. The biota is sparse, but essentially the same as that of the Barton Clay below. Terrestrial fossil material is similarly reduced and no mammals have been found. Towards the top of the sequence, the molluscs give an indication of shallowing waters with a change to brackish conditions, leading to the non-marine Totland Bay Member of Hordle (Hordwell) Cliff that succeeds conformably to the east.

The reptiles come from a number of levels, but have most frequently been obtained from horizons in the Lower and Middle Barton Beds, where they are commonly associated with shell-rich clays and silts. Burton (1933, p. 140) notes that in Horizon B (Lower Barton Beds), a 'grey sandy and glauconitic clay, four feet thick, . . . portions of the costals and marginals of ?*Argillochelys* sp. are fairly common, but as so happens in respect of such material in the Barton Clay, it is fragmentary and renders even generic determination somewhat speculative.' 'Vertebrae of fishes and remains of Chelonia in a fragmentary condition' were obtained from Horizon E ('Earthy' Bed, Middle Barton Beds), from a 'thin but persistent seam of *Ostrea* (*Ostrea* cf. *flabellula* Lamark). . .' that occurs at the base of the unit (Burton, 1929, p. 229). Burton (1933, p. 135) further notes ?*Argillochelys* from Horizon A1, the lowest unit of his Lower Barton Beds. A specimen of 'chelonoidean' carapace in the British Museum (Natural History) (BMNH R8358) is labelled 'Horizon B' (=*Pholadomya* Bed), the highest bed attributed to the Lower Barton Beds of Burton (1929).

From the Middle Barton Beds, Burton (1933, p. 143) records (from Horizon D) that 'Below the actual Corbula Bed occasional symphyses of chelonia occur; . . . encrusted with crystals of selenite cemented together by a ferruginous deposit'.

Although the majority of the Barton reptiles are turtles, other reptile groups are represented by rarer remains. Burton (1933, p. 140) noted: 'Occasionally, vertebrae referred to *Palaeophis* sp. from Horizon A3, in a thin ferruginous seam towards the base which is very seldom exposed.' Hooker (1972, p. 181) reported a 'lacertilian' humerus (BMNH R8580) after sieving a shelly seam high in Bed H (=*Chama*-Bed), the second unit of the Upper Barton Beds.

The reptile fossils are all preserved as isolated elements and some show signs of abrasion. The turtles are generally represented by carapace fragments, and the snake *Palaeophis* by its vertebrae.

Fauna

Many of the reptile remains from Barton Cliff are curated in the BMNH, and in the collection of Mr J. Athersuch.

	Numbers
Testudines: Cryptodira: Cheloniidae	
Argillochelys athersuchi Moody, 1980	
Type specimen: KP BA/19/VA	1
Argillochelys sp.	6
Eochelone brabantica Dollo, 1903	1
Puppigerus camperi Gray, 1831	1
Testudines: Cryptodira: Podocnemidae	
Podocnemis?	1
Testudines: Cryptodira: Trionychidae	
Trionyx incrassatus Owen, 1849	1

	Numbers
Trionyx planus Owen, 1849	1
Trionyx sp.	1
trionychid indet.	1
'chelonian'	4
Lepidosauria: Squamata: Sauria	
'lacertilian'	4
Lepidosauria: Squamata: Serpentes:	
Palaeophiidae	
Palaeophis sp.	3

Interpretation

The depositional environments of the Bartonian of the Hampshire Basin are divided into marine and non-marine provinces by Hooker (1986). The Barton Clay and Becton Sands formations in Christchurch Bay were deposited in three large cycles. The erosive base of each cycle has been interpreted as the result of a rapid marine transgression of a shelf sea, which then withdrew over a longer period, hence forming the rest of each cycle (Hooker, 1986). Marine indicators include the mineral glauconite, the trace fossil *Ophiomorpha*, and the shelly faunas of Foraminifera, bivalves and gastropods. These sediments seem to have been deposited in marine waters up to 100 m deep. Some terrestrial mammal fossils occur, as well as archaeocete whales, which were presumably preserved *in situ* (Hooker, 1986). The non-marine units occur in the Creechbarrow Limestone Formation, a lateral equivalent of the Barton Clay Formation, to be seen outside the GCR site.

The reptile fauna from Barton has never been studied in detail. Chance finds have been noted in various papers on the geology and fossil mammals from the site. Moody (1980a) reviewed the turtle fauna, noting *Puppigerus* sp. and a trionychid indet. from Barton Beds A1–3, *Eochelone brabantica*, *Puppigerus camperi*, *Argillochelys athersuchi* (Figure 9.5), ?*Trionyx planus*, trionychid indet., and *Allaeochelys* sp. from Barton Beds B–H, and *Puppigerus* sp. from Barton Beds I–K. Moody (1980b) described *Argillochelys athersuchi*, a new species from bed C in the Barton Clay Formation from Barton Cliff. Other species of *Argillochelys* are known from the London Clay (Early Eocene) of the Thames Valley, and the Late Palaeocene/Ypresian of Belgium (Moody, 1980b).

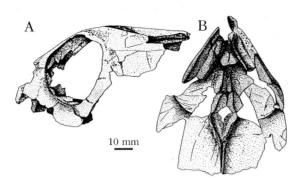

10 mm

Figure 9.5 The turtle *Argillochelys athersuchi* Moody, 1980, from the Late Eocene of Barton Cliff, partial skull in (A) lateral and (B) ventral views. After Moody (1980b).

Comparison with other localities

The perissodactyl mammal *Plagiolophus curtisi* from the Middle Barton Beds is shared with the Creechbarrow Limestone Formation. The mammal fauna of the Totland Bay Member of the Headon Hill Formation, which succeeds the Becton Sand Formation, correlates with the upper part of the Calcaire de Fons at Fons; thus the Upper Barton Beds may correlate with the Robiac unit below. The Lower Barton Beds may be Marinesian, perhaps partly equivalent to the Calcaire de St Ouen, since these lie above the Bracklesham Group that are well correlated with

the Auversian. The reptile fauna so far is insufficiently known to provide clear indications of relations. Its elements are known from a number of European faunas dating from Late Palaeocene to Late Eocene times.

Conclusions

Barton Cliff is the most productive British Mid Eocene site for reptiles, and source of a number of turtles, including one type specimen, as well as lizard and snake fossils. The site still yields abundant remains, and its potential has yet to be fully realized, giving it conservation value.

HORDLE CLIFF, HAMPSHIRE (SZ 253925–SZ 287915)

Highlights

Hordle Cliff has produced one of the richest assemblages of fossil reptiles and mammals from the Late Eocene in the world. The reptiles include nearly 40 species of turtles, crocodilians, lizards and snakes, and the specimens include the original named material of 15 species. New specimens come to light all the time.

Figure 9.6 The Lower Headon Beds of Hordle Cliff, looking towards Becton Bunny. (Photo: D.L. Harrison.)

Introduction

The Late Eocene (Priabonian) Totland Bay Member of the Headon Hill Formation (formerly the Lower Headon Beds or Headon Member; Figure 9.2) exposed at Hordle Cliff (Figure 9.6), a series of low cliffs between Becton Bunny and Milford-on-Sea, have produced an important assemblage of reptiles (Figure 9.7). A recent discovery of abundant squamate remains in the Mammal Bed has greatly enlarged the faunal list, rendering the Hordle herpetofauna equal in terms of diversity to the Late Eocene herpetofaunas of continental Europe. The section is usually masked by a thin covering of talus, and some parts are heavily slipped, but the relevant horizons remain accessible and may easily be cleared. The geology of Hordle Cliff has been described by Hastings (1848, 1852, 1853), Gardner *et al.* (1888), Curry (1958), Cray (1973), Milner *et al.* (1982) and Plint (1984).

The first vertebrate remains reported from the sections of Hordle (or Hordwell) Cliff were described from the extensive collections of Searles Wood and Barbara, Marchioness of Hastings which had been assembled during the late 1840s. These remains, including numerous specimens of mammals, fishes and reptiles (crocodilians, turtles, snakes, lizards), were initially reported by Wood (1844) and Charlesworth (1845). Wood (1846) listed and figured further material from Hordle Cliff, and Hastings (1852, 1853) reported the results of six years' further collecting, listing important finds of mammals, fishes, reptiles and birds. Subsequently, in 1855, the Hastings collection was acquired by the British Museum (Natural History).

The Hordle crocodilians were discussed further by Owen (1848), Pomel (1853), Meyer (1857), Huxley (1859e) and Lydekker (1882b, 1889h). The abundant turtles from Hordle Cliff were described by Owen and Bell (1849), Seeley (1876e), Baur (1889), Lydekker (1889b) and Hooley (1905), and the snake and lizard material was described by Owen (1850), Hastings (1852), Lydekker (1888a, 1888c), Hoffstetter (1942), Sullivan (1979), Rage and Ford (1980), Estes (1983) and Rage (1984).

A recent collection of small tetrapods from Hordle Cliff made between 1976 and 1981, containing numerous new specimens, has greatly expanded the faunal list. The specimens were obtained by Mr Roy Gardner of Fareham from the Mammal Bed, from the same locality which had produced some of the Hastings material. Milner *et al.* (1982) gave reports of the new finds and identified the occurrence of 16 new taxa of lizards and snakes, some of which were previously unknown from the Eocene of the British Isles.

Description

The stratigraphy of the Early Palaeogene rocks at Hordle Cliff has been described by Gardner *et al.* (1888) and Cray (1973). The following section is abridged from Cray (1973, p. 11). All the beds dip gently south-east at about 2.5°.

		Thickness (m)
Totland Bay Member		
('Lower Headon Beds')		
	Marl	seen 0.5
	Rodent Bed: *Limnaea* Marl with overlying dark clay (=Rodent Bed Marl)	0.25
	Unio Beds: grey clays with sandy layers	3.5
	Green clays	2.5
	Chara Bed: dark clays	1.4
	Blue and green clays	2.7
	Limnaea Limestone	0.4
13	Ironstone Bed	1.2
12	Crocodile Bed: sands	2.0
11	Rolled-Bone Bed: sand with abraded bones	0.3
10	Clay and sands	1.4
9(pars)	Leaf Bed: carbonaceous clay	1.0
9(pars)	Mammal Bed: clays, sands and sandy clays	3.5
8	Ironstone bed	0.4
7	Clays	1.1
6	Lignites	seen 1.4

The Totland Bay Member (Late Eocene) is included in the zone of the dinoflagellate *Wetzeliella perforata*. At Hordle Cliff the sediments form a series of low cliffs and slipped undercliffs between the east of Barton-on-Sea (just west of Becton Bunny) and Milford-on-Sea. The Mammal Bed, near the base, occurs beneath Plateau Gravel to the west of Becton Bunny, from where it may be traced as a distinct scar obliquely down the cliffs to reach sea-level just east of Long Mead End. Just east of Hordle House the highest unit in the Totland Bay Member, the *Limnaea* Marl and associated horizons, outcrops. The basal Colwell

Bay Member ('Middle Headon Beds' (pars)) are represented at Paddy's Gap by the occurrence of the Milford Marine Bed.

As has been established by Cray (1973) and Milner *et al.* (1982), among others, the Hordle reptiles were all found in the Totland Bay Member. The provenance of the early collections, however, is difficult to assess, the locality information provided by Hastings being merely 'Upper Eocene, Hordwell' or, in Lydekker's Catalogue (1888a, 1888b), 'from the Headon Beds of Hordwell'. The matrix on a number of specimens, although sparse, yields ambiguous information and cannot be used to demonstrate provenance with any degree of accuracy. Some of the specimens with adhering greenish-blue sandy clay may have come from the Mammal Bed, but other lithologies are undiagnostic. The accounts of Hastings (1848, 1852, 1853), however, indicate that most of the material came from two main horizons, the Mammal Bed and the Rodent Bed, and also from fossiliferous pockets within the Crocodile Bed.

The Rodent Bed (Hastings Bed 1), consisting predominantly of grey clays and marls, is limited in lateral extent, outcropping just to the east of Hordle House, and extending eastwards for some 275 m before wedging out. To the west the beds are absent, having been removed by recent erosion. The highest horizon of the Rodent Bed consists of clays, tinted pink and heavily altered by percolation from the overlying Plateau Gravel. These clays are underlain by a thin, dark, clayey sand which in turn rests on a comminuted shell bed, the *Limnaea* Marl.

Hastings (1852, p. 194) recorded an extensive vertebrate fauna from the Rodent Bed. The finds may be bracketed with the dark clayey marl on the basis of Hastings' detailed description of the host sediment and mention of the underlying *Limnaea* Marl. In her brief description of the fauna, Hastings (1852) recorded 'This band contains much debris, generally very compressed and fragile. You find here small rodent jaws, portions of carapace and a plastron of *Emys*, many teeth and bone fragments of crocodiles, some snake vertebrae, and rarely the teeth and bones of mammals' [translation]. Gardner *et al.* (1888, p. 596) also mention the occurrence of a large fauna from the dark clayey sand, listing 'serpents' vertebrae, rodents' teeth, etc.', but Tawney and Keeping (1883, p. 567), in their detailed stratigraphic account of Hordle Cliff, list only 'serpents' vertebrae' from the *Limnaea* Marl, without mention of the more abundant remains from the beds above.

Cray (1973, pp. 10–12) described the occurrence and preservation of the vertebrates: 'occasional rodent teeth and turtle fragments were recovered from the upper levels of the *Limnaea* marl, and the overlying dark sandy clay has yielded a moderate quantity of small-sized vertebrate debris . . . This material is always of very small size and evidently represents a current-sorted accumulation; all the large Headon Beds species are absent. All the specimens are fragmentary and . . . some of the material is water worn'.

The upper part of the Crocodile Bed is made up of fine, soft, white sands, but the lower layers are composed of more indurated sediments which are brownish in colour. The outcrop lies to the west of Hordle House, where the beds seem to rise from the base of the cliffs, and continue westwards until just west of Long Mead End. Hastings (1852, p. 198) noted crocodilians and the freshwater turtles *Trionyx* and '*Emys*' from the Crocodile Bed. *Diplocynodon hantoniensis*, collected by Wood in 1843 and described by Taylor in 1844, also appears to have come from this bed, in which the remains 'were embedded in the fine siliceous sand of which the freshwater deposit at Hordwell is chiefly composed'.

Hastings observed (1852, p. 197) that abundant shells invariably accompanied the vertebrate remains and recorded that the most richly fossiliferous level lay about 3 ft (*c.* 1 m) from the top of the bed, and that the middle of the outcrop, a little to the west of Hordle House was the most productive locality. Most material from the Crocodile Bed, however, appears to have been derived from isolated lenses rich in vertebrate remains, and such an origin is explicit in the earliest account by Hastings (1848, p. 63): 'the vertebrae and other bones of the Crocodile and *Paloplotherium* were found at intervals of from four inches to three feet apart to the westward of the heads . . . I must not omit likewise to state, that close to this crocodile's head (the whole group comprising a space of about six feet long by ten inches only in thickness, and following each other nearly in a straight line) were found the nearly entire shell of a fossil *Trionyx* . . . and the jaw, vertebrae, and scales of a fish of the order *Lepidosteus*'.

The Mammal Bed (*sensu* Curry, 1958; Cray, 1973), bed no. 9 of Tawney and Keeping (1883), and the upper part of bed 15 of Hastings, outcrops from beach level just west of Hordle House, westwards to Becton Bunny. Reptile material, although rare, was reported (Hastings, 1852) as coming from layers of white sand containing abundant remains of shells. Hastings (1852) provided a brief summary of the fauna listing '*Trionyx* and *Emys*, fragments of mammal jaws with teeth, fish vertebrae, occasional bones of birds, and some very small jaws, but no crocodiles' [translation]. The better-preserved remains appear to have come from the lowermost part of the Mammal Bed from bluish-green sandy clays, from which Hastings (1852, p. 201) recorded the dissociated remains of crocodilians, *Trionyx* and '*Emys*'. Seeley (1876e, p. 445) reported the remains of '*Emys*' from a horizon 'about 20 feet below the bed which yields the chief remains of *Crocodilus hastingsiae*, and about 10 feet above the brackish-water Upper Bagshot Beds, which are seen in the cliff rising westward at an angle of 3 degrees at Mead End', and therefore probably from the base of the Mammal Bed. Hastings (1852) also records remains from the next well-defined bed, of whitish brown sand with scattered bands of green clay, the upper half of which contained the same vertebrate material.

The specimens collected recently by Mr Gardner also derive from the Mammal Bed, from the stretch of Hordle Cliff sometimes referred to as Beacon Cliff, between Becton Bunny in the west and Long Mead End in the east (upper part of bed 15 of Hastings, 1852, Bed 9 of Tawney and Keeping, 1883). The material, consisting of many thousands of bones, all of small size, was found in numerous bone-bearing shelly pockets composed of pale-greenish, grey sand differing in some respects from the same level as described by Hastings (1852, p. 200).

Other horizons which have yielded reptile remains include the Rolled-Bone Bed. In 1852 (p. 199), Hastings reported finds of turtles and crocodilians from it. However, most of the specimens are generally highly abraded and cannot be identified precisely.

The Thin Shell Bed above the Lower Ironstone Band has yielded one of the largest collections of reptiles from Hordle. This bed occurs immediately above the ironstone band (numbered 8 in Tawney and Keepings' section), which is usually considered to mark the base of the Mammal Bed. Cray (1973, pp. 17–18), however, regards this unit as distinct on the basis of its mammal fauna, which is similar to that of a bed below the ironstone band. Hastings listed a wide range of taxa: 'You find here an equal quantity of snake and lizard vertebrae, some mammal teeth, rodent jaws, scales and

vertebrae of fish, crocodile debris, *Trionyx* and *Emys*, and more rarely larger and better preserved bones including astragalus and carpal bones' [translation]. A similar fauna to the above was mentioned by Hastings as occurring in the thin white sandy marl below the lower Ironstone Band, bed No. 7 of Tawney and Keeping (1883).

Fauna

The Hastings Collection is curated in the BMNH, and other material is held in CAMSM, OUM and YORMS. Repository numbers are only given for type specimens, but an estimate of the total numbers of each species preserved in these major collections is appended.

Numbers

Testudines: Cryptodira: Cheloniidae	
Argillochelys sp	1
Chelone sp.	1
Testudines: Cryptodira: Dermatemydidae	
Trachyaspis hantoniensis Lydekker, 1889	
Type specimen: BMNH R1443	1
Testudines: Cryptodira: Carettochelyidae	
Anosteira anglica Lydekker, 1889	
Type specimen: BMNH 33198 y,x	1
Testudines: Cryptodira: Trionychidae	
Aulacochelys (Trionyx) circumsulcata (Owen, 1849)	
Type specimen: BMNH 30404	2
Geoemyda (Nicoria) headonensis (Hooley, 1905)	
Type specimen: BMNH R1542	2
Geoemyda sp.	1
Trionyx barbarae Owen, 1849	
Type specimen BMNH 30409	2
Trionyx bowerbanki Lydekker, 1889	1
Trionyx henrici Owen, 1849	
Type specimen: BMNH 30406–7	11
Trionyx incrassatus Owen, 1849	
Type specimen: BMNH R1433	5
Trionyx planus Owen, 1849	
Type specimen: BMNH 30410,a	2
Trionyx rivosus Owen, 1849	
Type specimen BMNH 30405	1
Trionyx sp.	9+
Testudines: Cryptodira: Emydidae	
Ocadia crassa (Owen, 1849)	
Type specimen: CAMSM C20923	6+
Ocadia oweni (Lydekker, 1889)	
Type specimen: BMNH 36811	3

	Numbers
Ocadia sp.	1
Turtle indet.	100+
Archosauria: Crocodylia: Neosuchia:	
Eusuchia: Alligatoridae	
'*Crocodilus*' sp.	18
Diplocynodon hantoniensis	
(Wood, 1844)	
Type specimen: CAMSM ?unnumb.	*c.* 55
Diplocynodon sp.	*c.* 30
Lepidosauria: Squamata: Sauria	
Gekkonid	1
Necrosaurus sp.	5+
Ophisaurus sp.	50+
Anguine	5+
Glyptosaurinae *incertae sedis*	3
Plesiolacerta lydekkeri	
Hoffstetter, 1942	
Type specimen: BMNH 32840a	2
Lacertid	5+
Cordylid	50+
Squamata: Amphisbaenia: Amphisbaenidae	
Blanus sp.?	5+
Lepidosauria: Squamata: Serpentes	
Eoanilius cf. *E. europae*	1
Paleryx rhombifer Owen, 1850	
Type specimen: BMNH 25259	8
Paleryx sp.	3
Cadurcoboa sp.	1
Palaeophis	1
Calamagras sp.	50+
Platyspondylia sp.	1
cf. *Dunnophis*	5+
Vectophis wardi Rage and Ford, 1980	5+

Interpretation

Plint (1984) has interpreted the Hordle succession as representing a coastal environment, including littoral marine, barrier island shoreface, storm washover, and barrier flat, brackish lagoon, distributary channel and floodplain lake environments. The sequence indicates reducing salinity through time, and a transition towards river-dominated sedimentation in shallow floodplain lakes. Hooker (1992, p. 500) interprets the Hordle Mammal Bed as an open-forest subtropical setting. Wood (1844) and Charlesworth (1845) reported the first turtle finds from Hordle. The most abundant specimens are trionychids (softshelled turtles), such as the six species of *Trionyx* (Figure 9.7A) established by Owen and Bell

(1849). These were based on carapace remains, such as complete scutes and fragments bearing characteristic pustulose ornament, which are of limited use in modern taxonomic schemes which rely heavily on cranial characters. Meylan (1987) provides a cladistic classification of extant Trionychidae, but makes little reference to fossil taxa. Lydekker (1889h, pp. 53–4) established the genus *Aulacochelys* on the basis of carapace remains, and assigned *Trionyx circumsulcatus* Owen (1849) to it. However, Baur (1889) argued against this, pointing out that a free border on the costals, used by Lydekker to distinguish *A. circumsulcatus*, in fact occurs widely in the Trionychidae, and hence, that the genus must be regarded as invalid. Another trionychid was named *Geoemyda headonensis* by Hooley (1905), but was regarded as a *nomen vanum* by Młynarski (1976, p. 82), and was not mentioned by Moody (1980a).

Other turtles from Hordle belong to largely marine groups, typical of most British Tertiary sites (see Sheppey report). *Argillochelys* was a moderate-sized cheloniid with an estimated total length of 200 mm. *Trachyaspis hantoniensis*, a dermatemydid, is based on limited carapace remains showing a distinctive ornament (Lydekker, 1889h, p. 54). *Anosteira anglica*, established by Lydekker (1889h, p. 54) on the basis of carapace remains, is a carettochelyid. Młynarski (1976, p. 73) was uncertain of the systematic position of the species, while Moody (1980a) assigned it to *Allaeochelys*. *Ocadia crassa* Owen, 1849 (Figure 9.7B) and *O. oweni* Lydekker, 1889 are emydids, characterized by relatively thin unornamented scutes. The species '*Emys*', widely reported by the early authors, is probably referrable to this genus. Młynarski (1976) synonymized *O. oweni* with *O. crassa* and is not certain of their true affinities, while Moody (1980a) records both species of *Ocadia* from the Hordle Member.

The first recorded Hordle reptile was a crocodilian, apparently derived from the Crocodile Bed, and named *Alligator hantoniensis* by Wood (1844). The specimen, consisted of 'A great portion of the head . . . having nearly all the upper range of teeth (42 in number) remaining, along with the humerus, dermal scutae and other parts of the skeleton'. Pomel (1853, p. 383) correctly referred this species to the genus *Diplocynodon* on the basis of sharing with the French *D. ratelii* an expansion of the third lower tooth, which was nearly as much enlarged as the fourth. Meyer

(1857, p. 538) argued that *D. hastingsiae* and *D. ratelii* might be conspecific. Owen (1848) noted three more crocodilians, which he described as a gavial, an alligator and a crocodile. Huxley (1859e) described the dermal armour of *D. hastingsiae*. Woodward (1885) and Lydekker (1887b) reviewed the Hordle crocodilians, noting the synonymy between *A. hantoniensis* and *C. hastingsiae*. The material listed as *Diplocynodon* sp. and *Crocodilus* sp. also belongs to this taxon (A. R. Milner, pers. comm., 1994). *Diplocynodon hantoniensis* (Figure 9.7C, D) is a large alligator characterized by extensive development of ventral dermal armour, in addition to the distinctive features of the dentition.

The lizards from Hordle are now known from many hundreds of specimens, and eight or more genera are currently recognized as valid. The new collections include representatives of five lizard groups including a gekkonid, three anguids, two lacertids, a cordylid and an amphisbaenian (Milner *et al.*, 1982). The family Gekkonidae is represented by a single small dentary showing characteristic closure of the meckelian canal by an anterior downgrowth of the dentary. The Necrosauridae is represented by jaws and vertebrae assignable to *Necrosaurus*. Anguids are abundant in the assemblage, the commonest material belonging to a medium-sized, limbless anguine attributable to *Ophisaurus*. A smaller anguine, similar in size to the living *Anguis* (the slow worm), also occurs, but is distinguished from *Anguis* by the complete absence of neural spines on the vertebrae. Lacertids are abundantly

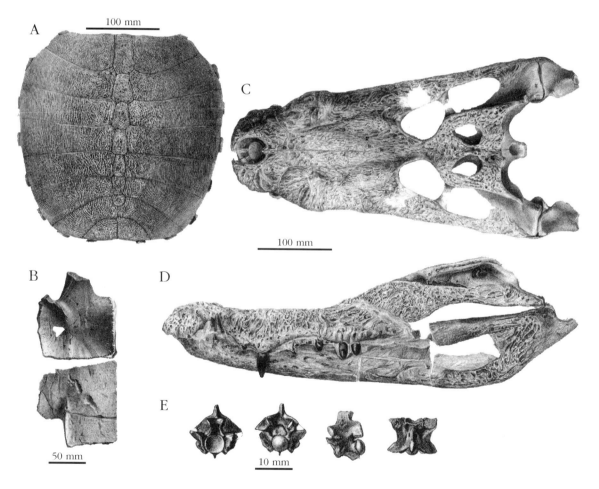

Figure 9.7 Typical reptiles of the Late Eocene Lower Headon Beds of Hordle Cliff. (A) The turtle *Trionyx henrici* Owen, 1849, carapace in dorsal view; (B) the turtle *Ocadia crassa* (Owen, 1849), internal views of plastron elements; (C) and (D) the crocodilian *Diplocynodon hantoniensis* (Wood, 1844), skull in (C) dorsal and (D) lateral views; (E) the snake *Paleryx rhombifer* Owen, 1850, mid-body vertebra in posterior, anterior, lateral, and dorsal views. (A) and (B) After Owen and Bell (1849); (C) and (D) after Owen (1850b); (E) after Owen (1850c).

represented by jaws and vertebrae, together with a few cranial and appendicular bones, and were the first lizards recognized from Hordle. The remains were originally referred to the supposed iguanid '*Iguana europeana*' by Lydekker (1888a, 1888c), but were reidentified as lacertid by Hoffstetter (1942), who designated one of the Hordle specimens the holotype of *Plesiolacerta lydekkeri* (Estes, 1983, p. 103). The other material, including vertebrae, was identified by Lydekker (1888a, 1888c) as belonging to the large limbed anguid *Placosaurus*. Sullivan (1979), however, argued that *P. rugosus*, the type species of *Placosaurus*, is indeterminate, but Estes (1983, p. 158) accepts the validity of the genus. Milner *et al.* (1982) refer *Placosaurus* to the anguid subfamily Glyptosaurinae *incertae sedis*. A large cordylid is represented by jaws and vertebrae. The burrowing lizards, the Amphisbaenidae (subfamily Amphisbaeninae), are represented by a vertebra, a maxilla, and some dentaries apparently similar to those of the extant *Blanus*.

The snake vertebrae collected by Wood and Hastings in the late 1840s were described as *Paleryx rhombifer* (Figure 9.7E) and *Paleryx depressus* by Owen (1850). Rage and Ford (1980) have questioned the validity of these assignments, pointing out that these species, designated on size differences in the vertebrae, are probably synonyms, since comparable modern taxa display similar size variations within a single species. The species *P. depressus* is consequently regarded as a junior synonym of *P. rhombifer* by Rage (1984, p. 20).

In the new collection snake remains are represented by vertebrae, occasional maxillae, palatines, pterygoids and dentaries, but only a few of the remains are suitable for taxonomic discussion. Besides distinctive remains of *Paleryx*, an erycine is the most abundant form present, known from caudal and trunk vertebrae with a complex morphology that most closely resemble those of *Calamagras*. Vertebrae bearing a distinct haemal keel are referred by Milner *et al.* (1982) to cf. *Dunnophis*. However, a haemal keel is unknown in other *Dunnophis* material. Two probable primitive caenophidians are present. The first, bearing tall neural spines on the trunk vertebrae, is referred to *Vectophis wardi*. The second form has a broad neural arch and a reduced neural spine. This may represent a new taxon, since no other comparable material is known. An aniliid is represented by an isolated dorsal vertebra. Several boids can be distinguished from the structure of their middle and posterior trunk vertebrae, and the aquatic snake '*Palaeophis*' is represented by somewhat limited remains of vertebrae, although this is a late record of this genus which should be checked (J.-C. Rage, pers. comm., 1993).

Comparison with other localities

The Totland Bay Member of Hordle Cliff is directly correlated with the top of the same unit (Insole and Daley, 1985) at Headon Hill, Isle of Wight, on the basis of their Late Eocene (Priabonian) mammal fauna and occurrence of calcareous nannoplankton zones NP17 (Barton Clay below) and NP19/20 (Curry *et al.*, 1978). The Hordle fauna includes a range of turtles, snakes, lizards and crocodilians, comparable to the Headon Hill reptiles (see Headon Hill report), however, differences are noted in the range and abundance of the taxa present between the two localities, with the Hordle fauna being at least quantitively different from that at Headon Hill. This may be a local ecological or taphonomic effect (Milner *et al.*, 1982).

The turtles are known from several other Late Palaeocene to Early Oligocene faunas in Britain and in continental Europe (Moody, 1974, 1980a). The fresh-water *Trionyx* and ?*Ocadia* and the crocodilian *Diplocynodon hantoniensis* are known from the Cliff End Member (Headon Hill Formation; Late Eocene) at Cliff End on the Isle of Wight (SZ 332893–SZ 335895) (Gamble, 1981, pp. 401–2; Moody, 1980a, pp. 23–4). The Late Eocene trionychids *T. henrici*, *T. marginatus* and *T. incrassatus* are known from the Bembridge Marls Member and *Trionyx* sp. is known from the Lower Hamstead Member (both Hamstead Formation) (Moody, 1980a, pp. 23–4).

Among the squamates, all of the Hordle families are found in the Eocene of France and Germany, and only the lizard families Agamidae and Helodermatidae and the snake family Typhlopidae, present in these successions, are not represented. The lizard *Placosaurus* is recorded from the Late Eocene Phosphorites du Quercy, France and Shara Murun Formation, Mongolia, as well as the Mid Eocene of the Geiseltal, Germany, and other Palaeocene to Oligocene sites in Europe and China (Estes, 1983, pp. 158–63). *Plesiolacerta* is known also from the Late Palaeocene of Dormaal, Belgium, the Late Eocene Phosphorites du Quercy, France and possibly from the Late Palaeocene of Germany (Estes, 1983, pp. 103–4). The snake *Paleryx* is known also from the Mid Eocene of the Geiseltal, Germany (Rage, 1984, p. 20), and the other less well-defined taxa are known from Palaeocene to Oligocene units in continental Europe.

The crocodilian *Diplocynodon* was named from the Aquitanian (Early Miocene) of Allier and other sites in France. Other supposed records of *D. hantoniensis* include a partial specimen from the Mid Eocene of the Geiseltal in Germany and a partial jaw from the Early Oligocene of Borken, Lower Hessen (Steel, 1973, p. 82). Other species of *Diplocynodon* have been reported from the Mid Eocene of France, Messel and the Geiseltal, Germany, Spain and Wyoming, USA, the Oligocene of France, the Miocene of Austria and the Mid Pliocene of Bulgaria, among others (Steel, 1973, pp. 81–4).

Conclusions

The Totland Bay Member at Hordle Cliff has yielded a rich fauna of reptiles of Late Eocene age. The locality, known since the early 19th century, has continued to produce abundant reptile remains and has recently produced an important collection of squamates. These recent finds, which include a variety of forms newly recorded from the British Palaeogene, indicate that the herpetofauna of this region during the Late Eocene was as diverse as those of continental Europe.

The conservation value lies in the richness of the reptile fauna and the continuing supply of new specimens.

HEADON HILL (ALUM BAY–TOTLAND), ISLE OF WIGHT (SZ 305855)

Highlights

Headon Hill has produced a good fauna of Late Eocene reptiles, and it is especially important for the specimens of lizards and snakes. These include several specimens of the glass lizard *Ophisaurus*, the first record in Britain, as well as three species of snakes.

Introduction

The Late Eocene (Priabonian) Headon Hill Formation in their type area in the degraded

coastal sections of Headon Hill, Isle of Wight (Figure 9.8), have produced in the recent past a good fauna of turtles, crocodilians, snakes and lizards (Figure 9.9). Large parts of the section are obscured by mud flows, but the relevant beds may easily be cleared for further excavation.

The Headon Hill Formation between Alum Bay and Totland has been described by Prestwich (1846), White (1921), Stinton (1971), Cray (1973), Daley and Edwards (1974), Daley and Insole (1984) and Insole and Daley (1985). Accounts of the reptilian faunas have been given by Cray (1973), Meszoely and Ford (1976) and Rage and Ford (1980), but there is as yet no complete overview.

Description

A generalized section of the Headon Hill Formation taken from the south-west corner of Headon Hill, based on Cray (1973) and Insole and Daley (1985), is:

	Thickness (m)
Cliff End Member (part of 'Upper Headon Beds')	
Clays and marls	seen to 6.6
Hatherwood Limestone Member (part of 'Upper Headon Beds')	
Limestones	2.8
Lignite (Lignite Bed)	0.7
Limestones	2.7
Linstone Chine Member (part of 'Upper Headon Beds'); white and grey sands (*Microchoerus* Bed at base)	0.8
Colwell Bay Member ('Middle Headon Beds')	
Blue-green clays and sands	2.0
Limnaea Limestone	0.2
Blue, green and brown sandy clays (*Venus* Bed)	*c.* 4.4
Sands, clays and lignites (*Neritina* Bed)	2.5
Totland Bay Member ('Lower Headon Beds')	
Limnaea limestone (How Ledge Limestone)	*c.* 2.0
Marls, clays, sands and lignites	4.6
Limnaea limestone	0.4
Green clays and pale sands	4.4
Limnaea limestone	0.8
Blue and green clays	1.0

Figure 9.8 Alum Cliff, at the southern end of the Headon Beds outcrop on Headon Hill, Isle of Wight. (Photo: M.J. Benton.)

	Thickness (m)
Limnaea limestone	0.25
Green sandy clays	0.7
Green clays	seen 1.1

Cray (1973, p. 24) mentions fragments of turtle bones and, more rarely, broken mammalian remains from a limestone immediately below the Lignite Bed, a horizon outcropping about a third of the way up the vertical cliff formed by the limestone on the south-west seaward face of Headon Hill. The limestone (which may also overlie the Lignite Bed in places) is very variable lithologically, and several subdivisions were recognized. The reptile-bearing lithology is a soft, impure, orange-coloured, marly limestone, rich in the shells of *Galba* sp. However, reptile debris, including dermal scutes of turtles (*'Emys'* sp.) and teeth of the crocodilian *Diplocynodon*, also occurs sporadically throughout the Lignite Bed and appears to come from all the lithologies, being represented both in friable lignite and in the shell marls (Cray, 1973, p. 25).

In the early 1970s large collections of reptiles (particularly squamates) and amphibians were obtained by Mr R.L.E. Ford from units in the Totland Bay Member, in particular from Bed HH2 (Bosma, 1974, fig. 9) beneath a unit of hard limestone named the 'How Ledge Limestone', from a series of green-grey clays. Two localities have yielded herpetofaunas from this stratum: in the undercliff at Headon Hill and in Totland Bay. The How Ledge Limestone occurs along the coast between Hatherwood Point and How Ledge, and it appears that the reptiles occur patchily beneath the entire length of the outcrop. The fossils are all represented by disarticulated, and frequently abraded and fragmented, elements, which indicate considerable predepositional disturbance.

Fauna

The main collections of fossil reptiles from Headon Hill are curated in the BMNH as well as the Museum National d'Histoire Naturelle (MNHN) Paris and the Stuttgarter Museum für Paläontologie (Ford collection). The collections include many mammal and amphibian taxa, as well as reptiles. Amphibians include palaeobatrachid and discoglossid frogs (Figure 9.9A).

Testudines: Cryptodira: Emydidae
    'Emys' sp.
Archosauria: Crocodylia: Neosuchia: Eusuchia
    Diplocynodon sp.
    Crocodilus sp.
Lepidosauria: Squamata: Sauria
    Scincomorph indet.
    Necrosaurus sp.
    Ophisaurus sp.
    Glyptosaurine indet.
Lepidosauria: Squamata: Serpentes
    Paleryx rhombifer Owen, 1850
    Vectophis wardi Rage and Ford, 1980
      Type specimen: MNHN CGB 27
    cf. *Dunnophis*

Interpretation

The Headon Hill Formation falls in the Headonian European mammal age and is equated with the upper part of this age, dated as late Late Eocene (Priabonian) by Curry *et al.* (1978). The environments are interpreted as floodplain and lagoonal, as for Hordle (q.v.), and the vertebrates are associated with closed, subtropical forests (Hooker, 1992). The squamates from the HH2 bed are associated with abundant amphibian remains, including three anurans (Discoglossidae indet., Palaeobatrachinae indet. and cf. *Eopelobates*) and rare salamanders such as 'cf. *Megalotriton*' (Rage and Ford, 1980).

The turtle *'Emys'* is represented by numerous thin unornamented scutes and appendicular bones. This genus may almost certainly be attributed to the well-known Eocene genus *Ocadia*, of which two forms have been noted from the Totland Bay Member by Moody (1980a, p. 24). A large number of isolated teeth have been collected from various parts of the section at Headon Hill; these are referred to the crocodilians *Diplocynodon* and *Crocodilus*.

The lizards from Headon Hill include members of the Necrosauridae (*Necrosaurus*), Scincomorpha and Anguidae (glyptosaurine, *Ophisaurus*), all of which are known from the Hordle Cliff section (see above). The faunal list from the HH2 horizon at Headon Hill is smaller than that from the Totland Bay Member at Hordle, lacking the gekkonid, cordylid and the lacertid *Plesiolacerta lydekkeri*.

The Scincomorpha are represented by several remains of dentary bones, but these are not entirely diagnostic. The dentaries possess a narrow meckelian groove that is shallow and restricted to the lower surface of the dentary anteriorly. The cylindrical teeth show pleurodont implantation, and (viewed laterally) are high relative to the average height of the dentary.

Ophisaurus (the glass lizard) is the most abundant lizard present, being represented by numerous distinctive remains of osteoscutes (or

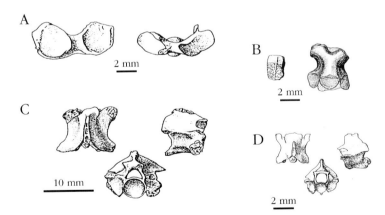

Figure 9.9 Typical reptiles and amphibians of the Late Eocene Lower Headon Beds of Headon Hill and Totland Bay. (A) A palaeobatrachid frog, fragmentary atlas; (B) the limbless lizard *Ophisaurus* sp., scute and trunk vertebra in ventral view; (C) the snake *Paleryx rhombifer* Owen, 1850, mid-trunk vertebra in dorsal, lateral, and anterior views; (D) the snake *Vectophis wardi* Rage and Ford, 1980, mid-trunk vertebra in dorsal, lateral and anterior views. (A), (C) and (D) After Rage and Ford (1980); (B) after Meszoely and Ford (1976).

osteoderms) and a few isolated dorsal vertebra (Meszoely and Ford, 1976; Rage and Ford, 1980; Figure 9.9B). The osteoscutes are flattened structures from the trunk and tail, bearing a smooth anterior 'gliding' surface and a flattened face with an ornament of irregularly branching grooves and ridges. Many of the osteoscutes, particularly those of the tail, carry a prominent median ridge. The osteoscutes and vertebrae of *Ophisaurus* show little morphological variation, and it has been hard to divide the genus into species (Rage and Ford, 1980). Meszoely and Ford (1976) suggested that the Headon Beds form was conspecific with *Ophisaurus ballensis* (Kuhn, 1940) from the Geiseltal deposits (Mid Eocene) near Halle, Germany, based on its European occurrence and Late Eocene age. This view was tentatively accepted by Rage and Ford (1980).

The anguine subfamily Glyptosaurinae is represented by two partly fragmented dorsal vertebrae and a caudal vertebra. These are larger than those of *Ophisaurus* and may be distinguished by the slightly concave ventral surface of the centra (a feature characteristic of limbed Anguidae). The necrosaurid *Necrosaurus* is represented by a single elongate posterior caudal vertebra, showing no fused haemapophyses, but two articular facets for the chevron and a groove on the ventral surface.

Of the snakes, the boid *Paleryx rhombifer* (Figure 9.9C), tropidophid cf. *Dunnophis* and caenophid *Vectophis wardi* are all represented by isolated remains of vertebrae. *Paleryx rhombifer* (represented by approximately 20 vertebrae) was regarded as congeneric with *Paleopython* from the Eocene of France by Lydekker (1888c), but Rage and Ford (1980) have argued that the two forms are distinct.

A small snake, represented by a number of isolated dorsal vertebrae, is referred by Rage and Ford (1980) to cf. *Dunnophis*. The genus is based on limited and damaged vertebral material from the Early Eocene of France and Belgium (Rage, 1984), and its precise relationships have been hard to establish. Over the years, this genus has been assigned to Serpentes *incertae sedis* or the Boidae (in particular the Tropidopheidae). These views have been disputed, but Rage and Ford (1980) suggested that the Isle of Wight form might provide a good morphological connection between 'typical' *Dunnophis* and the Boidae.

Vectophis wardi is a frequent element in the fauna, being represented by five vertebrae from Totland Bay and by about 60 vertebrae on Headon Hill (Figure 9.9D). This is a small alethinophidian snake with a distinctive vertebral morphology. The type specimen (MNHN CGB 27), collected from Totland Bay, consists of a single mid-trunk vertebra which carries a tall neural spine, a feature shared by several specimens from Hordle Cliff, which have consequently been referred to the species (Milner *et al.*, 1982, p. 152). Other features of the genus include a vaulted neural arch, robust neural spine, narrow centrum, mid and posterior trunk vertebrae which lack a hypophysis, a distinct and rather sharp haemal keel, grooves lying on either side of the haemal keel, absence of long prezygapophysial processes, and caudal vertebrae with pleurapophyses and haemapophyses. On the basis of these characters, Rage and Ford (1980) consider *Vectophis* as perhaps belonging to the Colubroidea, and as possibly a primitive member of this superfamily.

Comparison with other localities

Geographically and stratigraphically, the nearest comparable units to the Totland Bay Member at Headon Hill are the same stratigraphic unit at Hordle Cliff (SZ 253925–SZ 287915; see above), and the Fishbourne Member ('Osborne Beds') at Fishbourne (SZ 551927). In the 'Osborne Beds' shared faunal elements include *Ophisaurus* sp., *Paleryx rhombifer* (represented by one rounded and worn trunk vertebra) and cf. *Dunnophis*. The Erycinae cf. *Calamagras* and Erycinae unidentif. (Rage and Ford, 1980), present in these beds, do not occur in the Totland Bay Member. All of the reptiles recorded from Headon Hill are known from the directly correlative sequence at Hordle, but there are many genera known from Hordle that are absent on the Isle of Wight (see above), possibly the result of taphonomic differences (Milner *et al.*, 1982).

A dentary referred to a glyptosaurine lizard has been obtained elsewhere on the Isle of Wight, from the Bembridge Marls Member of the Bouldnor Formation (Early Oligocene) (BMNH R8716) (R. Estes, pers. comm. to Rage and Ford, 1980). Large-limbed Anguidae are represented in the Late Eocene of France by cranial osteoderms and other elements, named *Placosaurus rugosus* (Gervais, 1848–52), and also from Germany, where *Placotherium waltheri* (Weigelt, 1929) is known from deposits of Mid Eocene age. The status of these species, based mainly on external morphology of the osteoderms, is not clear and, although clearly belonging to the Glyptosaurinae, both forms are regarded by Sullivan (1979, pp. 43-4) as *nomina dubia*. In North America glyptosaurine lizards are represented by more complete remains bearing similar osteoderms, and numerous genera have been named, particularly from the Eocene and Oligocene (Sullivan, 1979; Estes, 1983).

The discovery of the anguid *Ophisaurus* from the Isle of Wight, extends the range of this genus from the Mid and Late Eocene of central Europe, to the British Isles. The genus is still extant and is confined to the eastern section of continental Europe.

Necrosaurus is known from the Late Eocene of France (*Necrosaurus cayluxi* Filhol, 1873) and from the Mid Eocene of Germany and latest Eocene and Early Oligocene of France (*N. eucarinatus* Kuhn, 1940). The genus is also known from the Paleocene of France and from the Early Oligocene of Belgium (Estes, 1983).

The snake *Dunnophis* is reported from the Early Eocene of France and Belgium, the Mid and Late Eocene of North America, the Late Eocene of France and the Early Oligocene of Belgium (Rage, 1984). As noted above, the closely related Totland Bay Member form cf. *Dunnophis*, may be phylogenetically intermediate between *Dunnophis* and the Tropidophiidae; in this sense, it is confined to the British Early Eocene.

Conclusions

Headon Hill is an important reptile site of Late Eocene age, unique for its record of the glass lizard *Ophisaurus*, a form known elsewhere in continental Europe from the Eocene to the present day. The type specimen of *Vectophis wardi* came from Headon Hill. The other snakes from Headon Hill, *Paleryx rhombifer* and cf. *Dunnophis* are of phylogenetic importance. The Headon Hill section offers great potential for future collecting, and it has been much less exploited than the equivalent-age units at Hordle Cliff (q.v.), hence its conservation value.

OLIGOCENE

The Oligocene deposits of the Isle of Wight have produced restricted, but important, reptile faunas. The finds are dominated by remains of freshwater turtles and crocodilians, but other elements include fully terrestrial forms including snakes. The better-documented localities include the following:

ISLE OF WIGHT: Thorness Bay (Bembridge Marls Member; Early Oligocene, Rupelian; SZ 455935; *Trionyx incrassatus, Trionyx* indet.; Hooker and Ward, 1980, p. 9; Daley, 1973, pp. 83-93); Gurnard Bay (=Gurnet Bay; Bembridge Marls Member, marine band; Early Oligocene, Rupelian; SZ 4795; *Trionyx incrassatus, T. circumsulcata*, '*Emys*' sp., trionychid indet., snake, *Diplocynodon hantoniensis*; Daley, 1973, Hooker and Ward, 1980, p. 9); Whitecliff Bay (Bembridge Marls Member; Early Oligocene, Rupelian; SZ 643864; turtle, *Trionyx*); Bembridge (Bembridge Marls Member; Early Oligocene, Rupelian; SZ 6588; *Trionyx* sp., *T. incrassatus* Owen, 1849, trionychid indet.; Hooker and Ward, 1980, p. 9); Bouldnor and Hamstead Cliffs (Bembridge Marls Member, Hamstead Member; Early Oligocene, Rupelian; SZ 391913; crocodilians, *Ocadia crassa, Trionyx* sp., *Diplocynodon hantoniensis, Diplocynodon* sp., *Paleryx* sp.); Yarmouth (Bembridge Marls Member; Early Oligocene, Rupelian; SZ 367899; trionychid indet.; Hooker and Ward, 1980, p. 8); Hamstead (=Hempstead; Hamstead Member, mottled clays and marls?; Early Oligocene, Rupelian; SZ 4091; *Trionyx* sp, 'chelonian', *Paleryx depressus, Crocodilus hastingsii, Diplocynodon hantoniensis*; Hooker and Ward, 1980, p. 9).

One GCR site has been selected for British Oligocene reptiles:

1. Bouldnor and Hamstead Cliffs, Isle of Wight (SZ 391913). Early Oligocene (Rupelian), Bembridge Marls Member, Hamstead Member, Bouldnor Formation.

BOULDNOR AND HAMSTEAD CLIFFS, ISLE OF WIGHT (SZ 391913)

Highlights

Bouldnor and Hamstead Cliffs is the only site in Britain for Oligocene reptiles, the fauna of five or six species of turtles, snakes and crocodilians is small, but important worldwide because of the general rarity of Oligocene reptiles sites everywhere.

Introduction

The Bembridge Limestone Formation and Bouldnor Formation (Figure 9.2) exposed at Bouldnor Cliff have produced the best fauna of British Oligocene reptiles. Large areas of the cliff are affected by landslips and debris flows, but exposures on the foreshore, visible at low water, are normally excellent and many new finds could be made.

The cliff sections at Hamstead and Bouldnor cliffs have been described by Forbes (1856), White (1921), Daley (1972, 1973), Daley and Edwards (1974) and Insole and Daley (1985).

Reptile remains have been noted sporadically by authors on the stratigraphy of the site, but there are no comprehensive descriptions. Hooker and Ward (1980) summarize the fauna, while Moody (1980a) gives some details of the turtles.

Description

At Bouldnor Cliff the whole of the Bouldnor Formation (Cranmore Member, Hamstead Member, Bembridge Marls Member; *c.* 87 m) and underlying Bembridge Limestone Formation are exposed. The Bembridge Limestone Formation, with the Late Eocene/Osborne Member below, occurs in the east of the section in Hamstead Ledge (SZ 401920), where three freshwater limestone beds are developed. West of Hamstead Point sections are seen in the top of the Bembridge Limestone Formation, the whole of the overlying Bembridge Marls Member and part of the Hamstead Member.

The Bembridge Marls Member here comprises 21.5 m of fresh- and brackish-water sediments, mainly clays and silts, and contain an abundant, but taxonomically restricted, molluscan fauna (Daley, 1972). The lower part of the unit occurs in the cliffs, but the sequence is best exposed on the foreshore and may be seen at low tide immediately to the west of Hamstead Ledge where numerous shell beds are developed in green or grey muds. All of the beds become visible at low water during the equinoctial spring tide. The Bembridge Oyster Bed (Forbes, 1856) (Bed HAM I of the Bembridge Marls Member), also seen at Whitecliff Bay, occurs at the base of the succession. The restricted assemblages include taxa which are regarded by modern analogy as brackish-water forms. The rest of the sequence is made up of grey to bluish-grey silts and clays deposited variously under fresh- and brackish-water conditions. Beds HAM XXIII–XXV (Daley, 1973) contain abundant monocotyledonous leaf fragments and the water-plant seeds *Brasenia* and *Stratiotes*, which occur in bands with the gastropods *Viviparus* and *Galba* (a pulmonate).

The Black Band, taken to mark the base of the Hamstead Member, occurs low in the cliffs about 200 m east of a line of posts. This comprises a carbonaceous mud and contains freshwater gastropods such as *Viviparus*. At the base of the unit, autochthonous root systems penetrate into the underlying bed. Another black, lignitic clay (the *Nematura* Bed) occurs somewhat higher up in the Hamstead Member succession, but contains a distinctive brackish-water molluscan fauna. The greater part of the Hamstead Member consists of grey-green and green muds with occasional dark-brown to black, laminated muds. However, these are much obscured by recent mudflows.

The succeeding Cranmore Member (9.2 m) is marked by a sudden change from brown-grey to bright green clay. The member occurs in the top of the cliff at the west of the exposure where it is capped by 'Plateau gravel'. The member is divided into the *Corbula* Beds (marine) and the *Cerithium* Beds (non-marine), which together consist of a mixture of grey, blue and black fossiliferous clays.

The section is based on White (1921, pp. 133–4, 140–1), Daley (1973), Daley and Edwards (1974), Daley and Insole (1984) and Insole and Daley (1985).

	Thickness (m)
Solent Group	
Bouldnor Formation	
Cranmore Member (Upper Hamstead Beds of White, 1921)	
Corbula Beds	5.8
Cerithium Beds	3.4
Hamstead Member (Lower Hamstead Beds of White, 1921)	
Green and mottled clays, with lignite beds and shell beds	*c.* 25
Water-Lily Bed: laminated lignite with seeds, palm leaves, water lily leaves and molluscs	0.6
Green and red marls (much obscured)	*c.* 2
White Band: green clays with white shell-marls	1.8

	Thickness (m)
Green clay with ironstone nodule band (much obscured)	10.8
Nematura Bed: black lignitic clay, full of gastropods	0.9
Green and black clays, with bivalves and gastropods	8.1
Black Band: lignite, full of *Paludina* and *Unio*	0.5
Bembridge Marls Member (Bed notation from Daley, 1973)	
HAM XXXI–XXXIV: green, red and mottled clays	10.2
HAM XXX: lignite with seeds and molluscs	0.6
HAM XXVI–XXIX: clays with seeds and molluscs	*c.* 5.0
HAM XXIII–XXV: lignite and clay, rich in water-plant seeds, leaf fragments, and gastropods	*c.* 2.0
HAM XX–XXII: freshwater clays and silts	1.6
HAM XIX: green clays and white marls, with bivalves	0.2
HAM XVI–XVIII: green mudstone and lignite band	15
HAM XV: black clay with gastropods	0.9
HAM XI–XIV: mudstones and siltstones, with bivalve band	3.3
HAM VI–X: grey and blue-green laminated clays, with brackish-water bivalves and gastropods	2.7
HAM V: greenish-grey clay with bands containing *M. acuta*, *Serpula* sp. and *Viviparus lentus*	0.3
HAM I–IV: grey and black clays with shelly partings and bands containing bivalves and gastropods; thin shell bed with *Ostrea* at the base	0.9
Bembridge Limestone Formation	

In the Bembridge Marls Member, vertebrate material is quite common, but usually comminuted. Fish vertebrae are mentioned by Daley (1972) as occurring commonly at many horizons, whereas fish scales and teeth are found in others. It is possible, but not proven, that the larger reptile and mammal remains may also derive from particular levels.

Fauna

Faunal lists are summarized by Hooker and Ward (1980) and Moody (1980a), but fuller accounts are not yet available. The specimen counts are based on collections in the BMNH and IWCMS.

	Numbers
Testudines: Cryptodira: Trionychidae	
Ocadia crassa (Owen, 1849)	1
Trionyx sp.	3
Lepidosauria: Squamata: Serpentes	
Paleryx sp.	1
Archosauria: Crocodylia: Neosuchia: Eusuchia: Alligatoridae	
Diplocynodon hantoniensis (Wood, 1944)	several
Diplocynodon sp.	several
crocodilian	several

Interpretation

Daley (1973) postulated three main environments of deposition for the Bembridge Marls Member. In the lower part the Bembridge Oyster Bed, which coincides with the main transgressive period of the Bembridge Marls Member, is interpreted as an estuarine deposit, since the sediments contain few primary sedimentary structures, and a fauna consisting predominantly of comminuted *Ostrea* shell debris indicates a considerable amount of water movement, in contrast to the fauna of lagoonal environments. Daley (1972) noted that the molluscan assemblages from this part of the succession are comparable with those of tropical and subtropical mudflats of the present day. The predominantly grey or blue-grey clays which form the bulk of the Bembridge Marls Member are interpreted as lagoonal in origin; the bivalves are commonly in life position, and some of the sediments exhibit varve-like lamination. On the basis of sedimentary evidence (ripples, irregular lamination, presence of lignite) the central and upper parts of the deposit are thought to represent floodplain and lacustrine deposits.

The trionychid turtles are represented by recent finds of complete carapaces and limited skull material (in IWCMS). The new trionychid carapaces fall into high- and low-domed types. The high-domed forms show great variation in the cross-sectional thickness of their shells, whereas in the low-domed forms the carapace is of uniform thickness. One of the specimens exhibits 'pathological' distortion of the rear dorsal surface of the carapace. The new skull material consists of a partial braincase with both quadrates attached. There are some associated postcranial remains of cervical vertebrae, but no carapace or lower jaw. New finds also include specimens of the crocodilian *Diplocynodon*, which had an estimated length of about 4 m.

Comparison with other localities

The Bembridge Marls Member has been correlated on the basis of mammals and charophytes with the Tongrian Stage in Belgium and the Ludian (Late Eocene) in the Paris Basin (Curry *et al.*, 1978). There has been some controversy over whether the beds should be included in the Eocene or Oligocene, but most British workers consider them as belonging to the latter.

Oligocene trionychids have been recorded from Monteviale in Italy (*Trionyx capellini*), Steiermark, Austria (*T. styriacus*), Catalonia (*T. marini*) and China (*T. gregaria*) (Młynarski, 1976, pp. 77-9). The crocodilian *Diplocynodon* has been reported especially from the Late Eocene Headon Beds of Hordle and Headon Hill (see above), as well as from sediments ranging in age from the Early Eocene to Miocene of Europe and the USA (see Headon report).

Oligocene reptiles are known from North America and Europe, but finds are much rarer than for those of Eocene or Miocene age. The famous Quercy deposits in France span the Eocene–Oligocene boundary, and thus equate in age with the Bembridge Marls. These have produced a diverse array of lizards, snakes, crocodilians and turtles. Other continental European Oligocene localities are spread as far afield as Spain and the Ukraine, and Germany and Italy. The mammals have been heavily studied, the reptiles less so. It is hard to make detailed comparisons with the Hamstead-Bouldnor locality until the reptiles from the site are more fully described.

Conclusions

Britain's only Oligocene reptile site, and one of the few of that age in the world. This international importance, plus the continuing opportunities for new finds provided by erosion, establish the site's conservation value.

PLEISTOCENE

Fossil reptiles from the British Pleistocene are presently known from interglacial sediments of Cromerian, Hoxnian, Ipswichian and Holocene age at a wide variety of localities, owing mainly to the result of a recent programme of research carried out by J.A. Holman in collaboration with A.J. Stuart and other workers during the 1980s and 1990s. The reptiles form part of herpetofaunas which are of value as reliable indicators of Pleistocene climates and environments. The finds also provide a valuable contribution to knowledge on the diversity and spread of reptiles through Pleistocene time, and demonstrate a link with climatic fluctuation. The localities are listed by county, and include only those that have produced reptiles (amphibian-only sites are not listed).

DEVON: Cow Cave, Chudleigh (SX 8679; *Anguis fragilis*); Happaway Cave, Torquay (Flandrian; *Natrix natrix*; Holman, 1987).

SOMERSET: Westbury-sub-Mendip (Cromerian; ?zone Cr IV; ; *Emys orbicularis, Coronella austriaca, Natrix natrix, Vipera berus*; Stuart, 1979; Holman, 1993).

SUSSEX: Selsey (Ipswichian; zones Ip Ib–IIb; *Emys orbicularis*; Stuart, 1979); Amey's Eartham Pit, Boxgrove (unnamed interglacial between Cromerian and Anglian; SU 920085; *Anguis fragilis, Lacerta* cf. *L. vivipara, Natrix natrix, Natrix* sp.; Holman, 1993).

KENT: Dierden's Pit, Ingress Vale, Swanscombe (Hoxnian; ?zone Ho III; TQ 6074; *Emys orbicularis, Natrix natrix*; Stuart, 1979; Holman, 1987); Ightham Fissure, near Ightham (Devensian-Holocene; TQ 5956; *Anguis fragilis, Natrix natrix, Vipera berus, Coronella austriaca*; Newton, 1894a; Holman, 1985).

SUFFOLK: Bobbitshole, Ipswich (Ipswichian; zones Ip Ia-IIa; *Emys orbicularis*; Stuart, 1979); Stoke Tunnel, Ipswich, Suffolk (Ipswichian; Stoke Tunnel 'Bone Bed', ?zone Ip IV; *Emys orbicularis*; Stuart, 1979); Harkstead (Ipswichian, ?zones Ip III–IV); *Emys orbicularis*; Stuart, 1979).

ESSEX: Cudmore Grove, East Mersea, Mersea Island (Hoxnian, Substage Ho IIIb, channel fill; TM 068146; *Emys orbicularis, Anguis fragilis, Lacerta vivipara, Lacerta* sp., *Elaphe longissima, Natrix maura* or *N. tessellata, Natrix natrix, Natrix* sp. indet., *Vipera berus*; Bridgland, 1987, p. 329; Holman *et al.*, 1990); Little Oakley (Cromerian Stage; Little Oakley Silts and Sands; TM 223294; *Emys orbicularis*; Bridgland, 1987, p. 321).

NORFOLK: Mundesley (Ipswichian; 'Forest Bed'; ?= Mundesley Sands, zones Ip Ib–IIb; TG 3136; *Emys orbicularis, Emys lutaria, Tropidonotus natrix*; Newton, 1862, 1879, 1882a; Woodward, 1880; Stuart, 1979); West Runton; Bacton (TG 1842; *Tropidonotus natrix, Vipera berus, Anguis fragilis*; Upper Freshwater Bed, Cromerian, W. Runton; *Natrix natrix, N. vipera*; Newton, 1882a, 1882b; Holman, 1993); Itteringham Gravel Pit (from Ipswichian interglacial bed; TG 139305; *Emys orbicularis, Natrix natrix*; Hallock *et al.*, 1990); East Wretham (Holocene; zone II (=VIIa, Atlantic); '*Emys lutaria*', *Emys orbicularis*; Newton, 1862; Woodward, 1880, Stuart; 1979); Swanton Morley (Ipswichian; zone Ip IIa; *Emys orbicularis*; Stuart, 1979).

LANCASHIRE: Dog Holes, Warton (Flandrian; SD 4128; *Anguis fragilis, ?Vipera, Natrix natrix*; Holman, 1987).

None of these sites could be selected as having a greater or lesser claim to be selected as a candidate GCR site to represent British Pleistocene reptiles. Indeed, several have been entirely worked out, and new ones are found when suitable sites are excavated.

References

A.B. (1757) [Further account of fossils.] *Gentleman's Magazine*, **27**, 122-3.

Adams-Tresman, S.M. (1987a) The Callovian (Middle Jurassic) marine crocodile *Metriorhynchus* from central England. *Palaeontology*, **30**, 179-94.

Adams-Tresman, S.M. (1987b) The Callovian (Middle Jurassic) teleosaurid marine crocodiles from central England. *Palaeontology*, **30**, 195-206.

Agassiz, L. (1844) Monographie des poissons fossiles du Vieux Grès Rouge ou système Devonien (Old Red Sandstone) des lles Britanniques et de Russie. Neuchâtel.

Aigner, T. (1980) Biofabrics and stratinomy of the Lower Kimmeridge Clay (Upper Jurassic, Dorset, England). *Neues Jahrbuch für Geologie und Paläontologie, Abhandlungen*, **159**, 324-38.

Allen, H.A. (1908) List of British Triassic fossils in the Warwick Museum. *Report of the British Association for the Advancement of Science*, 1908 (1907), 274-7.

Allen, P. (1949) Notes on Wealden bone-beds. *Proceedings of the Geologists' Association*, **60**, 275-83.

Allen, P. (1959) The Wealden environment: Anglo-Paris basin. *Philosophical Transactions of the Royal Society, Series B*, **242**, 283-346.

Allen, P. (1962) The Hastings Beds deltas: recent progress and Easter field trip meeting report. *Proceedings of the Geologists' Association*, **73**, 219-43.

Allen, P. (1976) Wealden of the Weald: a new model. *Proceedings of the Geologists' Association*, **86**, 390-436.

Allen, P. (1977) Wealden of the Weald - a new model; correspondence. *Proceedings of the Geologists' Association*, **87**, 433-42.

Allen, P. (1981) Pursuit of Wealden models. *Journal of the Geological Society of London*, **138**, 375-405.

Allen, P. (1989) Wealden research-ways ahead. *Proceedings of the Geologists' Association*, **100**, 529-64.

Allen, P. and Wimbledon, W.A. (1991) Correlation of NW European Purbeck-Wealden (nonmarine Lower Cretaceous) as seen from the English type areas. *Cretaceous Research*, **12**, 511-26.

Anderson, F.W. and Bazley, R.A.B. (1971) The Purbeck beds of the Weald (England). *Bulletin of the Geological Survey of Great Britain*, **34**, 173 pp.

Anderson, J.M. and Cruickshank, A.R.I. (1978) The biostratigraphy of the Permian and Triassic: Part 5. A review of the classification and distribution of Permo-Triassic tetrapods. *Palaeontologia Africana*, **21**, 15-44.

Andrews, C.W. (1897) Note on the cast of the brain-cavity of *Iguanodon*. *Annals and Magazine of Natural History, Series 6*, **19**, 585-91.

Andrews, C.W. (1910) *Descriptive Catalogue of the Marine Reptiles of the Oxford Clay, Part 1*. British Museum (Natural History), London.

Andrews, C.W. (1913) On the skull and part of the skeleton of a crocodile from the Middle Purbeck of Swanage, with a description of a new species (*Pholidosaurus laevis*) and a note on the skull of *Hylaeochampsa*. *Annals and Magazine of Natural History, Series 8*, **11**, 485-94.

Andrews, C.W. (1920) Note on two new species of fossil tortoises. *Annals and Magazine of Natural History, Series 9*, **5**, 145-50.

References

Andrews, C.W. (1921) On some remains of a theropodous dinosaur from the Lower Lias of Barrow-on-Soar. *Annals and Magazine of Natural History*, Series 9, **8**, 570-6.

Andrews, C.W. (1922) Description of a new plesiosaur from the Weald Clay of Berwick (Sussex). *Quarterly Journal of the Geological Society of London*, **78**, 285-98.

Andrews, J.E. (1985) Sedimentary facies of a late Bathonian regressive episode: the Kilmaluag and Skudiburgh formations of the Great Estuarine Group, Inner Hebrides, Scotland. *Journal of the Geological Society of London*, **142**, 1119-37.

Andrews, J.E. and Hudson, J.D. (1984) The first Jurassic dinosaur footprint from Scotland. *Scottish Journal of Geology*, **20**, 129-34.

Anonymous (1809) Petrified tortoise. *London, Edinburgh and Dublin Philosophical Magazine*, **33**, 501-2.

Anonymous (1834) Discovery of saurian bones in the magnesian conglomerate near Bristol. *London, Edinburgh and Dublin Philosophical Magazine, (3)*, **5**, 463.

Anonymous (1835) Saurian remains in the magnesian conglomerate of Bristol. *West of England Journal of Science and Literature*, **1**, 84-5.

Anonymous (1850a) Fossil foot-prints of Moray. *Elgin Courant*, 18th October, 1850, p. 2.

Anonymous (1850b) Geology – ancient foot-prints. *Elgin Courant*, 22nd November, 1850, p. 2.

Anonymous (1851) Geological discovery. *Elgin Courant*, 10th October, 1851, p. 3.

Anonymous (1864) Reptilian remains. *Elgin Courant*, 19th August, 1864, p. 5.

Anonymous (1866) *Telerpeton elginense*. *Elgin Courant*, 27th July, 1866, p. 5.

Anonymous (1891) 'A splendid addition to the Elgin fauna'. *Elgin Courant*, 16th June, 1891, p. 4.

Antia, D.D.J. (1979) Bone-beds: a review of their classification, occurrence, genesis, diagenesis, geochemistry, palaeoecology, weathering and microbiotas. *Mercian Geologist*, **7**, 93-174.

Arkell, W.J. (1931) The Upper Great Oolite, Bradford Beds and Forest Marble of south Oxfordshire and the succession of gastropod faunas in the Great Oolite. *Quarterly Journal of the Geological Society of London*, **87**, 563-629.

Arkell, W.J. (1933) *The Jurassic System in Great Britain*. Clarendon Press, Oxford, 681 pp.

Arkell, W.J. (1935) The Corallian beds of Dorset. *Proceedings of the Dorset Natural History and Archaeological Society*, **57**, 59-93.

Arkell, W.J. (1941) The Upper Oxford Clay at Purton, Wiltshire and the zones of the Lower Oxfordian. *Geological Magazine*, **78**, 161-72.

Arkell, W.J. (1947a) *The Geology of Oxford*. Clarendon Press, Oxford, 267 pp.

Arkell, W.J. (1947b) *Oxford Stone*. Faber and Faber, London, 185 pp.

Arkell, W.J. (1947c) *The Geology of the Country around Weymouth, Swanage, Corfe and Lulworth* (Sheets 341-3), Memoirs of the Geological Survey of the United Kingdom, HMSO, London, 386 pp.

Arkell, W.J. (1951-8) A monograph of English Bathonian ammonites. *Monographs of the Palaeontographical Society*, 264 pp.

Arkell, W.J. and Donovan, D.T. (1952) The Fuller's Earth of the Cotswolds and its relation to the Great Oolite. *Quarterly Journal of the Geological Society of London*, **107**, 227-53.

Aston, M.A. (1974) *Stonesfield Slate*. Department of Museum Services, Oxfordshire County Council, Publication No. 5, 85 pp.

Audley-Charles, M.G. (1970) Stratigraphical correlation of the Triassic rocks of the British Isles. *Quarterly Journal of the Geological Society of London*, **126**, 19-47.

Austen, J.H. (1852) *A Guide to the Geology of the Isle of Purbeck and the South Coast of Hampshire*. Blandford, London, 20 pp.

Bakewell, R. (1835) On the Maidstone fossil skeleton, in the museum of Gideon Mantell. *Magazine of Natural History*, **8**, 99-102.

Ballerstedt, M. (1914) Bemerkungen zu den älteren Berichten über Saurierfährten im Wealdensandstein und Behandlung einer neuen, aus 5 Fussabdrücken bestehenden Spur. *Centralblatt für Mineralogie, Geologie und Paläontologie, 1914*, 48-64.

Barker, D. (1966) Ostracods from the Portland and Purbeck beds of the Aylesbury district. *Bulletin of the British Museum (Natural History), Geology Series*, **11**, 458-87.

Barker, M.J. (1976) A stratigraphical, palaeoecological and biometrical study of some English Bathonian *Gastropoda* (especially *Nerineacea*). Unpublished PhD Thesis, University of Keele.

Barrow, G. (1908) In *The Geology of the Small Isles of Inverness-shire (Rum, Canna, Eigg, Muck, etc.)* (Sheet 60) (ed. A. Harker), Memoirs of the Geological Survey of the United Kingdom, HMSO, London, 210 pp.

Bate, R.H. (1978) The Jurassic. Part II Aalenian to Bathonian. In *A Stratigraphical Index of*

References

British Ostracoda (eds R.H. Bate and J.E. Robinson), *Geological Journal Special Issue*, **8**, Seel House Press, Liverpool, pp. 213-58

Baur, G. (1889) On '*Aulacochelys*' Lydekker and the systematic position of *Anosteira* Leidy and *Pseudotrionyx* Dollo. *Annals and Magazine of Natural History*, Series 6, **3**, 273-6.

Bayfield, T.G. (1864) Discovery of the skeleton of *Leiodon anceps* in the Chalk at Norwich. *Geological Magazine*, **1**, 296.

Beasley, H.C. (1896) An attempt to classify the footprints in the New Red Sandstone of this district. *Proceedings of the Liverpool Geological Society*, **7**, 391-409.

Beasley, H.C. (1898) Notes on examples of footprints and c., from the Trias in some provincial museums. *Proceedings of the Liverpool Geological Society*, **8**, 233-7.

Beasley, H.C. (1902) The fauna indicated in the Lower Keuper Sandstone of the neighbourhood of Liverpool. *Transactions of the Liverpool Biological Society*, **16**, 3-26.

Beasley, H.C. (1904) Report on footprints from the Trias - Part I. *Report of the British Association for the Advancement of Science*, 1904 (1903), 219-31.

Beasley, H.C. (1905) Report on footprints from the Trias - Part II. *Report of the British Association for the Advancement of Science*, 1905 (1904), 275-82.

Beasley, H.C. (1906) Notes on footprints from the Trias in the museum of the Warwickshire Natural History and Archaeological Society at Warwick. *Report of the British Association for the Advancement of Science*, 1906 (1905), 162-6.

Beckles, S.H. (1854) On the Ornithoidichnites of the Wealden. *Quarterly Journal of the Geological Society of London*, **10**, 456-64.

Beckles, S.H. (1856) On the lowest strata of the cliffs at Hastings. *Quarterly Journal of the Geological Society of London*, **12**, 288-92.

Beckles, S.H. (1859) On fossil foot-prints in the sandstone at Cummingstone. *Quarterly Journal of the Geological Society of London*, **15**, 461.

Beckles, S.H. (1862) On some natural casts of reptilian footprints in the Wealden Beds of the Isle of Wight and of Swanage. *Quarterly Journal of the Geological Society of London*, **18**, 443-7.

Bell, J., Holden, J., Pettigrew, T.H. and Sedman, K.W. (1979) The Marl Slate and Basal Permian Breccia at Middridge, County Durham. *Proceedings of the Yorkshire Geological Society*, **42**, 439-60.

Bensted, W.H. (1860) On the Kentish Ragstone as exhibited in the *Iguanodon* Quarry at Maidstone. *Proceedings of the Geologists' Association*, **1**, 57-60.

Benton, M.J. (1983a) Dinosaur success in the Triassic: a noncompetitive ecological model. *Quarterly Review of Biology*, **58**, 29-55.

Benton, M.J. (1983b) The age of the rhynchosaur. *New Scientist*, **98**, 9-13.

Benton, M.J. (1983c) Progressionism in the 1850s: Lyell, Owen, Mantell and the Elgin fossil reptile *Leptopleuron (Telerpeton)*. *Archives of Natural History*, **11**, 123-36.

Benton, M.J. (1983d) The Triassic reptile *Hyperodapedon* from Elgin: functional morphology and relationships. *Philosophical Transactions of the Royal Society of London, Series B*, **302**, 605-720.

Benton, M.J. (1984) Tooth form, growth and function in Triassic rhynchosaurs (Reptilia, Diapsida). *Palaeontology*, **27**, 737-76.

Benton, M.J. (1985) Classification and phylogeny of the diapsid reptiles. *Zoological Journal of the Linnean Society*, **84**, 97-164.

Benton, M.J. (1986a) More than one event in the late Triassic mass extinction. *Nature, London*, **321**, 857-61.

Benton, M.J. (1986b) The Late Triassic tetrapod extinction events. In *The Beginning of the Age of Dinosaurs; Faunal Change across the Triassic-Jurassic Boundary* (ed. K. Padian), Cambridge University Press, pp. 303-20.

Benton, M.J. (1988) British fossil reptile sites. In *The Use and Conservation of Palaeontological Sites* (eds P. Crowther and W.A. Wimbledon), *Special Papers in Palaeontology*, **40**, 73-84.

Benton, M.J. (1989) *On the Trail of the Dinosaurs*. Kingfisher, London; Crescent Books, New York, 144 pp.

Benton, M.J. (1990a) *Vertebrate Palaeontology; Biology and Evolution*. Chapman & Hall, London, 377 pp.

Benton, M.J. (1990b) Phylogeny of the major tetrapod groups: morphological data and divergence dates. *Journal of Molecular Evolution*, **30**, 409-24.

Benton, M.J. (1990c) The species of *Rhynchosaurus*, a rhynchosaur (Reptilia, Diapsida) from the Middle Triassic of England. *Philosophical Transactions of the Royal Society of London, Series B*, **328**, 213-306.

References

Benton, M.J. (1990d) *The Reign of the Reptiles.* Kingfisher, London; Crescent Books, New York, 144 pp.

Benton, M.J. (1991) What really happened in the Late Triassic? *Historical Biology,* **5**, 263-78.

Benton, M.J. (1993) Reptilia. In *The Fossil Record 2* (ed. M.J. Benton), Chapman & Hall, London, 681-715.

Benton, M.J. (1994a) Late Triassic terrestrial vertebrate extinctions: stratigraphic aspects and the record of the Germanic Basin. *Paleontologia Lombarda, Nuova Serie,* **2**, 19-38.

Benton, M.J. (1994b) Late Triassic to Jurassic extinctions among tetrapods: testing the pattern. In *In the Shadow of the Dinosaurs, Early Mesozoic Tetrapods* (eds N.C. Fraser and H.-D. Sues), Cambridge University Press, New York, 366-97.

Benton, M.J. and Clark, J. (1988) Archosaur phylogeny and the relationships of the Crocodylia. In *The Phylogeny and Classification of the Tetrapods* (ed. M.J. Benton), *Systematics Association Special Volume,* **35A**, Clarendon Press, Oxford, 295-338.

Benton, M.J. and Taylor, M.A. (1984) Marine reptiles from the Upper Lias (Lower Toarcian, Lower Jurassic) of the Yorkshire coast. *Proceedings of the Yorkshire Geological Society,* **44**, 399-429.

Benton, M.J. and Walker, A.D. (1981) The use of flexible synthetic rubbers for casts of complex fossils from natural moulds. *Geological Magazine,* **118**, 551-6.

Benton, M.J. and Walker, A.D. (1985) Palaeoecology, taphonomy and dating of Permo-Triassic reptiles from Elgin, north-east Scotland. *Palaeontology,* **28**, 207-34.

Benton, M.J. and Wimbledon, W.A. (1985) The conservation and use of fossil vertebrate sites: British fossil reptile sites. *Proceedings of the Geologists' Association,* **96**, 1-6.

Benton, M.J., Hart, M.B. and Clarey, T. (1993) A new rhynchosaur from the Middle Triassic of Devon. *Proceedings of the Ussher Society,* **8**, 167-71.

Benton, M.J., Warrington, G., Newell, A.J. and Spencer, P.S. (1994) A review of the British Mid Triassic tetrapod faunas. In *In the Shadow of the Dinosaurs, Early Mesozoic Tetrapods* (eds N.C. Fraser and H.-D. Sues), Cambridge University Press, New York, 131-60.

Binfield, W.R. and Binfield, H. (1854) On the occurrence of fossil insects in the Wealden strata of the Sussex Coast. *Quarterly Journal of the Geological Society of London,* **10**, 171-6.

Birkelund, T., Thusu, B. and Vigran, J. (1978) Jurassic-Cretaceous biostratigraphy of Norway, with comments on the British *Rasenia cymodoce* Zone. *Palaeontology,* **21**, 31-63.

Blake, C.C. (1863) On chelonian scutes from the Stonesfield Slate. *The Geologist,* **6**, 183-4.

Blake, J.F. (1875) On the Kimmeridge Clay of England. *Quarterly Journal of the Geological Society of London,* **31**, 196-233.

Blake, J.F. and Hudleston, W.H. (1877) The Corallian rocks of England. *Quarterly Journal of the Geological Society of London,* **33**, 206-405.

Blows, W.T. (1978) *Reptiles on the Rocks.* Isle of Wight County Council, Sandown, 60 pp.

Blows, W.T. (1982) A preliminary account of a new specimen *Polacanthus foxi* (Ankylosauria, Reptilia) from the Wealden of the Isle of Wight. *Proceedings of the Isle of Wight Natural History and Archaeological Society,* **7**, 303-6.

Blows, W.T. (1987) The armoured dinosaur *Polacanthus foxi* from the Lower Cretaceous of the Isle of Wight. *Palaeontology,* **30**, 557-80.

Bonaparte, J.F. (1969) Comments on early saurischians. *Zoological Journal of the Linnean Society,* **48**, 471-80.

Bonaparte, J.F. (1978) El Mesozoico de America del Sur y sus tetrapodos. *Opera Lilloana,* **26**, 596 pp.

Bone, D.A. (1986) The stratigraphy of the Reading Beds (Palaeocene), at Felpham, West Sussex. *Tertiary Research,* **8**, 17-32.

Bone, D.A. (1989) Temporary exposures in the Lower Palaeogene of the eastern Hampshire Basin (Chichester to Havant). *Proceedings of the Geologists' Association,* **100**, 147-59.

Boneham, B.F.W. and Forsey, G.F. (1991) Earliest stegosaur dinosaur. *Terra Abstracts,* **3**, 334.

Boneham, B.F.W. and Wyatt, R.J. (1993) The stratigraphical position of the Middle Jurassic (Bathonian) Stonesfield Slate of Stonesfield, Oxfordshire, UK. *Proceedings of the Geologists' Association,* **104**, 123-36.

Bosma, A.A. (1974) Rodent biostratigraphy of the Eocene, Oligocene transitional strata of the Isle of Wight. *Utrecht Micropalaeontological Bulletin, Special Publications,* **1**, 113 pp.

Boulenger, G.A. (1891) On British remains of

Homoeosaurus, with remarks on the classification of the Rhynchocephalia. *Proceedings of the Zoological Society of London*, 1891, 167-72.

Boulenger, G.A. (1903) On reptilian remains from the Trias of Elgin. *Philosophical Transactions of the Royal Society of London, Series B*, **196**, 175-89.

Boulenger, G.A. (1904a) On the characters and affinities of the Triassic reptile *Telerpeton elginense*. *Proceedings of the Zoological Society of London*, 1904, 470-80.

Boulenger, G.A. (1904b) A remarkable ichthyosaurian right anterior paddle. *Proceedings of the Zoological Society of London*, 1904, 424-6.

Bowerbank, J.S. (1846) On a new species of pterodactyl found in the Upper Chalk of Kent (*P. giganteus*). *Quarterly Journal of the Geological Society of London*, **2**, 7-8.

Bowerbank, J.S. (1848) Microscopical examination on the structure of the bones of *Pterodactylus giganteus* and other fossil animals. *Quarterly Journal of the Geological Society of London*, **4**, 2-10.

Bowerbank, J.S. (1852) On the pterodactyles of the Chalk Formation. *Proceedings of the Zoological Society of London*, **19**, 14-20.

Boyd, M.J. (1984) The Upper Carboniferous tetrapod assemblage from Newsham, Northumberland. *Palaeontology*, **37**, 367-92.

Boyd, M.J. (1985) A protorothyridid captorhinomorph (Reptilia) from the Upper Carboniferous of Newsham, Northumberland. *Palaeontology*, **28**, 393-9.

Brand, L. (1979) Field and laboratory studies on the Coconino Sandstone (Permian) vertebrate footprints and their paleoecological implications. *Palaeogeography, Palaeoclimatology, Palaeoecology*, **28**, 25-38.

Brickenden, L.B. (1850) Fossil foot-prints of Moray. *Elgin Courant*, 18 October, 1850, p. 2.

Brickenden, L.B. (1852) Notice of the discovery of reptilian foot-tracks and remains in the Old Red or Devonian strata of Moray. *Quarterly Journal of the Geological Society of London*, **8**, 97-100.

Bridgland, D.R. (1987) Report of Geologists' Association field meeting in north-east Essex, May 22nd-24th, 1987. *Proceedings of the Geologists' Association*, **99**, 315-33.

Bristow, C.R., Cox, B.M., Woods, M.A., Prudden, H.C., Sole, D., Edmunds, M. and Callomon, J.H. (1992) The geology of the A303 trunk road between Wincanton, Somerset and Mere, Wiltshire. *Proceedings of the Dorset Natural History and Archaeological Society*, **113**, 139-43.

Bristow, H.W. and Fisher, O. (1857) Comparative vertical sections of the Purbeck strata of Dorsetshire. *Geological Survey of the United Kingdom, Vertical Sections*, **22**.

Broadhurst, F.M. and Duffy, L. (1970) A plesiosaur in the Geology Department, University of Manchester. *Museums Journal*, **70**, 30-1.

Brookfield, M.E. (1978a) Revision of the stratigraphy of Permian and supposed Permian rocks of southern Scotland. *Geologische Rundschau*, **67**, 110-49.

Brookfield, M.E. (1978b) The lithostratigraphy of the Upper Oxfordian and Lower Kimmeridgian beds of south Dorset. *Proceedings of the Geologists' Association*, **89**, 1-32

Broom, R. (1913) On the South African pseudosuchian *Euparkeria* and allied genera. *Proceedings of the Zoological Society of London*, 1913, 619-33.

Brown, D.S. (1981) The English Upper Jurassic Plesiosauroidea (Reptilia) and a review of the phylogeny and classification of the Plesiosauria. *Bulletin of the British Museum (Natural History), Geology Series*, **35**, 253-347.

Brown, D.S. (1984) Discovery of a specimen of the plesiosaur *Colymbosaurus trochanterius* (Owen) on the island of Portland. *Proceedings of the Dorset Natural History and Archaeological Society*, **105**, 170.

Brown, D.S. (1990) Lincolnshire plesiosaur hunter discovers rich vertebrate fauna. *Geology Today*, **6**, 182.

Brown, D.S. and Keen, J.A. (1991) An extensive marine vertebrate fauna from the Kellaways Sand (Callovian, Middle Jurassic) of Lincolnshire. *Mercian Geologist*, **12**, 87-96.

Brown, D.S., Milner, A.C. and Taylor, M.A. (1986) New material of the plesiosaur *Kimmerosaurus langhami* Brown from the Kimmeridge Clay of Dorset. *Bulletin of the British Museum (Natural History), Geology Series*, **40**, 225-34.

Browne, H.B. (1946) *Chapters of Whitby History 1823-1946. The Story of Whitby Literary and Philosophical Society and of Whitby Museum*. Whitby Literary and Philosophical Society, Whitby.

Browne, M. (1889) *The Vertebrate Animals of Leicestershire and Rutland*. Midland Education

Co., Birmingham and Leicester, 223 pp.

Browne, M. (1894) On some vertebrate remains not hitherto recorded from the Rhaetic beds of Britain. *Report of the British Association for the Advancement of Science*, 1893 (1894), 748-9.

Buckland, W. (1824) Notice on the *Megalosaurus*, or great fossil lizard of Stonesfield. *Transactions of the Geological Society of London, Series 2*, **1**, 390-6.

Buckland, W. (1828) Note sur les traces de tortues observées dans le grès rouge. *Annales de Sciences Naturelles, Paris*, **13**, 85-6.

Buckland, W. (1829a) On the discovery of coprolites, or fossil faeces, in the Lias of Lyme Regis and in other formations. *Transactions of the Geological Society of London, Series 2*, **3**, 223-36.

Buckland, W. (1829b) On the discovery of a new species of pterodactyl in the Lias of Lyme Regis. *Transactions of the Geological Society of London, Series 2*, **3**, 217-22.

Buckland, W. (1829c) On the discovery of the bones of the *Iguanodon* and other large reptiles, in the Isle of Wight and Isle of Purbeck. *Proceedings of the Geological Society of London*, **1**, 159-60.

Buckland, W. (1835) On the discovery of the fossil bones of the *Iguanodon*, in the Iron Sand of the Wealden Formation in the Isle of Wight and in the Isle of Purbeck. *Transactions of the Geological Society of London, Series 2*, **3**, 425-32.

Buckland, W. (1837) On the occurrence of Keuper-Sandstone in the upper region of the New Red Sandstone formation or Poikilitic system in England and Wales. *Proceedings of the Geological Society of London*, **2**, 453-4.

Buckland, W. (1844) [President's address, 1839]. *Proceedings of the Ashmolean Society*, **16**, 5-7.

Buckland, W. and Conybeare, W.D. (1824) Observations on the south-western coal district of England. *Transactions of the Geological Society of London, Series 2*, **1**, 210-316.

Buckman, S.S. (1910) Certain Jurassic (Lias-Oolite) strata of south Dorset; and their correlation. *Quarterly Journal of the Geological Society of London*, **66**, 52-89.

Buckman, S.S. (1915) A palaeontological classification of the Jurassic rocks of the Whitby district; with a zonal table of Lias ammonites. In *The Geology of the Country between*

Whitby and Scarborough (Sheets 35 and 44), (Fox-Strangways, C. and Barrow, G.), Memoirs of the Geological Survey of the United Kingdom, HMSO, London, 59-102.

Buckman, S.S. (1925) *Type Ammonites*. Published by the author, London and Thame, 7 vols (1909-30).

Buffetaut, E. (1975) Sur l'anatomie et la position systématique de *Bernissartia fagesii* Dollo, L., 1883, crocodilien du Wealdien de Bernissart, Belgique. *Bulletin de l'Institut Royal des Sciences Naturelles de Belgique, Sciences de la Terre*, **51**(2), 1-20.

Buffetaut, E. (1982) Radiation évolutive, paléoecologie et biogéographie des crocodiliens mésosuchiens. *Mémoires de la Société Géologique de France*, **142**, 88 pp.

Buffetaut, E. (1983) The crocodilian *Theriosuchus* Owen, 1879 in the Wealden of England. *Bulletin of the British Museum (Natural History), Geology Series*, **37**, 93-7.

Buffetaut, E. (1989) New remains of the enigmatic dinosaur *Spinosaurus* from the Cretaceous of Morocco and the affinities between *Spinosaurus* and *Baryonyx*. *Neues Jahrbuch für Geologie und Paläontologie, Monatshefte, 1989*, 79-87.

Buffetaut, E. (1992) Remarks on the Cretaceous theropod dinosaurs *Spinosaurus* and *Baryonyx*. *Neues Jahrbuch für Geologie und Paläontologie, Monatshefte, 1992*, 88-96.

Buffetaut, E. and Ford, R.L.E. (1979) The crocodilian *Bernissartia* in the Wealden of the Isle of Wight. *Palaeontology*, **22**, 905-12.

Buffetaut, E. and Hutt, S. (1980) *Vectisuchus leptognathus* n. gen. n. sp., a slender-snouted goniopholid crocodilian from the Wealden of the Isle of Wight. *Neues Jahrbuch für Geologie und Paläontologie, Monatshefte, 1980*, 385-90.

Buffetaut, E., Cappetta, H., Gayet, M., Martin, M., Moody, R.T.J., Rage, J.-C., Taquet, P. and Wellnhofer, P. (1981) Les vertébrés de la partie moyenne du Crétacé en Europe. *Cretaceous Research*, **2**, 275-81.

Buffetaut, E., Cuny, G. and Le Loeuff, J. (1991) French dinosaurs: the best record in Europe? *Modern Geology*, **16**, 17-42.

Burckhardt, R. (1900) On *Hyperodapedon gordoni*. *Geological Magazine*, Decade 4, **7**, 486-92, 529-35.

Burton, E.St.J. (1929) The horizons of Bryozoa (Polyzoa) in the Upper Eocene beds of Hampshire. *Quarterly Journal of the*

Geological Society of London, **85**, 223-39.

Burton, E.St.J. (1933) Faunal horizons in the Barton Beds in Hampshire. *Proceedings of the Geologists' Association,* **44**, 131-66.

Callomon, J.H. and Cope, J.C.W. (1971) The stratigraphy and ammonite succession of the Oxford and Kimmeridge Clays in the Warlingham Borehole. *Bulletin of the Geological Survey of Great Britain*, **36**, 147-76.

Carpenter, K. and Currie, P.J. (1990) *Dinosaur Systematics; Approaches and Perspectives.* Cambridge University Press, Cambridge, 318 pp.

Carroll, R.L. (1964) The earliest reptiles. *Journal of the Linnean Society, Zoology*, **45**, 61-83.

Carroll, R.L. (1978) Permo-Triassic 'lizards' from the Karoo System. Part II. A gliding reptile from the Upper Permian of Madagascar. *Palaeontologia Africana*, **21**, 143-59.

Carroll, R.L. (1988) *Vertebrate Paleontology and Evolution.* Freeman, San Francisco, 698 pp.

Carte, A. and Baily, W.H. (1863) Description of a new species of *Plesiosaurus* from the Lias near Whitby, Yorkshire. *Journal of the Royal Dublin Society*, **4**, 160-70.

Carter, H.J. (1888) On some vertebrate remains in the Triassic strata of the south coast of Devonshire between Budleigh Salterton and Sidmouth. *Quarterly Journal of the Geological Society of London*, **44**, 318-19.

Casey, R. (1959) Field meeting at Wrotham and the Maidstone by-pass. *Proceedings of the Geologists' Association*, **70**, 206-9.

Casey, R. (1961) The stratigraphical palaeontology of the Lower Greensand. *Palaeontology*, **3**, 487-621.

Casier, E. (1978) *Les Iguanodons de Bernissart*, 2nd edn., Institut Royal des Sciences Naturelles Belgiques, Bruxelles, 166 pp.

Chapman, W. (1758) An account of the fossil bones of an alligator, found on the sea-shore, near Whitby in Yorkshire. *Philosophical Transactions of the Royal Society*, **50**, 688-91.

Charig, A.J. (1979) *A New Look at the Dinosaurs.* Heinemann, London, 160 pp.

Charig, A.J. (1980) A diplodocid sauropod from the Lower Cretaceous of England. In *Aspects of Vertebrate History* (ed. L.L. Jacobs), Museum of North Arizona Press, Flagstaff, pp. 231-44.

Charig, A.J. and Milner, A.C. (1986) *Baryonyx*, a remarkable new theropod dinosaur. *Nature, London*, **324**, 359-61.

Charig, A.J. and Milner, A.C. (1990) The systematic position of *Baryonyx walkeri*, in the light of Guthier's reclassification of the Theropoda. In *Dinosaur Systematics; Approaches and Perspectives* (eds K. Carpenter and P.J. Currie), Cambridge University Press, Cambridge, pp. 127-40.

Charig, A.J. and Newman, B.H. (1962) Footprints in the Purbeck. *New Scientist*, **14**, 234-5.

Charlesworth, E. (1837) [Fossil crocodile at Whitby.] *Magazine of Natural History, New Series*, **1**, 531-53.

Charlesworth, E. (1845) On the discovery, by Searles Wood, of an alligator in the freshwater cliff at Hordwell, associated with extinct Mammalia. *Report of the British Association for the Advancement of Science*, 1844 (1845), p. 50.

Charlesworth, E. (1855) [Notice on new vertebrate fossils.] *Report of the British Association for the Advancement of Science*, 1854 (1855), p. 80.

Chatterjee, S. (1974) A rhynchosaur from the Upper Triassic Maleri Formation of India. *Philosphical Transactions of the Royal Society of London, Series B*, **267**, 209-61.

Chatterjee, S. (1980) The evolution of rhynchosaurs. *Mémoires de la Société Géologique de France, Nouveau Série*, **139**, 57-65.

Clark, J.M. (in press) Cranial anatomy of *Protosuchus richardsoni* (Brown) and two new protosuchids and the relationships of the 'Protosuchia' (Archosauria: Crocodylomorpha). *Bulletin of the American Museum of Natural History*.

Clark, J.M. and Norell, M.A. (1992) The Early Cretaceous crocodylomorph *Hylaeochampsa vectiana* from the Wealden of the Isle of Wight. *American Museum Novitates*, **3032**, 1-19.

Clarke, J. and Etches, S. (1992) Predation amongst Jurassic marine reptiles. *Proceedings of the Dorset Natural History and Archaeological Society*, **113**, 202-5.

Clayden, A.W. (1908a) Note on the discovery of footprints in the 'Lower Sandstones' of the Exeter district. *Report and Transactions of the Devonshire Association for the Advancement of Science*, **40**, 172-3.

Clayden, A.W. (1908b) On the occurrence of footprints in the Lower Sandstones of the Exeter district. *Quarterly Journal of the Geological Society of London*, **64**, 496-500.

Clemens, W.A. and Lees, P.M. (1971) A review of

References

English Early Cretaceous mammals. In *Early Mammals* (eds D.M. Kermack and K.A. Kermack), *Zoological Journal of the Linnean Society*, **50**, Supplement, 1 Academic Press, London, pp. 117-30.

Clements, R. G. (1993) Type-section of the Purbeck Limestone Group, Durlston Bay, Swanage, Dorset. *Proceedings of the Dorset Natural History and Archaeological Society*, **114**, 181-206.

Clemmensen, L. (1987) Complex star dunes and associated aeolian bedforms, Hopeman Sandstone (Permo-Triassic), Moray Firth Basin, Scotland. In *Desert Sediments: Ancient and Modern* (eds L. Frostick and I. Reid), *Geological Society Special Publication*, **35**, Geological Society, London, pp. 213-31.

Cluver, M.A. and King, G.M. (1983) A reassessment of the relationships of Permian Dicynodontia (Reptilia, Therapsida) and a new classification of dicynodonts. *Annals of the South African Museum*, **91**, 195-273.

Colbert, E.H. (1970) The Triassic gliding reptile *Icarosaurus*. *Bulletin of the American Museum of Natural History*, **143**, 85-142.

Colbert, E.H. and Morales, M. (1991) *Evolution of the Vertebrates; a History of the Backboned Animals through Time*, 4th edn., Wiley-Liss, New York, 470 pp.

Collins, J.I. (1970) The chelonian *Rhinochelys* Seeley from the Upper Cretaceous of England and France. *Palaeontology,* **13,** 355-78.

Conybeare, W.D. (1821) Notice of the discovery of a new fossil animal, forming a link between the *Ichthyosaurus* and the crocodile, together with general remarks on the osteology of the *Ichthyosaurus*. *Transactions of the Geological Society of London*, **5**, 559-94.

Conybeare, W.D. (1822) Additional notices on the fossil genera *Ichthyosaurus* and *Plesiosaurus*. *Transactions of the Geological Society of London*, **(2), 1**, 103-23.

Conybeare, W.D. (1824) Additional notes on the fossil genera *Ichthyosaurus* and *Plesiosaurus*. *Transactions of the Geological Society of London*, **(2), 1**, 381-9.

Cookson, I.C. and Hughes, N.F. (1964) Microplankton from the Cambridge Greensand (mid Cretaceous). *Palaeontology*, **7**, 37-59.

Coombs, W.P. (1978) The families of the ornithischian dinosaur order Ankylosauria. *Palaeontology*, **21**, 143-70.

Coombs, W.P. and Maryanska, M. (1990) Ankylosauria. In *The Dinosauria* (eds D.B. Weishampel, P. Dodson and H. Osmólska), University of California Press, Berkeley, pp. 456-83.

Coombs, W.P. Jr, Weishampel, D.B. and Witmer, L.M. (1990) Basal Thyreophora. In *The Dinosauria* (eds D.B. Weishampel, P. Dodson and H. Osmólska), University of California Press, Berkeley, pp. 427-34.

Cope, J.C.W. (1967) The palaeontology and stratigraphy of the lower part of the Upper Kimmeridge Clay of Dorset. *Bulletin of the British Museum (Natural History), Geology Series*, **15**, 79 pp.

Cope, J.C.W. (1974) Ichthyosaur remains from the Oxford Clay. *Proceedings of the Dorset Natural History and Archaeological Society*, **95**, p. 106.

Cope, J.C.W. (1978) The ammonite faunas and stratigraphy of the upper part of the Upper Kimmeridge Clay of Dorset. *Palaeontology*, **21**, 469-533.

Cope, J.C.W. (1993) The Bolonian Stage: an old answer to an old problem. *Newsletters in Stratigraphy*, **28**, 156-63.

Cope, J.C.W. and Wimbledon, W.A. (1973) Ammonite faunas of the uppermost Kimmeridge Clay, the Portland Sand and the Portland Stone of Dorset. *Proceedings of the Ussher Society*, **2**, 593-8.

Cope, J.C.W., Getty, T.A., Howarth, M.K., Morton, N. and Torrens, H.S. (1980a) A correlation of Jurassic rocks in the British Isles. Part one: Introduction and Lower Jurassic. *Geological Society of London, Special Report*, **14**, 73 pp.

Cope, J.C.W., Duff, K.L., Parsons, C.F., Torrens, H.S., Wimbledon, W.A. and Wright, J.K. (1980b) A correlation of Jurassic rocks in the British Isles. Part two: Middle and Upper Jurassic. *Geological Society of London, Special Report*, **15**, 109 pp.

Cox, B.M. and Gallois, R.W. (1981) The stratigraphy of the Kimmeridge Clay of the Dorset type area and its correlation with some other Kimmeridgian sequences. *Report of the Institute of Geological Sciences*, **80**, 43 pp.

Cox, L.R. (1925) The fauna of the Basal Shell Bed of the Portland Stone, Isle of Portland. *Proceedings of the Dorset Natural History and Antiquarian Field Club*, **46**, 113-72.

Cray, P.E. (1973) Marsupialia, Insectivora, Primates, Creodonta and Carnivora from the Headon Beds (Upper Eocene) of southern England. *Bulletin of the British Museum (Natural History), Geology Series*, **23**, 102 pp.

References

Cruickshank, A.R.I. and Keyser, A.W. (1984) Remarks on the genus *Geikia* Newton, 1893 and its relationships with other dicynodonts (Reptilia: Therapsida). *Transactions of the South African Geological Society*, **87**, 35-9.

Cruickshank, A.R.I. and Taylor, M.A. (1993) A plesiosaur from the Linksfield erratic (Rhaetian, Upper Triassic) near Elgin, Morayshire. *Scottish Journal of Geology*, **29**, 191-6.

Crush, P.J. (1980) An early terrestrial crocodile from South Wales. Unpublished PhD Thesis, University College, London, 357 pp.

Crush, P.J. (1984) A late Upper Triassic sphenosuchid crocodilian from Wales. *Palaeontology*, **27**, 131-57.

Cuny, G., Buffetaut, E., Cappetta, H., Martin, M., Mazin, J.-M. and Rose, J.-M. (1991) Nouveaux restes de vertébrés du Jurassique terminal du Boulonnais (nord de la France). *Neues Jahrbuch für Geologie und Paläontologie, Abhandlungen*, **180**, 323-47.

Curry, D. (1958) The Barton area. In *Geology of the Southampton Area* (eds D. Curry and D.E. Wisden), Geologists' Association, London, pp. 8-12.

Curry, D., Adams, C.G., Boulter, M.C., Dilley, F.C., Eames, F.E., Funnell, B.M. and Wells, M.K. (1978) A correlation of Tertiary rocks in the British Isles. *Geological Society of London, Special Report*, **12**, 72 pp.

Cuvier, G. (1824) *Recherches sur les ossemens fossiles, où l'on rétablit les caractères de plusieurs animaux dont les révolutions du globe ont détruit les espèces*, 2nd edn., Dulour et d'Ocagne, Paris, 5 vols.

Daley, B. (1972) Macroinvertebrate assemblages from the Bembridge Marls (Oligocene) of the Isle of Wight, England and their environmental significance. *Palaeogeography, Palaeoclimatology, Palaeoecology*, **11**, 11-32.

Daley, B. (1973) The palaeoenvironment of the Bembridge Marls (Oligocene) of the Isle of Wight, Hampshire. *Proceedings of the Geologists' Association*, **84**, 83-93.

Daley, B. and Edwards, N. (1974) Weekend field meeting: the Upper Eocene-Lower Oligocene beds of the Isle of Wight. *Proceedings of the Geologists' Association*, **85**, 281-92.

Daley, B. and Insole, A. (1984) The Isle of Wight. *Geologists' Association Guide*, **25**, 34 pp.

Daley, B. and Stewart, D.J. (1979) Weekend field meeting: the Wealden Group in the Isle of Wight. 17-19 June 1977. *Proceedings of the Geologists' Association*, **90**, 51-4.

Damon, R. (1884) *Handbook of the Geology of Weymouth, Portland and Coast of Dorsetshire, from Swanage to Bridport-on-Sea*, 2nd edn., R.F. Damon, Weymouth; E. Stanford, London, 250 pp.

Davis, A.G. (1936) The London Clay of Sheppey and the location of its fossils. *Proceedings of the Geologists' Association*, **47**, 328-45.

Davis, A.G. (1937) Additional notes on the geology of Sheppey. *Proceedings of the Geologists' Association*, **48**, 77-81.

Dean, W.T. (1954) Notes on part of the Upper Lias succession at Blea Wyke, Yorkshire. *Proceedings of the Yorkshire Geological Society*, **29**, 161-79.

Deeming, D.C., Halstead, L.B., Manabe, M. and Unwin, D.M. (1993) An ichthyosaur embryo from the Lower Lias (Jurassic: Hettangian) of Somerset, England, with comments on the reproductive biology of ichthyosaurs. *Modern Geology*, **18**, 423-42.

De la Beche, H.T. (1820) Some additional remarks concerning several species of *Proteosaurus* which have been discovered. *Annals of Philosophy, London*, **15**, 57.

De la Beche, H.T. and Conybeare, W.D. (1821) Notice of the discovery of a new fossil animal, forming a link between the *Ichthyosaurus* and crocodile, together with general remarks on the osteology of the *Ichthyosaurus*. *Transactions of the Geological Society of London*, **5**, 559-94.

Delair, J.B. (1958) The Mesozoic reptiles of Dorset: Part 1. *Proceedings of the Dorset Natural History and Archaeological Society*, **79**, 47-72.

Delair, J.B. (1959) The Mesozoic reptiles of Dorset: Part 2. *Proceedings of the Dorset Natural History and Archaeological Society*, **80**, 52-90.

Delair, J.B. (1960) The Mesozoic reptiles of Dorset: Part 3. *Proceedings of the Dorset Natural History and Archaeological Society*, **81**, 59-85.

Delair, J.B. (1963) Notes on Purbeck footprints, with descriptions of two hitherto unknown forms from Dorset. *Proceedings of the Dorset Natural History and Archaeological Society*, **84**, 92-100.

Delair, J.B. (1966) New records of dinosaurs and other fossil reptiles from Dorset. *Proceedings of the Dorset Natural History and Archaeological Society*, **87**, 57-66.

Delair, J.B. (1969a) A history of the early discover-

References

ies of Liassic ichthyosaurs in Dorset and Somerset (1779–1835). *Proceedings of the Dorset Natural History and Archaeological Society*, **90**, 115–27.

Delair, J.B. (1969b) The first record of the occurrence of ichthyosaurs in the Purbeck. *Proceedings of the Dorset Natural History and Archaeological Society*, **90**, 128–32.

Delair, J.B. (1973) The dinosaurs of Wiltshire. *Wiltshire Archaeological and Natural History Magazine*, **68**, 1–7.

Delair, J.B. (1982a) New and little-known Jurassic reptiles from Wiltshire. *Wiltshire Archaeological and Natural History Magazine*, **76**, 155–64.

Delair, J.B. (1982b) Multiple dinosaur trackways from the Isle of Purbeck. *Proceedings of the Dorset Natural History and Archaeological Society*, **102**, 65–7.

Delair, J.B. (1982c) Notes on an armoured dinosaur from Barnes High, Isle of Wight. *Proceedings of the Isle of Wight Natural History and Archaeological Society*, **7**, 297–302.

Delair, J.B. (1986) Some little known Jurassic ichthyosaurs from Dorset. *Proceedings of the Dorset Natural History and Archaeological Society*, **107**, 127–34.

Delair, J.B. (1987) An unusual ichthyosaurian forelimb from Rodwell, Dorset. *Proceedings of the Dorset Natural History and Archaeological Society*, **108**, 210–12.

Delair, J.B. (1989) A history of dinosaur footprint discoveries in the British Wealden. In *Dinosaur Tracks and Traces* (eds D.D. Gillette and M.G. Lockley), Cambridge University Press, Cambridge, pp. 19–25.

Delair, J.B. (1992) The occurrence of megalosaurs in the Portlandian of Dorset. *Proceedings of the Dorset Natural History and Archaeological Society*, **113**, 196.

Delair, J.B. and Brown, P.A. (1975) Worbarrow Bay footprints. *Proceedings of the Dorset Natural History and Archaeological Society*, **96**, 14–16.

Delair, J.B. and Lander, A.D. (1973) A short history of the discovery of reptilian footprints in the Purbeck beds of Dorset, with notes on their stratigraphical distribution. *Proceedings of the Dorset Natural History and Archaeological Society*, **94**, 17–20.

Delair, J.B. and Sarjeant, W.A.S. (1975) The earliest discoveries of dinosaurs. *Isis*, **66**, 5–25.

Delair, J.B. and Sarjeant, W.A.S. (1985) History and bibliography of the study of fossil vertebrate footprints in the British Isles: supplement 1973–1983. *Palaeogeography, Palaeoclimatology, Palaeoecology*, **49**, 123–60.

De Rance, C.E. (1868) On the Albian, or Gault, of Folkestone. *Geological Magazine*, **5**, 163–71.

Deslongchamps, E.E-. (1869) Mémoire sur les téléosauriens de la Normandie. *Bulletin de la Société Linnéenne de Normandie, Série 2*, **3**, 124–221.

Dibley, G.E. (1900) Zonal features of the Chalk pits in the Rochester, Gravesend and Croydon areas. *Proceedings of the Geologists' Association*, **16**, 484–96.

Dibley, G.E. (1904) Excursion to Holborough and Burham. *Proceedings of the Geologists' Association*, **18**, 474–5.

Dibley, G.E. (1907) Excursion to Rochester, Wouldham and Blue Bell Hill. *Proceedings of the Geologists' Association*, **20**, 178–81.

Dibley, G.E. (1918) Additional notes on the Chalk of the Medway Valley, Gravesend, west Kent, north-east Surrey and Grays (Essex). *Proceedings of the Geologists' Association*, **29**, 68–93.

Dibley, G.E. and Spath, L.F. (1926) Report on an excursion to Burham and Aylesford, Kent. *Proceedings of the Geologists' Association*, **37**, 432–3.

Dines, H.G., Holmes, S.C.A. and Robbie, J.A. (1954) *Geology of the Country around Chatham* (Sheet 272), Memoirs of the Geological Survey of Great Britain, HMSO, London, 157 pp.

Dixon, F. (1850) *The Geology and Fossils of the Tertiary and Cretaceous Formations of Sussex*. Longman, Brown, Green and Longmans, London, 422 pp.

Dodson, P., Behrensmeyer, A.K., Bakker, R.T. and McIntosh, J.S. (1980) Taphonomy and paleoecology of the dinosaur beds of the Jurassic Morrison Formation. *Paleobiology*, **6**, 208–32.

Dong, Z.M. (1990) Stegosaurs of Asia. In *Dinosaur Systematics: Approaches and Perspectives* (eds K. Carpenter and P.J. Currie), Cambridge University Press, Cambridge, pp. 255–68.

Douglas, J.A. and Arkell, W.J. (1928) The stratigraphical distribution of the Cornbrash. I. The south-western area. *Quarterly Journal of the Geological Society of London*, **84**, 117–78.

Douglas, J.A. and Arkell, W.J. (1932) The stratigraphical distribution of the Cornbrash. II. The north-eastern area. *Quarterly Journal of the Geological Society of London*, **88**, 112-32.

Douglas, J.A. and Arkell, W.J. (1935) On a new section of fossiliferous Upper Cornbrash of north-eastern facies at Enslow Bridge, near Oxford. *Quarterly Journal of the Geological Society of London*, **91**, 318-22.

Drake, H. and Sheppard, T. (1909) Classified list of organic remains from the rocks of the East Riding of Yorkshire. *Proceedings of the Yorkshire Geological Society*, **17**, 4-71.

Duff, K.L. (1975) Palaeoecology of a bituminous shale: the Lower Oxford Clay of Central England. *Palaeontology*, **18**, 443-82.

Duff, K.L., McKirdy, A.P. and Harley, M.J. (eds) (1985) *New Sites for Old. A Students' Guide to the Geology of the East Mendips*. Nature Conservancy Council, Peterborough, 192 pp.

Duff, P. (1842) *Sketch of the Geology of Moray*. Forsyth and Young, Elgin.

Duffin, C. (1979a) The Moore collections of Upper Liassic crocodiles: a history. *Geological Curators' Group Newsletter*, **2**, 235-52.

Duffin, C. (1979b) *Pelagosaurus* (Mesosuchia, Crocodylia) from the English Toarcian. *Neues Jahrbuch für Geologie und Paläontologie, Monatshefte*, 1979, 475-85.

Duffin, C.J. (1980) The Upper Triassic section at Chilcompton, Somerset, with notes on the Rhaetic of the Mendips in general. *Mercian Geologist*, **7**, 251-68.

Dunn, J. (1831) On a large species of *Plesiosaurus* in the Scarborough Museum. *Proceedings of the Geological Society of London*, **1**, 336-7.

Edmonds, E.A. and Dinham, C.H. (1965) *Geology of the Country around Huntingdon and Biggleswade* (Sheets 187 and 204N), Memoirs of the Geological Survey of the United Kingdom, HMSO, London, 90 pp.

Edwards, R.A. and Freshney, E.C. (1987) Lithostratigraphical classification of the Hampshire Basin Palaeogene deposits (Reading Formation to Headon Formation). *Tertiary Research*, **8**, 43-73.

Ensom, P.C. (1977) A therapsid tooth from the Forest Marble (Middle Jurassic) of Dorset. *Proceedings of the Geologists' Association*, **88**, 201-5.

Ensom, P.C. (1982a) Dinosaur footprints at 19 Townsend Road, Swanage. *Proceedings of the Dorset Natural History and Archaeological Society*, **103**, 141.

Ensom, P.C. (1982b) *Ichnites* spp. from Worbarrow Tout, near West Lulworth. *Proceedings of the Dorset Natural History and Archaeological Society*, **103**, 141.

Ensom, P.C. (1983) *Ichnites* spp. from the Chief Beef Bed and Broken Shell Limestone, Durlston Beds, Purbeck Limestone Formation, Durlston Bay, Swanage. *Proceedings of the Dorset Natural History and Archaeological Society*, **104**, 201.

Ensom, P.C. (1984a) *Purbeckopus pentadactylus* Delair. *Proceedings of the Dorset Natural History and Archaeological Society*, **105**, 166.

Ensom, P.C. (1984b) *Ichnites* spp. in Durlston Bay and on Worbarrow Tout. *Proceedings of the Dorset Natural History and Archaeological Society*, **105**, 166-7.

Ensom, P.C. (1985a) An annotated section of the Purbeck Limestone Formation at Worbarrow Tout, Dorset. *Proceedings of the Dorset Natural History and Archaeological Society*, **106**, 87-91.

Ensom, P.C. (1985b) A correction and additions to the distribution of *Ichnites* spp. in the Purbeck Limestone Formation of Worbarrow Tout and Durlston Bay, Dorset. *Proceedings of the Dorset Natural History and Archaeological Society*, **106**, 166-7.

Ensom, P.C. (1986a) *Purbeckopus pentadactylus* Delair; a figured specimen rediscovered. *Proceedings of the Dorset Natural History and Archaeological Society*, **107**, 183.

Ensom, P.C. (1986b) *Ichnites* sp. from the Upper *Cypris* Clays and Shales Member (Purbeck Limestone Formation), near Harman's Cross, Dorset. *Proceedings of the Dorset Natural History and Archaeological Society*, **107**, 183.

Ensom, P.C. (1987a) Scelidosaur remains from the Lower Lias of Dorset. *Proceedings of the Dorset Natural History and Archaeological Society*, **108**, 203-5.

Ensom, P.C. (1987b) A remarkable new vertebrate site in the Purbeck Limestone Formation of the Isle of Purbeck. *Proceedings of the Dorset Natural History and Archaeological Society*, **108**, 205-6.

Ensom, P.C. (1987c) Notes on *Ichnites* spp. in the Purbeck Limestone Formation, Dorset. *Proceedings of the Dorset Natural History and Archaeological Society*, **108**, 206.

Ensom, P.C. (1987d) Dinosaur tracks in Dorset. *Geology Today*, **3**, 182-3.

Ensom, P.C. (1988) Excavation at Sunnydown Farm, Langton Matravers, Dorset: amphibians discovered in the Purbeck Limestone

References

Formation. *Proceedings of the Dorset Natural History and Archaeological Society*, **109**, 148-50.

Ensom, P.C. (1989a) New scelidosaur remains from the Lower Lias of Dorset. *Proceedings of the Dorset Natural History and Archaeological Society*, **110**, 165, 167.

Ensom, P.C. (1989b) *Plesiosaurus* sp. from the Middle Lias of the Dorset coast. *Proceedings of the Dorset Natural History and Archaeological Society*, **110**, 167.

Ensom, P.C. (1989c) Sunnydown Farm sauropod footprint site. *Proceedings of the Dorset Natural History and Archaeological Society*, **110**, 167-8.

Ensom, P.C. (1990) [Correction to 'A remarkable vertebrate site in the Purbeck Limestone Formation of the Isle of Purbeck'.] *Proceedings of the Dorset Natural History and Archaeological Society*, **111**, 133.

Ensom, P.C., Evans, S.E. and Milner, A.R. (1991) Amphibians and reptiles from the Purbeck Limestone Formation (Upper Jurassic) of Dorset. In *Fifth Symposium on Mesozoic Terrestrial Ecosystems and Biota* (eds Z. Kielan-Jaworowska, N. Heintz and H.A. Nakrem), *Contributions from the Paleontological Museum, University of Oslo*, **364**, 19-20.

Estes, R. (1983) Sauria terrestria, Amphisbaenia. *Handbuch der Paläoherpetologie*, **10A**, 249 pp.

Estes, R., de Queiroz, K. and Gauthier, J. (1988) Phylogenetic relationships within Squamata. In *Phylogenetic Relationships of the Lizard Families* (eds R. Estes and G. Pregill), Essays commemorating Charles L. Camp, Stanford University Press, Stanford, pp. 119-281.

Etheridge, R. (1867) On the stratigraphical position of *Acanthopholis horridus* (Huxley). *Geological Magazine*, **4**, 67-9.

Etheridge, R. (1868) Physical structure of west Somerset and north Devon and the palaeontological value of Devonian fossils. *Quarterly Journal of the Geological Society of London*, **23**, 251-2, 568-98.

Etheridge, R. (1870) On the geological position and geographical distribution of the reptilian or Dolomitic Conglomerate of the Bristol area. *Quarterly Journal of the Geological Society of London*, **26**, 174-92.

Etheridge, R. (1872) On the physical structure and organic remains of the Penarth (Rhaetic) Beds of Penarth and Lavernock. *Transactions of the Cardiff Naturalists' Society*, 1872, 39-64.

Evans, J. and Kemp, T.S. (1975) The cranial morphology of a new Lower Cretaceous turtle from southern England. *Palaeontology*, **18**, 25-40.

Evans, J. and Kemp, T.S. (1976) A new turtle skull from the Purbeckian of England and a note on the early dichotomies of cryptodire turtles. *Palaeontology*, **19**, 317-24.

Evans, S.E. (1980) The skull of a new eosuchian reptile from the Lower Jurassic of South Wales. *Zoological Journal of the Linnean Society*, **70**, 203-64.

Evans, S.E. (1981) The postcranial skeleton of the Lower Jurassic eosuchian *Gephyrosaurus bridensis*. *Zoological Journal of the Linnean Society*, **73**, 81-116.

Evans, S.E. (1982) The gliding reptiles of the Upper Permian. *Zoological Journal of the Linnean Society*, **76**, 97-123.

Evans, S.E. (1984) The classification of the Lepidosauria. *Zoological Journal of the Linnean Society*, **82**, 87-100.

Evans, S.E. (1988a) The early history and relationships of the Diapsida. In *The Phylogeny and Classification of the Tetrapods* (ed. M.J. Benton), *Systematics Association Special Volume*, **35A**, Clarendon Press, Oxford, pp. 221-60.

Evans, S.E. (1988b) The Upper Permian reptile *Adelosaurus* from Durham. *Palaeontology*, **31**, 957-64.

Evans, S.E. (1989) New material of *Cteniogenys* (Reptilia: Diapsida) and a reassessment of the systematic position of the genus. *Neues Jahrbuch für Geologie und Paläontologie, Monatshefte*, 1989, 577-89.

Evans, S.E. (1990) The skull of *Cteniogenys*, a choristodere (Reptilia: Archosauromorpha) from the Middle Jurassic of Oxfordshire. *Zoological Journal of the Linnean Society*, **99**, 205-37.

Evans, S.E. (1991) The postcranial skeleton of the choristodere *Cteniogenys* (Reptilia: Diapsida) from the Middle Jurassic of England. *Geobios*, **24**, 187-99.

Evans, S.E. (1992a) A sphenodontian (Reptilia: Lepidosauria) from the Middle Jurassic of England. *Neues Jahrbuch für Geologie und Paläontologie, Monatshefte*, 1992, 449-57.

Evans, S.E. (1992b) Small reptiles and amphibians from the Forest Marble (Middle Jurassic) of Dorset. *Proceedings of the Dorset Natural History and Archaeological Society*, **113**, 201-2.

Evans, S.E. (1992c) A sphenodontid jaw (Reptilia: Lepidosauria) from the Upper Jurassic of Dorset. *Proceedings of the Dorset Natural History and Archaeological Society*, **113**, 199-200.

Evans, S.E. and Haubold, H. (1987) A review of the Upper Permian genera *Coelurosauravus, Weigeltisaurus* and *Gracilisaurus* (Reptilia: Diapsida). *Zoological Journal of the Linnean Society*, **90**, 275-303.

Evans, S.E. and King, M.S. (1993) A new specimen of *Protorosaurus* (Reptilia: Diapsida) from the Marl Slate (late Permian) of Britain. *Proceedings of the Yorkshire Geological Society*, **49**, 229-34.

Evans, S.E. and Milner, A.R. (1991) Middle Jurassic microvertebrate faunas from the British Isles. In *Fifth Symposium on Mesozoic Terrestrial Ecosystems and Biota* (eds Z. Kielan-Jaworowska, N. Heintz and H.A. Nakrem), *Contributions from the Paleontological Museum, University of Oslo*, **364**, 21-2.

Evans, S.E. and Milner, A.R. (1994) Middle Jurassic microvertebrate faunas from the British Isles. In *In the Shadow of the Dinosaurs, Early Mesozoic Tetrapods* (eds N.C. Fraser and H.-D. Sues), Columbia University Press, New York, 303-21.

Evans, S.E., Milner, A.R. and Mussett, F. (1988) The earliest known salamanders (Amphibia: Caudata): a record from the Middle Jurassic of England. *Geobios*, **21**, 539-52.

Evans, S.E., Milner, A.R. and Mussett, F. (1990) A discoglossid frog from the Middle Jurassic of England. *Palaeontology*, **33**, 299-311.

Fitton, W.H. (1836) Observations on some of the strata between the Chalk and the Oxford Oolite, in the south-east of England. *Transactions of the Geological Society of London, Series 2*, **4**, 103-388.

Fitton, W.H. (1847) A stratigraphical account of the section from Atherfield to Rocken End, on the south-west coast of the Isle of Wight. *Quarterly Journal of the Geological Society of London*, **3**, 289-325.

Forbes, E. (1856) On the Tertiary fluvio-marine formation of the Isle of Wight. *Memoirs of the Geological Survey of the United Kingdom*, **10**, 162 pp.

Forster, S.C. and Warrington, G. (1985) Geochronology of the Carboniferous, Permian and Triassic. *Geological Society of London, Memoir* **10**, 99-113.

Fox, W. (1866) Another new Wealden reptile. *Geological Magazine*, **3**, p. 383.

Fox, W. (1869) On the skull and bones of an *Iguanodon. Report of the British Association for the Advancement of Science*, 1868 (1869), 64-5.

Fox-Strangways, C. (1892) *The Jurassic Rocks of Britain. Volumes I, II. Yorkshire.* Memoirs of the Geological Survey of the United Kingdom, HMSO, London, 570 + 250 pp.

Fox-Strangways, C. (1903) *The Geology of the Country near Leicester* (Sheet 156), Memoirs of the Geological Survey of England and Wales, HMSO, London.

Fraser, N.C. (1982) A new rhynchocephalian from the British Upper Trias. *Palaeontology*, **25**, 709-25.

Fraser, N.C. (1985) Vertebrate faunas from Mesozoic fissure deposits of southwest Britain. *Modern Geology*, **9**, 273-300.

Fraser, N.C. (1986) New Triassic sphenodontids from southwest England and a review of their classification. *Palaeontology*, **29**, 165-86.

Fraser, N.C. (1988a) Rare tetrapod remains from the Late Triassic fissure infillings of Cromhall Quarry, Avon. *Palaeontology*, **31**, 567-76.

Fraser, N.C. (1988b) Latest Triassic terrestrial vertebrates and their biostratigraphy. *Modern Geology*, **13**, 125-40.

Fraser, N.C. (1988c) The osteology and relationships of *Clevosaurus* (Reptilia: Sphenodontida). *Philosophical Transactions of the Royal Society of London, Series B*, **321**, 125-78.

Fraser, N.C. (1994) Assemblages of small tetrapods from British Late Triassic fissure deposits. In *In The Shadow of the Dinosaurs, Early Mesozoic Tetrapods* (eds N.C. Fraser and H.-D. Sues), Columbia University Press, New York.

Fraser, N.C. and Benton, M.J. (1989) The Triassic reptiles *Brachyrhinodon* and *Polysphenodon* and the relationships of the spenodontids. *Zoological Journal of the Linnean Society*, **96**, 413-45.

Fraser, N.C. and Sues, H.-D. (eds) (1994) *In The Shadow of the Dinosaurs, Early Mesozoic Tetrapods.* Columbia University Press, New York.

Fraser, N.C. and Unwin, D.M. (1990) Pterosaur remains from the Upper Triassic of Britain. *Neues Jahrbuch für Geologie und Paläontologie, Monatshefte*, 1990, 272-82.

Fraser, N.C. and Walkden, G.M. (1983) The ecology of a Late Triassic reptile assemblage from

References

Gloucestershire, England. *Palaeogeography, Palaeoclimatology, Palaeoecology*, **42**, 341-65.

Fraser, N.C. and Walkden, G.M. (1984) The post-cranial skeleton of *Planocephalosaurus robinsonae*. *Palaeontology*, **27**, 575-95.

Fraser, N.C., Walkden, G.M. and Stewart, V. (1985) The first pre-Rhaetic therian mammal. *Nature, London*, **314**, 161-3.

Freeman, E.F. (1976) Mammal teeth from the Forest Marble (Middle Jurassic) of Oxfordshire, England. *Science*, **194**, 1053-5.

Freeman, E.F. (1979) A Middle Jurassic mammal bed from Oxfordshire. *Palaeontology*, **22**, 135-66.

Gaffney, E.S. (1975a) A taxonomic revision of the Jurassic turtles *Portlandemys* and *Plesiochelys*. *American Museum Novitates*, **2574**, 1-19.

Gaffney, E.S. (1975b) A phylogeny and classification of the higher categories of turtles. *Bulletin of the American Museum of Natural History*, **155**, 387-436.

Gaffney, E.S. (1976) Cranial morphology of the European Jurassic turtles *Portlandemys* and *Plesiochelys*. *Bulletin of the American Museum of Natural History*, **157**, 487-544.

Gaffney, E.S. (1979a) Comparative cranial morphology of recent and fossil turtles. *Bulletin of the American Museum of Natural History*, **164**, 65-376.

Gaffney, E.S. (1979b) The Jurassic turtles of North America. *Bulletin of the American Museum of Natural History*, **162**, 91-136.

Gaffney, E.S. (1980) Phylogenetic relationships of the major groups of amniotes. In *The Terrestrial Environment and the Origin of Land Vertebrates* (ed. A.L. Panchen), *Systematics Association Special Volume*, **15**, Academic Press, London, pp. 593-610.

Gaffney, E.S. and Meylan, P.A. (1988) A phylogeny of turtles. In *The Phylogeny and Classification of the Tetrapods. Volume 1. Amphibians, Reptiles, Birds* (ed. M.J. Benton), *Systematics Association Special Volume*, **35A**, Clarendon Press, Oxford, pp. 157-219.

Gallois, R.W. (1988) *Geology of the Country around Ely* (Sheet 173), Memoirs of the British Geological Survey, HMSO, London, 116 pp.

Gallois, R.W. and Cox, B.M. (1976) The stratigraphy of the Lower Kimmeridge Clay of eastern England. *Proceedings of the Yorkshire Geological Society*, **41**, 13-26.

Galton, P.M. (1969) The pelvic musculature of the dinosaur *Hypsilophodon*. *Postilla*, **131**, 1-64.

Galton, P.M. (1971a) *Hypsilophodon*, the cursorial non-arboreal dinosaur. *Nature*, **231**, 159-61.

Galton, P.M. (1971b) The mode of life of *Hypsilophodon*, the supposedly arboreal ornithopod dinosaur. *Lethaia*, **4**, 453-65.

Galton, P.M. (1971c) A primitive dome-headed dinosaur (Ornithischia: Pachycephalosauridae) from the Lower Cretaceous of England and the function of the dome in pachycephalosaurids. *Journal of Paleontology*, **45**, 40-7.

Galton, P.M. (1973) A femur of a small theropod dinosaur from the Lower Cretaceous of England. *Journal of Paleontology*, **47**, 996-1001.

Galton, P.M. (1974) The ornithischian dinosaur *Hypsilophodon* from the Wealden of the Isle of Wight. *Bulletin of the British Museum (Natural History), Geology*, **25**, 152 pp.

Galton, P.M. (1975) English hypsilophodontid dinosaurs (Reptilia: Ornithischia). *Palaeontology*, **18**, 741-52.

Galton, P.M. (1976a) The dinosaur *Vectisaurus valdensis* (Ornithischia: Iguanodontidae) from the Lower Cretaceous of England. *Journal of Palaeontology*, **50**, 976-84.

Galton, P.M. (1976b) *Iliosuchus*, a Jurassic dinosaur from Oxfordshire and Utah. *Palaeontology*, **19**, 587-9.

Galton, P.M. (1977) The ornithopod dinosaur *Dryosaurus* and a Laurasia-Gondwanaland connection in the Upper Jurassic. *Nature*, **268**, 230-2.

Galton, P.M. (1978) Fabrosauridae, the basal family of ornithischian dinosaurs (Reptilia: Ornithopoda). *Paläontologische Zeitschrift*, **52**, 138-59.

Galton, P.M. (1980a) Armoured dinosaurs (Ornithischia: Ankylosauria) from the Middle and Upper Jurassic of England. *Geobios*, **13**, 825-37.

Galton, P.M. (1980b) European Jurassic ornithopod dinosaurs of the families Hypsilophodontidae and Iguanodontidae. *Neues Jahrbuch für Geologie und Paläontologie, Abhandlungen*, **160**, 73-95.

Galton, P.M. (1980c) *Dryosaurus* and *Camptosaurus*, intercontinental genera of Upper Jurassic ornithopod dinosaurs. *Mémoires de la Société Géologique de France, Nouvelle Série*, **139**, 103-8.

Galton, P.M. (1981a) A juvenile stegosaurian dinosaur *Astrodon pusillus*, from the Upper Jurassic of Portugal, with comments on Upper

References

Jurassic and Lower Cretaceous biogeography. *Journal of Vertebrate Paleontology*, **1**, 245-56.

Galton, P.M. (1981b) *Craterosaurus pottonensis* Seeley, a stegosaurian dinosaur from the Lower Cretaceous of England and a review of Cretaceous stegosaurs. *Neues Jahrbuch für Geologie und Paläontologie, Abhandlungen*, **161**, 28-46.

Galton, P.M. (1983a) Armoured dinosaurs (Ornithischia: Ankylosauria) from the Middle and Upper Jurassic of Europe. *Palaeontographica, Abteilung A*, **182**, 1-25.

Galton, P.M. (1983b) *Sarcolestes leedsi*, an ankylosaurian dinosaur from the Middle Jurassic of England. *Neues Jahrbuch für Geologie und Paläontologie*, 1983, 141-55.

Galton, P.M. (1983c) A juvenile stegosaurian dinosaur, *Omosaurus phillipsi* Seeley from the Oxfordian (Upper Jurassic) of England. *Geobios*, **16**, 95-101.

Galton, P.M. (1985a) The poposaurid thecodontian *Teratosaurus suevicus* von Meyer, plus referred specimens mostly based on prosauropod dinosaurs from the Middle Stubensandstein (Upper Triassic) of Nordwürttemberg. *Stuttgarter Beiträge zur Naturkunde, Serie B*, **116**, 1-29.

Galton, P.M. (1985b) British plated dinosaurs (Ornithischia, Stegosauridae). *Journal of Vertebrate Paleontology*, **5**, 211-54.

Galton, P.M. (1985c) Notes on the Melanorosauridae, a family of large prosauropod dinosaurs (Saurischia: Sauropodomorpha). *Geobios*, **19**, 671-6.

Galton, P.M. (1990) Basal Sauropodomorpha - prosauropods. In *The Dinosauria* (eds D.B. Weishampel, P. Dodson and H. Osmólska), University of California Press, Berkeley, pp. 320-44.

Galton, P.M. and Cluver, M.A. (1976) *Anchisaurus capensi* (Broom) and a revision of the Anchisauridae (Reptilia, Saurischia). *Annals of the South African Museum*, **69**, 121-59.

Galton, P.M. and Jensen, J.A. (1979) Remains of ornithopod dinosaurs from the Lower Cretaceous of North America. *Brigham Young University Geological Studies*, **25**, 1-10.

Galton, P.M. and Powell, H.P. (1980) The ornithischian dinosaur *Camptosaurus prestwichii* from the Upper Jurassic of England. *Palaeontology*, **23**, 411-43.

Galton, P.M. and Powell, H.P. (1983) Stegosaurian dinosaurs from the Bathonian (Middle Jurassic) of England, the earliest record of the Family Stegosauridae. *Geobios*, **16**, 219-29.

Galton, P.M. and Taquet, P. (1982) *Valdosaurus*, a hypsilophodontid dinosaur from the Lower Cretaceous of Europe and Africa. *Geobios*, **15**, 147-59.

Galton, P.M., Brun, R. and Rioult, M. (1980) Skeleton of the stegosaurian dinosaur *Lexovisaurus* from the lower part of the Middle Callovian (Middle Jurassic) of Argences (Calvados), Normandy. *Bulletin trimestrielle de la Société Géologique de Normandie et Amis du Muséum du Havre*, **67**(4), 39-60.

Gamble, H.J. (1981) Observations on a vertebrate fauna from the Osborne Member, Solent Formation (Upper Eocene), of Cliff End, near Colwell Bay, Isle of Wight. *Proceedings of the Isle of Wight Natural History and Archaeological Society*, 7, 397-404.

Gardiner, B.G. (1982) Tetrapod classification. *Zoological Journal of the Linnean Society*, **74**, 207-32.

Gardiner, C.I. (1935) Recent discoveries of reptilian remains in the Chipping Norton Limestone. *Proceedings of the Cotteswolds Naturalists' Field Club*, **25**, p. 222.

Gardiner, C.I. (1937) Reptile-bearing Oolite, Stow. *Report of the British Association for the Advancement of Science*, 1936, p. 296.

Gardiner, C.I. (1938) Reptile-bearing Oolite, Stow. *Report of the British Association for the Advancement of Science*, 1937, p. 290.

Gardner, J.S., Keeping, H. and Monckton, H.W. (1888) The Upper Eocene, comprising the Barton and Upper Bagshot Formation. *Quarterly Journal of the Geological Society of London*, **44**, 578-635.

Gauthier, J.A. (1986) Saurischian monophyly and the origin of birds. *Memoir of the California Academy of Sciences*, **8**, 56 pp.

Gauthier, J.A., Estes, R. and de Queiroz, K. (1988c) A phylogenetic analysis of Lepidosauromorpha. In *Phylogenetic Relationships of the Lizard Families* (eds R. Estes and G. Pregill), Stanford University Press, Stanford, California, pp. 15-98.

Gauthier, J.A., Kluge, A.G. and Rowe, T. (1988a) The early evolution of the Amniota. In *The Phylogeny and Classification of the Tetrapods. Volume 1. Amphibians, Reptiles, Birds* (ed. M.J. Benton), *Systematics Association Special Volume*, **35A**, Clarendon Press, Oxford, pp. 103-55.

References

Gauthier, J.A., Kluge, A.G. and Rowe, T. (1988b) Amniote phylogeny and the importance of fossils. *Cladistics*, **4**, 105-208.

Geiger, M.E. and Hopping, C.A. (1968) Triassic stratigraphy of the southern North Sea Basin. *Philosophical Transactions of the Royal Society, Series B*, **254**, 1-36.

Gibson, T.F. (1858) Notice of the discovery of a large femur of the *Iguanodon* in the Weald Clay at Sandown Bay, Isle of Wight. *Quarterly Journal of the Geological Society of London*, **14**, 175-6.

Gilmore, C.W. (1928) Fossil lizards of North America. *Memoirs of the National Academy of Sciences*, **22**, 169 pp.

Glennie, K.W. and Buller, A.T. (1983) The Permian Weissliegend of NW Europe: the partial deformation of aeolian sand dunes caused by the Zechstein transgression. *Sedimentary Geology*, **35**, 43-81.

Goodrich, E.S. (1894) On the fossil Mammalia from the Stonesfield Slate. *Quarterly Journal of the Microscopical Society, New Series*, **35**, 407-32.

Gordon, G. (1859) On the geology of the lower or northern part of the province of Moray: its history, present state of inquiry and points for future examination. *Edinburgh New Philosophical Journal, New Series*, **9**, 14-60.

Gordon, G. (1892) The reptiliferous sandstones of Elgin (with map). *Transactions of the Geological Society of Edinburgh*, **6**, 241-5.

Grierson, J. (1828) On footsteps before the Flood, in a specimen of red sandstone. *Edinburgh Journal of Science*, **8**, 130-4.

Hallock, L.A., Holman, J.A. and Warren, M.R. (1990) Herpetofauna of the Ipswichian Interglacial Bed (Late Pleistocene) of the Itteringham Gravel Pit, Norfolk, England. *Journal of Herpetology*, **24**, 33-9.

Halstead, L.B. and Nicoll, P.G. (1971) Fossilized caves of Mendip. *Studies in Speleology*, **2**, 93-102.

Hamilton, D. (1977) Aust Cliff. In *Geological Excursions in the Bristol District* (ed. R.J.G. Savage), University of Bristol, pp. 110-18.

Hancock, A. and Howse, R. (1870a) On a new labyrinthodont amphibian from the Magnesian Limestone of Middridge, Durham. *Quarterly Journal of the Geological Society of London*, **26**, 556-64.

Hancock, A. and Howse, R. (1870b) On *Proterosaurus speneri* von Meyer and a new species, *Proterosaurus huxleyi*, from the Marl-Slate of Middridge, Durham. *Quarterly Journal of the Geological Society of London*, **26**, 565-72.

Hardenbol, J. and Berggren, W.A. (1978) A new Paleogene numerical time scale. *American Association of Petroleum Geologists, Studies in Geology*, **6**, 213-34.

Harkness, R. (1850) On the position of the impressions of footsteps in the Bunter Sandstone of Dumfries-shire. *Annals and Magazine of Natural History, Series 2*, **6**, 203-8.

Harkness, R. (1851) Notice of some new footsteps in the Bunter Sandstone of Dumfries-shire. *Annals and Magazine of Natural History, Series 2*, **8**, 90-5.

Harkness, R. (1864) On the reptiliferous sandstones and the footprint-bearing strata of the north-east of Scotland. *Quarterly Journal of the Geological Society of London*, **20**, 429-43.

Harland, W.B., Armstrong, R.L., Cox, A.V., Craig, L.E., Smith, A.G. and Smith, D.G. (1990) *A Geologic Time Scale 1989*. Cambridge University Press, Cambridge, 263 pp.

Harris, J.P. and Hudson, J.D. (1980) Lithostratigraphy of the Great Estuarine Group (Middle Jurassic), Inner Hebrides. *Scottish Journal of Geology*, **16**, 231-50.

Hastings, B. (1848) On the freshwater Eocene beds of the Hordle Cliff, Hampshire. *Report of the British Association for the Advancement of Science*, 1847, 63-4.

Hastings, B. (1852) Description géologique des falaises d'Hordle et sur la côte de Hampshire, en Angleterre. *Bulletin de la Société Géologique de France, Série 2*, **9**, 191-203.

Hastings, B. (1853) On the Tertiary beds of Hordwell, Hampshire. *London, Edinburgh and Dublin Philosophical Magazine, Series 4*, **6**, 1-11.

Haubold, H. (1971) Ichnia amphibiorum et reptiliorum. In *Handbuch der Paläoherpetologie* (ed. O. Kuhn), **18**, Gustav Fischer, Stuttgart, 124 pp.

Haubold, H. (1986) Archosaur footprints at the terrestrial Triassic-Jurassic boundary. In *The Beginning of the Age of Dinosaurs; Faunal Change across the Triassic-Jurassic Boundary* (ed. K. Padian), Cambridge University Press, pp. 189-201.

Haubold, H. (1990) Ein neuer Dinosaurier (Ornithischia, Thyreophora) aus dem Unteren Jura des nördlichen Mitteleuropa. *Révue de Paléobiologie*, **9**, 149-77.

Haubold, H. and Sarjeant, W.A.S. (1973)

Tetrapodenfährten aus den Keele und Enville Groups (Permokarbon: Stefan und Autun) von Shropshire und South Staffordshire, Grossbritannien. *Zeitschrift für Geologischen Wissenschaft*, **8**, 895-933.

Haubold, H. and Schaumberg, G. (1985) *Die Fossilien des Kupferschiefers*. Neue Brehm-Bücherei, 333, Ziemsen, Witteberg Lutherstadt, 223 pp.

Hauff, B. (1921) Untersuchung der Fossil-fundstätten von Holzmaden im Posidonienschiefer des oberen Lias Württembergs. *Palaeontographica*, **64**, 1-42.

Hawkins, T. (1840) *The Book of the Great Sea-Dragons, Ichthyosauri and Plesiosauri, Gedolim Taninim of Moses*. W. Pickering, London, 27 pp.

Heaton, M.J. and Reisz, R.R. (1986) Phylogenetic relationships of captorhinomorph reptiles. *Canadian Journal of Earth Sciences*, **23**, 402-18.

Hemingway, J.E. (1974) Jurassic. In *The Geology and Mineral Resources of Yorkshire* (eds D.H. Rayner and J.E. Hemingway), Yorkshire Geological Society, pp. 161-224.

Henson, M.R. (1970) The Triassic rocks of south Devon. *Proceedings of the Ussher Society*, **2**, 172-7.

Hickling, G. (1909) British Permian footprints. *Memoirs and Proceedings of the Manchester Literary and Philosophical Society*, **53** (22), 31 pp.

Hill, C., Jarzembowski, E.A. and Batten, D.J. (1992) A new Wealden plant. 7. In *Early Cretaceous Environments*. Postgraduate Research Institute for Sedimentology, University of Reading [abstracts].

Hoffstetter, R. (1942) Sur les restes de Sauria du Nummulitique Européen rapportés à la famille des Iguanidae. *Bulletin du Museum National d'Histoire Naturelle*, Série 2, **14**, 233-40.

Hoffstetter, R. (1967) Coup d'oeil sur les sauriens (= lacertiliens) des couches de Purbeck (Jurassique supérieur d'Angleterre). In *Problèmes Actuelles de Paléontologie (Évolution des Vertébrés). Colloques Internationaux du Centre National de la Recherche Scientifique*, **163**, CNRS, Paris, pp. 349-71.

Holloway, S. (1985) Triassic: Sherwood Sandstone Group (excluding the Kinnerton Sandstone Formation and the Lenton Sandstone Formation). In *Atlas of Onshore Sedimentary Basins in England and Wales: post-Carboniferous Tectonics and Stratigraphy* (ed. A. Whittaker), Blackie, Glasgow, pp. 31-3.

Holloway, S., Milodowski, A.E., Strong, G.E. and Warrington, G. (1989) The Sherwood Sandstone Group (Triassic) of the Wessex Basin, southern England. *Proceedings of the Geologists' Association*, **100**, 383-94.

Holman, J.A. (1979) A review of North American Tertiary snakes. *Publications of the Museum, Michigan State University, Paleontology Series*, **1**, 203-60.

Holman, J.A. (1985) Herpetofauna of the Late Pleistocene fissures near Ightham, Kent. *Herpetological Journal*, **1**, 26-32.

Holman, J.A. (1987) Additional records of British Pleistocene amphibians and reptiles. *British Herpetological Society Bulletin*, **19**, 18-20.

Holman, J.A. (1993) The Boxgrove, England, Middle Pleistocene herpetofauna: paleogeographic, evolutionary, stratigraphic and paleoecological relationships. *Historical Biology*, **6**, 263-79.

Holman, J.A., Stuart, A.J. and Clayden, J.D. (1990) A Middle Pleistocene herpetofauna from Cudmore Grove, Essex, England and its paleogeographic and paleoclimatic implications. *Journal of Vertebrate Paleontology*, **10**, 86-94.

Home, E. (1814) Some account of the fossil remains of an animal more nearly allied to fishes than to any other classes of animals. *Philosophical Transactions of the Royal Society*, **104**, 571-7.

Home, E. (1816) Some further account of the fossil remains of an animal of which a description was given in 1814. *Philosophical Transactions of the Royal Society*, **106**, 318-21.

Home, E. (1818) Additional facts respecting the fossil remains of an animal, on the subject of which two papers have already been printed in the Philosophical Transactions. *Philosophical Transactions of the Royal Society*, **108**, 24-32.

Home, E. (1819a) An account of the fossil skeleton of the *Proteosaurus*. *Philosophical Transactions of the Royal Society*, **109**, 209-11.

Home, E. (1819b) Reasons for giving the name *Proteosaurus* to the fossil skeleton which has been described. *Philosophical Transactions of the Royal Society*, **109**, 212-16.

Hooker, J.J. (1972) The first land mammals from the marine Barton Beds (Upper Eocene) of

References

Hampshire. *Proceedings of the Geologists' Association*, **83**, 179-84.

Hooker, J.J. (1986) Mammals from the Bartonian (middle-late Eocene) of the Hampshire Basin, southern England. *Bulletin of the British Museum (Natural History), Geology Series*, **39**, 191-478.

Hooker, J.J. (1991) The sequence of mammals in the Thanetian and Ypresian of the London and Belgian Basins. Location of the Palaeocene-Eocene boundary. *Newsletters in Stratigraphy*, **25**, 75-90.

Hooker, J.J. (1992) British mammalian palaeocommunities across the Eocene-Oligocene transition and their environmental implications. In *Eocene-Oligocene Climatic and Biotic Evolution*, (eds D.R. Prothero and W.A. Berggren), Princeton University Press, Princeton, New Jersey, pp. 494-515.

Hooker, J.J. and Ward, D.J. (1980) List of localities. *Tertiary Research*, **3**, 3-12.

Hooley, R.W. (1900) Note on a tortoise from the Wealden of the Isle of Wight. *Geological Magazine*, Decade 4, **7**, 263-5.

Hooley, R.W. (1905) On a new tortoise from the Lower Headon beds of Horwell, *Nicoria headonensis*, sp. nov. *Geological Magazine*, Decade 5, **2**, 66-8.

Hooley, R.W. (1907) On the skull and greater portion of the skeleton of *Goniopholis crassidens* from the Wealden Shales of Atherfield (Isle of Wight). *Quarterly Journal of the Geological Society of London*, **63**, 50-63.

Hooley, R.W. (1912) On the discovery of remains of *Iguanodon mantelli* in the Wealden Beds of Brighstone Bay, Isle of Wight. *Geological Magazine*, Decade 5, **9**, 444-9.

Hooley, R.W. (1913) On the skeleton of *Ornithodesmus latidens*; an ornithosaur from the Wealden Shales of Atherfield. *Quarterly Journal of the Geological Society of London*, **69**, 372-422.

Hooley, R.W. (1925) On the skeleton of *Iguanodon atherfieldensis* sp. nov., from the Wealden Shales of the Isle of Wight. *Quarterly Journal of the Geological Society of London*, **81**, 1-61.

Horwood, A.P. (1909) Bibliographical notes upon the flora and fauna of the British Keuper. *Report of the British Association for the Advancement of Science*, 1909 (1908), 158-62.

Horwood, A.R. (1916) The Upper Trias of Leicestershire, Part 2. *Geological Magazine*, Decade 6, **3**, 360, 411, 456.

Hotton, N.C., III, MacLean, P.D., Roth, J.J. and Roth, E.C. (1986) *The Ecology and Biology of Mammal-like Reptiles*. Smithsonian, Washington, DC, 326 pp.

House, M.R. (1990) *Geology of the Dorset Coast*. The Geologists' Association, London, 162 pp.

Howarth, M.K. (1955) Domerian of the Yorkshire coast. *Proceedings of the Yorkshire Geological Society*, **30**, 147-75.

Howarth, M.K. (1962) The Jet Rock Series and the Alum Shale Series of the Yorkshire coast. *Proceedings of the Yorkshire Geological Society*, **33**, 381-421.

Howarth, M.K. (1973) The stratigraphy and ammonite fauna of the Upper Liassic Grey Shales of the Yorkshire coast. *Bulletin of the British Museum (Natural History), Geology*, **24**, 235-77.

Howell, H.H. (ed.) (1859) *The Geology of the Warwickshire Coalfield and the Permian Rocks and Trias of the Surrounding District*. Memoirs of the Geological Survey of the United Kingdom, London, 57 pp.

Howse, S.C.B. (1986) On the cervical vertebrae of the Pterodactyloidea (Reptilia: Archosauria). *Zoological Journal of the Linnean Society*, **88**, 307-28.

Howse, S.C.B. and Milner, A.R. (1993) *Ornithodesmus* - a maniraptoran theropod dinosaur from the Lower Cretaceous of the Isle of Wight, England. *Palaeontology*, **36**, 425-37.

Hudlestone, W.H. (1887) Excursion to Aylesbury. *Proceedings of the Geologists' Association*, **10**, 166-72.

Hudson, J.D. (1962) The stratigraphy of the Great Estuarine Series (Middle Jurassic) of the Inner Hebrides. *Transactions of the Edinburgh Geological Society*, **19**, 139-65.

Hudson, J.D. (1966) Hugh Miller's Reptile Bed and the Mytilus Shales, Middle Jurassic, Isle of Eigg, Scotland. *Scottish Journal of Geology*, **2**, 265-81.

Hudson, J.D. and Morton, N. (1969) Excursion Guide No. 4: Guide for Western Scotland. In *International Field Symposium on the British Jurassic* (ed. H.S. Torrens), University of Keele, Keele, D1-D47.

Huene, E. von (1935) Ein Rhynchocephale aus dem Rhät (*Pachystropheus* n.g.). *Neues Jahrbuch für Mineralogie, Geologie, und Paläontologie, Abhandlungen*, **74**, 441-7.

Huene, F. von (1902) Übersicht über die Reptilien der Trias. *Geologische und Paläontologische*

Abhandlungen, Neue Folge, **6**, 1-84.

Huene, F. von (1906) Über das Hinterhaupt von *Megalosaurus bucklandi* aus Stonesfield. *Neues Jahrbuch für Mineralogie, Geologie, und Paläontologie*, 1906, 1-12.

Huene, F. von (1908a) Neue und verkannte Pelycosaurier-Reste aus Europa. *Centralblatt für Mineralogie, Geologie, und Paläontologie*, 1908, 431-4.

Huene, F. von (1908b) Die Dinosaurier der europäischen Triasformation mit Berucksichtigung der aussereuropäischen Vorkommnisse. *Geologische und Paläontologische Abhandlungen (Supplement-Band)*, **1**, 419 pp.

Huene, F. von (1908c) Note on two sections in the Lower Keuper Sandstone of Guy's Cliff, Warwick. *Geological Magazine*, Decade 5, **5**, 100-2.

Huene, F. von (1908d) Eine Zusammenstellung über die englische Trias und das Alter ihrer Fossilien. *Centralblatt für Mineralogie, Geologie, und Paläontologie*, 1908, 9-17.

Huene, F. von (1908e) On phytosaurian remains from the Magnesian Conglomerate of Bristol (*Rileya platyodon*). *Annals and Magazine of Natural History*, Series 8, **1**, 228-30.

Huene, F. von (1908f) On the age of the reptile faunas contained in the Magnesian Conglomerate at Bristol and in the Elgin sandstone. *Geological Magazine*, Decade 5, **5**, 99-100.

Huene, F. von (1910a) Ein primitiver Dinosaurier aus der mittleren Trias von Elgin. *Geologische und Paläontologische Abhandlungen, Neue Folge*, **8**, 315-22.

Huene, F. von (1910b) Über einen echten Rhynchocephalen aus der Trias von Elgin, *Brachyrhinodon taylori*. *Neues Jahrbuch für Mineralogie, Geologie, und Paläontologie*, 1910 (2), 29-62.

Huene, F. von (1910c) Über den ältesten Rest von *Omosaurus (Dacentrurus)* im englischen Dogger. *Neues Jahrbuch für Mineralogie, Geologie, und Paläontologie*, 1910 (1), 75-8.

Huene, F. von (1912a) Die Cotylosaurier der Trias. *Palaeontographica*, **59**, 69-102.

Huene, F. von (1912b) Die zweite Fund des Rhynchocephalen *Brachyrhinodon* in Elgin. *Neues Jahrbuch für Mineralogie, Geologie, und Paläontologie*, 1912 (I), 51-7.

Huene, F. von (1912c) Der Unterkiefer eines riesigen Ichthyosauriers aus dem englischen Rhät. *Centralblatt für Mineralogie, Geologie, und Paläontologie*, 1912, 61-3.

Huene, F. von (1913) Über die reptilfuhrenden Sandsteine bei Elgin in Schottland. *Centralblatt für Mineralogie, Geologie, und Paläontologie*, 1913, 617-23.

Huene, F. von (1914) Beiträge zur Geschichte der Archosaurier. *Geologische und Paläontologische Abhandlungen, Neue Folge*, **13**, 1-53.

Huene, F. von (1920) Ein *Telerpeton* mit gut erhaltenem Schadel. *Centralblatt für Mineralogie, Geologie, und Paläontologie*, 1920, 189-92.

Huene, F. von (1922) Die Ichthyosaurier des Lias und ihre Zusammenhänge. *Monographien zur Geologie und Paläontologie*, **1**, 114 pp.

Huene, F. von (1923) Carnivorous Saurischia in Europe since the Triassic. *Bulletin of the Geological Society of America*, **34**, 449-58.

Huene, F. von (1926) The carnivorous Saurischia in the Jura and Cretaceous formations principally in Europe. *Revista del Museo de La Plata*, **29**, 35-167.

Huene, F. von (1929a) Über Rhynchosaurier und andere Reptilien aus den Gondwana-ablagerungen Südamerikas. *Geologische und Paläontologische Abhandlungen, Neue Folge*, **17**, 1-62.

Huene, F. von (1929b) Los Saurisquios y Ornithisquios del Cretáceo Argentino. *Anales del Museo de La Plata (2a)*, **3**, 196 pp.

Huene, F. von (1932) Die fossile Reptil-Ordnung Saurischia, ihre Entwicklung und Geschichte. *Monographien zur Geologie und Paläontologie*, **4**, 361 pp.

Huene, F. von (1938) *Stenaulorhynchus*, ein Rhynchosauride der ostafrikanischen Obertrias. *Nova Acta Leopoldina, Neue Folge*, **6**, 83-121.

Huene, F. von (1939) Die Verwandtshaftgeschichte der Rhynchosauriden. *Physis, Augsburg*, **14**, 499-523.

Huene, F. von (1942) *Die Fossilen Reptilien des Südamerikanischen Gondwanalandes*. C.H. Beck, München.

Huene, F. von (1956) *Paläontologie und Phylogenie der Niederen Tetrapoden*. G. Fischer, Jena, 716 pp.

Hughes, B. (1968) The tarsus of rhynchocephalian reptiles. *Journal of Zoology*, **156**, 457-81.

Hughes, N.F. and McDougall, A.D. (1990) New Wealden correlation for the Wessex Basin. *Proceedings of the Geologists' Association*, **101**, 85-90.

Hulke, J.W. (1869a) Note on a large saurian

humerus from the Kimmeridge Clay of the Dorset coast. *Quarterly Journal of the Geological Society of London,* **25**, 386-9.

Hulke, J.W. (1869b) Note on some fossil remains of a gavial-like saurian from Kimmeridge Bay, collected by J.C. Mansel, Esq., establishing its identity with Cuvier's deuxième gavial d'Honfleur, tête à museau plus court (*Steneosaurus rostro-minor* of Geoffroy St Hilaire, 1825) and with Quenstedt's *Dakosaurus. Quarterly Journal of the Geological Society of London,* **25**, 390-401.

Hulke, J.W. (1870a) Note on a crocodilian skull from Kimmeridge Bay, Dorset. *Quarterly Journal of the Geological Society of London,* **26**, 167-72.

Hulke, J.W. (1870b) Note on some teeth associated with two fragments of a jaw from Kimmeridge Bay. *Quarterly Journal of the Geological Society of London,* **26**, 172-4.

Hulke, J.W. (1870c) Note on some plesiosaurian remains obtained by J.C. Mansel, in Kimmeridge Bay, Dorset. *Quarterly Journal of the Geological Society of London,* **26**, 611-22.

Hulke, J.W. (1870d) Note on a new and undescribed Wealden vertebra. *Quarterly Journal of the Geological Society of London,* **26**, 318-24.

Hulke, J.W. (1871a) Note on an *Ichthyosaurus (I. enthekiodon)* from Kimmeridge Bay, Dorset. *Quarterly Journal of the Geological Society of London,* **27**, 440-1.

Hulke, J.W. (1871b) Note on a fragment of a teleosaurian snout from Kimmeridge Bay, Dorset. *Quarterly Journal of the Geological Society of London,* **27**, 442-3.

Hulke, J.W. (1871c) Note on a large reptilian skull from Brooke, Isle of Wight, probably dinosaurian and referable to the genus *Iguanodon. Quarterly Journal of the Geological Society of London,* **27**, 199-206.

Hulke, J.W. (1872) Note on some ichthyosaurian remains from Kimmeridge Bay, Dorset. *Quarterly Journal of the Geological Society of London,* **28**, 34-5.

Hulke, J.W. (1873) Contribution to the anatomy of *Hypsilophodon foxii. Quarterly Journal of the Geological Society of London,* **29**, 522-31.

Hulke, J.W. (1874a) Note on a very large saurian limb-bone adapted for progression upon land, from the Kimmeridge Clay of Weymouth, Dorset. *Quarterly Journal of the Geological Society of London,* **30**, 16-18.

Hulke, J.W. (1874b) Note on an astragalus of *Iguanodon mantelli. Quarterly Journal of the Geological Society of London,* **30**, 24-6.

Hulke, J.W. (1874c) Note on a reptilian tibia and humerus (probably of *Hylaeosaurus*) from the Wealden formation in the Isle of Wight. *Quarterly Journal of the Geological Society of London,* **30**, 516-20.

Hulke, J.W. (1874d) Supplemental note on the anatomy of *Hypsilophodon foxii. Quarterly Journal of the Geological Society of London,* **30**, 18-23.

Hulke, J.W. (1874e) Note on a modified form of dinosaurian ilium, hitherto reputed scapula. *Quarterly Journal of the Geological Society of London,* **30**, 521-8.

Hulke, J.W. (1876) Appendix to 'Note on a modified form of dinosaurian ilium, hitherto reputed scapula'. *Quarterly Journal of the Geological Society of London,* **32**, 364-6.

Hulke, J.W. (1878) Note on an os articulare, presumably that of *Iguanodon mantelli. Quarterly Journal of the Geological Society of London,* **34**, 744-7.

Hulke, J.W. (1879a) Note (3d) on (*Eucamerotus* Hulke) *Ornithopsis,* H.G. Seeley, =*Bothriospondylus magnus,* Owen, =*Chondrosteosaurus magnus,* Owen. *Quarterly Journal of the Geological Society of London,* **35**, 752-62.

Hulke, J.W. (1879b) *Vectisaurus valdensis,* a new Wealden dinosaur. *Quarterly Journal of the Geological Society of London,* **35**, 421-4.

Hulke, J.W. (1880a) *Iguanodon prestwichii,* a new species from the Kimmeridge Clay, distinguished from *Iguanodon mantelli* of the Wealden Formation in the S.E. of England and Isle of Wight by differences in the shape of the vertebral centra, by fewer than five sacral vertebrae, by the simpler character of its tooth-serrature, etc., founded on numerous fossil remains lately discovered at Cumnor, near Oxford. *Quarterly Journal of the Geological Society of London,* **36**, 433-56.

Hulke, J.W. (1880b) Supplementary note on the vertebrae of *Ornithopsis,* Seeley, =*Eucamerotus,* Hulke. *Quarterly Journal of the Geological Society of London,* **36**, 31-4.

Hulke, J.W. (1882a) Description of some *Iguanodon* remains indicating a new species, *I. seelyi. Quarterly Journal of the Geological Society of London,* **38**, 135-44.

Hulke, J.W. (1882b) *Polacanthus foxii,* a large undescribed dinosaur from the Wealden formation in the Isle of Wight. *Philosophical Transactions of the Royal Society of London,*

172, 653-67.

Hulke, J.W. (1882c) An attempt at a complete osteology of *Hypsilophodon foxii*. *Philosophical Transactions of the Royal Society of London*, **173**, 1035-62.

Hulke, J.W. (1882d) Note on the os pubis and ischium of *Ornithopsis eucamerotus*. *Quarterly Journal of the Geological Society of London*, **38**, 372-6.

Hulke, J.W. (1885) Note on the sternal apparatus in *Iguanodon*. *Quarterly Journal of the Geological Society of London*, **41**, 473-5.

Hulke, J.W. (1887) Note on some dinosaurian remains in the collection of A. Leeds, Esq., of Eyebury, Northamptonshire. *Quarterly Journal of the Geological Society of London*, **43**, 695-702.

Hull, E. (1859) *The Geology of the Country around Woodstock, Oxfordshire* (Sheet 45, S.W.), Memoirs of the Geological Survey of the United Kingdom, 30 pp.

Hull, E. (1860) On the Blenheim iron-ore and the thickness of the formations below the Great Oolite at Stonesfield, Oxfordshire. *The Geologist*, **3**, 303-5; and *Report of the British Association for the Advancement of Science*, 1860 (1861), 81-3.

Hull, E. (1869) *The Triassic and Permian Rocks of the Midland Counties of England*. Memoirs of the Geological Survey of the United Kingdom, 127 pp.

Hull, E. (1892) A comparison of the Red Rocks of the South Devon coast with those of the Midland and Western counties. *Quarterly Journal of the Geological Society of London*, **48**, 60-7.

Hunt, A.P. and Lucas, S.G. (1991a) The *Paleorhinus* Biochron and the correlation of the nonmarine Upper Triassic of Pangaea. *Palaeontology*, **34**, 487-501.

Hunt, A.P. and Lucas, S.G. (1991b) A new rhynchosaur from the Upper Triassic of west Texas, U.S.A. and the biochronology of Late Triassic rhynchosaurs. *Palaeontology*, **34**, 927-38.

Hutchinson, P.O. (1879) Fossil plant, discovered near Sidmouth. *Transactions of the Devonshire Association for the Advancement of Science, Literature and Art*, **11**, 383-5.

Hutchinson, P.O. (1906) Geological section of the cliffs to the west and east of Sidmouth, Devon. *Report of the British Association for the Advancement of Science*, 1906 (1905), 168-70.

Hutt, S., Simmonds, K. and Hullman, G. (1989) Predatory dinosaurs from the Isle of Wight. *Proceedings of the Isle of Wight Natural History and Archaeological Society*, **9**, 137-46.

Huxley, T.H. (1859a) Postscript [to Murchison (1859)]. *Quarterly Journal of the Geological Society of London*, **15**, 435-6.

Huxley, T.H. (1859b) On the *Stagonolepis robertsoni*; and on the recently discovered footmarks in the sandstones of Cummingstone. *Quarterly Journal of the Geological Society of London*, **15**, 440-60.

Huxley, T.H. (1859c) On a fragment of a lower jaw of a large labyrinthodont from Cubbington. In *The Geology of the Warwickshire Coalfield and the Permian Rocks and Trias of the Surrounding District* (ed. H. H. Howell), Memoirs of the Geological Survey of the United Kingdom, London, pp. 56-7.

Huxley, T.H. (1859d) On *Rhamphorhynchus bucklandi*, a pterosaurian from the Stonesfield Slate. *Quarterly Journal of the Geological Society of London*, **15**, 658-70.

Huxley, T.H. (1859e) On the dermal armour of *Crocodilus hastingsiae*. *Quarterly Journal of the Geological Society of London*, **15**, 678-80.

Huxley, T.H. (1867a) On a new specimen of *Telerpeton elginense*. *Quarterly Journal of the Geological Society of London*, **23**, 77-84.

Huxley, T.H. (1867b) On *Acanthopholis horridus*, a new reptile from the Chalk-marl. *Geological Magazine*, **4**, 65-7.

Huxley, T.H. (1869) On *Hyperodapedon*. *Quarterly Journal of the Geological Society of London*, **25**, 138-52.

Huxley, T.H. (1870a) On the classification of the Dinosauria, with observations on the Dinosauria of the Trias. *Quarterly Journal of the Geological Society of London*, **26**, 32-50.

Huxley, T.H. (1870b) On *Hypsilophodon foxii*, a new dinosaurian from the Wealden of the Isle of Wight. *Quarterly Journal of the Geological Society of London*, **26**, 3-12.

Huxley, T.H. (1875) On *Stagonolepis robertsoni* and on the evolution of the Crocodilia. *Quarterly Journal of the Geological Society of London*, **31**, 423-38.

Huxley, T.H. (1877) The crocodilian remains found in the Elgin Sandstones, with remarks on the ichnites of Cummingstone. *Memoirs of the Geological Survey of the United Kingdom, Monograph*, **3**, 52 pp.

References

Huxley, T.H. (1887) Further observations upon *Hyperodapedon gordoni. Quarterly Journal of the Geological Society of London*, **43**, 675-94.

Insole, A. (1980) New dinosaur remains from the Wealden. *Proceedings of the Isle of Wight Natural History and Archaeological Society*, **7**, 201.

Insole, A. (1982) The habitat of the Wealden dinosaurs. *Journal of the Portsmouth and District Natural History Society*, **3**, 80-7.

Insole, A. and Daley, B. (1985) A revision of the lithostratigraphical nomenclature of the Late Eocene and Early Oligocene strata of the Hampshire Basin, southern England. *Tertiary Research*, **7**, 67-100.

Insole, A.N. and Hutt, S. (1994) The palaeoecology of the dinosaurs of the Wessex Formation (Wealdon Group, Early Cretaceous). Isle of Wight, Southern England. *Zoological Journal of the Linnean Society*, **112**, 135-50.

Ireland, R.J., Pollard, J.E., Steel, R.S. and Thompson, D.B. (1978) Intertidal sediments and trace fossils from the Waterstones (Scythian-Anisian?) at Daresbury, Cheshire. *Proceedings of the Yorkshire Geological Society*, **41**, 399-436.

Irving, A. (1888) The red-rock series of the Devon coast-section. *Quarterly Journal of the Geological Society of London*, **44**, 149-63.

Irving, A. (1892) Supplementary note to the paper on the 'Red rocks of the Devon coast-section'. *Quarterly Journal of the Geological Society of London*, **48**, 68-80.

Irving, A. (1893) The base of the Keuper formation in Devon. *Quarterly Journal of the Geological Society of London*, **49**, 79-83.

Ivimey-Cook, H.C. (1974) The Permian and Triassic deposits of Wales. In *The Upper Palaeozoic and post-Palaeozoic Rocks of Wales* (ed. T.R. Owen), University of Wales Press, Cardiff, pp. 295-321.

Jaeger, G.F. (1828) *Über die fossile Reptilien, welche in Württemberg aufgefunden worden sind.* Metzler, Stuttgart, 46 pp.

Jarzembowski, E.A. (1976) Insect fossils from the Wealden of the Weald. *Proceedings of the Geologists' Association*, **87**, 433-46.

Jarzembowski, E.A. (1991a) The Weald Clay of the Weald: a report of 1988/89 field meetings. *Proceedings of the Geologists' Association*, **102**, 83-92.

Jarzembowski, E.A. (1991b) New insects from the Weald Clay of the Weald. *Proceedings of the Geologists' Association*, **102**, 93-108.

Joffe, J. (1967) The 'dwarf' crocodiles of the Purbeck Formation, Dorset: a reappraisal. *Palaeontology*, **10**, 629-39.

Johnson, M.R.W. (1950) The fauna of the Rhaetic Beds in south Nottinghamshire. *Geological Magazine*, **87**, 116-20.

Judd, J.W. (1873) The Secondary rocks of Scotland. *Quarterly Journal of the Geological Society of London*, **29**, 97-195.

Judd, J.W. (1885) The presence of the remains of *Dicynodon* in the Triassic sandstone of Elgin. *Nature*, **32**, p. 573.

Judd, J.W. (1886a) On the relation of the reptiliferous sandstone of Elgin to the Upper Old Red Sandstone. *Proceedings of the Royal Society of London, Series B*, **39**, 394-404.

Judd, J.W. (1886b) Section C: Presidential address. *Reports of the British Association for the Advancement of Science, Reports and Transactions*, 1885, 994-1013.

Jukes-Browne, A.J. (1896) On a collection of fossils from the Upper Greensand in the Dorset County Museum. *Proceedings of the Dorset Natural History and Archaeological Field Club*, **17**, 96-108.

Jukes-Browne, A.J. and Hill, W. (1900) *The Cretaceous Rocks of Britain. Volume I. The Gault and Upper Greensand of England.* Memoirs of the Geological Survey of the United Kingdom, HMSO, London, 499 pp.

Jukes-Browne, A.J. and Hill, W. (1903) *The Cretaceous Rocks of Britain. Volume II. The Lower and Middle Chalk of England.* Memoirs of the Geological Survey of the United Kingdom, HMSO, London, 568 pp.

Jukes-Browne, A.J. and Hill, W. (1904) *The Cretaceous Rocks of Britain. Volume III. The Upper Chalk of England.* Memoirs of the Geological Survey of the United Kingdom, HMSO, London, 566 pp.

Keeping, W. (1883) *The Fossils and Palaeontological Affinities of the Neocomian Deposits of Upware and Brickhill (Cambridgeshire and Bedfordshire).* Cambridge University Press, Cambridge, 167 pp.

Kemp, T.S. (1982) *Mammal-like Reptiles and the Origin of Mammals.* Academic Press, London, 363 pp.

Kennedy, W.J. (1969) The correlation of the Lower Chalk of south-east England. *Proceedings of the Geologists' Association*, **80**, 459-560.

Kent, P.E. (1968) The Rhaetic beds. In *The*

References

Geology of the East Midlands (eds P.C. Sylvester-Bradley and T.D. Ford), Leicester University Press, Leicester, pp. 174-87.

Kent, P.E. (1970) Problems of the Rhaetic in the East Midlands. *Mercian Geologist*, **3**, 361-71.

Kermack, D. (1984) New prosauropod material from South Wales. *Zoological Journal of the Linnean Society*, **82**, 101-17.

Kermack, K.A., Mussett, F. and Rigney, H.W. (1973) The lower jaw of *Morganucodon*. *Zoological Journal of the Linnean Society*, **53**, 87-175.

Kermack, K.A., Lee, A.J., Lees, P.M. and Mussett, F. (1987) A new docodont from the Forest Marble. *Zoological Journal of the Linnean Society*, **89**, 1-39.

Kerth, M. and Hailwood, E.A. (1988) Magnetostratigraphy of the Lower Cretaceous Vectis Formation (Wealden Group) on the Isle of Wight, Southern England. *Journal of the Geological Society, London*, **145**, 351-60.

King, C. (1970) The biostratigraphy of the London Clay in the London district. *Tertiary Times*, **1**, 13-15.

King, C. (1981) The stratigraphy of the London Clay and associated deposits. *Tertiary Research Special Paper*, **6**, 158 pp.

King, C. (1984) The stratigraphy of the London Clay Formation and Virginia Water Formation in the coastal sections of the Isle of Sheppey (Kent, England). *Tertiary Research*, **5**, 121-58.

King, G.M. (1988) Anomodontia. In *Handbuch der Paläoherpetologie, 17C* (ed. P. Wellnhofer), Gustav Fischer, Stuttgart, pp. 1-174.

King, M.J. (in prep.) Tetrapod footprints from the Middle Triassic of England. Unpublished PhD Thesis, University of Bristol.

Kitchener, A. (1987) Function of Claws' claws. *Nature, London*, **325**, 114.

Klein, G. de V. (1965) Dynamic significance of primary structures in the Middle Jurassic Great Oolite Series, southern England. *Special Publication of the Society of Economic Paleontologists and Mineralogists*, **12**, 173-91.

Krantz, R. (1972) Die Sponge-Gravels von Faringdon (England). *Neues Jahrbuch für Geologie und Paläontologie, Abhandlungen*, **140**, 207-31.

Krebs, B. (1967) Der Jura-Krokodilier *Machimosaurus* H. v. Meyer. *Paläontologisches Zeitschrift*, **41**, 46-59.

Krebs, B. (1969) *Ctenosauriscus koeneni* (v. Huene), die Pseudosuchier und die Buntsandstein Reptilien. *Eclogae Geologicae Helvetiae*, **62**, 697-714.

Krebs, B. (1976) Pseudosuchia. In *Handbuch der Paläoherpetologie 13* (ed. O. Kuhn), Gustav Fischer, Stuttgart, pp. 40-98.

Kuhn, O. (1969) Proganosauria, Bolosauria, Placodontia, Araeoscelidia, Trilophosauria, Weigeltisauria, Millerosauria, Rhynchocephalia, Protorosauria. In *Handbuch der Paläoherpetologie, 9* (ed. O. Kuhn), Gustav Fischer, Stuttgart, pp. 1-74.

Kühne, W.G. (1956) *The Liassic Therapsid Oligokyphus*. British Museum (Natural History), London, 149 pp.

Lake, R.D. and Shephard-Thorn, E.R. (1987) *Geology of the Country around Hastings and Dungeness* (Sheets 320 and 321), Memoirs of the British Geological Survey, HMSO, London, 81 pp.

Laming, D.J.C. (1966) Imbrications, palaeocurrents and other sedimentary features in the Lower New Red Sandstone, Devonshire, England. *Journal of Sedimentary Petrology*, **36**, 940-59.

Laming, D.J.C. (1968) New Red Sandstone stratigraphy in Devon and west Somerset. *Proceedings of the Ussher Society*, **2**, 23-5.

Laming, D.J.C. (1982) The New Red Sandstone. In *The Geology of Devon* (eds E.M. Durrance and D.J.C. Laming), Exeter University Press, Exeter, pp. 148-78.

Lang, W.D. (1914) The geology of the Charmouth cliffs, beach and foreshore. *Proceedings of the Geologists' Association*, **25**, 293-360.

Lang, W.D. (1924) The Blue Lias of the Devon and Dorset coasts. *Proceedings of the Geologists' Association*, **35**, 169-85.

Lang, W.D. (1932) The Lower Lias of Charmouth and the Vale of Marshwood. *Proceedings of the Geologists' Association*, **43**, 97-126.

Lang, W.D. (1939) Mary Anning (1799-1847) and the pioneer geologists of Lyme. *Proceedings of the Dorset Natural History and Archaeological Society*, **60**, 142-64.

Lang, W.D. (1947) James Harrison of Charmouth, geologist (1819-1864). *Proceedings of the Dorset Natural History and Archaeological Society*, **68**, 103-18.

Lang, W.D. and Spath, L.F. (1926) The Black Marl of Black Ven and Stonebarrow, in the Lias of the Dorset coast. *Quarterly Journal of the Geological Society of London*, **82**, 144-87.

Lang, W.D., Spath, L.F. and Richardson, W.A.

(1923) Shales with Beef: a sequence in the Lower Lias of the Dorset coast. *Quarterly Journal of the Geological Society of London*, **79**, 47-99.

Lang, W.D., Spath, L.F., Cox, L.R. and Muir-Wood, H.M. (1928) The Belemnite Marls of Charmouth, a series in the Lias of the Dorset coast. *Quarterly Journal of the Geological Society of London*, **84**, 179-257.

Lavis, H.J. (1876) On the Triassic strata which are exposed in the cliff sections near Sidmouth and a note on the occurrence of an ossiferous zone containing bones of a *Labyrinthodon*. *Quarterly Journal of the Geological Society of London*, **32**, 274-7.

Leeds, E.T. (1956) *The Leeds Collection of Fossil Reptiles from the Oxford Clay of Peterborough*. Blackwell, Oxford, 104 pp.

Leonard, A.J., Moore, A.G. and Selwood, E.B. (1982) Ventifacts from a deflation surface marking the top of the Budleigh Salterton Pebble Beds, east Devon. *Proceedings of the Ussher Society*, **5**, 333-9.

Lhuyd, E. (1699) *Eduardi Luidii apud Oxonienses Cimeliarchae Ashmoleani Lithophylacii Britannici Ichnographia*. MC, London, 145 pp.

Lillegraven, J.A., Kielan-Jaworowska, Z. and Clemens, W.A. (1979) *Mesozoic Mammals: the First Two-thirds of Mammalian History*. University of California Press, Berkeley, 311 pp.

Linn, J. (1886) Memoir of Sheet 95. [MS, never published; details in Peacock *et al.*, 1968.]

Lorsong, J.A., Clarey, T.J. and Atkinson, C.D. (eds) (1990) Lithofacies architecture of sandy braided stream deposits in the Otter Sandstone, U.K. 138. (ed.). *13th International Sedimentological Congress, Nottingham, England, Abstracts of Posters*. International Sedimentological Union, Utrecht.

Lucas, S.G. and Hunt, A.P. (1990) The oldest mammal. *New Mexico Journal of Science*, **30**, 41-9.

Lydekker, R. (1887a) On certain dinosaurian vertebrae from the Cretaceous of India and the Isle of Wight. *Quarterly Journal of the Geological Society of London*, **43**, 156-60.

Lydekker, R. (1887b) Note on the Hordwell and other crocodilians. *Geological Magazine*, Decade 3, **4**, 307-12.

Lydekker, R. (1888a) *Catalogue of the Fossil Reptilia and Amphibia in the British Museum (Natural History). Part I. The Orders Ornithosauria, Crocodilia, Dinosauria, Squamata, Rhynchocephalia and Proterosauria*. British Museum (Natural History), London, 309 pp.

Lydekker, R. (1888b) Note on a new Wealden iguanodont and other dinosaurs. *Quarterly Journal of the Geological Society of London*, **44**, 46-61.

Lydekker, R. (1888c) Notes on Tertiary Lacertilia and Ophidia. *Geological Magazine*, Decade 3, **5**, 110-13.

Lydekker, R. (1889a) *Catalogue of the Fossil Reptilia and Amphibia in the British Museum (Natural History). Part II. The Orders Ichthyopterygia and Sauropterygia*. British Museum (Natural History), London, 307 pp.

Lydekker, R. (1889b) *Catalogue of the Fossil Reptilia and Amphibia in the British Museum (Natural History). Part III. The Order Chelonia*. British Museum (Natural History), London, 239 pp.

Lydekker, R. (1889c) On the remains and affinities of five genera of Mesozoic reptiles. *Quarterly Journal of the Geological Society of London*, **45**, 41-59.

Lydekker, R. (1889d) On remains of Eocene and Mesozoic Chelonia and a tooth of ?*Ornithopsis*. *Quarterly Journal of the Geological Society of London*, **45**, 227-46.

Lydekker, R. (1889e) Notes on some points in the nomenclature of fossil reptiles and amphibians, with preliminary notices on two new species. *Geological Magazine*, Decade 3, **6**, 325-6.

Lydekker, R. (1889f) Notes on new and other dinosaurian remains. *Geological Magazine*, Decade 3, **6**, 352-6.

Lydekker, R. (1889g) On a skull of the chelonian genus *Lytoloma*. *Proceedings of the Zoological Society of London*, 1889, 60-6.

Lydekker, R. (1889h) Preliminary notice of new fossil Chelonia. *Annals and Magazine of Natural History, Series 6*, **3**, 53-4.

Lydekker, R. (1890a) *Catalogue of the Fossil Reptilia and Amphibia in the British Museum (Natural History). Part IV. The Orders Anomodontia, Ecaudata, Caudata and Labyrinthodontia; and supplement*. British Museum (Natural History), London, 295 pp.

Lydekker, R. (1890b) Contributions to our knowledge of the dinosaurs of the Wealden and sauropterygians of the Purbeck and Oxford

Clay. *Quarterly Journal of the Geological Society of London*, 46, 36-53.

Lydekker, R. (1890c) On remains of small sauropodous dinosaurs from the Wealden. *Quarterly Journal of the Geological Society of London*, 46, 182-5.

Lydekker, R. (1890d) On a peculiar horn-like dinosaurian bone from the Wealden. *Quarterly Journal of the Geological Society of London*, 46, 185-6.

Lydekker, R. (1891) On certain ornithosaurian and dinosaurian remains. *Quarterly Journal of the Geological Society of London*, 47, 41-4.

Lydekker, R. (1892) Note on two dinosaurian foot-bones from the Wealden. *Quarterly Journal of the Geological Society of London*, 48, 375-6.

Lydekker, R. (1893a) On two dinosaurian teeth from Aylesbury. *Quarterly Journal of the Geological Society of London*, 49, 566-8.

Lydekker, R. (1893b) On a sauropodous dinosaurian vertebra from the Wealden of Hastings. *Quarterly Journal of the Geological Society of London*, 49, 276-80.

Lyell, C. (1852) *A Manual of Elementary Geology*, 4th edn., Murray, London, 512 pp.

Macfadyen, W.A. (1970) *Geological Highlights of the West Country*. Butterworths, London, 296 pp.

Mackeson, H.B. (1840) Note on the discovery of some portions of a large saurian near the bottom of the Lower Greensand in the vicinity of Hythe. *Proceedings of the Geological Society of London*, 3, p. 325.

Mackie, S.J. (1863) The reptiles of the Chalk. *The Geologist*, 6, 266-8.

Mackie, W. (1902) The pebble-band of the Elgin Trias and its wind-worn pebbles. *Report of the British Association for the Advancement of Science, Reports and Transactions*, 1901, 650-1.

Macquaker, J. (1994) Palaeoenvironmental significance of 'bone-beds' in organic-rich mudstone successions: an example from the Upper Triassic of south-west Britain. *Zoological Journal of the Linnean Society*, 112, 285-301.

Mader, D. (1990) *Palaeoecology of the Flora in Buntsandstein and Keuper in the Triassic of Middle Europe. Volume 1. Buntsandstein.* Gustav Fischer, Stuttgart, 1231 pp.

Mader, D. and Laming, D.J.C. (1985) Braidplain and alluvial-fan environmental history and climatological evolution controlling origin and destruction of aeolian dune fields and governing overprinting of sand seas and river plains

by calcrete pedogenesis in the Permian and Triassic of south Devon (England). In *Aspects of Fluvial Sedimentation in the Lower Triassic Buntsandstein of Europe* (ed. D. Mader), Springer-Verlag, Berlin, pp. 519-28.

Mansel-Pleydell, J.C. (1888) Fossil reptiles of Dorset. *Proceedings of the Dorset Natural History and Antiquarian Field Club*, 9, 1-40.

Mantell, G.A. (1822) *The Fossils of the South Downs; or, Illustrations of the Geology of Sussex*. Lupton Relfe, London, 327 pp.

Mantell, G.A. (1825) Notice on the *Iguanodon*, a newly discovered fossil reptile, from the sandstone of Tilgate Forest, in Sussex. *Philosophical Transactions of the Royal Society*, 115, 179-86.

Mantell, G.A. (1827) *Illustrations of the Geology of Sussex: Containing a General View of the Geological Relations of the South-eastern Part of England; with Figures and Descriptions of the Fossils of the Tilgate Forest*. Lupton Relfe, London, 92 pp.

Mantell, G.A. (1833) Observations on the remains of the *Iguanodon* and other fossil reptiles, of the strata of Tilgate Forest in Sussex. *Proceedings of the Geological Society of London*, 1, 410-11.

Mantell, G.A. (1837) On the structure of the fossil saurians. *Loudon's Magazine of Natural History*, 10, 281, 341.

Mantell, G.A. (1842) On the fossil remains of turtles, discovered in the Chalk Formation of the south-east of England. *Philosophical Transactions of the Royal Society*, 131, 153-8.

Mantell, G.A. (1844) *The Medals of Creation*. London, 876 pp.

Mantell, G.A. (1846) Notes on the Wealden strata of the Isle of Wight, with an account of the bones of *Iguanodons* and other reptiles discovered at Brook Point and Sandown Bay. *Quarterly Journal of the Geological Society of London*, 2, 91-6.

Mantell, G.A. (1847) *Geological Excursions Round the Isle of Wight and Along the Adjacent Coast of Dorsetshire; Illustrative of the Most Interesting Geological Phenomena and Organic Remains*. S.H.G. Bohn, London, 428 pp.

Mantell, G.A. (1849) Additional observations on the osteology of the *Iguanodon* and *Hylaeosaurus*. *Philosophical Transactions of the Royal Society*, 139, 271-305.

Mantell, G.A. (1850a) On the *Pelorosaurus*; an

undescribed gigantic terrestrial reptile, whose remains are associated with those of the *Iguanodon* and other saurians in the strata of the Tilgate Forest, in Sussex. *Philosophical Transactions of the Royal Society*, **140**, 379-90.

Mantell, G.A. (1850b) On a dorsal dermal spine of the *Hylaeosaurus*, recently discovered in the strata of Tilgate Forest. *Philosophical Transactions of the Royal Society*, **140**, 391-2.

Mantell, G.A. (1852) Description of the *Telerpeton elginense* and observations on supposed fossil ova of batrachians in the Lower Devonian strata of Forfarshire. *Quarterly Journal of the Geological Society of London*, **8**, 100-9.

Mantell, G.A. (1854) *Geological Excursions Round the Isle Of Wight and Along the Adjacent Coast of Dorsetshire; Illustrative of the Most Interesting Geological Phenomena and Organic Remains*, 3rd edn., S.H.G. Bohn, London, 356 pp.

Marshall, J.E.A. and Whiteside, D.I. (1980) Marine influences in the Triassic 'uplands'. *Nature, London*, **287**, 627-8.

Martill, D.M. (1985a) Plesiosaur discovery in Scotland. *Geology Today*, **1**, 162.

Martill, D.M. (1985b) The preservation of marine vertebrates in the Lower Oxford Clay (Jurassic) of central England. *Philosophical Transactions of the Royal Society, Series B*, **311**, 155-65.

Martill, D.M. (1986) The stratigraphic distribution of fossil vertebrates in the Oxford Clay of England. *Mercian Geologist*, **10**, 161-86.

Martill, D.M. (1988) A review of terrestrial vertebrate fossils of the Oxford Clay (Callovian-Oxfordian) of England. *Mercian Geologist*, **11**, 171-90.

Martill, D.M. (1990) New plesiosaur find in Oxford Clay. *Geology Today*, **6**, 6-7.

Martill, D.M. (1991) Organically preserved dinosaur skin: taphonomic and biological implications. *Modern Geology*, **16**, 61-8.

Martill, D.M. (1992) New marine reptile find in the Oxford Clay. *Earth science conservation*, No. **30**, 20-1.

Martill, D.M. and Dawn, A. (1986) Fossil vertebrates from new exposures of the Westbury Formation (Upper Triassic) at Newark, Nottinghamshire. *Mercian Geologist*, **10**, 127-33.

Martill, D.M. and Hudson, J.D. (eds) (1991) *Fossils of the Oxford Clay*. Palaeontological Association, London, 286 pp.

Martill, D.M., Taylor, M.A. and Duff, K.L. (1994) The trophic structure of the biota of the Peterborough Member, Oxford Clay Formation (Jurassic), UK. *Journal of the Geological Society of London*, **151**, 173-94.

Martin, J., Frey, E. and Riess, J. (1986) Soft tissue preservation in ichthyosaurs and a stratigraphic review of the Lower Hettangian of Barrow-upon-Soar, Leicestershire. *Transactions of Leicester Literary and Philosophical Society*, **80**, 58-72.

Martin, J.C. (1860) A ramble among the fossiliferous beds of Moray. [9 pp.; unpublished MS held by Elgin Museum.]

Maryanska, M. (1990) Pachycephalosauria. In *The Dinosauria* (eds D.B. Weishampel, P. Dodson and H. Osmólska), University of California Press, Berkeley, pp. 564-77.

Massare, J.A. and Callaway, J.M. (1990) The affinities and ecology of Triassic ichthyosaurs. *Geological Society of America Bulletin*, **102**, 406-16.

Maxwell, W.D. (1991) The pareiasaur *Elginia* from Elgin, north-east Scotland and the Late Permian extinction event. Unpublished PhD Thesis, Queen's University of Belfast.

McGowan, C. (1972) The systematics of Cretaceous ichthyosaurs, with particular reference to the material from North America. *Contributions in Geology, University of Wyoming*, **11**, 9-29.

McGowan, C. (1973a) The cranial morphology of the lower Liassic latipinnate ichthyosaurs of England. *Bulletin of the British Museum (Natural History), Geology*, **24**, 109 pp.

McGowan, C. (1973b) Differential growth in three ichthyosaurs: *Ichthyosaurus communis, I. breviceps* and *Stenopterygius quadriscissus* (Reptilia: Ichthyosauria). *Life Science Contributions of the Royal Ontario Museum*, **93**, 21 pp.

McGowan, C. (1974a) A revision of the longipinnate ichthyosaurs of the Lower Jurassic of England, with descriptions of two new species (Reptilia: Ichthyosauria). *Life Science Contributions of the Royal Ontario Museum*, **97**, 37 pp.

McGowan, C. (1974b) A revision of the latipinnate ichthyosaurs of the Lower Jurassic of England (Reptilia: Ichthyosauria). *Life Science Contributions of the Royal Ontario Museum*, **100**, 30 pp.

References

McGowan, C. (1976) The description and phenetic relationships of a new ichthyosaur genus from the Upper Jurassic of England. *Canadian Journal of Earth Sciences*, **76**, 668-83.

McGowan, C. (1978) Further evidence for the wide geographical distribution of ichthyosaur taxa (Reptilia: Ichthyosauria). *Journal of Paleontology*, **52**, 1155-62.

McGowan, C. (1979) A revision of the Lower Jurassic ichthyosaurs of Germany with descriptions of two new species. *Palaeontographica, Abteilung A*, **166**, 93-135.

McGowan, C. (1986) A putative ancestor for the swordfish-like ichthyosaur *Eurhinosaurus*. *Nature, London*, **322**, 454-6.

McGowan, C. (1989a) *Leptopterygius tenuirostris* and other long-snouted ichthyosaurs from the English Lower Lias. *Palaeontology*, **32**, 409-27.

McGowan, C. (1989b) Computed tomography reveals further details of *Excalibosaurus*, a putative ancestor for the swordfish-like ichthyosaur *Eurhinosaurus*. *Journal of Vertebrate Paleontology*, **9**, 269-81.

McGowan, C. (1989c) The ichthyosaurian tail-bend: A verification problem facilitated by computed tomography. *Palaeobiology*, **15**, 429-36.

McGowan, C. (1993) A new species of large, long-snouted ichthyosaur from the English Lower Lias. *Canadian Journal of Earth Sciences*, **30**, 1197-1204.

McIntosh, J.S. (1990) Sauropoda. In *The Dinosauria* (eds D.B. Weishampel, P. Dodson and H. Osmólska), University of California Press, Berkeley, pp. 345-401.

McKeever, P.J. (1990) Studies on the sedimentology and palaeoecology of the Permian of Scotland. Unpublished PhD Thesis, Queen's University of Belfast.

McKeever, P.J. (1991) Trackway preservation in aeolian sandstones from the Permian of Scotland. *Geology*, **19**, 726-9.

McKerrow, W.S. and Baden-Powell, D.F.W. (1953) Easter field meeting, 1952: The Jurassic rocks of Oxfordshire and their superficial deposits. *Proceedings of the Geologists' Association*, **64**, 88-98.

McKerrow, W.S. and Baker, S. (1988) Field meeting to Charlbury and Stonesfield, Oxfordshire. *Proceedings of the Geologists' Association*, **99**, 61-5.

McKerrow, W.S., Johnson, R.T. and Jakobson, M.E. (1969) Palaeoecological studies in the Great Oolite at Kirtlington, Oxfordshire. *Palaeontology*, **12**, 56-83.

Melmore, S. (1930) A description of the type-specimen of *Ichthyosaurus crassimanus* Blake. *Annals and Magazine of Natural History, Series 10*, **6**, 615-19.

Melmore, S. (1931) A reptilian egg from the Lias of Whitby. *Proceedings of the Yorkshire Philosophical Society*, 1930, 3-5.

Meszoely, C.A. and Ford, R.E. (1976) Eocene glass-lizard *Ophisaurus* (Anguidae) from the British Isles. *Copeia*, 1976, 407-8.

Metcalf, S.J., Vaughan, R.F., Benton, M.J., Cole, J., Simms, M.J. and Dartnall, D.L. (1992) A new Bathonian (Middle Jurassic) microvertebrate site, within the Chipping Norton Limestone Formation at Hornsleasow Quarry, Gloucestershire. *Proceedings of the Geologists' Association*, **103**, 321-42.

Metcalfe, A.T. (1884) On further discoveries of vertebrate remains in the Triassic strata of the south coast of Devonshire between Budleigh Salterton and Sidmouth. *Quarterly Journal of the Geological Society of London*, **40**, 257-62.

Meyer, H. von (1857) Beiträge zur näheren Kenntniss fossiler Reptilien. *Neues Jahrbuch für Mineralogie, Geologie, und Paläontologie*, 1857, 532-43.

Meylan, P.A. (1987) The phylogenetic relationships of soft-shelled turtles (Family Trionychidae). *Bulletin of the American Museum of Natural History*, **186**, 101 pp.

Miall, L.C. (1874) On the remains of Labyrinthodonta from the Keuper Sandstone of Warwick, preserved in the Warwick Museum. *Quarterly Journal of the Geological Society of London*, **30**, 417-35.

Miller, H. (1858) *The Cruise of The Betsey*. Nimmo, Edinburgh, 486 pp.

Mills, D.A.C. and Hull, J.H. (1976) *Geology of the Country around Barnard Castle* (Sheet 32), Memoirs of the Geological Survey of the United Kingdom, HMSO, London, 385 pp.

Milner, A.C., Milner, A.R. and Estes, R. (1982) Amphibians and squamates from the Upper Eocene of Hordle Cliff, Hampshire - a preliminary report. *Tertiary Research*, **4**, 149-54.

Milner, A.R. (1987) The Westphalian tetrapod fauna; some aspects of its geography and ecology. *Journal of the Geological Society, London*, **144**, 495-506.

Milner, A.R., Gardiner, B.G., Fraser, N.C. and Taylor, M.A. (1990) Vertebrates from the Middle Triassic Otter Sandstone Formation of

Devon. *Palaeontology*, **33**, 873-92.

Mlynarski, M. (1976) Testudines. *Handbuch der Paläoherpetologie*, **7**, 130 pp.

Molnar, R.E. (1990) Problematic Theropoda: 'carnosaurs'. In *The Dinosauria* (eds D.B. Weishampel, P. Dodson and H. Osmólska), University of California Press, Berkeley, pp. 169-209.

Molnar, R.E., Kurzanov, S.M. and Dong Z. (1990) Carnosauria. In *The Dinosauria* (eds D.B. Weishampel, P. Dodson and H. Osmólska), University of California Press, Berkeley, pp. 306-17.

Moody, R.T.J. (1968) A turtle, *Eochelys crassicosta* (Owen), from the London Clay of the Isle of Sheppey. *Proceedings of the Geologists' Association*, **79**, 129-40.

Moody, R.T.J. (1974) The taxonomy and morphology of *Puppigerus camperi* (Gray), an Eocene sea-turtle from northern Europe. *Bulletin of the British Museum (Natural History), Geology*, **25**, 153-86.

Moody, R.T.J. (1980a) The distribution of turtles in the British Palaeogene. *Tertiary Research*, **3**, 21-4.

Moody, R.T.J. (1980b) Notes on some European Palaeogene turtles. *Tertiary Research*, **2**, 161-8.

Moody, R.T.J. and Walker, C.A. (1970) A new trionychid turtle from the British Lower Eocene. *Palaeontology*, **13**, 503-10.

Mook, C.C. (1955) Two new genera of Eocene crocodilians. *American Museum Novitates*, **1727**, 1-4.

Moore, C. (1881) On abnormal geological deposits in the Bristol district. *Quarterly Journal of the Geological Society of London*, **37**, 67-82.

Morales, M. (1987) Terrestrial fauna and flora from the Triassic Moenkopi Formation of the southwestern United States. *Journal of the Arizona-Nevada Academy of Sciences*, **22**, 1-19.

Morris, J. (1843) *A Catalogue of British Fossils*. London, 222 pp.

Morris, J. (1856) General sketch of the geology of Hartwell. *London University Magazine*, 1856, p. 102.

Murchison, R.I. (1829) Geological sketch of the north-western extremity of Sussex and the adjoining parts of Hants and Surrey. *Transactions of the Geological Society of London, Series 2*, **2**, 97-107.

Murchison, R.I. (1839) *The Silurian System*. John Murray, London, 768 pp.

Murchison, R.I. (1859) On the sandstones of Morayshire (Elgin and c.) containing reptilian remains; and on their relations to the Old Red Sandstone of that country. *Quarterly Journal of the Geological Society of London*, **15**, 419-39.

Murchison, R.I. (1867) *Siluria*, 4th edn., John Murray, London.

Murchison, R.I. and Strickland, H.E. (1840) On the upper formations of the New Red Sandstone System in Gloucestershire, Worcestershire and Warwickshire; etc. *Transactions of the Geological Society of London*, **(2) 5**, 331-48.

Murry, P.A. and Long, R.A. (1989) Geology and paleontology of the Chinle Formation, Petrified Forest National Park and vicinity, Arizona and a discussion of vertebrate fossils of the southwestern Upper Triassic. In *Dawn of the Age of Dinosaurs in the American Southwest* (eds S.G. Lucas and A.P. Hunt), New Mexico Museum of Natural History, Albuquerque, pp. 29-64.

Newman, B.H. (1968) The Jurassic dinosaur *Scelidosaurus harrisoni*, Owen. *Palaeontology*, **11**, 40-3.

Newman, B.H. (1990) A dinosaur trackway from the Purbeck beds of Swanage, England. *Palaeontologia africana*, **27**, 97-100.

Newton, A. (1862) On the discovery of ancient remains of *Emys lutaria* in Norfolk. *Annals and Magazine of Natural History, Series 3*, **4**, 224-8.

Newton, E.T. (1878) Notes on a crocodilian jaw from the Corallian rocks of Weymouth. *Quarterly Journal of the Geological Society of London*, **34**, 398-400.

Newton, E.T. (1879) Note on some fossil remains of *Emys lutaria* from the Norfolk coast. *Geological Magazine*, Decade 2, **6**, 304-6.

Newton, E.T. (1882a) *The Vertebrata of the Forest Bed Series of Norfolk and Suffolk*. Memoirs of the Geological Survey of the United Kingdom, 1882, 143 pp.

Newton, E.T. (1882b) Notes on the Vertebrata of the pre-glacial Forest Bed Series of the east of England. *Geological Magazine*, Decade 2, **9**, 7-9.

Newton, E.T. (1888) On the skull, brain and auditory organ of a new species of pterosaurian (*Scaphognathus purdoni*) from the Upper Lias near Whitby, Yorkshire. *Philosophical Transactions of the Royal Society of London, Series B*, **179**, 503-37.

Newton, E.T. (1893) On some new reptiles from

References

the Elgin sandstones. *Philosophical Transactions of the Royal Society of London, Series B*, **184**, 431-503.

Newton, E.T. (1894a) The vertebrate fauna collected by Mr Lewis Abbott from the fissures near Ightham, Kent. *Quarterly Journal of the Geological Society of London*, **50**, 188-210.

Newton, E.T. (1894b) Reptiles from the Elgin sandstone. Description of two new genera. *Philosophical Transactions of the Royal Society of London, Series B*, **185**, 573-607.

Newton, E.T. (1899) On a megalosauroid jaw from Rhaetic beds near Bridgend. *Quarterly Journal of the Geological Society of London*, **55**, 89-96.

Nopcsa, B.F. (1905a) Notes on British dinosaurs. Part I. *Hypsilophodon. Geological Magazine*, Decade 5, **2**, 203-8.

Nopcsa, F. (1905b) Notes on British dinosaurs. Part II. *Polacanthus. Geological Magazine*, Decade 5, **2**, 241-50.

Nopcsa, F. (1911) Notes on British dinosaurs. Part IV. *Stegosaurus priscus. Geological Magazine*, Decade 5, **8**, 109-15, 145-53.

Nopcsa, F. (1912) Notes on British dinosaurs. Part V. *Craterosaurus* (Seeley). *Geological Magazine*, Decade 5, **9**, 481-4.

Nopcsa, F. (1928) Palaeontological notes on reptiles. *Geologica Hungarica, Series Palaeontologia*, **1** (1), 1-84.

Norell, M.A. and Clark, J.M. (1990) A reanalysis of *Bernissartia fagesii*, with comments on its phylogenetic position and its bearing on the origin and diagnosis of the Eusuchia. *Bulletin de l'Institut Royal des Sciences Naturelles de Belgique, Sciences de la Terre*, **60**, 115-28.

Norman, D.B. (1980) On the ornithischian dinosaur *Iguanodon bernissartensis* from the Lower Cretaceous of Bernissart (Belgium). *Mémoire de l'Institut Royal des Sciences Naturelles de Belgique*, **178**, 103 pp.

Norman, D.B. (1984) On the cranial morphology and evolution of ornithopod dinosaurs. In *The Structure, Development and Evolution of Reptiles* (ed. M.W.J. Ferguson), *Zoological Society of London Symposia*, **52**, Academic Press, London, 521-47.

Norman, D.B. (1985) *The Illustrated Encyclopedia of Dinosaurs*. Salamander, London, 208 pp.

Norman, D.B. (1986) On the anatomy of *Iguanodon atherfieldensis* (Ornithischia: Ornithopoda). *Bulletin de l'Institut Royal des Sciences Naturelles de Belgique*, **56**, 281-372.

Norman, D.B. (1987) On the history of the discovery of fossils at Bernissart in Belgium. *Archives of Natural History*, **14**, 59-75.

Norman, D.B. (1990a) Problematic Theropoda: 'coelurosaurs'. In *The Dinosauria* (eds D.B. Weishampel, P. Dodson and H. Osmólska), University of California Press, Berkeley, pp. 280-305.

Norman, D.B. (1990b) A review of *Vectisaurus valdensis*, with comments on the family Iguanodontidae. In *Dinosaur Systematics; Approaches and Perspectives* (eds K. Carpenter and P.J. Currie), Cambridge University Press, Cambridge, pp. 147-61.

Norman, D.B. and Weishampel, D.B. (1990) Iguanodontidae and related ornithopods. In *The Dinosauria* (eds D.B. Weishampel, P. Dodson and H. Osmólska), University of California Press, Berkeley, pp. 510-33.

Norman, D.B., Hilpert, K.-H. and Hölder, H. (1987) Die Wirbeltierfauna von Nehden (Sauerland), Westdeutschland. *Geologie und Paläontologie Westphalens*, **8**, 1-77.

Nunn, J.F. (1990) A new tridactyl footprint impression in Durlston Bay, Swanage. *Proceedings of the Dorset Natural History and Archaeological Society*, **111**, 133-4.

Nunn, J.F. (1992) A geological map of Purbeck beds in the northern part of Durlston Bay. *Proceedings of the Dorset Natural History and Archaeological Society*, **113**, 145-8.

Odin, G.S., Curry, D. and Hunziker, J.C. (1978) Radiometric dates from the NW European glauconites and the Palaeogene time scale. *Journal of the Geological Society of London*, **135**, 481-97.

Odling, M. (1913) The Bathonian rocks of the Oxford District. *Quarterly Journal of the Geological Society of London*, **69**, 484-513.

Old, R.A., Sumbler, M.G. and Ambrose, K. (1987) *Geology of the Country around Warwick* (Sheet 184), Memoirs of the British Geological Survey, HMSO, London, 93 pp.

Olsen, P.E. and Baird, D. (1986) The ichnogenus *Atreipus* and its significance for Triassic biostratigraphy. In *The Beginning of the Age of Dinosaurs; Faunal Change Across the Triassic-Jurassic boundary* (ed. K. Padian), Cambridge University Press, pp. 61-87.

Olsen, P.E. and Galton, P.M. (1977) Triassic-Jurassic tetrapod extinctions: are they real? *Science*, **197**, 983-6.

Ostrom, J.H. (1970) Stratigraphy and paleontology of the Cloverly Formation (lower Cretaceous) of the Bighorn Basin area, Wyoming and Montana. *Bulletin of the Peabody Museum of Natural History*, **35**, 234 pp.

Owen, H.G. (1971) Middle Albian stratigraphy in the Anglo-Paris Basin. *Bulletin of the British Museum (Natural History), Geology Series*, Supplement, **8**, 164 pp.

Owen, H.G. (1976) The stratigraphy of the Gault and Upper Greensand of the Weald. *Proceedings of the Geologists' Association*, **86**, 475-98.

Owen, R. (1841a) Report on British fossil reptiles. Part I. *Report of the British Association for the Advancement of Science*, 1839, 43-126.

Owen, R. (1841b) *Odontography; or a Treatise on the Comparative Anatomy of the Teeth*. Hippolyte Baillière, London, 655 pp.

Owen, R. (1841c) Description of some remains of a gigantic crocodilian saurian, probably marine, from the Lower Greensand at Hythe; and of the teeth from the same formation at Maidstone, referable to the genus *Polyptychodon*. *Proceedings of the Geological Society of London*, **3**, 449-52.

Owen, R. (1841d) Description of the remains of six species of marine turtles (*Chelones*) from the London Clay of Sheppey and Harwich. *Proceedings of the Geological Society*, **3**, 570-8.

Owen, R. (1841e) Description of some ophidiolites (*Palaeophis toliapicus*) from the London Clay at Sheppey, indicative of an extinct species of serpent. *Transactions of the Geological Society of London, Series 2*, **6**, 209-10.

Owen, R. (1842a) Description of parts of the skeleton and teeth of five species of the genus *Labyrinthodon* (*Lab. leptognathus, Lab. pachygnathus* and *Lab. ventricosus*, from the Coton-end and Cubbington Quarries of the Lower Warwick Sandstone; *Lab. jaegeri*, from Guy's Cliff, Warwick; and *Lab. scutulatus*, from Leamington); with remarks on the probable identity of the *Cheirotherium* with this genus of extinct batrachians. *Transactions of the Geological Society of London, Series 2*, **6**, 515-43.

Owen, R. (1842b) Report on British fossil reptiles. Part II. *Report of the British Association for the Advancement of Science*, 1841, 60-204.

Owen, R. (1842c) Description of an extinct lacertian reptile, *Rhynchosaurus articeps,* Owen, of which the bones and foot-prints characterize the upper New Red Sandstone at Grinsill, near Shrewsbury. *Transactions of the Cambridge Philosophical Society*, (**2**) **7**, 355-69.

Owen, R. (1842d) On the teeth of a species of *Labyrinthodon (Mastodonsaurus)* of Jaeger, common to the German Keuper Formation and the Lower Sandstone of Warwick and Leamington. *Transactions of the Geological Society of London, Series 2*, **6**, 503-13.

Owen, R. (1842e) Description of the remains of a bird, tortoise and lizard from the Chalk. *Transactions of the Geological Society of London, Series 2*, **6**, 411-13.

Owen, R. (1846) *A History of British Fossil Mammals and Birds*. London, 560 pp.

Owen, R. (1848) On the fossils obtained by the Marchioness of Hastings from the freshwater Eocene beds of the Hordle Cliffs. *Report of the British Association for the Advancement of Science*, 1847, 65-6.

Owen, R. (1850a) Monograph on the fossil Reptilia of the London Clay, and of the Bracklesham and other Tertiary beds. Part I. Chelonia (*Platemys*). *Palaeontographical Society (Monographs)*, **3**, 1-4.

Owen, R. (1850b) Monograph on the fossil Reptilia of the London Clay, and of the Bracklesham and other Tertiary beds. Part II. Crocodilia (*Crocodilus*, etc.). *Palaeontographical Society (Monographs)*, **3**, 5-50.

Owen, R. (1850c) Monograph on the fossil Reptilia of the London Clay, and of the Bracklesham and other Tertiary beds. Part III. Ophidia (*Palaeophis*, etc.). *Palaeontographical Society (Monographs)*, **3**, 51-68.

Owen, R. (1851a) Vertebrate air-breathing life in the Old Red Sandstone. *Literary Gazette*, 1851, p. 900.

Owen, R. (1851b) Monograph of the fossil Reptilia of the Cretaceous formations. *Palaeontographical Society (Monographs)*, **5**, 1-118.

Owen, R. (1852a) On a new species of pterodactyle (*Pterodactylus compressirostris,* Owen) from the Chalk, with some remarks on the nomenclature of the previously described species. *Proceedings of the Zoological Society of London*, **19**, 21-34.

Owen, R. (1853) Monograph on the fossil Reptilia of the Wealden and Purbeck formations. Part I. Chelonia (*Pleurosternon* etc.) (Purbeck). *Palaeontographical Society (Monographs)*, **7**, 1-12.

References

Owen, R. (1854) On some fossil reptilian and mammalian remains from the Purbecks. *Quarterly Journal of the Geological Society of London*, **10**, 420-33.

Owen, R. (1855a) Notice of some new reptilian fossils from the Purbeck beds near Swanage. *Quarterly Journal of the Geological Society of London*, **11**, 123-4.

Owen, R. (1855b) Monograph on the fossil Reptilia of the Wealden and Purbeck formations. Part II. Dinosauria (*Iguanodon*) (Wealden). *Palaeontographical Society (Monographs)*, **8**, 1-54.

Owen, R. (1857) On the affinities of *Stereognathus ooliticus* (Charlesworth), a mammal from the Oolitic slate of Stonesfield. *Quarterly Journal of the Geological Society of London*, **13**, 1-11.

Owen, R. (1858) Note on the bones of the hindfoot of the *Iguanodon*, discovered and exhibited by S.H. Beckles. *Quarterly Journal of the Geological Society of London*, **14**, 174-5.

Owen, R. (1859a) Note on the affinities of *Rhynchosaurus*. *Annals and Magazine of Natural History, Series 3*, **4**, 237-8.

Owen, R. (1859b) Monograph on the fossil Reptilia of the Wealden and Purbeck formations. Supplement no. II. Crocodilia (*Streptospondylus*, etc.) (Wealden). *Palaeontographical Society (Monographs)*, **11**, 20-44.

Owen, R. (1859c) On remains of new and gigantic species of pterodactyle (*Pter. fittoni and Pter. sedgwickii*) from the Upper Greensand near Cambridge. *Report of the British Association for the Advancement of Science*, 1858, 98-103.

Owen, R. (1860) Note on some remains of *Polyptychodon* from Dorking. *Quarterly Journal of the Geological Society of London*, **16**, 262-3.

Owen, R. (1861a) A monograph of the fossil Reptilia of the Liassic formations. Part I. A monograph of a fossil dinosaur (*Scelidosaurus harrisonii*, Owen) of the Lower Lias. *Palaeontographical Society (Monographs)*, **13**, 14 pp.

Owen, R. (1861b) Monograph on the fossil Reptilia of the Cretaceous formations. Supplement no. III. Pterosauria (*Pterodactylus*) and Sauropterygia (*Polyptychodon*). *Palaeontographical Society (Monographs)*, **12**, 25 pp.

Owen, R. (1861c) Monograph on the fossil Reptilia of the Wealden and Purbeck formations. Part V. Lacertilia (*Nuthetes*, etc.) (Purbeck). *Palaeontographical Society (Monographs)*, **12**, 31-9.

Owen, R. (1863a) Notice of a skull and parts of the skeleton of *Rhynchosaurus articeps*. *Philosophical Transactions of the Royal Society of London*, **152**, 466-7.

Owen, R. (1863b) A monograph of the fossil Reptilia of the Liassic formations. Part II. *Scelidosaurus harrisonii* continued. *Palaeontographical Society (Monographs)*, **15**, 1-26.

Owen, R. (1864) Monograph on the fossil Reptilia of the Wealden and Purbeck formations. Supplement no. III. Dinosauria (*Iguanodon*) (Wealden). *Palaeontographical Society, Monographs*, **16**, 19-21.

Owen, R. (1865) A monograph of the fossil Reptilia of the Liassic formations. Part III. *Plesiosaurus, Dimorphodon* and *Ichthyosaurus*. Part I. Sauropterygia. *Palaeontographical Society (Monographs)*, **17**, 1-40.

Owen, R. (1869) A monograph on the Reptilia of the Kimmeridge Clay and Portland Stone. No. III. Containing *Pliosaurus grandis, P. trochanterius* and *P. portlandicus*. *Palaeontographical Society (Monographs)*, **22**, 1-12.

Owen, R. (1870) A monograph of the fossil Reptilia of the Liassic formations. Part III. *Plesiosaurus, Dimorphodon* and *Ichthyosaurus*. Part II. Order Pterosauria. *Palaeontographical Society (Monographs)*, **23**, 41-81.

Owen, R. (1871) Monograph of the fossil Mammalia of the Mesozoic formations. *Palaeontographical Society (Monographs)*, **24**, 1-115.

Owen, R. (1874a) Monograph of the fossil Reptilia of the Mesozoic formations. Part I. *Pterosauria*. *Palaeontographical Society (Monographs)*, **27**, 1-14.

Owen, R. (1874b) Monograph on the fossil Reptilia of the Wealden and Purbeck formations. Supplement no. V. *Iguanodon* (Wealden). *Palaeontographical Society (Monographs)*, **27**, 1-18.

Owen, R. (1874c) Monograph on the fossil Reptilia of the Wealden and Purbeck formations. Supplement no. VI. *Hylaeochampsa* (Wealden). *Palaeontographical Society (Monographs)*, **27**, 1-7.

References

Owen, R. (1876) Monograph on the fossil Reptilia of the Wealden and Purbeck formations. Supplement no. VII. Crocodilia (*Poikilopleuron*) and Dinosauria? (*Chondrosteosaurus*) (Wealden). *Palaeontographical Society (Monographs)*, **29**, 1-7.

Owen, R. (1878a) Monograph on the fossil Reptilia of the Wealden and Purbeck formations. Supplement no. VIII. Crocodilia (*Goniopholis, Petrosuchus* and *Suchosaurus*). *Palaeontographical Society (Monographs)*, **32**, 1-15.

Owen, R. (1878b) On the fossils called 'granicones'; being a contribution to the histology of the exo-skeleton in 'Reptilia'. *Journal of the Royal Microscopical Society*, **1**, 233-6.

Owen, R. (1879a) Monograph on the fossil Reptilia of the Wealden and Purbeck formations. Supplement no. IX. Crocodilia (*Goniopholis, Brachydectes, Nannosuchus, Theriosuchus* and *Nuthetes*). *Palaeontographical Society (Monographs)*, **33**, 1-19.

Owen, R. (1879b) On the association of dwarf crocodiles (*Nannosuchus* and *Theriosuchus pusillus*, e.g.) with diminutive mammals of the Purbeck shales. *Quarterly Journal of the Geological Society of London*, **35**, 148-55.

Owen, R. (1881) [Description of *Plastremys lata*.] In Parkinson, C. (1881), 370-1.

Owen, R. (1884a) On the cranial and vertebral characters of the crocodilian genus *Plesiosuchus*, Owen. *Quarterly Journal of the Geological Society of London*, **40**, 153-9.

Owen, R. (1884b) *A History of British Fossil Reptiles*. Cassell and Co., London (1849-1884).

Owen, R. and Bell, T. (1849) Monograph on the fossil Reptilia of the London Clay. Part I. Chelonia. *Palaeontographical Society (Monographs)*, **2**, 1-76.

Padgham, R.C. (1972) Field meeting to the Folkestone Beds (Lower Greensand) of West Surrey. *Proceedings of the Geologists' Association*, **83**, 355-9.

Padian, K. (1983) Osteology and functional morphology of *Dimorphodon macronyx* (Buckland) (Pterosauria: Rhamphorhynchoidea) based on new material in the Yale Peabody Museum. *Postilla*, **189**, 1-44.

Padian, K. (1984) A functional analysis of walking and flying in pterosaurs. *Paleobiology*, **9**, 218-39.

Palmer, C.P. (1972) A revision of the zonal classification of the Lower Lias of the Dorset coast in south west England. *Newsletters in Stratigraphy*, **2**, 45-54.

Palmer, C.P. (1988) The Kimmeridgian fauna associated with the Portland plesiosaur. *Proceedings of the Dorset Natural History and Archaeological Society*, **109**, 109-12.

Palmer, T.J. (1973) Field Meeting in the Great Oolite of Oxfordshire. *Proceedings of the Geologists' Association*, **84**, 53-64.

Palmer, T.J. (1979) The Hampen Marly and White Limestone formations: Florida-type carbonate lagoons in the Jurassic of Central England. *Palaeontology*, **22**, 189-228.

Parish, W. (1833) [Collection of fossils made by Mr Parish during the last summer at St Leonards.] *Proceedings of the Geological Society of London*, **2**, 3-4.

Parkinson, C. (1881) Upper Greensand and chloritic marl, Isle of Wight. *Quarterly Journal of the Geological Society of London*, **37**, 370-5.

Parkinson, J. (1811) Observations on some of the strata in the neighbourhood of London and on the fossil remains contained in them. *Transactions of the Geological Society of London*, **1**, 324-54.

Parsons, T.S. and Williams, E.E. (1961) Two Jurassic turtle skulls: a morphological study. *Bulletin of the Museum of Comparative Zoology, Harvard*, **125**, 43-107.

Paton, R.L. (1974a) Capitosauroid labyrinthodonts from the Trias of England. *Palaeontology*, **17**, 253-89.

Paton, R.L. (1974b) Lower Permian pelycosaurs from the English Midlands. *Palaeontology*, **17**, 541-52.

Patterson, C. (1966) British Wealden sharks. *Bulletin of the British Museum (Natural History), Geology Series*, **11**, 281-350.

Pattison, J., Smith, D.B. and Warrington, G. (1973) A review of late Permian biostratigraphy in the British Isles. In *The Permian and Triassic Systems and their Mutual Boundary* (eds A.V. Logan and L.V. Mills), *Memoirs of the Canadian Society of Petroleum Geologists*, **2**, 220-60.

Peacock, J.D. (1966) Contorted beds in the Permo-Triassic aeolian sandstones of Morayshire. *Bulletin of the Geological Survey of Great Britain*, **24**, 157-62.

Peacock, J.D., Berridge, N.G., Harris, A.L. and May, F. (1968) *The Geology of the Elgin District* (Sheet 95), Memoirs of the Geological Survey of Great Britain, HMSO, Edinburgh.

Peake, N.B. and Hancock, J.M. (1978) The Upper Cretaceous of Norfolk. In *The Geology of Norfolk* (eds G.P. Larwood and B.M. Funnell), Paramoudra Club, Norwich, pp. 293-339.

Pereda-Suberbiola, J. (1991) Nouvelle évidence d'une connexion terrestre entre Europe et Amérique du Nord au Crétacé inférieur: *Hoplitosaurus*, synonyme de *Polacanthus* (Ornithischia: Ankylosauria). *Comptes Rendus de l'Académie des Sciences, Paris, Série II*, **313**, 971-6.

Pereda-Suberbiola, J. (1993) *Hylaeosaurus, Polacanthus*, and the systematics and stratigraphy of Wealden armoured dinosaurs. *Geological Magazine*, **130**, 767-81.

Persson, P.O. (1963) A revision of the classification of the Plesiosauria with a synopsis of the stratigraphical and geographical distribution of the group. *Lunds Universitets Årsskrift. Ny Följd. Andra Avdelningen, 2*, **59**, 1-60.

Pettigrew, T.H. (1979) A gliding reptile from the Upper Permian of north east England. *Nature*, **281**, 297-8.

Phillips, J. (1853) Report of the council of the Yorkshire Philosophical Society, etc. *Annual Report of the Yorkshire Philosophical Society* 1852, **7-8**, 19-20.

Phillips, J. (1854) On a new *Plesiosaurus* in the York Museum. *Report of the British Association for the Advancement of Science*, 1853, p. 54.

Phillips, J. (1860) Notice of some sections of the strata near Oxford. No. I. The Great Oolite in the valley of the Cherwell. *Quarterly Journal of the Geological Society of London*, **16**, 115-19.

Phillips, J. (1871) *Geology of Oxford and the Valley of the Thames*. Clarendon Press, Oxford, 523 pp.

Phillips, J. G. (1886) The Elgin sandstones. *British Association for the Advancement of Science, Report and Transactions*, 1885, 1023-4.

Phizackerley, P.H. (1951) A revision of the Teleosauridae in the Oxford University Museum (Natural History). *Annals and Magazine of Natural History, Series 12*, **4**, 1169-92.

Platt, J. (1758) An account of the fossil thigh-bone of a large animal, dug up at Stonesfield, near Woodstock, in Oxfordshire. *Philosophical Transactions of the Royal Society*, **50**, 524-7.

Plint, A.G. (1984) A regressive coastal sequence from the Upper Eocene of Hampshire, southern England. *Sedimentology*, **31**, 213-25.

Plot, R. (1677) *The Natural History of Oxfordshire, being an Essay toward the Natural History of England*, Theatre, Oxford, 358 pp.

Pocock, R. and Wray, D.A. (1925) *The Geology of the Country around Wem* (Sheet 138), Memoirs of the Geological Survey of England and Wales, HMSO, London, 125 pp.

Pollard, J.E. (1968) The gastric contents of an ichthyosaur from the Lower Lias of Lyme Regis. *Palaeontology*, **11**, 376-88.

Pomel, A. (1853) Catalogue méthodique et descriptif des vertébrés fossiles découverts dans les bassins de la Loire et de l'Allier. *Annales de Science, Littérature et Industrie d'Auvergne*, **25**, 337-80; **26**, 81-229.

Powell, H.P. (1988) A megalosaurid dinosaur jawbone from the Kimmeridge Clay of the seabed of West Bay, Dorset. *Proceedings of the Dorset Natural History and Archaeological Society*, **109**, 105-7.

Powell, J.H. (1984) Lithostratigraphical nomenclature of the Lias Group in the Yorkshire Basin. *Proceedings of the Yorkshire Geological Society*, **45**, 51-7.

Prestwich, J. (1846) On the Tertiary or supracretaceous formations of the Isle of Wight as exhibited in the sections at Alum Bay and White Cliff Bay. *Quarterly Journal of the Geological Society of London*, **2**, 223-59.

Prestwich, J. (1879) On the discovery of a species of *Iguanodon* in the Kimmeridge Clay near Oxford; and a notice of a very fossiliferous band of the Shotover Sands. *Geological Magazine*, Decade 2, **6**, 193-5.

Prestwich, J. (1880) Notes on the occurrence of a new species of *Iguanodon* in a brick-pit of the Kimmeridge Clay at Cumnor Hurst, three miles WSW of Oxford. *Quarterly Journal of the Geological Society of London*, **36**, 430-2.

Price, F.G.H. (1875) On Gault of Folkestone. *Quarterly Journal of the Geological Society of London*, **30**, 342-66.

Pringle, J. (1923) On the concealed Mesozoic rocks in south-west Norfolk. *Summary of Progress of the Geological Survey, London, for 1922*, HMSO, London, 126-39.

Pringle, J. (1926) *The Geology of the Country around Oxford* (Special Oxford Sheet), Memoirs of the Geological Survey of the United Kingdom, 191 pp.

References

Prothero, D.R. and Estes, R. (1980) Late Jurassic lizards from Como Bluff, Wyoming and their palaeobiogeographic significance. *Nature, London*, **286**, 484-6.

Purvis, K. and Wright, V.P. (1991) Calcretes related to phreatophytic vegetation from the Middle Triassic Otter Sandstone of south-west England. *Sedimentology*, **38**, 539-51.

Pyrah, B.J. (1979) Catalogue of type and figured fossils in the Yorkshire Museum: Part 4. *Proceedings of the Yorkshire Geological Society*, **42**, 415-37.

Radley, J.D. (1991) Palaeoecology and deposition of Portlandian (Upper Jurassic) strata at the Bugle Pit, Hartwell, Buckinghamshire. *Proceedings of the Geologists' Association*, **102**, 241-9.

Radley, J. (1993) Excavation of a sauropod dinosaur on the Isle of Wight. *Geology Today*, **9**, 167-8.

Radley, J. and Hutt, S. (1993) The Isle of Wight sauropod. *Earth science conservation*, No. **33**, 10-2.

Rage, J.-C. (1984) Serpentes. *Handbuch der Paläoherpetologie*, **11**, 80 pp.

Rage, J.-C. and Ford, R.L.E. (1980) Amphibians and squamates from the Upper Eocene of the Isle of Wight. *Tertiary Research*, **3**, 47-60.

Rawson, P.F., Curry, D., Dilley, F.C., Hancock, J.M., Kennedy, W.J., Neale, J.W., Wood, C.J. and Worssam, B.C. (1978) A correlation of Cretaceous rocks of the British Isles. *Geological Society of London, Special Report*, **9**, 70 pp.

Reid, C. (1899) *Geology of the Country around Dorchester* (Sheet 328), Memoirs of the Geological Survey of the United Kingdom, HMSO, London, 52 pp.

Reid, C. and Strahan, A. (1889) *The Geology of the Isle of Wight*. Memoirs of the Geological Survey of the United Kingdom, HMSO, London, 349 pp.

Reid, R.E.H. (1987) Claws' claws. *Nature, London*, **325**, 487.

Reisz, R. R. (1986) Pelycosauria. *Handbuch der Paläoherpetologie*, **17A**, 102 pp.

Reynolds, S.H. (1939) On a collection of reptilian bones from the Oolite of Stow-on-the-Wold, Gloucestershire. *Geological Magazine*, **76**, 193-214.

Reynolds, S.H. (1946) The Aust section. *Proceedings of the Cotteswold Naturalists' Field Club*, **29**, 29-39.

Richardson, L. (1909) The Rhaetic section at Wigston. *Geological Magazine*, Decade 5, **6**, 366-70.

Richardson, L. (1911a) On the sections of Forest Marble and Great Oolite on the railway between Cirencester and Chedworth, Gloucestershire. *Proceedings of the Geologists' Association*, **22**, 95-115.

Richardson, L. (1911b) On the Rhaetic and contiguous deposits of west, mid and east Somerset. *Quarterly Journal of the Geological Society of London*, **67**, 1-72.

Richardson, L. (1929) *The Country around Moreton-in-Marsh* (Sheet 217), Memoirs of the Geological Survey of England and Wales, HMSO, London, 162 pp.

Richardson, L., Arkell, W.J. and Dines, H.G. (1946) *Geology of the Country around Witney* (Sheet 236), Memoirs of the Geological Survey of the United Kingdom, HMSO, London, 150 pp.

Riley, H. and Stutchbury, S. (1840) A description of various fossil remains of three distinct saurian animals, recently discovered in the Magnesian Conglomerate near Bristol. *Transactions of the Geological Society of London*, **5**, 349-57.

Rivett, W.H.E. (1953) Saurian remains from the Weald Clay at Ockley, Surrey. *South Eastern Naturalist*, **58**, 36-7.

Rivett, W.H.E. (1956) Reptilian bones from the Weald Clay. *Proceedings of the Geological Society of London*, **1540**, 110-11.

Rixon, A.E. (1968) The development of the remains of a small *Scelidosaurus* from a Lias nodule. *Museums Journal*, **67**, 315-27.

Robinson, P.L. (1957a) The Mesozoic fissures of the Bristol Channel area and their vertebrate faunas. *Zoological Journal of the Linnean Society*, **43**, 260-82.

Robinson, P.L. (1957b) An unusual sauropsid dentition. *Zoological Journal of the Linnean Society*, **43**, 282-93.

Robinson, P.L. (1962) Gliding lizards from the Upper Keuper of Great Britain. *Proceedings of the Geological Society of London*, **1601**, 137-46.

Robinson, P.L. (1967) The evolution of the Lacertilia. *Colloques Internationaux du Centre National des Recherches Scientifiques*, **163**, 394-407.

Robinson, P.L. (1971) A problem of faunal replacement on Permo-Triassic continents. *Palaeontology*, **14**, 131-53.

Robinson, P.L. (1973) A problematic reptile from

the British Upper Trias. *Journal of the Geological Society of London*, **129**, 457-79.

Robinson, P.L., Kermack, K.A. and Joysey, K.A. (1952) Exhibition of a new Upper Triassic land fauna from Slickstones Quarry, Gloucestershire. *Proceedings of the Geological Society of London*, **1485**, 86-7.

Romer, A.S. (1956) *Osteology of the Reptiles*. University of Chicago Press, Chicago, 772 pp.

Romer, A.S. (1966) *Vertebrate Paleontology*, 3rd edn., University of Chicago Press, Chicago, 468 pp.

Rowe, T. (1980) The morphology, affinities and age of the dicynodont reptile *Geikia elginensis*. In *Aspects of Vertebrate History* (ed. L. Jacobs), Museum of Northern Arizona Press, Flagstaff, 269-94.

Russell, D.A. (1967) Systematics and morphology of American mosasaurs (Reptilia, Sauria). *Bulletin of the Peabody Museum of Natural History*, **23**, 237 pp.

Salfeld, H. (1914) Die Gliederung des oberen Jura in Nordwesteuropa. *Neues Jahrbuch für Geologie und Paläontologie, Abhandlungen*, **37**, 125-246.

Sanders, W. (1876) On certain large bones in Rhaetic bone beds at Aust Cliff, near Bristol. *Report of the British Association for the Advancement of Science*, 1875, 80-1.

Sarjeant, W.A.S. (1974) A history and bibliography of the study of fossil vertebrate footprints in the British Isles. *Palaeogeography, Palaeoclimatology, Palaeoecology*, **16**, 265-378.

Savage, R.J.G. (1958) Pliosaur from Portland. *Proceedings of the Bristol Naturalists' Society*, **29**, 379-80.

Savage, R.J.G. (1963) The Witts collection of Stonesfield Slate fossils. *Proceedings of the Cotteswold Naturalists' Field Club*, **33**, 177-82.

Savage, R.J.G. (1971) Tritylodontid *incertae sedis*. *Proceedings of the Bristol Naturalists' Society*, **32**, 80-3.

Savage, R.J.G. (1977) The Mesozoic strata of the Mendip Hills. In *Geological Excursions in the Bristol District* (ed. R.J.G. Savage), University of Bristol, Bristol, pp. 85-100.

Savage, R.J.G. (1984) Mid-Jurassic mammals from Scotland. In *Third Symposium on Mesozoic Terrestrial Ecosystems* (eds W.-E. Reif and F. Westphal), Attempto, Tübingen, pp. 211-13.

Savage, R.J.G. (1993) Vertebrate fissure faunas with special reference to Bristol Channel

Mesozoic faunas. *Journal of the Geological Society of London*, **150**, 1025-34.

Savage, R.J.G. and Large, N.F. (1966) On *Birgeria acuminata* and the absence of labyrinthodonts from the Rhaetic. *Palaeontology*, **9**, 135-41.

Savage, R.J.G. and Waldman, M. (1966) *Oligokyphus* from Holwell Quarry, Somerset. *Proceedings of the Bristol Naturalists' Society*, **31**, 185-92.

Sedgwick, A. (1829) On the geological relations and internal structure of the Magnesian Limestone and the lower portions of the New Red Sandstone Series in their range through Nottinghamshire, Derbyshire, Yorkshire and Durham, to the southern extremity of Northumberland. *Transactions of the Geological Society of London*, **(2) 3**, 37-124.

Seeley, H.G. (1865a) On *Plesiosaurus macropterus*, a new species from the Lias of Whitby. *Annals and Magazine of Natural History, Series 3*, **15**, 49-53.

Seeley, H.G. (1865b) Note to a paper on *Plesiosaurus macropterus*. *Annals and Magazine of Natural History, Series 3*, **15**, 232-3.

Seeley, H.G. (1869a) *Index to the Fossil Remains of Aves, Ornithosauria and Reptilia, from the Secondary System of Strata Arranged in the Woodwardian Museum of the University of Cambridge*. Deighton, Bell and Co., Cambridge.

Seeley, H.G. (1869b) Discovery of *Dakosaurus* in England. *Geological Magazine*, **6**, 188-9.

Seeley, H.G. (1870a) On *Ornithopsis*, a gigantic animal of the pterodactyle kind from the Wealden. *Annals and Magazine of Natural History, Series 4*, **5**, 279-83.

Seeley, H.G. (1870b) *The Ornithosauria, an Elementary Study of the Bones of Pterodactyles, Made from Fossil Remains Found in the Cambridge Upper Greensand and Arranged in the Woodwardian Museum of the University of Cambridge*. Deighton, Bell and Co., Cambridge, 135 pp.

Seeley, H.G. (1871) Note on some chelonian remains from the London Clay. *Annals and Magazine of Natural History, Series 4*, **8**, 227-33.

Seeley, H.G. (1873) On *Cetarthrosaurus walkeri* (Seeley), an ichthyosaurian from the Cambridge Upper Greensand. *Quarterly Journal of the Geological Society of London*, **29**, 505-7.

References

Seeley, H.G. (1874a) On the base of a large lacertian cranium from the Potton Sands, presumably dinosaurian. *Quarterly Journal of the Geological Society of London*, **30**, 690-2.

Seeley, H.G. (1874b) Note on some of the generic modifications of the plesiosaurian pectoral arch. *Quarterly Journal of the Geological Society of London*, **30**, 436-49.

Seeley, H.G. (1874c) On cervical and dorsal vertebrae of *Crocodilus cantabrigiensis* (Seeley) from the Cambridge Upper Greensand. *Quarterly Journal of the Geological Society of London*, **30**, 693-5.

Seeley, H.G. (1875a) On an ornithosaurian (*Doratorhynchus validus*) from the Purbeck limestone of Langton near Swanage. *Quarterly Journal of the Geological Society of London*, **31**, 465-8.

Seeley, H.G. (1875b) On the maxillary bone of a new dinosaur (*Priodontognathus phillipsii*) contained in the Woodwardian Museum of the University of Cambridge. *Quarterly Journal of the Geological Society of London*, **31**, 439-43.

Seeley, H.G. (1875c) On the axis of a dinosaur from the Wealden of Brook in the Isle of Wight, probably referable to the *Iguanodon*. *Quarterly Journal of the Geological Society of London*, **31**, 461-4.

Seeley, H.G. (1876a) On the posterior portion of a lower jaw of *Labyrinthodon* (*L. lavisi*) from the Trias of Sidmouth. *Quarterly Journal of the Geological Society of London*, **32**, 278-84.

Seeley, H.G. (1876b) On an associated series of cervical and dorsal vertebrae of *Polyptychodon*, from the Cambridge Upper Greensand, in the Woodwardian Museum of the University of Cambridge. *Quarterly Journal of the Geological Society of London*, **32**, 433-6.

Seeley, H.G. (1876c) On *Crocodilus icenicus*, a second and larger species of crocodile from the Cambridge Upper Greensand, contained in the Woodwardian Museum of the University of Cambridge. *Quarterly Journal of the Geological Society of London*, **32**, 437-9.

Seeley, H.G. (1876d) On *Macrurosaurus semnus*, a long-tailed animal with procoelous vertebrae from the Cambridge Upper Greensand, preserved in the Woodwardian Museum of the University of Cambridge. *Quarterly Journal of the Geological Society of London*, **32**, 440-4.

Seeley, H.G. (1876e) On remains of *Emys horwellensis* (Seeley) from the Lower Hordwell Cliff, contained in the Woodwardian Museum

of the University of Cambridge. *Quarterly Journal of the Geological Society of London*, **32**, 445-50.

Seeley, H.G. (1877) On *Mauisaurus gardneri* (Seeley), an elasmosaurian from the base of the Gault at Folkestone. *Quarterly Journal of the Geological Society of London*, **33**, 541-7.

Seeley, H.G. (1879) On the Dinosauria of the Cambridge Greensand. *Quarterly Journal of the Geological Society of London*, **35**, 591-636.

Seeley, H.G. (1880) On the skull of an *Ichthyosaurus* from the Lias of Whitby, apparently indicating a new species (*I. zetlandicus*, Seeley), preserved in the Woodwardian Museum of the University of Cambridge. *Quarterly Journal of the Geological Society of London*, **36**, 635-64.

Seeley, H.G. (1882a) On a remarkable dinosaurian coracoid from the Wealden of Brook in the Isle of Wight, preserved in the Woodwardian Museum of Cambridge, probably referable to *Ornithopsis*. *Quarterly Journal of the Geological Society of London*, **38**, 367-71.

Seeley, H.G. (1882b) On *Thecospondylus horneri*, a new dinosaur from the Hastings Sand, indicated by the sacrum and the neural canal of the sacral region. *Quarterly Journal of the Geological Society of London*, **38**, 457-60.

Seeley, H.G. (1883) On the dorsal region of the vertebral column of a new dinosaur (indicating a new genus, *Sphenospondylus*), from the Wealden of Brook in the Isle of Wight, preserved in the Woodwardian Museum of the University of Cambridge. *Quarterly Journal of the Geological Society of London*, **39**, 55-61.

Seeley, H.G. (1887a) On the mode of development of the young in *Plesiosaurus*. *Geological Magazine*, Decade 3, **4**, 562-3.

Seeley, H.G. (1887b) On a sacrum, apparently indicating a new type of bird, *Ornithodesmus cluniculus* Seeley, from the Wealden of Brook. *Quarterly Journal of the Geological Society of London*, **43**, 206-11.

Seeley, H.G. (1887c) On *Aristosuchus pusillus* (Owen), being further notes on the fossils described by Sir R. Owen as *Poikilopleuron pusillus*, Owen. *Quarterly Journal of the Geological Society of London*, **43**, 221-8.

Seeley, H.G. (1887d) On the reputed clavicles and interclavicles of *Iguanodon*. *Geological Magazine*, Decade 3, **4**, 561-2.

Seeley, H.G. (1887e) On *Heterosuchus valdensis* Seeley, a procadian crocodile from the

References

Hastings Sand of Hastings. *Quarterly Journal of the Geological Society of London*, **43**, 212-15.

Seeley, H.G. (1888a) [Letter stating that he had neither written nor proof-read Seeley, 1887d.] *Geological Magazine*, Decade 3, **5**, 45-6.

Seeley, H.G. (1888b) On the mode of development of the young *Plesiosaurus*. *Report of the British Association for the Advancement of Science*, 1887, 697-8.

Seeley, H.G. (1888c) On *Cumnoria*, an iguanodont genus founded upon the *Iguanodon prestwichii*, Hulke. *Report of the British Association for the Advancement of Science*, 1887, p. 698.

Seeley, H.G. (1888d) On *Thecospondylus daviesi* (Seeley), with some remarks on the classification of the dinosaurs. *Quarterly Journal of the Geological Society of London*, **44**, 79-87.

Seeley, H.G. (1892) On the os pubis of *Polacanthus foxii*. *Quarterly Journal of the Geological Society of London*, **48**, 81-5.

Seeley, H.G. (1893a) On a reptilian tooth with two roots. *Annals and Magazine of Natural History, Series 6*, **12**, 227-30.

Seeley, H.G. (1893b) Supplemental note on a double-rooted tooth from the Purbeck beds in the British Museum. *Annals and Magazine of Natural History, Series 6*, **12**, 274-6.

Seeley, H.G. (1895) On *Thecodontosaurus* and *Palaeosaurus*. *Annals and Magazine of Natural History, Series 6*, **15**, 144-63.

Seeley, H.G. (1896) On a pyritous concretion from the Lias of Whitby. *Annual Report of the Yorkshire Philosophical Society*, 1895, 20-9.

Seeley, H.G. (1898) On large terrestrial saurians from the Rhaetic Beds of Wedmore Hill, described as *Avalonia sanfordi* and *Picrodon herveyi*. *Geological Magazine*, Decade 4, **5**, 1-6.

Seeley, H.G. (1901) *Dragons of the Air; an Account of Extinct Flying Reptiles*. Methuen, London, 239 pp.

Seiffert, J. (1973) Upper Jurassic lizards from central Portugal. *Serviços Geológicos de Portugal, Memória*, **22**, 85 pp.

Sellwood, B.W. and McKerrow, W.S. (1974) Depositional environments in the lower part of the Great Oolite Group of Oxfordshire and North Gloucestershire. *Proceedings of the Geologists' Association*, **85**, 189-210.

Selwood, E.B., Edwards, R.A., Simpson, S., Chesher, J.A., Hamblin, R.J.O., Henson, M.R., Riddolls, B.W. and Waters, R.A. (1984) *Geology of the Country around Newton Abbott* (Sheet 339), Memoirs of the British Geological Survey, HMSO, London, 212 pp.

Sereno, P.C. (1986) Phylogeny of the bird-hipped dinosaurs (Order Ornithischia). *National Geographic Research*, **2**, 234-56.

Sereno, P.C. (1991a) *Lesothosaurus*, 'fabrosaurids' and the early evolution of Ornithischia. *Journal of Vertebrate Paleontology*, **11**, 168-97.

Sereno, P.C. (1991b) Basal archosaurs: phylogenetic relationships and functional implications. *Journal of Vertebrate Paleontology, Supplement*, **11**, 53 pp.

Sereno, P.C. and Arcucci, A.D. (1990) The monophyly of crurotarsal archosaurs and the origin of bird and crocodile ankle joints. *Neues Jahrbuch für Geologie und Paläontologie, Abhandlungen*, **180**, 21-52.

Short, A.R. (1904) A description of some Rhaetic sections in the Bristol district, with considerations on the mode of deposition of the Rhaetic Series. *Quarterly Journal of the Geological Society of London*, **60**, 170-93.

Shotton, F.W. (1929) The geology of the country around Kenilworth (Warwickshire). *Quarterly Journal of the Geological Society of London*, **85**, 167-220.

Simms, M.J. (1990) Triassic palaeokarst in Britain. *Cave Science*, **17**, 93-101.

Simms, M.J. and Ruffell, A.H. (1989) Synchroneity of climatic change and extinctions in the late Triassic. *Geology*, **17**, 265-8.

Simms, M.J. and Ruffell, A.H. (1990) Climatic and biotic change in the late Triassic. *Journal of the Geological Society of London*, **147**, 321-7.

Simpson, G.G. (1928) *A Catalogue of Mesozoic Mammalia in the Geological Department of the British Museum of Natural History*. British Museum (Natural History), London, 215 pp.

Simpson, M. (1884) *The Fossils of the Yorkshire Lias Described from Nature*. 2nd edn., Wheldon, London, 268 pp.

Simpson, M.I. (1985) The stratigraphy of the Atherfield Clay Formation (Lower Aptian, Lower Cretaceous) at the type and other localities in southern England. *Proceedings of the Geologists' Association*, **96**, 23-45.

Sinclair, J. (1791-9) *The Statistical Account of Scotland*, 21 vols, W. Creech, Edinburgh.

Smart, J.G.O., Bisson, G. and Worssam, B.C. (1966) *Geology of the Country around Canterbury and Folkestone* (Sheets 289, 303,

References

306), Memoirs of the Geological Survey of the United Kingdom, HMSO, London, 337 pp.

Smith, D.B. (1989) The Late Permian palaeogeography of north-east England. *Proceedings of the Yorkshire Geological Society*, **47**, 285-312.

Smith, D.B. and Taylor, J.C.M. (1992) Permian. In *Atlas of Palaeogeography and Lithofacies* (eds J.C.W. Cope, J.K. Ingham and P.F. Rawson), *Geological Society of London Memoir*, **13**, Bath, pp. 87-96.

Smith, D.B., Brunstrom, R.G.W., Manning, P.I., Simpson, S. and Shotton, F.W. (1974) A correlation of Permian rocks in the British Isles. *Journal of the Geological Society, London*, **130**, 1-45.

Smith, G.V. (1884) On further discoveries of the footprints of vertebrate animals in the Lower New Red Sandstone of Penrith. *Quarterly Journal of the Geological Society of London*, **40**, 479-81.

Smith, S.A. (1990) The sedimentology and accretionary styles of an ancient gravel-bed stream: the Budleigh Salterton Pebble Beds (Lower Triassic), southwest England. *Sedimentary Geology*, **67**, 199-219.

Smith, S.A. and Edwards, R.A. (1991) Regional sedimentological variations in Lower Triassic fluvial conglomerates (Budleigh Salterton Pebble Beds), southwest England: some implications for palaeogeography and basin evolution. *Geological Journal*, **26**, 65-83.

Smithson, T.R. (1989) The earliest known reptile. *Nature, London*, **342**, 676-8.

Smithson, T.R. and Rolfe, W.D.I. (1991) *Westlothiana* gen. nov.: naming the earliest known reptile. *Scottish Journal of Geology*, **26**, 137-8.

Smithson, T.R., Carroll, R.L., Panchen, A.L. and Andrews, S.M. (1994) *Westlothiana lizziae* from the Viséan of East Kirkton, West Lothian, Scotland and the amniote stem. *Transactions of the Royal Society of Edinburgh, Earth Sciences*, **84**, 383-412.

Sollas, W.J. (1879) On some three-toed footprints from the Triassic conglomerate of South Wales. *Quarterly Journal of the Geological Society of London*, **35**, 511-16.

Sollas, W.J. (1881) On a new species of *Plesiosaurus (P. Conybeari)* from the Lower Lias of Charmouth, with observations on *P. megacephalus*, Stutchbury and *E. brachycephalus*, Owen. *Quarterly Journal of the Geological Society of London*, **37**, 440-81.

Spath, L.F. (1923-43) A monograph of the Ammonoidea of the Gault. *Palaeontographical Society (Monographs)*, parts 1-16, 787 pp.

Spencer, P.S. (1994) The early inter-relationships and morphology of Amniota. Unpublished PhD Thesis, University of Bristol.

Spencer, P.S. and Isaac, K.P. (1983) Triassic vertebrates from the Otter Sandstone Formation of Devon, England. *Proceedings of the Geologists' Association*, **94**, 267-9.

Steel, R. (1970) Saurischia. *Handbuch der Paläoherpetologie*, **14**, 87 pp.

Steel, R. (1973) Crocodylia. *Handbuch der Paläoherpetologie*, **16**, 116 pp.

Stewart, D.J. (1978) The sedimentology of the Wealden Group of the Isle of Wight. Unpublished PhD Thesis, University of Reading.

Stewart, D.J. (1981a) A meander-belt sandstone of the Lower Cretaceous of southern England. *Sedimentology*, **28**, 1-20.

Stewart, D.J. (1981b) A field guide to the Wealden Group of the Hastings area and the Isle of Wight (3.1-3.31). In *Field Guides to Modern and Ancient Fluvial Systems in Britain and Spain* (ed. T. Elliott), University of Keele, Keele.

Stewart, D.J. (1983) Possible suspended load channel deposits from the Wealden Group (Lower Cretaceous) of southern England. *Special Publication of the International Association of Sedimentologists*, **6**, 369-84.

Stewart, D.J., Ruffel, A., Wach, G. and Goldring, R. (1991) Lagoonal sedimentation and fluctuation salinities in the Vectis Formation (Wealden Group, Lower Cretaceous) of the Isle of Wight, south England. *Sedimentary Geology*, **72**, 117-34.

Stinton, F.C. (1971) Easter field meeting to the Isle of Wight. *Proceedings of the Geologists' Association*, **82**, 403-10.

Storrs, G.W. (1991) Anatomy and relationships of *Corosaurus alcovensis* (Diapsida: Sauropterygia) and the Triassic Alcova Limestone of Wyoming. *Bulletin of the Peabody Museum of Natural History*, **44**, 151 pp.

Storrs, G.W. (1994) Fossil vertebrate faunas of the British Rhaetian (latest Triassic). *Zoological Journal of the Linnean Society*, **112**, 217-60.

Storrs, G.W. and Gower, D.J. (1993) The earliest possible choristodere (Diapsida) and gaps in the fossil record of semi-aquatic reptiles.

References

Journal of the Geological Society of London, **150**, 1103-7.

Strahan, A. (1898) *The Geology of the Isle of Purbeck and Weymouth* (Sheets 341-3), Memoirs of the Geological Survey of the United Kingdom, HMSO, London, 196 pp.

Strickland, H.E. (1841) On the occurrence of the Bristol Bone-Bed on the Lower Lias near Tewkesbury. *Proceedings of the Geological Society of London*, **3**, 585-8.

Stuart, A.J. (1979) Pleistocene occurrences of the European pond tortoise (*Emys orbicularis L.*) in Britain. *Boreas*, **8**, 359-71.

Stutchbury, S. (1850) [A large cylindrical bone found by Mr Thompson in the 'bone bed' of Aust Cliff on the Severn.] *Report of the British Association for the Advancement of Science*, 1849, p. 67.

Sues, H.-D. and Norman, D.B. (1990) Hypsilophodontidae, *Tenontosaurus*, Dryosauridae. In *The Dinosauria* (eds D.B. Weishampel, P. Dodson and H. Osmólska), University of California Press, Berkeley, pp. 498-509.

Sullivan, R. (1979) Revision of the Paleogene genus *Glyptosaurus* (Reptilia, Anguidae). *Bulletin of the American Museum of Natural History*, **163**, 72 pp.

Sumbler, M.G. (1984) The stratigraphy of the Bathonian White Limestone and Forest Marble Formations of Oxfordshire. *Proceedings of the Geologists' Association*, **95**, 51-64.

Sweeting, G.S. (1925) The geology of the country around Crowhurst, Sussex. *Proceedings of the Geologists' Association*, **36**, 406-18.

Swinton, W.E. (1930) Preliminary account of a new genus and species of plesiosaur. *Annals and Magazine of Natural History*, Series 10, **6**, 206-9.

Swinton, W.E. (1934) *The Dinosaurs; a Short History of a Great Group of Extinct Reptiles*. Murby, London, 233 pp.

Swinton, W.E. (1936a) The dinosaurs of the Isle of Wight. *Proceedings of the Geologists' Association*, **47**, 204-20.

Swinton, W.E. (1936b) Notes on the osteology of *Hypsilophodon* and on the Family Hypsilophodontidae. *Proceedings of the Zoological Society of London*, 1936, 555-78.

Swinton, W.E. (1939) A new Triassic rhynchocephalian from Gloucestershire. *Annals and Magazine of Natural History*, Series 11, **4**, 591-4.

Swinton, W.E. (1970) *The Dinosaurs*. George Allen and Unwin, London, 331 pp.

Sykes, J.H. (1977) British Rhaetic bone-beds. *Mercian Geologist*, **6**, 197-239.

Sykes, J.H., Cargill, J.S. and Kent, P.E. (1970) The stratigraphy and palaeontology of the Rhaetic Beds (Rhaetian: Upper Triassic) of Barnstone, Nottinghamshire. *Mercian Geologist*, **3**, 235-46.

Tagart, E. (1846) On markings in the Hastings Sand Beds near Hastings, supposed to be footprints of birds. *Quarterly Journal of the Geological Society of London*, **2**, 267.

Tarlo, L.B. (1958) The scapula of *Pliosaurus macromerus* Phillips. *Palaeontology*, **1**, 193-9.

Tarlo, L.B. (1959a) A new Middle Triassic reptile fauna from fissures in the Middle Devonian limestones of Poland. *Proceedings of the Geological Society of London*, **1568**, 63-4.

Tarlo, L.B. (1959b) *Pliosaurus brachyspondylus* (Owen) from the Kimmeridge Clay. *Palaeontology*, **1**, 283-91.

Tarlo, L.B. (1959c) *Stretosaurus* gen. nov., a giant pliosaur from the Kimmeridge Clay. *Palaeontology*, **2**, 39-55.

Tarlo, L.B. (1960) A review of the Upper Jurassic pliosaurs. *Bulletin of the British Museum (Natural History), Geology Series*, **4**, 145-89.

Tarlo, L.B. (1962) Ancient animals of the upland. *New Scientist*, **15** (294), 32-4.

Tate, R. and Blake, J.F. (1876) *The Yorkshire Lias*. Van Voorst, London, 475 pp.

Tawney, E.B. and Keeping, H. (1883) On the section at Hordwell cliffs, from the top of the Lower Headon to the base of the Upper Bagshot Sands. *Quarterly Journal of the Geological Society of London*, **39**, 566-74.

Taylor, J.H. (1963) Sedimentary features of an ancient deltaic complex: the Wealden rocks of south-east England. *Sedimentology*, **2**, 2-28.

Taylor, M.A. (1986) The Lyme Regis (Philpot) Museum: the history, problems and prospects of a small museum and its geological collection. *Geological Curator*, **4**, 309-17.

Taylor, M.A. (1992a) Functional anatomy of the head of the large aquatic predator *Rhomaleosaurus zetlandicus* (Plesiosauria, Reptilia) from the Toarcian (Lower Jurassic) of Yorkshire, England. *Philosophical Transactions of the Royal Society*, **335**, 247-80.

Taylor, M.A. (1992b) Taxonomy and taphonomy of *Rhomaleosaurus zetlandicus* (Plesiosauria, Reptilia) from the Toarcian (Lower Jurassic) of the Yorkshire coast. *Proceedings of the*

References

Yorkshire Geological Society, **49**, 49-55.

Taylor, M.A. and Benton, M.J. (1986) Reptiles from the Upper Kimmeridge Clay (Kimmeridgian, Upper Jurassic) of the vicinity of Egmont Bight. *Proceedings of the Dorset Natural History and Archaeological Society*, **107**, 121-5.

Taylor, M.A. and Cruickshank, A.R.I. (1993) A plesiosaur from the Linksfield erratic (Rhaetian, Upper Triassic) near Elgin, Morayshire. *Scottish Journal of Geology*, **29**, 191-6.

Taylor, M.A. and Torrens, H.S. (1987) Mary Anning's forgotten fossil: the fish *Squaloraja* from the Lias of Lyme Regis. *Proceedings of the Dorset Natural History and Archaeological Society*, **108**, 135-48.

Taylor, W. (1894) [Note on Cutties Hillock reptiles.] *Natural Science*, **4**, 472.

Taylor, W. (1920) A new locality for Triassic reptiles, with notes on the Trias found in the parishes of Urquhart and Lhanbryde, Morayshire. *Transactions of the Geological Society of Edinburgh*, **11**, 11-13.

Thomas, T.H. (1879) Tridactyl uniserial ichnolites in the Trias at Newton Nottage, near Porthcawl, Glamorganshire. *Reports of the Transactions of the Cardiff Naturalists' Society*, **10**, 77-91.

Thompson, D.B. (1969) Dome-shaped aeolian dunes in the Frodsham Member of the so-called 'Keuper' Sandstone Formation (Scythian-?Anisian: Triassic) at Frodsham, Cheshire (England). *Sedimentary Geology*, **3**, 263-89.

Thompson, D.B. (1970a) The stratigraphy of the so-called Keuper Sandstone Formation (Scythian-?Anisian) in the Permo-Triassic Cheshire Basin. *Quarterly Journal of the Geological Society of London*, **126**, 151-81.

Thompson, D.B. (1970b) Sedimentation of the Triassic (Scythian) Red Pebbly Sandstones in the Cheshire Basin and its margins. *Geological Journal*, **7**, 183-216.

Thompson, D.B. (1985) *Field Excursion to the Permo-Triassic of the Cheshire-East Irish Sea-Needwood and Stafford Basins*. Poroperm-Geochem Ltd, Chester, 160 pp.

Thulborn, R.A. (1977) Relationships of the Lower Jurassic dinosaur *Scelidosaurus harrisonii*. *Journal of Paleontology*, **51**, 725-39.

Thulborn, R.A. (1982) Liassic plesiosaur embryos reinterpreted as shrimp burrows. *Palaeontology*, **25**, 351-9.

Topley, W. (1875) *Geology of the Weald*. Memoirs of the Geological Survey of England and Wales, HMSO, London, 503 pp.

Torrens, H.S. (1968) The Great Oolite Series. In *The Geology of the East Midlands* (eds P.C. Sylvester-Bradley and T.D. Ford), University Press, Leicester, pp. 227-63.

Torrens, H.S.(ed.) (1969a) *International Field Symposium on the British Jurassic*. University of Keele, Keele.

Torrens, H.S. (1969b) The stratigraphical distribution of Bathonian ammonites in Central England. *Geological Magazine*, **106**, 63-76.

Torrens, H.S. (1974) Standard zones of the Bathonian. *Mémoires du Bureau des Recherches Géologiques et Minières*, **75**, 581-604.

Torrens, H.S. and Taylor, M.A. (1990) Collections, collectors and museums of note. No. 55, Geological collectors and museums in Cheltenham 1810-1988: a case history and its lessons. *The Geological Curator*, **5**, 175-213.

Traquair, R.H. (1886) Preliminary note on a new fossil reptile recently discovered at New Spynie, near Elgin. *British Association for the Advancement of Science, Reports and Transactions*, 1885, 1024-5.

Tresise, G.R. (1989) *The Invisible Dinosaur*. National Museums and Galleries on Merseyside, Liverpool, 32 pp.

Tresise, G.R. (1991) The Storeton Quarry discoveries of Triassic vertebrate footprints, 1838: John Cunningham's account. *Geological Curator*, **5**, 225-9.

Tresise, G.R. (1993) Triassic vertebrate footprints from Cheshire, England: localities and lithologies. *Modern Geology*, **18**, 407-17.

Tucker, M.E. (1977) The marginal Triassic deposits of South Wales: continental facies and palaeogeography. *Geological Journal*, **12**, 169-88.

Tucker, M.E. and Benton, M.J. (1982) Triassic environments, climates and reptile evolution. *Palaeogeography, Palaeoclimatology, Palaeoecology*, **40**, 361-79.

Tucker, M.E. and Burchette, T.P. (1977) Triassic dinosaur footprints from South Wales; their context and preservation. *Palaeogeography, Palaeoclimatology, Palaeoecology*, **22**, 195-208.

Tylor, A. (1862) On the footprint of an *Iguanodon*, lately found at Hastings. *Quarterly Journal of the Geological Society of London*, **18**, 247-53.

References

Unwin, D.M. (1988a) A new pterosaur from the Kimmeridge Clay of Kimmeridge, Dorset. *Proceedings of the Dorset Natural History and Archaeological Society*, **109**, 150-3.

Unwin, D.M. (1988b) New remains of the pterosaur *Dimorphodon* (Pterosauria: Rhamphorhynchoidea) and the terrestrial ability of early pterosaurs. *Modern Geology*, **13**, 57-68.

Unwin, D.M. (1991) The morphology, systematics and evolutionary history of pterosaurs from the Cretaceous Cambridge Greensand of England. Unpublished PhD Thesis, University of Reading.

Unwin, D.M. (in prep) *Phylogeny of the Pterosaurs*.

Urlichs, M. 1977. The Lower Jurassic in south-western Germany. *Stuttgarter Beiträge zu Naturkunde, Serie B*, **24**, 1-41.

Ussher, W.A.E. (1876) On the Triassic rocks of Somerset and Devon. *Quarterly Journal of the Geological Society of London*, **32**, 367-94.

Vaughan, R.F. (1989) *The Excavation at Hornsleasow Quarry: an Interim Report Prepared for the Nature Conservancy Council*. City Museum and Art Gallery, Gloucester, 65 pp.

Vaughn, P.P. (1955) The Permian reptile *Araeoscelis*. *Bulletin of the Museum of Comparative Zoology, Harvard*, **113**, 305-469.

Waagen, W. (1865) *Versuch einer Allgemeinen Classification der Schichten des Oberen Jura*. Hermann Manz, München.

Wach, G.D. and Ruffell, A.H. (1991) Sedimentology and sequence stratigraphy of a Lower Cretaceous tide and storm-dominated clastic succession, Isle of Wight and SE England. *13th International Sedimentological Congress, Nottingham, U.K., 1990, Field Guide*, British Sedimentological Research Group, Cambridge, **4**, 100 pp.

Waldman, M. (1974) Megalosaurids from the Bajocian (Middle Jurassic) of Dorset. *Palaeontology*, **17**, 325-39.

Waldman, M. and Evans, S.E. (1994) Lepidosauromorph reptiles from the Middle Jurassic of Skye. *Zoological Journal of the Linnean Society*, **112**, 135-50.

Waldman, M. and Savage, R.J.G. (1972) The first Jurassic mammal from Scotland. *Journal of the Geological Society of London*, **128**, 119-25.

Walford, E.A. (1895) Stonesfield Slate - Report of the Committee. *Report of the British Association for the Advancement of Science*, 1894, 304-6.

Walford, E.A. (1896) Stonesfield Slate - Second report of the Committee. *Report of the British Association for the Advancement of Science*, 1895, 414-15.

Walford, E.A. (1897) Stonesfield Slate - Third and final report of the Committee. *Report of the British Association for the Advancement of Science*, 1896, p. 356.

Walkden, G.M. and Fraser, N.C. (1993) Late Triassic fissure sediments and vertebrate faunas: environmental change and faunal succession at the Cromhall SSSI, south west Britain. *Modern Geology*, **18**, 511-35.

Walkden, G.M. and Oppé, E.F. (1969) In the footsteps of dinosaurs. *Amateur Geologist*, **3(2)**, 1-18.

Walkden, G.M., Fraser, N.C. and Muir, J. (1987) A new specimen of *Steneosaurus* (Mesosuchia, Crocodilia) from the Toarcian of the Yorkshire coast. *Proceedings of the Yorkshire Geological Society*, **46**, 279-87.

Walker, A.D. (1961) Triassic reptiles from the Elgin area: *Stagonolepis, Dasygnathus* and their allies. *Philosophical Transactions of the Royal Society of London, Series B*, **244**, 103-204.

Walker, A.D. (1964) Triassic reptiles from the Elgin area: *Ornithosuchus* and the origin of carnosaurs. *Philosophical Transactions of the Royal Society of London, Series B*, **248**, 53-134.

Walker, A.D. (1966) *Elachistosuchus*, a Triassic rhynchocephalian from Germany. *Nature*, **211**, 583-5.

Walker, A.D. (1969) The reptile fauna of the 'Lower Keuper' Sandstone. *Geological Magazine*, **106**, 470-6.

Walker, A.D. (1970a) Discussion contributions. *Quarterly Journal of the Geological Society of London*, **126**, 217-18.

Walker, A.D. (1970b) A revision of the Jurassic reptile *Hallopus victor* (Marsh), with remarks on the classification of crocodiles. *Philosophical Transactions of the Royal Society of London, Series B*, **257**, 323-72.

Walker, A.D. (1973) The age of the Cuttie's Hillock Sandstone (Permo-Triassic) of the Elgin area. *Scottish Journal of Geology*, **9**, 177-83.

Walker, C.A. and Moody, R.T.J. (1974) A new trionychid turtle from the Lower Eocene of Kent. *Palaeontology*, **17**, 901-7.

References

Walker, C.A. and Moody, R.T.J. (1985) Redescription of *Eurycephalochelys*, a trionychid turtle from the Lower Eocene of England. *Bulletin of the British Museum (Natural History), Geology Series*, **38**, 373–80.

Walker, J.F. (1867) On a new phosphatic deposit near Upware in Cambridgeshire. *Report of the British Association for the Advancement of Science*, 1866, p. 73.

Wall, W.P. and Galton, P.M. (1979) Notes on pachycephalosaurid dinosaurs (Reptilia: Ornithischia) from North America, with comments on their status as ornithopods. *Canadian Journal of Earth Science*, **16**, 1176–86.

Walrond, L.F.J. (1976) *The Stroud Dinosaur: recent Developments at Stroud and District Museum*. Museums Journal.

Ward, D.J. (1979) The Lower London Tertiary (Palaeocene) succession at Herne Bay, Kent. *Report of the Institute of Geological Sciences*, **78/10**, 12 pp.

Ward, T.O. (1840) On the foot-prints and ripple-marks of the New Red Sandstone of Grinshill Hill, Shropshire. *Report of the British Association for the Advancement of Science*, 1840 (1839), 75–6.

Ward, T.O. (1841) The *Labyrinthodon*. *Salopian Journal*, 28 April 1841, p. 2.

Ward, T.O. (1874) Note on the *Rhynchosaurus articeps*, Owen. *Nature*, **11**, p. 8.

Warrington, G. (1967) Correlation of the Keuper Series of the Triassic by miospores. *Nature*, **214**, 1323–4.

Warrington, G. (1970a) The 'Keuper' Series of the British Trias in the northern Irish Sea and neighbouring areas. *Nature*, **226**, 254–6.

Warrington, G. (1970b) The stratigraphy and palaeontology of the 'Keuper' Series of the central Midlands of England. *Quarterly Journal of the Geological Society of London*, **126**, 183–223.

Warrington, G. (1971) Palynology of the New Red Sandstone of the south Devon coast. *Proceedings of the Ussher Society*, **2**, 307–14.

Warrington, G. (1974a) Trias. In *The Geology and Mineral Resources of Yorkshire* (eds D.H. Rayner and J.E. Hemingway), Yorkshire Geological Society, York, pp. 145–60.

Warrington, G. (1974b) Les évaporites du Trias Britannique. *Bulletin de la Société Géologique de France, 7me Série*, **16**, 708–23.

Warrington, G. (1978) Palynology of the Keuper, Westbury and Cotham beds and the White Lias of the Withycombe Farm Borehole. *Bulletin of the Geological Survey of Great Britain*, **68**, 22–8.

Warrington, G. and Ivimey-Cook, H.C. (1992) Triassic. In *Atlas of Palaeogeography and Lithofacies* (eds J.C.W. Cope, J.K. Ingham and P.F. Rawson), *Geological Society of London Memoir*, **13**, Bath, pp. 97–106.

Warrington, G. and Scrivener, R.C. (1988) Late Permian fossils from Devon: regional geological implications. *Proceedings of the Ussher Society*, **7**, 95–6.

Warrington, G. and Scrivener, R.C. (1990) The Permian of Devon, England. *Review of Palaeobotany and Palynology*, **66**, 263–72.

Warrington, G., Audley-Charles, M.G., Elliott, R.E., Evans, W. B., Ivimey-Cook, H.C., Kent, P.E., Robinson, P.L., Shotton, F.W. and Taylor, F.M. (1980) A correlation of Triassic rocks in the British Isles. *Geological Society of London, Special Report*, **13**, 78 pp.

Watson, D.M.S. (1909a) On some reptilian remains from the Trias of Lossiemouth (Elgin). *Quarterly Journal of the Geological Society of London*, **65**, p. 440.

Watson, D.M.S. (1909b) The 'Trias' of Moray. *Geological Magazine*, Decade 5, **6**, 102–7.

Watson, D.M.S. (1909c) A preliminary note on two new genera of Upper Liassic plesiosaurs. *Memoirs and Proceedings of the Manchester Literary and Philosophical Society*, **54 (4)**, 28 pp.

Watson, D.M.S. (1910a) On a skull of *Rhynchosaurus* in the Manchester Museum. *Report of the British Association for the Advancement of Science*, 1910 (1909), 155–8.

Watson, D.M.S. (1910b) Upper Liassic Reptilia. Part II. The Sauropterygia of the Whitby Museum. *Memoirs and Proceedings of the Manchester Literary and Philosophical Society*, **54 (11)**, 13 pp.

Watson, D.M.S. (1911a) The Upper Liassic Reptilia. Part III. *Microcleidus macropterus* (Seeley) and the limbs of *Microcleidus homalospondylus* (Owen). *Memoirs and Proceedings of the Manchester Literary and Philosophical Society*, **55 (17)**, 9 pp.

Watson, D.M.S. (1911b) Notes on some British Mesozoic crocodiles. *Memoirs and Proceedings of the Manchester Literary and Philosophical Society*, **55 (18)**, 13 pp.

Watson, D.M.S. (1914) *Broomia perplexa* gen. et sp. nov., a fossil reptile from South Africa.

References

Proceedings of the Zoological Society of London, 1914, 995–1010.

Watson, D.M.S. and Hickling, G. (1914) On the Triassic and Permian rocks of Moray. *Geological Magazine*, Decade 6, **1**, 399–402.

Weishampel, D.B. (1990) Dinosaurian distribution. In *The Dinosauria* (eds D.B. Weishampel, P. Dodson and H. Osmólska), University of California Press, Berkeley, pp. 63–139.

Weishampel, D.B., Dodson, P. and Osmólska, H. (eds) (1990) *The Dinosauria*. University of California Press, Berkeley, 733 pp.

Welles, S.P. and Estes, R. (1969) *Hadrokkosaurus bradyi* from the Upper Moenkopi Formation of Arizona. *University of California Publications in the Geological Sciences*, **84**, 56 pp.

Wellnhofer, P. (1978) Pterosauria. *Handbuch der Paläoherpetologie*, **19**, 82 pp.

Wellnhofer, P. (1991) *The Illustrated Encyclopedia of Pterosaurs*. Salamander Books Ltd, London, 192 pp.

West, I.M. (1988) Notes on some Purbeck sediments associated with the dinosaur footprints at Sunnydown Farm, near Langton Matravers, Dorset. *Proceedings of the Dorset Natural History and Archaeological Society*, **109**, 153–4.

West, I.M. and El-Shahat, A. (1985) Dinosaur footprints and early cementation of Purbeck bivalve beds. *Proceedings of the Dorset Natural History and Archaeological Society*, **106**, 169–70.

West, I.M., Shearman, D.J. and Pugh, M.E. (1969) Whitsun field meeting in the Weymouth area. *Proceedings of the Geologists' Association*, **80**, 331–40.

Westphal, F. (1961) Zur Systematik der deutschen und englischen Lias-Krokodilier. *Neues Jahrbuch für Geologie und Paläontologie, Abhandlungen*, **113**, 207–18.

Westphal, F. (1962) Die Krokodilier des deutschen und englischen oberen Lias. *Palaeontographica, Abteilung A*, **118**, 23–118.

Westphal, F. (1976) Phytosauria. *Handbuch der Paläoherpetologie*, **13**, 99–120.

Whitaker, W. (1869) On the succession of beds in the 'New Red' on the south coast of Devon and on the locality of a new specimen of *Hyperodapedon*. *Quarterly Journal of the Geological Society of London*, **25**, 152–8.

Whitaker, W., Woodward, H.B., Bennett, F.J.,

Skertchley, S.B.J. and Jukes-Browne, A.J. (1891) *The Geology of Parts of Cambridgeshire and Suffolk* (Sheet 51 NE and NW), Memoirs of the Geological Survey of the United Kingdom, HMSO, London, 127 pp.

White, E.I. (1931) *The Vertebrate Faunas of the English Eocene: Volume 1. From the Thanet Sands to the Basement Bed of the London Clay*. British Museum (Natural History), London, 123 pp.

White, H.J.O. (1921) *A Short Account of the Geology of the Isle of Wight*, Memoirs of the Geological Survey of the United Kingdom, HMSO, London, 219 pp.

White, H.J.O. (1923) *The Geology of the Country South and West of Shaftesbury* (Sheet 313), Memoirs of the Geological Survey of the United Kingdom, HMSO, London, 112 pp.

White, H.J.O. (1924) *The Geology of the Country around Brighton and Worthing* (Sheets 318 and 333), Memoirs of the Geological Survey of the United Kingdom, HMSO, London, 114 pp.

White, H.J.O. (1928) *The Geology of the Country near Hastings and Dungeness* (Sheets 320 and 321), Memoirs of the Geological Survey of the United Kingdom, HMSO, London, 104 pp.

White, T.E. (1940) Holotype of *Plesiosaurus longirostris* Blake and classification of the plesiosaurs. *Journal of Paleontology*, **14**, 451–67.

Whiteside, D.I. (1983) A fissure fauna from Avon. Unpublished PhD Thesis, University of Bristol, 216 pp.

Whiteside, D.I. (1986) The head skeleton of the Rhaetian sphenodontid *Diphydontosaurus avonis* gen. et sp. nov. and the modernizing of a living fossil. *Philosophical Transactions of the Royal Society of London, Series B*, **312**, 379–430.

Whiteside, D.I. and Robinson, D. (1983) A glauconitic clay-mineral from a speleological deposit of Late Triassic age. *Palaeogeography, Palaeoclimatology, Palaeoecology*, **41**, 81–5.

Wickes, W.H. (1904) The Rhaetic Bone-beds. *Proceedings of the Bristol Naturalists' Society*, 1904, 213–27.

Wild, R. (1978a) Die Flugsaurier (Reptilia, Pterosauria) aus der Oberen Trias von Cene bei Bergamo. *Bollettino della Società Paleontologia Italiana*, **17**, 176–256.

Wild, R. (1978b) Ein Sauropoden-Rest (Reptilia, Saurischia) aus dem Posidonicnschiefer (Lias, Toarcium) von Holzmaden. *Stuttgarter Beiträge zur Naturkunde, Serie B*, **41**, 1–15.

Wild, R. (1980) *Tanystropheus* (Reptilia: Squamata) and its importance for stratigraphy. *Mémoires de la Société Géologique de France, Nouvelle Série,* **109**, 201-6.

Wild, R. (1983) A new pterosaur (Reptilia, Pterosauria) from the Upper Triassic (Norian) of Friuli, Italy. *Gortania - Atti del Museo Friulano di Storia Naturale,* **5**, 45-62.

Williams, D. (1973) The sedimentology and petrology of the New Red Sandstone of the Elgin basin, north-east Scotland. Unpublished PhD Thesis, University of Hull.

Wills, L.J. (1907) On some fossiliferous Keuper rocks at Bromsgrove (Worcestershire). *Geological Magazine,* Decade 5, **4**, 28-34.

Wills, L.J. (1908) Note on the fossils from the Lower Keuper of Bromsgrove. *Report of the British Association for the Advancement of Science,* 1908 (1907), 312-13.

Wills, L.J. (1910) On the fossiliferous Lower Keuper rocks of Worcestershire. *Proceedings of the Geologists' Association,* **21**, 249-331.

Wills, L.J. (1916) The structure of the lower jaw of Triassic labyrinthodonts. *Proceedings of the Birmingham Natural History and Philosophical Society,* **14**, 1-16.

Wills, L.J. and Sarjeant, W.A.S. (1970) Fossil vertebrate and invertebrate tracks from boreholes through the Bunter Series (Triassic) of Worcestershire. *Mercian Geologist,* **3**, 399-413.

Wimbledon, W.A. and Hunt, C.O. (1983) The Portland–Purbeck junction (Portlandian–Berriasian) in the Weald and correlation of latest Jurassic–early Cretaceous rocks in southern England. *Geological Magazine,* **120**, 267-80.

Wood, S.V. (1844) Record of the discovery of an alligator with several new Mammalia in the freshwater strata at Hordwell. *Annals and Magazine of Natural History,* **14**, 319-51.

Wood, S.V. (1846) On the discovery of an alligator and several new Mammalia in the Hordwell Cliff; with observations upon the geological phenomena at that locality. *London Geological Journal,* **1**, 1-7, 117-22.

Woodhams, K.E. and Hines, J.S. (1989) Dinosaur footprints from the Lower Cretaceous of East Sussex, England. In *Dinosaur Tracks and Traces* (eds D.D. Gillette and M.G. Lockley), Cambridge University Press, Cambridge, pp. 301-8.

Woodward, A.S. (1885) On the literature and nomenclature of British fossil Crocodilia. *Geological Magazine,* Decade 3, **2**, 496-510.

Woodward, A.S. (1888) A synopsis of the vertebrate fossils of the English Chalk. *Proceedings of the Geologists' Association,* **10**, 273-338.

Woodward, A.S. (1895) Note on megalosaurian teeth discovered by Mr J. Alstone in the Portlandian of Aylesbury. *Proceedings of the Geologists' Association,* **14**, 31-2.

Woodward, A.S. (1904) On two new labyrinthodont skulls of the genera *Capitosaurus* and *Aphaneramma. Proceedings of the Zoological Society of London,* 1904 (2), 170-6.

Woodward, A.S. (1906) Note on some portions of mosasaurian jaws obtained by Mr G.E. Dibley from the Middle Chalk of Cuxton, Kent. *Proceedings of the Geologists' Association,* **19**, 185-7.

Woodward, A.S. (1907a) On *Rhynchosaurus articeps* (Owen). *Report of the British Association for the Advancement of Science,* 1907 (1906), 293-9.

Woodward, A.S. (1907b) On a new dinosaurian reptile (*Scleromochlus taylori,* gen. et sp. nov.) from the Trias of Lossiemouth, Elgin. *Quarterly Journal of the Geological Society of London,* **63**, 140-6.

Woodward, A.S. (1908a) On a mandible of *Labyrinthodon leptognathus,* Owen. *Report of the British Association for the Advancement of Science,* 1908 (1907), 298-300.

Woodward, A.S. (1908b) Note on a megalosaurian tibia from the Lower Lias of Wilmcote, Warwickshire. *Annals and Magazine of Natural History,* Series 8, **1**, 257-65.

Woodward, A.S. (1908c) Note on *Dinodocus mackesoni,* a cetiosaurian from the Lower Greensand of Kent. *Geological Magazine,* Decade 5, **5**, 204-6.

Woodward, A.S. (1910) On a skull of *Megalosaurus* from the Great Oolite of Minchinhampton (Gloucestershire). *Quarterly Journal of the Geological Society of London,* **66**, 111-15.

Woodward, A.S. and Sherborn, C.D. (1890) *A Catalogue of British Fossil Vertebrata.* Dulau, London, 396 pp.

Woodward, H.B. (1880) Discovery of remains of *Emys lutaria* in the Mundesley river-bed. *Transactions of the Norfolk and Norwich Naturalists' Society,* **3**, 36-7.

Woodward, H.B. (1893) *The Jurassic Rocks of Britain. Volume III. The Lias of England and*

Wales (Yorkshire Excepted). HMSO, London, 399 pp.

Woodward, H.B. (1894) *The Jurassic Rocks of Britain. Volume IV. The Lower Oolitic Rocks of England (Yorkshire Excepted)*. HMSO, London, 628 pp.

Woodward, H.B. (1895) *The Jurassic Rocks of Britain. Volume V. The Middle and Upper Oolitic Rocks of England (Yorkshire Excepted)*. HMSO, London, 499 pp.

Woodward, H.B. and Ussher, W.A.E. (1911) *The Geology of the Country near Sidmouth and Lyme Regis* (Sheets 326 and 340), Memoirs of the Geological Survey of the United Kingdom, HMSO, London, 96 pp.

Woodward, J. (1728) *An Attempt Towards a Natural History of the Fossils of England*, 2 vols, F. Fayram *et al.*, London.

Wooler, ('Mr') (1758) A description of the fossil skeleton of an animal found in the Alum Rock near Whitby. *Philosophical Transactions of the Royal Society*, **50**, 786-90.

Worssam, B.C. (1963) *Geology of the Country around Maidstone* (Sheet 288), Memoirs of the Geological Survey of the United Kingdom, HMSO, London, 152 pp.

Wright, J.K. (1986) The Upper Oxford Clay at Furzy Cliff, Dorset: stratigraphy, palaeoenvironment and ammonite fauna. *Proceedings of the Geologists' Association*, **97**, 221-8.

Wright, T. (1852) An account of the section from Round Tower Point to Alum Bay, on the northwest coast of the Isle of Wight. *Proceedings of the Cotteswold Naturalists' Field Club*, **1**, 87-100.

Wright, V.P., Marriott, S.B. and Vanstone, S.D. (1991) A 'reg' palaeosol from the Lower Triassic of south Devon: stratigraphic and palaeoclimatic implications. *Geological Magazine*, **128**, 517-23.

Wrigley, A.G. (1924) Faunal divisions of the London Clay. *Proceedings of the Geologists' Association*, **35**, 245-59.

Young, B. and Lake, R.D. (1988) *Geology of the Country around Brighton and Worthing* (Sheets 318 and 333), Memoirs of the British Geological Survey, HMSO, London, 115 pp.

Young, G. (1820) Account of a singular fossil skeleton, discovered at Whitby, in February 1819. *Memoirs of the Wernerian Natural History Society*, **3**, 450-7.

Young, G. (1825) Account of a fossil crocodile recently discovered in the Alum-Shale near Whitby. *Edinburgh New Philosophical Journal*, **13**, 76-81.

Young, G. and Bird, J. (1822) *A Geological Survey of the Yorkshire Coast*. Clark, Whitby, 335 pp.

Young, G. and Bird, J. (1828) *A Geological Survey of the Yorkshire Coast*. 2nd edn., Kirby, Whitby, 366 pp.

Zangerl, R. (1971) Two toxochelyid sea turtles from the Londian Sands of Erquellinnes (Hainaut) of Belgium. *Mémoires de l'Institut Royal des Sciences Naturelles de Belgique*, **169**, 12 pp.

Glossary

This glossary provides simple explanations of the more important technical and arcane terms used in the Introductions to the chapters and in the Highlights and Conclusions of Chapters 2 to 9. These explanations do not pretend to be scientific definitions but are intended to help the general reader. Stratigraphical terms are omitted as they are given context within the tables and figures.

Acanthodian: 'spiny', a member of an extinct class of Palaeozoic primitive jawed fish, the so-called 'spiny sharks', which occupied both marine and fresh waters (Silurian – Permian).

Aeolian: sediments carried and deposited by wind.

Aetosaur: a member of the only plant-eating and armoured suborder of the extinct **thecodont** (basal **archosaurs**) quadruped reptiles of the late Triassic.

Allantois: an embryonic membranous sac which acts as an organ of respiration / nutrition / excretion.

Ammonite zone: a stratigraphically restricted unit of sedimentary rocks defined by its fossil content, most usefully by species of narrowly defined temporal range and named after a single characteristic species. Here, the term refers to the **cephalopod** ammonites, which are particularly useful zonal fossils in the Mesozoic because of their rapid evolution and widespread distribution.

Ammonoid: 'Jupiter form', a member of an extinct group of marine **cephalopods**, whose nearest living relative is the *Nautilus* and is generally characterised by a coiled shell, regularly partitioned into chambers (Devonian – end Cretaceous).

Amniota: 'foetal membrane', a group of craniates including reptiles, birds and mammals having an amnion (foetal membrane) around the embryo.

Amphisbaenid: 'fabulous serpent having a head at each end', a member of a highly specialised extant group of terrestrial **squamate** reptiles, limbless and worm-like with a wedge-shaped skull adapted for burrowing (mid Cretaceous – Recent).

Anapsid: 'without an arch', a member of a group of reptiles characterised by having no openings in the skull behind the eye, including turtles, tortoises, and extinct groups such as procolophonids and captorhinids.

Ankylosaur: 'fused lizard', a member of a group of **ornithischian** quadrupedal armoured dinosaurs with a horny beak for plant eating and the neck, shoulders and back encased in a tough skin reinforced with bony plates, spikes and a tail armed with a bony club (mid – end Cretaceous).

Archosaur: 'ruling lizard', a member of a major grouping of **diapsid** reptiles including the extinct **dinosaurs**, **pterosaurs**, **thecodontians** and living crocodiles.

Astragalus: a vertebrate ankle bone.

Atoposaurid: 'unusual lizard', a member of a group of very small terrestrial crocodilians with reduced armour and short broad skulls (late Jurassic – mid Cretaceous).

Azhdarchid: a member of a late Cretaceous group of pterodactyl pterosaurs.

Belemnite: 'javelin form', a member of an extinct group of **cephalopod** marine molluscs related to squids, having an internal solid calcium carbonate 'bullet shaped' and posterior skeletal element (predominantly

Glossary

Jurassic – Cretaceous but with problematic earlier Carboniferous and Triassic forms and a later questionable Tertiary form).

Biostratigraphy: the subdivision and correlation of sedimentary strata based on their fossil content.

Bioturbated: sediment that has been churned up or disturbed by the activities of organisms especially burrowing. The patterns of disturbance (trace fossils) are often characteristic for particular groups of organisms.

Bone-bed: a stratigraphically restricted sedimentary accumulation and concentration of bones, or other vertebrate remains such as teeth or scales often worn by transport and associated with fluviatile deposition, especially channel-lag deposits or marine near-shore conglomerates. Bone-beds may reflect a lack of other coarse grained sediment or a mass / catastrophic extinction event and occasionally are an economic source of phosphates. They represent an important source of palaeontological and geological information.

Calcaneum: the relatively large heel bone of the foot in vertebrates.

Capitosaurid: 'head lizard', a member of a group of stereospondyl labyrinthodont amphibians with flattened skulls, some of which were 'crocodile-like' and reached considerable size (Triassic).

Captorhinomorph: a member of a 'primitive' group of small to medium sized **anapsid** stem reptiles of carnivorous habit and with a small pineal opening in the skull (late Carboniferous – Permian).

Carnosaur: 'meat-eating lizards', a member of a group of large carnivorous **saurischian** dinosaurs.

Cephalopod: 'head foot', a member of a class of marine molluscs including octopus, squid, nautiloids and ammonoids, having a well developed head surrounded by tentacles and a large mantle cavity opening to the exterior by a siphon. Some secrete a chambered shell, e.g. nautiloids and ammonoids, and have an excellent fossil record (Cambrian – Recent).

Chelonian: a member of a large group of **anapsid** reptiles including the turtles and tortoises, having a short broad body protected by dorsal and ventral shields composed of bony plates overlain by epidermal plates of tortoiseshell (late Triassic – Recent).

Choristodere: 'separate neck', a member of an enigmatic fresh water, fish-eating crocodile-like **diapsid** group (late Triassic – Tertiary).

Cladistic analysis: an attempt to characterise natural groupings of organisms by means of a search for shared derived characters.

Cladogram: a branched tree-like classification diagram produced by **cladistic analysis**.

Clastic: fragmental sediment composed mainly of particles derived from pre-existing rocks or minerals, including organic remains (designated as bioclastic).

Cleidoic: 'closed', referring to the egg of amniotes enclosed within a protective outer coating or shell and a complex system of membranes around the embryo.

Coccolith: 'berry stone', a member of a palaeontologically important group of unicellular flagellate and planktonic marine microorganisms producing a calcium carbonate skeleton made up of a series of plates. The Cretaceous chalk limestone is often largely made up of coccolith skeletons (late Triassic – Recent).

Conchostracan: 'shelled shell', a member of a group of freshwater crustaceans in which the body is contained within a chitinous bivalved shell (Devonian to Recent).

Coprolite: petrified or fossil faecal material which may contain identifiable food remains and occasionally abundant enough to be a source of phosphate.

Cranial: referring to that part of the skull which encloses the brain.

Crocodilian: a member of an ancient but extant group of quadrupedal **archosaurs**, which are often armoured, have front legs shorter than hind and an elongate body and laterally flattened tail for swimming (late Triassic – Recent).

Cryptoclidid: 'hidden key', a member of a group of marine long necked **plesiosaurs** (late Jurassic – end Cretaceous).

Cycad: a member of a group of gymnosperms having a palm-like appearance with massive stems, which may be short or tree-like, pinnate leaves and sporophylls in cones (Permian – Recent).

Cynodont: 'dog tooth', a member of a group of advanced **synapsid** mammal-like reptiles of the Triassic.

Dermochelid: 'skin turtle', a member of an extant group of marine, ocean living leathery turtles, in which there is almost no connected dermal skeleton but a series of small bony plates studding the skin of the back (Tertiary – Recent).

Diachronous: relating to sedimentary or stratigraphic units where the environmental or facies boundaries cut across the time boundaries in the succession of deposition. Diachronism reflects the migration of a geological event through time so that the sediment produced by that event is not everywhere the same age.

Diapsid: 'two arches', a member of a major grouping of extant reptiles, which includes the dinosaurs, extinct marine reptiles, crocodiles, lizards, snakes and the descendent birds, characterised by a pair of openings in the skull immediately behind the eye socket (late Carboniferous – Recent).

Diagenesis: the sum of all changes at 'normal' surface pressures and temperatures, i.e. chemical, physical and biological, which an unconsolidated sediment undergoes after deposition.

Dicynodont: 'double dog tooth', a member of a curious, specialised 'pig-like' group of herbivorous **therapsids** with reduced numbers of teeth (Permian – Triassic).

Dinocephalian: 'terrible head', a member of a short-lived group of large **therapsid** mammal-like reptiles, both herbivorous and carnivorous (late Permian).

Dinoflagellate: 'rotating whip', a member of a large and diverse group of aquatic unicellular micro-organisms, loosely placed with the algae, which swim by means of flagellae and some of which are covered with cellulose plates that can be preserved in the fossil record (mid Triassic – Recent).

Dinosaur: 'terrible lizard', a member of an extinct, diverse and particular group of land-living **archosaur** reptiles with an erect gait that flourished from the late Triassic to the end of the Cretaceous.

Discoglossid: 'round flat tongue', a member of a group of primitive frogs (anurans) with a long fossil record (late Jurassic – Recent) and both aquatic and terrestrial representatives.

Elasmosaurid: 'elongate lizard', a member of a marine group of sauropterygian diapsid reptiles along with the **pliosaurs** and **plesiosaurs**. They had a very long neck with more than seventy vertebrae, small head and paddle shaped limbs (mid Jurassic – end Cretaceous).

Emydid: 'fresh water turtle', a member of an extant group of marsh dwelling (fresh water / terrestrial) turtles with a relatively abundant fossil record (Tertiary – Recent).

Epeiric: produced by large scale uplift or subsidence of continental crust without the severe deformation of rocks associated with orogeny.

Epicontinental: located on a continent or the surrounding continental shelf.

Euryapsid: 'wide arch', a member of a group of reptiles characterised by a single upper opening in the skull behind the eye, now thought in this instance to be an artificial grouping of extinct marine reptiles that are modified Mesozoic **diapsids**.

Eurypterid: a member of an extinct group of Palaeozoic aquatic arthropods resembling scorpions.

Eusuchian: 'true crocodile', a member of the large extant group of crocodilians which retain the dorsal armour, have a well developed secondary palate and an extensive fossil record (mid Cretaceous – Recent).

Evaporite: a sediment deposited from a solution by evaporation of the solvent – normally water, which may be sea or fresh water. A wide range of mineral salts may be precipitated depending on the original composition of the solvent, e.g. carbonates, sulphates, chlorides and mixed with other types of sediment, often finely laminated.

Facies: a particular sedimentary deposit or part of a stratigraphic unit with sediment related characteristics which clearly distinguish it from other parts of the unit.

Fissures: cavities, often formed by solution of limestone host rock, infilled with relatively younger deposits, which may be of particular interest when they contain fossils, especially microvertebrates that are not preserved elsewhere.

Foraminiferan: 'carrying an opening', a member of a group of small unicellular aquatic organisms which secrete a coiled shell of various materials; often very abundant in marine waters with representatives that are benthic and planktonic (Cambrian – Recent).

Gastrocentrous: a distinctive type of vertebra with centra formed by pairs of interventralia.

Geosaur: see **metriorhynchid.**

Glauconitic: containing the diagenetic (growing in place) mineral glauconite, a complex green-coloured hydrous potassium iron silicate which is sufficiently common in some shallow water marine sediments to give them an overall green colouration, e.g. Cretaceous Greensands.

Goniopholidid: 'angle scale', a member of a number of groups of crocodilians with both fresh water and terrestrial representatives (late Jurassic - end Cretaceous).

Gorgonopsian: a member of a late Permian group of moderate to large sized carnivorous **therapsid** mammal-like reptiles with 'sabreteeth'.

Gymnosperm: 'naked seed', a member of a major division of the plant kingdom, consisting of woody plants with alternation of generations and seeds produced on the surface of the sporophylls and not enclosed in an ovary, e.g. seed ferns and conifers (late Devonian - Recent).

Hadrosaur: 'big lizard', a member of a group of large plant eating **ornithischian ornithopod** dinosaurs, both bipedal and quarupedal, having horny beaks and batteries of cheek teeth (mid - end Cretaceous).

Halite: common salt, NaCl, a naturally occurring mineral particularly associated with evaporite deposits from sea water.

Haematothermia: 'heated blood', a grouping of birds and mammals with a presumed shared ancestor in the Triassic.

Herpetofauna: a fauna of reptiles.

Hypsilophodontid: 'high ridge tooth', a member of a group of **ornithopod ornithischian** dinosaurs, small (2 metres high) fast moving bipedal plant eaters with both horny beaks and self sharpening cheek teeth, short arms, long legs and feet and a stiff tail (late Jurassic - end Cretaceous).

Ichnofauna: 'track fauna', an assemblage of trace fossils, records of life in sediments disturbed by the activity of organisms, e.g. worm burrows or foot prints (see **bioturbated)**.

Ichthyosaur: 'fish lizard', a highly specialised marine reptile of the Mesozoic Era, well adapted for swimming with a streamlined body and paddle shaped limbs (Triassic - late Cretaceous).

Inoceramid: 'strong clay pot', a member of a large group of extinct pterioid marine bivalves, which have been used for biostratigraphical subdivision (Triassic - end Cretaceous).

Intertemporal: a paired membrane bone of the braincase.

Kuehneosaur: 'Kuehne's lizard', a member of a remarkable late Triassic group of terrestrial **lepidosaurs**, small early lizard forms with the ribs highly modified so that they could be projected out sideways as a pair of horizontal 'sails' for gliding, in a mode similar to that seen in the modern lizard *Draco*.

Labyrinthodont: 'labyrinth tooth', a member of a large extinct grouping of primitive amphibians, which included the first land vertebrates and are characterised by teeth with complex infolding of the dentine (late Devonian - Triassic).

Lepidosaur: 'scaly reptiles', a member of a diverse group of largely terrestrial **diapsid** lizards (pleurosaurs, a marine exception) which includes living representatives, e.g. the tuatara (Triassic - Recent).

Lignite: a brown coal formed from peat under moderate pressure having a low calorific value, typically of Tertiary age.

Lissamphibia: 'smooth both lives', a subclass which includes all living amphibians with reduced or absent scales and skin respiration i.e. anurans, urodeles and apodans.

Lithostratigraphy: the organisation and division of strata into units and their correlation based entirely upon their lithological (rock compositional) characteristics.

Mammal-like reptile: a member of a large extinct group of synapsid reptiles, including the pelycosaurs and therapsids and which the mammals evolved, of particular importance during the Permian and Triassic (late Carboniferous - late mid Jurassic).

Maniraptor: 'hand robber', a member of a newly recognised grouping of small carnivorous **theropod** dinosaurs, including birds, based on **cladistic analysis**.

Mass extinction: a heightened rate of extinction as recorded in the fossil record by the termination of a significant number of species lineages over a relatively short period of time (in geological terms), reflecting a biotic crisis that may have a variety of causes e.g. a change in sea level or climate.

Megalosaur: 'great lizard', a member of a group of terrestrial **theropod carnosaurs**, large heavily built bipedal carnivores. They had large heads, short strong necks, saw-edged teeth, strong arms and powerful legs with formidable claws and long tails (late Triassic - mid Cretaceous)

Glossary

Metriorhynchid: 'moderate nose', a member of an extinct group of Jurassic to Cretaceous crocodiles that was highly adapted to an aquatic mode of life, having paddle shaped limbs and tail fins. The group includes both marine and fresh water forms.

Micrite: the fine-grained microcrystalline carbonate matrix of limestones much of which is chemically precipitated as a lime mud but also may include a significant proportion of organic derived mud.

Microvertebrate: literally the small fossil remains of vertebrates, whether they be of juveniles of a large species or just a small species. Such fossil remains tend to be disarticulated teeth and bones, and are usuallly size-sorted and deposited together by the processes of transport and deposition, especially by water currents.

Monophyletic: a natural taxonomic group that includes all descendants of a single common ancestor, e.g. the Amniota, which includes the reptiles, birds and mammals.

Mosasaur: 'Meuse lizard', a member of a group of large marine predatory **squamate lepidosaurs**, having large heads and jaws, short powerful necks, bodies and elongate flattened tails for swimming and paddle shaped limbs for steering and balance (late Cretaceous).

Nodosaur: 'node lizard', a member of a group of quadrupedal armoured **ornithischian** dinosaurs with alternate rows of large and small plates on the back and flanks (late Jurassic – end Cretaceous).

Nothosaur: 'false lizard', a member of a group of long necked **diapsid** lizard-like aquatic reptiles, up to 4 metres long that flourished in Triassic seas.

Notosuchian: 'back crocodile', a member of a group of stratigraphically restricted (Coniacian, Upper Cretaceous) terrestrial armourless crocodilians.

Oolite: a sedimentary rock, usually a limestone made up of small (1 mm – 1 cm) ovoid accretionary bodies cemented together. The ovoids resemble fish eggs but are formed by the precipitation of layers of calcium carbonate concentrically arranged around a nucleus e.g. a sand grain, as it is rolled around on the sea floor by waves and currents, especially in shallow tropical and subtropical seas.

Ophthalmosaur: 'eye lizard', a member of a group of Jurassic marine **ichthyosaurs.**

Ornithischian: 'bird hipped dinosaurs', a member of a major grouping of diapsid plant eating dinosaurs with a 'bird-like' pelvis, horny 'bill' and leaf shaped teeth, e.g. **stegosaurs** (Triassic – end Cretaceous).

Ornithopod: 'bird-feet', a bipedal or quadrupedal plant eating and bird-hipped (ornithischian) dinosaur (Jurassic – end Cretaceous).

Ornithosuchid: 'bird crocodiles' carnivorous bipedal and quadrupedal members of a family of Triassic thecodonts (Upper Triassic).

Orogeny: a process of mountain building during which the rocks and sediments of a particular area of a continent(s) is deformed and uplifted to form mountain belts. Although these processes take a long time they can be distinguished as recognisable and discrete phases in Earth history and are named accordingly, e.g. Variscan orogeny.

Ostracod: 'shell like', a member of a group of small crustaceans having a bivalved shell around the body. Throughout their long geological history (Cambrian – Recent) they have diversified into a wide range of aquatic ecological niches both on land and at sea.

Pachycephalosaur: 'thick skulled lizard', a member of a group of large plant eating **ornithischian** dinosaurs having high domed, thick bony brain cases with both bipedal and quadrupedal representatives (mid – end Cretaceous).

Palynology: the study of plant spores and pollen and their distribution, which has proved to be of considerable biostratigraphic use.

Palynomorph: a microscopic, resistant walled organic body found in palynological preparations, including both plant derived bodies such as spores and pollen and also other acid resistant remains such as acritarchs and chitinozoans.

Pangaea: a supercontinent formed by plate collision of all continents in the late Permian.

Paraphyletic: arising from a single common ancestor but not including all descendants, e.g. Class Reptilia, which does not include the descendent birds and mammals.

Pareiasaurs: a member of an extinct primitive and relatively short lived group of large plant-eating **anapsid** reptiles in which the limbs were positioned close to the body and bore the weight more vertically than in earlier reptiles (late Permian).

Parietal: one of a pair of bones forming the roof of the vertebrate braincase.

Glossary

Pelomedusid: 'clay snake head', a member of an archaic but extant group of marine and fresh water pleurodire turtles which are capable of withdrawing their heads into their carapaces by means of a sideways movement (mid Cretaceous – Recent).

Pelycosaur: 'sail-finned lizards', a mammal-like reptile of the Carboniferous and Permian, some of which have distinctive 'sail fins' on their backs.

Peneplain: a landscape surface with greatly reduced features as the result of prolonged weathering and erosion.

Perleidid: a member of a group of bony 'ray-finned' fish with ganoid scales of Triassic age.

Pes: technical name for the vertebrate foot.

Pholidosaurid: 'scale lizard', a member of an extinct group of long snouted crocodilians that included both fresh water and terrestrial forms (late Jurassic – mid Cretaceous).

Phreatic: relating to the water table, here referring to solution cavities opened up in limestone by the underground rise and fall of the water table.

Phytosaur: so called 'plant lizard', a member of a group of Triassic crocodile like **thecodontians**, which were in fact carnivorous fish eaters (piscivores).

Playa: the flat dry bottom of a desert basin, often the bed of an ephemeral lake and underlain by evaporites.

Plesiosaur: 'near lizard', a predatory marine reptile of the Mesozoic Era, swimming with flipper shaped limbs, a long neck and relatively small head (total length up to 12 metres).

Pleurosaur: 'side lizard', a member of an extinct marine group of **lepidosaurian diapsid** lizards (Jurassic – early Cretaceous).

Pleurosternid: 'side chest', a member of a group of **anapsid** turtles (testudines) with both fresh water and marine representatives (late Jurassic – early Tertiary).

Pliosaur: 'more lizard', a member of a short necked group of predatory marine Mesozoic reptiles, swimming with flipper shaped limbs.

Postorbital: a bone forming part of the posterior wall of the eye socket in vertebrates.

Prolacertiform: 'early lizard shape', a member of a group of Triassic **diapsids**.

Prolocophonids: a member of an extinct group of small primitive plant eating **anapsid** reptiles (late Permian – late Triassic).

Prosauropod: 'first lizard foot', a member of a group of **saurischian sauropod** plant eating dinosaurs with long necks and tails and both bipedal and quadrupedal forms (late Triassic – early Jurassic).

Protobranch: 'first gills', a member of a 'primitive' group of marine bivalve molluscs with a very long fossil record (early Cambrian – Recent) that commonly occupy mud substrates and feed by extracting organic material from the mud, e.g. the nuculids.

Protorothyridid: 'first door', a member of one of the 'stem' or 'basal' reptile groups.

Pterodactyloid: 'winged finger', a member of a large group of **pterosaurs** including the pterodactyls and pteranodontids (late Jurassic – end Cretaceous).

Pterosaur: 'winged lizards', members of an order of Jurassic and Cretaceous **archosaur** reptiles capable of flight, having a membranous wing supported by an elongate fourth finger.

Quadrate: the bone with which the lower jaw articulates in birds, reptiles, amphibians and most fish.

Rauisuchian: 'Rau's crocodile', terrestrial quadrupedal **thecodont** reptiles of the Triassic.

Red beds: sedimentary deposits that are predominantly red in colour, generally as a result of abundant iron oxides, which often reflect deposition in an oxidising situation, e.g. in an arid terrestrial environment and may be associated with evaporites.

Regressive: referring to the retreat of the sea from land areas as the result of a fall in sea level or elevation of the landmass.

Reptile: 'creeping animals', a member of a large class of amniote vertebrates, having a long fossil history extending back to the Carboniferous, with a dry, waterproof horny skin of scales, plates or scutes, functional lungs, a four chambered heart and laying eggs fertilised inside the female's body.

Rhynchosaur: 'snout lizard', a member of a group of squat 'pig-like' late Triassic **diapsids** with curious hooked 'beaks'.

Rhythmic sequence: a regularly banded vertical sequence of sediments, reflecting rhythmic changes in the supply of sediment often related to seasonal changes, e.g. the varved couplets of silt and clay in glacial lakes.

Rudist: 'stirring rod', a member of an unusual and varied group of extinct marine cemented bivalve molluscs (also known as hippuritoids), which flourished in the shallow tropical seas of the Tethyan area and in places formed reef like clusters. Some had thick cone shaped shells up to a metre long, whilst others had coiled 'snail like' shells (late Jurassic – end Cretaceous).

Sabkha: a halite encrusted surface of salt flats, which are often developed just inland parallel to dry hot tropical coastlines, where periodic flooding by the sea is evaporated with precipitation of various **evaporite** minerals and laminae of dried algae.

Saurischian: 'lizard hip', a member of a major grouping of Mesozoic **archosaur** reptiles with a characteristic pelvic structure in which the pubis is long and points forward and down from the hip socket. Includes both bipedal carnivores (**therapods**) and very large quadrupedal herbivores (**sauropods**).

Sauropod: 'lizard feet', a member of a large group of **saurischian** dinosaurs, many of which were very large quadrupedal plant eaters with small heads, long necks and tails and bulky bodies, e.g. brachiosaurs (late Triassic – end Cretaceous).

Sauropterygian: 'lizard paddle', a member of an extinct Mesozoic group of amphibious and marine reptiles, e.g. **plesiosaurs**.

Scute: an external scale as seen in reptiles and fish.

Sebecosuchian: a member of an aberrant extinct group of giant terrestrial predatory crocodilians having high narrow skulls (Tertiary).

Sedimentary cycle: a regularly repeated sequence of environmental changes which are reflected in a repeated vertical succession of deposits.

Sphenodontid: 'wedge tooth', a member of a large group of **lepidosaurian diapsid** reptiles including the sphenodonts and **pleurosaurs** (Triassic – Recent).

Sphenopsid: 'wedge appearance', a member of an ancient group of pteridophyte plants, commonly called 'horsetails' (Devonian – Recent).

Squamate: 'scaly', a member of a large group of **lepidosaur** reptiles that includes the lizards and snakes (late Jurassic – Recent).

Squamosal: a membrane bone forming part of the side wall of the vertebrate skull.

SSSI: Site of Special Scientific Interest.

Stegosaur: 'roof lizard', a member of a group of quadrupedal **ornithischian** plant eating dinosaurs with rows of plates or spines arising from the neck, back and tail (late Jurassic – late Cretaceous).

Steneosaur: 'narrow lizard', a member of an extinct group of Jurassic to early Cretaceous marine crocodiles.

Supratemporal: a bone in the upper temporal region of the braincase.

Synapsid: 'union', a member of one of the 'stem' or 'basal' reptile groups to which the mammals are distantly related, characterised by a single opening low down on the skull behind the eye socket.

Tabular: a bone posterior to the parietal in the braincase of some vertebrates.

Teleosaur: 'complete lizard', a member of an extinct group of Jurassic to early Cretaceous marine crocodiles which includes the steneosaurs.

Temnospondyl: 'cut vertebrae', a member of an extinct group of labyrinthodont **tetrapod** amphibians that lived from Carboniferous to Triassic times.

Temporal opening: an important characteristic opening in the skull behind the eye and used to distinguish major groups of reptiles, e.g. anapsids (no opening), diapsids (two openings), euryapsids (single upper opening) and synapsids (single lower opening).

Tetrapod: developmentally four footed vertebrate including amphibians, reptiles and mammals.

Thalassemyid: 'sea power', a member of a group of marine amphibious Jurassic turtles (testudines).

'Thecodontian': 'socket toothed', a member of a primitive group of early **archosaurs** (late Permian) from which the more advanced **archosaurs**, e.g. pterosaurs, dinosaurs and crocodiles evolved in the Triassic.

Therapsid: 'attendant', a member of a group of mammal-like reptiles of the late Permian and early Triassic.

Therian: 'small animal', a member of a group of mammals whose living members are viviparous and have a distinctive molar tooth pattern and a spiral cochlea.

Theropod: 'beast foot', a member of a large group of **saurischian** bipedal and largely carnivorous dinosaurs, e.g. carnosaurs (late Triassic – end Cretaceous).

Thyreophoran: 'shield bearer', a member of a large group of armoured **ornithischian** dinosaurs which includes the **stegosaurs** and **ankylosaurs** (Jurassic – end Cretaceous).

Trilophosaur: 'three ridged lizard', archosauro-morph Triassic beaked reptiles with broad, sharp, shearing cheek teeth.

Trionychid: a member of an extant group of soft shelled cryptodire turtles with both fresh water and terrestrial representatives, which have lost the horny scutes and bear only thin dermal plates beneath the skin giving increased buoyancy (mid Cretaceous – Recent).

Tritylodontid: 'three knobbed tooth', a member of an extinct group of terrestrial **cynodont** reptiles with skulls close to the mammal condition including well differentiated teeth and a well developed secondary palate (late Triassic – late Jurassic).

Turtle: see **chelonian**.

Zone: see **ammonite zone**.

Fossil index

Page numbers in **bold** type refer to figures, page numbers in *italic* type refer to tables

Fossil index

Fossil index

Fossil index

Fossil index

Fossil index

Fossil index

Fossil index

General index

Page numbers in **bold** type refer to figures and page numbers in *italic* type refer to tables.

General index

General index

General index

General index